COUNTERFACTUALS AND CAUSAL INFERENCE

Second Edition

In this completely revised and expanded second edition of *Counterfactuals and Causal Inference*, the essential features of the counterfactual approach to observational data analysis are presented with examples from the social, demographic, and health sciences. Alternative estimation techniques are first introduced using both the potential outcome model and causal graphs; after which conditioning techniques, such as matching and regression, are presented from a potential outcomes perspective. For research scenarios in which important determinants of causal exposure are unobserved, alternative techniques, such as instrumental variable estimators, longitudinal methods, and estimation via causal mechanisms, are then presented. The importance of causal effect heterogeneity is stressed throughout the book, and the need for deep causal explanation via mechanisms is discussed.

Stephen L. Morgan is the Bloomberg Distinguished Professor of Sociology and Education at Johns Hopkins University. He was previously the Jan Rock Zubrow '77 Professor in the Social Sciences and the director of the Center for the Study of Inequality at Cornell University. His current areas of interest include social stratification, the sociology of education, and quantitative methodology. He has published *On the Edge of Commitment: Educational Attainment and Race in the United States* (2005) and, as editor, the *Handbook of Causal Analysis for Social Research* (2013).

Christopher Winship is the Diker-Tishman Professor of Sociology and a member of the senior faculty of Harvard's Kennedy School of Government. Prior to coming to Harvard in 1992, he was Professor of Sociology and Statistics and by courtesy Economics at Northwestern University. His research focuses on statistical models for causal inference, most recently mechanisms and endogenous selection; how black clergy in Boston have worked with police to reduce youth violence; the effects of education on mental ability; pragmatism as the basis for a theory of action; the implications of advances in cognitive psychology for sociology; and sociological approaches to how individuals understand justice. Since 1995 he has been editor of *Sociological Methods and Research*.

Analytical Methods for Social Research

Analytical Methods for Social Research presents texts on empirical and formal methods for the social sciences. Volumes in the series address both the theoretical underpinnings of analytical techniques as well as their application in social research. Some series volumes are broad in scope, cutting across a number of disciplines. Others focus mainly on methodological applications within specific fields such as political science, sociology, demography, and public health. The series serves a mix of students and researchers in the social sciences and statistics.

Series Editors:

R. Michael Alvarez, California Institute of Technology
Nathaniel L. Beck, New York University
Stephen L. Morgan, Johns Hopkins University
Lawrence L. Wu, New York University

Other Titles in the Series:

Time Series Analysis for the Social Sciences, by Janet M. Box-Steffensmeier, John R. Freeman, Matthew Perry Hitt, and Jon C. W. Pevehouse

Event History Modeling: A Guide for Social Scientists, by Janet M. Box-Steffensmeier and Bradford S. Jones

Ecological Inference: New Methodological Strategies, edited by Gary King, Ori Rosen, and Martin A. Tanner

Spatial Models of Parliamentary Voting, by Keith T. Poole

Essential Mathematics for Political and Social Research, by Jeff Gill

Political Game Theory: An Introduction, by Nolan McCarty and Adam Meirowitz

Data Analysis Using Regression and Multilevel/Hierarchical Models, by Andrew Gelman and Jennifer Hill

Counterfactuals and Causal Inference

Methods and Principles for Social Research

Second Edition

STEPHEN L. MORGAN

Johns Hopkins University

CHRISTOPHER WINSHIP

Harvard University

CAMBRIDGE
UNIVERSITY PRESS

CAMBRIDGE
UNIVERSITY PRESS

University Printing House, Cambridge CB2 8BS, United Kingdom

One Liberty Plaza, 20th Floor, New York, NY 10006, USA

477 Williamstown Road, Port Melbourne, VIC 3207, Australia

314–321, 3rd Floor, Plot 3, Splendor Forum, Jasola District Centre, New Delhi – 110025, India

79 Anson Road, #06–04/06, Singapore 079906

Cambridge University Press is part of the University of Cambridge.

It furthers the University's mission by disseminating knowledge in the pursuit of
education, learning and research at the highest international levels of excellence.

www.cambridge.org
Information on this title: www.cambridge.org/9781107065079

First published 2007
Second edition 2015
Reprinted with corrections 2015
12th printing 2020

Printed in the United Kingdom by TJ Books Limited, Padstow Cornwall

A catalogue record for this publication is available from the British Library.

Library of Congress Cataloging in Publication Data
Morgan, Stephen L. (Stephen Lawrence), 1971–
Counterfactuals and causal inference : methods and principles for social research /
Stephen L. Morgan, Christopher Winship.
pages   cm. – (Analytical methods for social research)
Revised edition of the authors' Counterfactuals and causal inference, published in 2007.
Includes bibliographical references and index.
ISBN 978-1-107-06507-9 (hardback) – ISBN 978-1-107-69416-3 (paperback)
1. Social sciences–Research.   2. Social sciences–Methodology.
3. Causation.   I. Winship, Christopher.   II. Title.
H62.M646 2015
300.72–dc23   2014033205

ISBN 978-1-107-06507-9 Hardback
ISBN 978-1-107-69416-3 Paperback

To my wife, Sydney, my son, Vinny, and my daughter, Beatrix
– Steve Morgan

To my wife, Nancy, and my sons, David and Michael
– Chris Winship

Contents

Figures

Tables

Acknowledgments for First Edition

Without yet knowing it, we began to write this book in 1997 when collaborating on a paper for the 1999 volume of the *Annual Review of Sociology*, titled "The Estimation of Causal Effects from Observational Data." We benefited from many helpful comments in the preparation of that manuscript, and we were pleased that many of our colleagues found it to be a useful introduction to a literature that we were, at the time, still working to understand ourselves. Since then, considerable progress in the potential outcomes and counterfactual modeling literature has been achieved, which led us into long discussions of the utility of writing a more comprehensive introduction. In the end, our motivation to learn even more of the literature was the decisive factor.

We thank Richard Berk, Felix Elwert, George Farkas, Glenn Firebaugh, Jeremy Freese, Andrew Gelman, Gary King, Trond Petersen, David Weakliem, and Kim Weeden for reading some or all of the penultimate draft of the book. We also thank the anonymous reviewer recruited by Cambridge University Press. The insightful comments of all of these readers helped tremendously. We also thank our students at Cornell and Harvard, from whom we have learned much in the course of learning and then presenting this material to them. Their comments and questions were more valuable than they are probably aware.

Finally, we thank Kelly Andronicos and Jenny Todd at Cornell University for assistance with the preparation of the manuscript, as well as Larry Wu and Ed Parsons at Cambridge University Press, Project Manager Peter Katsirubas at Aptara, Inc., and Victoria Danahy at In Other Words.

Acknowledgments for Second Edition

We thank all of the students in our classes at Cornell and at Harvard, as well as those who have attended presentations of the new material in this second edition at other universities. Your excellent questions over the years have shaped this book more than you may realize.

For their generosity and willingness to read and comment on substantial portions of this second edition, we thank Weihua An, Neal Beck, Richard Berk, David Bills, Ken Bollen (and his students), Andy Cherlin, Tom DiPrete, Felix Elwert, Markus Gangl, Guanglei Hong, Mike Hout, Tim Liao, Scott Lynch, Isaac Reed, Matt Salganik, Jasjeet Sekhon, Peter Steiner, Jessica Su, Steve Vaisey, Tyler VanderWeele, David Weakliem, and Hui Zheng. In addition, we thank John Cawley and Dan Lichter for pointing us to relevant literature in health economics and demography.

We also thank Cornell University and Harvard University for the sabbatical support that allowed us to begin the writing of this second edition. Morgan thanks Collegio Carlo Alberto for providing a restful and stimulating environment for work from January through June 2013.

Part I: Causality and Empirical Research in the Social Sciences

Chapter 1

Introduction

Do charter schools increase the test scores of elementary school students? If so, how large are the gains in comparison to those that could be realized by implementing alternative educational reforms? Does obtaining a college degree increase an individual's labor market earnings? If so, is this particular effect large relative to the earnings gains that could be achieved only through on-the-job training? Did the use of a butterfly ballot in some Florida counties in the 2000 presidential election cost Al Gore votes? If so, was the number of miscast votes sufficiently large to have altered the election outcome?

At their core, these types of questions are simple cause-and-effect questions of the form, Does X cause Y? If X causes Y, how large is the effect of X on Y? Is the size of this effect large relative to the effects of other causes of Y?

Simple cause-and-effect questions are the motivation for much research in the social, demographic, and health sciences, even though definitive answers to cause-and-effect questions may not always be possible to formulate given the constraints that researchers face in collecting data and evaluating alternative explanations. Even so, there is reason for optimism about our current and future abilities to effectively address cause-and-effect questions. Over the past four decades, a counterfactual model of causality has been developed and refined, and as a result a unified framework for the prosecution of causal questions is now available. With this book, we aim to convince more social scientists to apply this model to the core empirical questions of the social sciences and to applied research questions of public importance.

In this introductory chapter, we provide a skeletal précis of the main features of the potential outcome model, which is a core piece of the more general counterfactual approach to observational data analysis that we present in this book. We then offer a brief and selective history of causal analysis in quantitatively oriented observational social science. We develop some background on the examples that we will draw on throughout the book, and we conclude with an introduction to directed graphs for systems of causal relationships.

1.1 The Potential Outcome Model of Causal Inference

With its origins in early work on experimental design by Neyman (1990[1923], 1935), Fisher (1935), Cochran and Cox (1950), Kempthorne (1952), and Cox (1958), the potential outcome model of causal inference was formalized in a series of papers by the statistician Donald Rubin (1974, 1977, 1978, 1980a, 1981, 1986, 1990). The name "potential outcome" is a reference to the potential yields from Neyman's work in agricultural statistics (see Gelman and Meng 2004; Rubin 2005). The model also has roots in the economics literature (Roy 1951; Quandt 1972), with important subsequent work by James Heckman (see Heckman 1974, 1978, 1979, 1989, 1992), Charles Manski (1995), and others. The model is now dominant in both statistics and economics, and it is being used with increasing frequency across the basic and applied social and health sciences.

The core of the potential outcome model is simple. Suppose that each individual in a population of interest can be exposed to two alternative states of a cause. Each state is characterized by a distinct set of conditions, exposure to which potentially affects an outcome of interest, such as labor market earnings or scores on a standardized mathematics test. If the outcome is earnings, the population of interest could be adults between the ages of 30 and 50, and the two states could be whether or not an individual has obtained a college degree. Alternatively, if the outcome is a mathematics test score, the population of interest could be high school seniors, and the two states could be whether or not a student has taken a course in trigonometry. For the potential outcome model, these alternative causal states are referred to as alternative treatments. When only two treatments are considered, they are referred to as treatment and control. Throughout this book, we will conform to this convention.

The key assumption of the model is that each individual in the population of interest has a potential outcome under each treatment state, even though each individual can be observed in only one treatment state at any point in time. For example, for the causal effect of having a college degree rather than only a high school diploma on subsequent earnings, adults who have completed only high school diplomas have theoretical what-if earnings under the state "have a college degree," and adults who have completed college degrees have theoretical what-if earnings under the state "have only a high school diploma." These what-if potential outcomes are counterfactual in the sense that they exist in theory but are not observed.

Formalizing this conceptualization for a two-state treatment, the potential outcomes of each individual are defined as the true values of the outcome of interest that would result from exposure to the alternative causal states. The potential outcomes of each individual i are y_i^1 and y_i^0, where the superscript 1 signifies the treatment state and the superscript 0 signifies the control state. Because both y_i^1 and y_i^0 exist in theory for each individual, an individual-level causal effect can be defined as some contrast between y_i^1 and y_i^0, usually the simple difference $y_i^1 - y_i^0$. Because it is impossible to observe both y_i^1 and y_i^0 for any individual, causal effects cannot be observed or directly calculated at the individual level.[1]

---

[1] The only generally effective strategy for estimating individual-level causal effects is a crossover design, in which individuals are exposed to two alternative treatments in succession and with enough

By necessity, a researcher must analyze an observed outcome variable Y that takes on values y_i for each individual i that are equal to y_i^1 for those in the treatment state and y_i^0 for those in the control state. We usually refer to those in the treatment state as the treatment group and those in the control state as the control group.[2] Accordingly, y_i^0 is an unobservable counterfactual outcome for each individual i in the treatment group, and y_i^1 is an unobservable counterfactual outcome for each individual i in the control group.

In the potential outcome modeling tradition, attention is focused on estimating various average causal effects, by analysis of the values y_i, for groups of individuals defined by specific characteristics. To do so effectively, the process by which individuals of different types are exposed to the cause of interest must be modeled. Doing so involves introducing defendable assumptions that allow for the estimation of the average unobservable counterfactual values for specific groups of individuals. If the assumptions are defendable, and a suitable method for constructing an average contrast from the data is chosen, then the resulting average difference in the values of y_i can be given a causal interpretation.

The potential outcome model is one core piece of the more general counterfactual approach to causal analysis that we will present in this book. Another core piece, which we will introduce at the end of this chapter, is the directed graph approach to causal analysis, most closely associated with the work of the computer scientist Judea Pearl. We will use Pearl's work extensively in our presentation, drawing on his 2000 book, *Causality: Models, Reasoning, and Inference* (2nd edition, 2009), as well as related literature.

A counterfactual account of causation also exists in philosophy, which began with the seminal 1973 article by David Lewis, titled "Causation." It is related to the counterfactual model of causal analysis that we will present in this book, but the philosophical version, as implied by the title of Lewis' original article, aims to be a general model of causation.[3] As noted by the philosopher James Woodward in his 2003 book,

---

time elapsed in between exposures such that the effects of the cause have had time to dissipate (see Rothman, Greenland, and Lash 2008). Obviously, such a design can be attempted only when a researcher has control over the allocation of the treatments and only when the treatment effects are sufficiently ephemeral. These conditions rarely exist for the causal questions that interest social scientists.

[2] We assume that, for observational data analysis, an underlying causal exposure mechanism exists in the population, and thus the distribution of individuals across the treatment and control states exists separately from the observation and sampling process. Accordingly, the treatment and control groups exist in the population, even though we typically observe only samples of them in the observed data. We will not require that the labels "treatment group" and "control group" refer only to the observed treatment and control groups.

[3] In this tradition, causation is defined with reference to counterfactual dependence (or, as is sometimes written, the "ancestral" to counterfactual dependence). Accordingly, and at the risk of a great deal of oversimplification, the counterfactual account in philosophy maintains that (in most cases) it is proper to declare that, for events c and e, c causes e if (1) c and e both occur and (2) if c had not occurred and all else remained the same, then e would not have occurred. The primary challenge of the approach is to define the counterfactual scenario in which c does not occur (which Lewis did by imagining a limited "divergence miracle" that prevents c from occurring in a closest possible hypothetical world where all else is the same except that c does not occur). The approach differs substantially from the regularity-based theories of causality that dominated metaphysics through the 1960s, based on relations of entailment from covering law models. For a collection of essays in

Making Things Happen: A Theory of Causal Explanation, the counterfactual account of causation championed by Lewis and his students has not been influenced to any substantial degree by the potential outcomes version of counterfactual modeling used by statisticians, social scientists, and other empirical researchers. However, Woodward and other philosophers have engaged the directed graph approach to causality with considerable energy in the past decade. We will discuss the broader philosophical literature in Chapters 2, 10, and 13, as it does have some implications for social science practice and the pursuit of explanation more generally.

1.2   Causal Analysis and Observational Social Science

The challenges of using observational data to justify causal claims are considerable. In this section, we present a selective history of the literature on these challenges, focusing on the varied usage of experimental language in observational social science. We will also consider the growth of survey research and the shift toward outcome-equation-based motivations of causal analysis that led to the widespread usage of regression estimators. Many useful discussions of these developments exist, and our presentation here is not meant to be complete.[4] We review only the literature that is relevant for explaining the connections between the counterfactual approach and other traditions of quantitatively oriented analysis that are of interest to us here.

1.2.1   Experimental Language in Observational Social Science

Although the common definition of the word experiment is broad, in the social sciences it is most closely associated with randomized experimental designs, such as the double-blind clinical trials that have revolutionized the biomedical sciences and the routine small-scale experiments that psychology professors perform on their own students.[5]

---

philosophy on counterfactuals and causation, see Collins, Hall, and Paul (2004). For a penetrating examination of the counterfactual model in philosophy and its rivals, see Paul and Hall (2013). The counterfactual model that we consider in this book is close to what Paul and Hall label the "causal model" approach to causation, which they consider one of four variants of counterfactual modeling.

[4]To gain a more complete appreciation of the expansive literature on causality in the social sciences, see, for sociology, Barringer, Leahey, and Eliason (2013), Berk (1988, 2004, 2008), Blossfeld (2009), Bollen (1989), Bollen and Pearl (2013), Firebaugh (2008), Fox (2008), Gangl (2010), Goldthorpe (2007), Harding and Seefeldt (2013), Lieberson (1985), Marini and Singer (1988), Morgan (2013), Rohwer (2010), Singer and Marini (1987), Smith (1990, 2003, 2013), Sobel (1995, 1996, 2000), and Treiman (2009). For economics, see Angrist and Krueger (1999), Angrist and Pischke (2009, 2010), Heckman (2000, 2005, 2008b, 2010), Imbens and Wooldridge (2009), Keane (2010), Lee (2005), Manski (1994, 1995, 2003), Moffitt (2003), Pratt and Schlaifer (1984), and Rosenzweig and Wolpin (2000). For political science, see Brady and Collier (2010), Druckman, Kuklinski, and Lupia (2011), Dunning (2012), Gerber and Green (2012), Gerring (2007), Goertz and Mahoney (2012), King, Keohane, and Verba (1994), Morton and Williams (2010), and Sekhon (2009). For applied evaluation and policy analysis, see Shadish, Cook, and Campbell (2001), and especially for education research, see Murnane and Willett (2011)

[5]The *Oxford English Dictionary* provides the scientific definition of experiment: "An action or operation undertaken in order to discover something unknown, to test a hypothesis, or establish or illustrate some known truth" and also provides source references from as early as 1362.

Randomized experiments have their origins in the work of statistician Ronald A. Fisher during the 1920s, which then diffused throughout various research communities via his widely read 1935 book, *The Design of Experiments*.

Statisticians David Cox and Nancy Reid (2000) offer a definition of an experiment that focuses on the investigator's deliberate control and that allows for a clear juxtaposition with an observational study:

> The word *experiment* is used in a quite precise sense to mean an investigation where the system under study is under the control of the investigator. This means that the individuals or material investigated, the nature of the treatments or manipulations under study and the measurement procedures used are all selected, in their important features at least, by the investigator.
>
> By contrast in an observational study some of these features, and in particular the allocation of individuals to treatment groups, are outside the investigator's control. (Cox and Reid 2000:1)

We will maintain this basic distinction throughout this book. We will argue in this section that the potential outcome model of causality that we introduced in the last section is valuable precisely because it helps researchers to stipulate assumptions, evaluate alternative data analysis techniques, and think carefully about the process of causal exposure. Its success is a direct result of the language of potential outcomes, which permits the analyst to conceptualize observational studies as if they were experimental designs controlled by someone other than the researcher – quite often, the subjects of the research. In this section, we offer a brief discussion of other important attempts to use experimental language in observational social science and that succeeded to varying degrees.

Samuel A. Stouffer, the sociologist and pioneering public opinion survey analyst, argued that "the progress of social science depends on the development of limited theories – of considerable but still limited generality – from which prediction can be made to new concrete instances" (Stouffer 1962[1948]:5). Stouffer argued that, when testing alternative ideas, "it is essential that we always keep in mind the model of a controlled experiment, even if in practice we may have to deviate from an ideal model" (Stouffer 1950:356). He followed this practice over his career, from his 1930 dissertation that compared experimental with case study methods of investigating attitudes, to his leadership of the team that produced *The American Soldier* during World War II (see Stouffer 1949), and in his 1955 classic *Communism, Conformity, and Civil Liberties*.

On his death, and in celebration of a posthumous collection of his essays, Stouffer was praised for his career of survey research and attendant explanatory success. The demographer Philip Hauser noted that Stouffer "had a hand in major developments in virtually every aspect of the sample survey – sampling procedures, problem definition, questionnaire design, field and operating procedures, and analytic methods" (Hauser 1962:333). Arnold Rose (1962:720) declared, "Probably no sociologist was so ingenious in manipulating data statistically to determine whether one hypothesis or another could be considered as verified." And Herbert Hyman portrayed Stouffer's method of tabular analysis in charming detail:

> While the vitality with which he attacked a table had to be observed in
> action, the characteristic strategy he employed was so calculating that
> one can sense it from reading the many printed examples.... Multivariate
> analysis for him was almost a way of life. Starting with a simple cross-
> tabulation, the relationship observed was elaborated by the introduction
> of a third variable or test factor, leading to a clarification of the original
> relationship.... But there was a special flavor to the way Sam handled it.
> With him, the love of a table was undying. Three variables weren't enough.
> Four, five, six, even seven variables were introduced, until that simple thing
> of beauty, that original little table, became one of those monstrous crea-
> tures at the first sight of which a timid student would fall out of love with
> our profession forever. (Hyman 1962:324–25)

Stouffer's method was to conceive of the experiment that he wished he could have
conducted and then to work backwards by stratifying a sample of the population
of interest into subgroups until he felt comfortable that the remaining differences in
the outcome could no longer be easily attributed to systematic differences within the
subgroups. He never lost sight of the population of interest, and he appears to have
always regarded his straightforward conclusions as the best among plausible answers.
Thus, as he said, "Though we cannot always design neat experiments when we want to,
we can at least keep the experimental model in front of our eyes and behave cautiously"
(Stouffer 1950:359).

Not all attempts to incorporate experimental language into observational social
science were as well received. Most notably in sociology, F. Stuart Chapin had earlier
argued explicitly for an experimental orientation to nearly all of sociological research,
but while turning the definition of an experiment in a direction that agitated others.
For Chapin, a valid experiment did not require that the researcher obtain control over
the treatment to be evaluated, only that observation of a causal process be conducted
in controlled conditions (see Chapin 1932, 1947). He thus considered what he called
"ex post facto experiments" to be the solution to the inferential problems of the social
sciences, and he advocated matching designs to select subsets of seemingly equivalent
individuals from those who were and were not exposed to the treatment of interest.
In so doing, however, he proposed to ignore the incomparable, unmatched individuals,
thereby losing sight of the population that Stouffer, the survey analyst, always kept
in the foreground.

Chapin thereby ran afoul of emergent techniques of statistical inference, and he
suffered attacks from his natural allies in quantitative analysis. The statistician Oscar
Kempthorne, whose 1952 book *The Design and Analysis of Experiments* would later
become a classic, dismissed Chapin's work completely. In a review of Chapin's 1947
book, *Experimental Designs in Sociological Research*, Kempthorne wrote:

> The usage of the word "experimental design" is well established by now
> to mean a plan for performing a comparative experiment. This implies
> that various treatments are actually applied by the investigator and are
> not just treatments that happened to have been applied to particular units
> for some reason, known or unknown, before the "experiment" was planned.
> This condition rules out practically all of the experiments and experimental
> designs discussed by the author. (Kempthorne 1948:491)

Chapin's colleagues in sociology and demography were often just as unforgiving. Nathan Keyfitz (1948:260), for example, chastised Chapin for ignoring the population of interest and accused him of using terms such as "experimental design" merely to "lend the support of their prestige."

In spite of the backlash against Chapin, in the end he has a recognizable legacy in observational data analysis. The matching techniques he advocated will be discussed in Chapter 5. They have been reborn in the new literature, in part because the population of interest has been brought back to the foreground. But there is an even more direct legacy. Many of Chapin's so-called experiments were soon taken up, elaborated, and analyzed by the psychologist Donald T. Campbell and his colleagues under the milder and more general name of "quasi-experiments."[6]

The first widely read presentation of Campbell's perspective emerged in 1963 (see Campbell and Stanley 1966[1963]), in which quasi-experiments were discussed alongside randomized and fully controlled experimental trials, with an evaluation of their relative strengths and weaknesses in alternative settings. In the subsequent decade, Campbell's work with his colleagues moved closer toward observational research, culminating in the volume by Cook and Campbell (1979), *Quasi-Experimentation: Design & Analysis Issues for Field Settings*, wherein a whole menu of quasi-experiments was described and analyzed: from the sort of ex post case-control matching studies advocated by Chapin (but relabeled more generally as nonequivalent group designs) to novel proposals for regression discontinuity and interrupted time series designs (which we will discuss in Chapter 11). For Cook and Campbell, the term quasi-experiment refers to "experiments that have treatments, outcome measures, and experimental units, but do not use random assignment to create the comparisons from which treatment-caused change is inferred" (Cook and Campbell 1979:6).[7] And, rather than advocate for a reorientation of a whole discipline as Chapin had, they pitched the approach as a guide for field studies, especially program evaluation studies of controlled interventions. Nonetheless, the ideas were widely influential throughout the social sciences, as they succeeded in bringing a tamed experimental language to the foreground in a way that permitted broad assessments of the strengths and weaknesses of alternative study designs and data analysis techniques.

1.2.2 "The Age of Regression"

Even though the quasi-experiment tradition swept through the program evaluation community and gained many readers elsewhere, it lost out in the core social science disciplines to regression-equation-based motivations of observational data analysis,

---

[6]In his first publication on quasi-experiments, Campbell (1957) aligned himself with Stouffer's perspective on the utility of experimental language, and in particular Stouffer (1950). Chapin is treated roughly by Campbell and Stanley (1966[1963]:70), even though his ex post facto design is identified as "one of the most extended efforts toward quasi-experimental design."

[7]Notice that Cook and Campbell's definition of quasi-experiments here is, in fact, consistent with the definition of an experiment laid out by Cox and Reid, which we cited earlier in this section. For that definition of an experiment, control is essential but randomization is not. The text of Cook and Campbell (1979) equivocates somewhat on these issues, but it is clear that their intent was to discuss controlled experiments for which randomization is infeasible and which they then label quasi-experiments.

under the influence at first of researchers who promoted regression modeling from
a path-modeling orientation. In sociology, Hubert Blalock and Otis Dudley Duncan
are usually credited with introducing the techniques, first via Blalock's 1964[1961]
book *Causal Inferences in Nonexperimental Research* and then later via Duncan's
1966 article, "Path Analysis: Sociological Examples," which was published as the lead
article in that year's *American Journal of Sociology*.[8] In both presentations, caution
was stressed. Blalock discussed carefully the differences between randomized experiments and observational survey research. Duncan stated explicitly in his abstract that
"path analysis focuses on the problem of interpretation and does not purport to be
a method for discovering causes," and he concluded his article with a long quotation
from Sewall Wright attesting to the same point.

A confluence of developments then pushed path models toward widespread usage
and then basic regression modeling toward near complete dominance of observational
research in some areas of social science.[9] In sociology, the most important impetus
was the immediate substantive payoff to the techniques. *The American Occupational
Structure*, which Duncan cowrote with Peter Blau and published in 1967, offered new
decompositions of the putative causal effects of parental background and individuals'
own characteristics on later educational and occupational attainment. By pushing
social stratification research into new terrain, their book transformed a core subfield of
the discipline of sociology, leading to major theoretical and methodological redirections
of many existing lines of scholarship.[10]

Researchers seemed to then ignore many of the cautionary statements of Blalock,
Duncan, and others. In their defense, it should be noted that Blalock's guidance was
confusing at times. When introducing regression equations in his 1961 book, specified
as $Y_i = a + bX_i + e_i$, where X is the causal variable of interest and Y is the outcome
variable of interest, Blalock stated the matter correctly and clearly:

> What if there existed a major determinant of Y, not explicitly contained
> in the regression equation, which was in fact correlated with some of the

---

[8]Goldberger (1972) and Heckman (2000) offer a history of usage in economics, which begins before
the history we offer for sociology. The biologist Sewall Wright (1925, 1934) deserves credit for the
earliest developments (see Bollen and Pearl 2013; Pearl 2009).

[9]Regression estimation of systems of linear causal models with observed variables should be
regarded as a restricted form of the more general structural equation modeling approach that Blalock,
Duncan, and others introduced into sociology (see Bollen and Pearl 2013 for an explanation, especially
their debunking of the myth "SEM and Regression Are Essentially Equivalent"). In these early and
very influential pieces, Blalock and Duncan considered only linear causal models with observed variables. Duncan (1966:7) clarified: "As a statistical technique, therefore, neither path analysis nor the
Blalock-Simon procedure adds anything to conventional regression analysis as applied recursively to
generate a system of equations. As a pattern of interpretation, however, path analysis is invaluable in
making explicit the rationale for a set of regression calculations." The literature moved quickly from
the late 1960s to consider overidentified models (see Duncan, Haller, and Portes 1968; Hauser and
Goldberger 1971), after which a general latent variable structural equation model was developed (see
Bollen 1989). Nonetheless, the pattern of practice that elevated regression to its dominant position,
we maintain, was shaped by these early pieces, as well as later work on interpreting the regression
coefficients estimated for basic linear path models with observed variables (e.g., Alwin and Hauser
1975).

[10]For example, compare the methods (and substantive motivations) in Sewell (1964), with its nonparametric table standardization techniques, to Sewell, Haller, and Portes (1969), with its path model
of the entire stratification process.

independent variables X_i? Clearly, it would be contributing to the error term in a manner so as to make the errors systematically related to these particular X_i. If we were in a position to bring this unknown variable into the regression equation, we would find that at least some of the regression coefficients (slopes) would be changed. This is obviously an unsatisfactory state of affairs, making it nearly impossible to state accurate scientific generalizations. (Blalock 1964[1961]:47)

At other points in his book, however, Blalock characterized the same issue in ways that encouraged more permissive practice. He wrote at several points that the goal of causal inference should not be too easily sacrificed:

> Since it will always be possible that some unknown forces may be operating to disturb a given causal relationship, or to lead us to believe a causal relationship exists when in fact it does not, the only way we can make causal inferences at all is to make simplifying assumptions about such disturbing influences. (Blalock 1964[1961]:13)

> We shall assume that error terms are uncorrelated with each other and with any of the independent variables in a given equation.... In non-experimental studies involving nonisolated systems, this kind of assumption is likely to be unrealistic. This means that disturbing influences must be explicitly brought into the model. But at some point one must stop and make the simplifying assumption that variables left out do not produce confounding influences. Otherwise, causal inferences cannot be made. (Blalock 1964[1961]:176)

And, even though Blalock was clear that regression coefficients are estimated quantities (and more fundamentally that the causal models that give regression equations their specifications are subject to simplifying assumptions that may be unrealistic), he still wrote about the resulting coefficients and equations in ways that would surely have excited readers interested in powerful new ways to gain insight from observational data:

> It is the regression coefficients which give us the laws of science. (Blalock 1964[1961]:51)

> In causal analyses our aim is to focus on causal laws as represented by regression equations and their coefficients. (Blalock 1964[1961]:177)[11]

Finally, in the concluding section of his book, he suggested,

> The method for making causal inferences may be applied to models based on a priori reasoning, or it may be used in exploratory fashion to arrive at models which give closer and closer approximations to the data. (Blalock 1964[1961]:179)

---

[11] When using this language of causal laws, Blalock was arguing for the comparative value of metric regression coefficients, in contrast to standardized regression coefficients, and probably using this causal law language with implicit reference to the covering law model of explanation that was dominant in philosophy of science at the time. He was not claiming that regression is a method for the discovery of causal laws. We maintain, nonetheless, that this sort of language had the potential to mislead readers.

Given this type of guidance, it is not hard to imagine practitioners offering up exploration-enhanced causal models, enabled by unrealistic simplifying assumptions, and then writing about their regression coefficients as if causal laws had been uncovered.

Duncan never failed to mention that assumptions about causal relationships must be grounded in theory and cannot be revealed by data. Yet, as Abbott (2001[1998]:115) notes, "Duncan was explicit in [*The American Occupational Structure*] ... about the extreme assumptions necessary for the analysis, but repeatedly urged the reader to bear with him while he tried something out to see what could be learned." What Duncan learned transformed the field, and it was thus hard to ignore the potential power of the techniques to move the literature.

Duncan's 1975 methodological text, *Introduction to Structural Equation Models*, is appropriately restrained, with many fine discussions that echo the caution in the abstract of his 1966 article. Yet he also encouraged widespread application of regression techniques to estimate causal effects, and at times he gave the impression that researchers should just get on with it, as he did in *The American Occupational Structure*. For example, in his chapter 8, titled "Specification Error," Duncan noted that "it would require no elaborate sophistry to show that we will never have the 'right' model in any absolute sense" (Duncan 1975:101). But he then continued:

> As the term will be used here, analysis of specification error relates to a rhetorical strategy in which we suggest a model as the "true" one for sake of argument, determine how our working model [the model that has been estimated] differs from it and what the consequences of the difference(s) are, and thereby get some sense of how important the mistakes we will inevitably make may be. Sometimes it is possible to secure genuine comfort by this route. (Duncan 1975:101–2)

As is widely known, Duncan later criticized the widespread usage of regression analysis, both in his 1984 book *Notes on Social Measurement: Historical and Critical* and in private communication, in which he reminded many inside and outside of sociology of his long-standing cautionary perspective (see Xie 2007).

Finally, the emergent ease with which regression models could be estimated with new computing power was important as well. No longer would Stouffer have needed to concentrate on a seven-way cross-tabulation. Researchers could instead estimate and then interpret only a few estimated regression slopes, rather than attempt to make sense of the hundred or so cells that Stouffer often generated by subdivision of the sample. Aage Sørensen has given the most memorable indictment of the consequences of this revolution in computing power:

> With the advent of the high-speed computer, we certainly could study the relationships among many more variables than before. More importantly, we could compute precise quantitative measures of the strength of these relationships. The revolution in quantitative sociology was a revolution in statistical productivity. Social scientists could now calculate almost everything with little manual labor and in very short periods of time. Unfortunately, the sociological workers involved in this revolution lost control

of their ability to see the relationship between theory and evidence. Sociologists became alienated from their sociological species being. (Sørensen 1998:241)

As this quotation intimates, enthusiasm for regression approaches to causal inference had declined dramatically by the mid-1990s. Naive usage of regression modeling was blamed for nearly all the ills of sociology, everything from stripping temporality and context from the mainstream (see Abbott 2001 for a collections of essays), the suppression of attention to explanatory mechanisms (see Goldthorpe 2001 and Hedstrm 2005), the denial of causal complexity (see Ragin 1987, 2008), and the destruction of mathematical sociology (Sørensen 1998).

It is unfair to lay so much at the feet of least squares formulas, and we will argue later that regression can be put to work quite sensibly in the pursuit of causal questions. However, the critique of regression-based practice was largely on target. For causal analysis, the rise of regression led to a focus on estimated equations for outcomes, rather than careful thinking about how the data in hand differ from what would have been generated by the ideal experiments one might wish to have conducted. This sacrifice of attention to experimental thinking might have been reasonable if the outcome-equation tradition had led researchers to specify and then carefully investigate the plausibility of alternative explanatory mechanisms that generate the outcomes of the equations. But, instead, it seems that researchers all too often chose not to develop fully articulated mechanisms that generate outcomes and chose to simply act as if estimated regression equations somehow mimic appreciably well (by a process not amenable to much analysis) the experiments that researchers might otherwise have wished to undertake.

Largely independent of these critiques, the potential outcome model for observational data analysis has achieved success in the past two decades in the social sciences because it brings experimental language back into observational data analysis. But it does so in the way that Stouffer used it: as a framework in which to ask carefully constructed what-if questions that lay bare the limitations of observational data and the need to clearly articulate assumptions that are believable because they are grounded in theory that is defendable.

Consider the motivating questions in the opening paragraph of this book, which we posed to draw in readers who are not yet utilizing the counterfactual model. These questions are stated in the traditional style of

Does X cause Y?

If X causes Y, how large is the effect of X on Y?

The counterfactual model encourages the formulation of more precise causal questions with clear counterfactual contrasts, such as

If individuals with $X = x'$ had instead had $X = x''$, how much would their value for Y have changed?

For example, rather than ask the question "Does obtaining a college degree increase an individual's labor market earnings?," the counterfactual model encourages researchers to ask two questions:

> 1. If high school graduates had instead obtained college degrees, how much would their labor market earnings have changed?

> 2. If college graduates had only obtained high school diplomas, how much would their labor market earnings have changed?

For these two particular questions, the empirical literature suggests that the answers differ in magnitude, suggesting that important heterogeneity exists in the individual-level causal effects that underlie them. Such differences are of theoretical interest to researchers and of practical value to policymakers, and they are obscured by general cause-and-effect questions without clear and specific counterfactual states. This book is grounded on the position that social scientists ought to use a conceptual and methodological framework that encourages the asking and answering of rigorously posed questions such as these.

1.3   Examples Used Throughout the Book

In this section, we offer background on the main substantive examples that we will draw on throughout the book when discussing the methods and approach abstractly and then when demonstrating particular empirical analysis strategies. We first present five broad foundational examples that have guided empirical research in the social, demographic, and health sciences for decades. We then present eight narrower examples that are at the frontier of current research, all of which can be addressed using the counterfactual model.

1.3.1   Broad Examples from the Social, Demographic, and Health Sciences

We first outline three prominent classic examples that, in spite of their distinct disciplinary origins, are related to each other: (1) the causal effects of family background and mental ability on educational attainment, (2) the causal effects of educational attainment and mental ability on earnings, and (3) the causal effects of family background, educational attainment, and earnings on political participation. These examples are classic and wide-ranging, having been developed, respectively, in the formative years of observational data analysis in sociology, economics, and political science. We then present two broad examples from the interdisciplinary social, demographic, and health sciences: (4) the causal effects of family background and life course events on fertility patterns and (5) the causal effects of socioeconomic status on health and mortality.

The Causal Effects of Family Background and Intelligence on Educational Attainment

In the status attainment tradition in sociology, as pioneered by Blau and Duncan (1967), family background and mental ability are considered to be ultimate causes of educational attainment. This claim is grounded on the purported existence of a specific causal mechanism that relates individuals' expectations and aspirations for the future to the social contexts that generate them. This particular explanation is most often identified with the Wisconsin model of status attainment, which was based on early analyses of the Wisconsin Longitudinal Survey (see Sewell, Haller, and Portes 1969; Sewell, Haller, and Ohlendorf 1970).

According to the original Wisconsin model, the joint effects of high school students' family backgrounds and mental abilities on their eventual educational attainments can be completely explained by the expectations that others hold of them. In particular, significant others – parents, teachers, and peers – define expectations based on students' family background and observable academic performance. Students then internalize the expectations crafted by their significant others. In the process, the expectations become individuals' own aspirations, which then compel achievement motivation.

The implicit theory of the Wisconsin model maintains that students are compelled to follow their own aspirations. Accordingly, the model is powerfully simple, as it implies that significant others can increase high school students' future educational attainments merely by increasing their own expectations of them.[12] Critics of this status attainment perspective argued that structural constraints embedded in the opportunity structure of society should be at the center of all models of educational attainment, and hence that concepts such as aspirations and expectations offer little or no explanatory power. Pierre Bourdieu (1973) dismissed all work that asserts that associations between aspirations and attainments are causal. Rather, for Bourdieu, the unequal opportunity structures of society "determine aspirations by determining the extent to which they can be satisfied" (Bourdieu 1973:83). And, as such, aspirations have no autonomous explanatory power because they are nothing other than alternative indicators of structural opportunities and resulting attainment.

Research on the relationships between family background and educational attainment is now vast, especially in economics and sociology (see Ehrenberg 2004; Morgan 2005; Stevens, Armstrong, and Arum 2008). Scholars disagree on the effects of the resource constraints imposed by disadvantaged family backgrounds, the role of individual choices in response to incentives, the importance of beliefs about the opportunity structure, and features of the institutions that must be navigated (see Breen and Johnson 2005; Heckman 2008a; Holmlund, Lindahl, and Plug 2011; Hoxby 2004; Jackson 2013; Morgan, Leenman, Todd, and Weeden 2013).

The Causal Effects of Educational Attainment and Mental Ability on Earnings

The economic theory of human capital maintains that education has a causal effect on the subsequent labor market earnings of individuals. The theory presupposes that educational training provides skills that increase the potential productivity of workers. Because productivity is prized in the labor market, firms are willing to pay educated workers more.

These claims are largely accepted within economics, but considerable debate remains over the size of the causal effect of education. In reflecting on the first edition of his

---

[12]See Hauser, Warren, Huang, and Carter (2000) for the latest update of the original model and Sewell, Hauser, Warren, and Hauser (2004) for a review of the entire research tradition.

book *Human Capital*, which was published in 1964, Gary Becker wrote nearly 30 years later:

> Education and training are the most important investments in human cap-
> ital. My book showed, and so have many other studies since then, that high
> school and college education in the United States greatly raise a person's
> income, even after netting out direct and indirect costs of schooling, and
> after adjusting for the better family backgrounds and greater abilities of
> more educated people. Similar evidence is now available for many points in
> time from over one hundred countries with different cultures and economic
> systems. (Becker 1993[1964]:17)

The complication, hinted at in this quotation, is that economists also accept that mental ability enhances productivity. Thus, because those with relatively high ability are assumed to be more likely to obtain higher educational degrees, the highly educated are presumed to have higher innate ability and higher natural rates of productivity. As a result, some portion of the purported causal effect of education on earnings may instead reflect innate ability rather than any productivity-enhancing skills provided by educational institutions (see Willis and Rosen 1979).

The degree of "ability bias" in standard estimates of the causal effect of education on earnings has remained one of the largest causal controversies in the social sciences since the 1970s (see Card 1999). Scholars continue to disagree on the magnitude of biases in traditional estimates, and debate has also developed on the variability of returns that may be related to underlying cognitive ability and other individual characteristics (see Cunha and Heckman 2007; Brand and Xie 2010; Carneiro, Heckman, and Vytlacil 2011; Hout 2012).

The Causal Effects of Family Background, Educational Attainment, and Earnings on Political Participation

The socioeconomic status model of political participation asserts that education, occupational attainment, and income predict strongly most measures of political participation (see Verba and Nie 1972). Critics of this model maintain instead that political interests and engagement determine political participation, and these are merely correlated with the main dimensions of socioeconomic status.[13] In other words, those who have a predilection to participate in politics are likely to show commitment to other institutions, such as the educational system.

Verba, Schlozman, and Brady (1995) later elaborated the socioeconomic status model, focusing on the contingent causal processes that they argue generate patterns of participation through the resources conferred by socioeconomic position. They claim that

> interest, information, efficacy, and partisan intensity provide the desire,
> knowledge, and self-assurance that impel people to be engaged by poli-
> tics. But time, money, and skills provide the wherewithal without which

---

[13]This interest model of participation has an equally long lineage. Lazarsfeld, Berelson, and Gaudet (1955[1948]:157) write that, in their local sample, "the difference in deliberate non-voting between people with more or less education can be completely accounted for by the notion of interest."

> engagement is meaningless. It is not sufficient to know and care about politics. If wishes were resources, then beggars would participate. (Verba et al. 1995:355–56)

They reach this conclusion through a series of regression models that predict political participation. They use temporal order to specify causal order, and they then claim to eliminate alternative theories that emphasize political interests and engagement by showing that these variables have relatively weak predictive power in their models.

Moreover, they identify education as the single strongest cause of political participation. Beyond generating the crucial resources of time, money, and civic skills, education shapes preadult experiences and transmits differences in family background (see Verba et al. 1995, figure 15.1). Education emerges as the most powerful cause of engagement because it has the largest net association with measures of political participation.

Nie, Junn, and Stehlik-Barry (1996) then built on the models of Verba and his colleagues, specifying in detail the causal pathways linking education to political participation. For this work, the effects of education, family income, and occupational prominence (again, the three basic dimensions of socioeconomic status) on voting frequency are mediated by verbal proficiency, organizational membership, and social network centrality. Nie et al. (1996:76) note that these variables "almost fully explain the original bivariate relationship between education and frequency of voting."

Current research continues to investigate these relationships, especially the effect of education on participation. Some studies have focused attention on relative education (e.g., Tenn 2005), while others have questioned whether the effect is genuine at all (Berinsky and Lenz 2011; Highton 2009; Kam and Palmer 2008, 2011). As often happens in causal controversies, the latter research prompted additional effort to restore the original claim (Henderson and Chatfield 2011; Mayer 2011; Sondheimer and Green 2010).[14]

The Causal Effects of Family Background and Life Course Events on Fertility Patterns

Through the analysis of population trends in the nineteenth and early twentieth centuries, demographers developed the concept of a "demographic transition," during which countries move in comparatively short periods of time from population regimes of high fertility and high mortality toward those of low fertility and low mortality (see Davis 1945; Thompson 1949; Notestein 1950). Ronald Lee characterizes the worldwide march of country-specific transitions as "Three Centuries of Fundamental Change":

> Before the start of the demographic transition, life was short, births were many, growth was slow and the population was young. During the transition, first mortality and then fertility declined, causing population growth

---

[14]Perhaps the most widely known recent political participation studies are those that do not focus on the effects of stable background characteristics of individuals. Research on "Get Out the Vote" operations shows, for example, that social pressure is surprisingly effective (Gerber, Green, and Larimer 2008; Davenport, Gerber, Green et al. 2010), while attack advertisements broadcast on television are less effective than many have assumed (Krasno and Green 2008).

rates first to accelerate and then to slow again, moving toward low fertility, long life and an old population. The transition began around 1800 with declining mortality in Europe. It has now spread to all parts of the world and is projected to be completed by 2100. (Lee 2003:167)

Although these trends have inspired many strands of literature in demography, we will use as our broad example the interconnected causes that are typically examined for differences in fertility rates, both over time and across groups, focusing mostly on the literature on fertility in the first group of industrializing countries to experience the demographic transition.[15]

Early studies document differences in fertility rates by immigrant status, geographic region, urban–rural location, and social class (see Dinkel 1952; Karpinos 1938; Notestein 1933; Rose 1942; Thompson 1948; Whelpton 1932), often using causal language, for example, "the birth rate for married couples not separating during the migration was higher after they came to the United States than it would have been had they remained in Italy" (Rose 1942:621). Most of this literature is concerned with how fertility differences vary with family background, typically across social classes as measured by the occupations of husbands. Analyzing trends during a time when the eugenics movement had not yet fully receded, Notestein (1933:22) concludes:

> At present the white collar classes are not reproducing rapidly enough to maintain equal permanent replacement, but the unskilled laborer class and the agricultural population appear to be reproducing more rapidly than is required to maintain their numbers. (Notestein 1933:33)

Notestein offers empirical models to support the position that some of these differences can be attributed to the effects of age at marriage, but he also notes the possibility that "in the young marriages of the professional class fertility was purposefully and effectively controlled even prior to 1910" (Notestein 1933:25).

Although this early literature considers the timing of marriage and the possibility of differential patterns of overt birth control, it does not consider the full range of gender-related mechanisms that with hindsight were obviously at play. After correctives were offered in subsequent literature, newly available individual-level data have allowed demographers to model the consequences of changing rates of female labor force participation (see Brewster and Rindfuss 2000 for a review), changes in birth control practices (Westoff and Bumpass 1973; Goldin and Katz 2002), and some of the deeper contingencies of related life course events, such as how age and marital status at first birth structure later fertility (Bumpass, Rindfuss, and Janosik 1978; see also Morgan and Rindfuss 1999).

Studies that examine family background differences in fertility, as indexed by social class, are less common in current research, having been supplanted by the study of the relationship between educational attainment and fertility, as conditioned by family

---

[15]Although there is considerable debate on the role of causal analysis in demographic research (see Duncan 2008; Engelhardt, Kohler, and Prskawetz 2009; Moffitt 2005; Ní Bhrolcháin and Dyson 2007; Smith 1989, 2009, 2013; Xie 2011), we do not see demography as inherently different than the other domains of observational research considered in this book. However, it may be the case that demography has a special additional burden of documenting social and demographic patterns that do not depend on any particular causal assumptions.

background of origin (see Brand and Davis 2011; Musick, England, Edgington, and Kangas 2009). This more contemporary literature also takes full account of the distinct patterns of marital and nonmarital fertility (see Musick 2002; Musick, England, Edgington, and Kangas 2009; Seltzer, Bachrach, Bianchi et al. 2005; Wu 1996, 2008; Wu and Wolfe 2001).

The Causal Effects of Socioeconomic Status on Health and Mortality

In reaction to the traditional focus of epidemiology on the proximate direct causes of disease and mortality, a group of social epidemiologists has advanced the case that socioeconomic status should be considered a fundamental cause of health disparities across the life course, generating robust and recurrent associations between socioeconomic status and both health and mortality (Link and Phelan 1995; Phelan, Link, Diez-Roux et al. 2004; Phelan, Link, and Tehranifar 2010). Here, the three traditional dimensions of socioeconomic status – education, income, and occupation – are all considered to be active (see Adler and Newman 2002). Lutfey and Freese characterize the nature of the posited causal relationship:

> If an explanatory variable is a fundamental cause of an outcome, then the association cannot be successfully reduced to a set of more proximate, intervening causes because the association persists even while the relative influence of various proximate mechanisms changes. (Lutfey and Freese 2005:1328)

Socioeconomic status is therefore a fundamental cause, according to this perspective, because it is a paramount distal determinant of health that activates alternative and replaceable causal pathways, such as those that arise through differential access to quality health care, knowledge about health innovations, and propensity to engage in risky health behaviors (see Cawley and Ruhm 2012; Fiscella, Franks, Gold, and Clancy 2000; Pampel, Krueger, and Denney 2010). Thus, while health may improve on average for all, socioeconomic disparities may persist, or even grow, because those who are disadvantaged by low education, low income, or lack of employment are less able to take advantage of improvements in health care.

Each of these five broad examples, as noted earlier, is concerned with relationships that unfold over the life course of the majority of individuals in most industrialized societies. As such, these examples encompass some of the most important early substantive scholarship in sociology, economics, and political science as well as the latest frontiers of research at the interdisciplinary nexus of social, demographic, and health science. At the same time, however, these examples pose some fundamental challenges for causal analysis: measurement complications and potential nonmanipulability of the causes of interest. Each of these deserves some comment before the narrower and less complicated examples that follow are introduced.

First, the purported causal and outcome variables in these models are sometimes highly abstract and internally differentiated. Consider the political science example. Political participation takes many forms, from volunteer work to financial giving and voting. Each of these, in turn, is itself heterogeneous, given that individuals can contribute episodically and vote in only some elections. Furthermore, family background and socioeconomic status include at least three underlying dimensions: family income, parental education, and occupational position. But other dimensions of advantage, such as wealth and family structure, must also be considered, as these are thought to be determinants of both an individual's educational attainment and also the resources that supposedly enable political participation.[16]

Scholars who pursue analysis of these causal effects must therefore devote substantial energy to the development of measurement scales. Although very important to consider, in this book we will not discuss measurement issues so that we can focus closely on causal effect estimation strategies. But, of course, it should always be remembered that, in the absence of agreement on issues of how to measure causes and their outcomes, few causal controversies can be resolved, no matter what estimation strategy seems best to adopt.

Second, most of these examples examine causal effects for individual characteristics that are not easily manipulable through external intervention. Or, more to the point, even when they are manipulable, any such induced variation may differ fundamentally from the naturally occurring (or socially determined) variation with which the models are most directly concerned. For example, family background could be manipulated by somehow convincing a sample of middle-class and working-class parents to exchange their children at particular well-chosen ages, but the subsequent outcomes of this induced variation may not correspond to the family background differences that the original models attempt to use as explanatory differences.

As we will discuss later, whether nonmanipulability of a cause presents a challenge to an observational data analyst is a topic of continuing debate in the methodological and philosophical literature. We will discuss this complication at several points in this book, including a section in the concluding Chapter 13 (see pages 439–441), where we argue that critics have overemphasized this concern. But, given that the measurement and manipulability concerns of these broad examples present challenges at some level, we also draw on more narrow examples throughout the book, as we discuss in the next section. For these examples, measurement is generally less controversial and potential manipulability is more plausible (and in some cases is completely straightforward).

1.3.2   Narrow and Specific Examples

In this section, we present additional examples that we will use at multiple points throughout the book: the causal effects of neighborhood of residence and father absence on child development, educational performance, and deviance in adolescence; the causal effects of Catholic schooling, school vouchers, and charter schools on learning; the causal effect of worker training on earnings; the causal effects of risky health behaviors and peer relationships on obesity and mortality; and the causal effects of alternative voting technology on valid voting and election outcomes. These additional

---

[16]Moreover, education as a cause is somewhat ungainly as well. For economists who wish to study the effects of learned skills on labor market earnings, simple variables measuring years of education obtained are oversimplified representations of human capital.

examples are more specific, and often more recent, versions of the five broad examples presented in the last section.

The Causal Effects of Neighborhood of Residence on Educational Performance, Deviance, and Youth Development

Working within the broad tradition of research on educational attainment and the transition to adulthood, social scientists have investigated the effects of neighborhood of residence since at least the 1980s (see Jencks and Mayer 1990 for a review of the early research). Reflecting on his four decades of research on neighborhoods, William Julius Wilson writes of one such possible effect:

> [O]ne of the significant arguments in *When Work Disappears* is that a neighborhood in which people are poor and working is significantly different from a neighborhood in which people are poor and jobless. Jobless neighborhoods create special problems, exacerbating conditions that reinforce racial stereotypes and prejudices. High rates of joblessness trigger other problems in the neighborhood ranging from crime, gang violence, and drug trafficking to family breakups and other disruptions in the organization of family life. (Wilson 2011:10)

The effects of neighborhoods have proven persistently difficult to estimate. Individuals make systematic but constrained residential choices, and analysts rarely have sufficient information to model their choices effectively. Furthermore, neighborhoods have many characteristics, and individuals living within them can be influenced to varying degrees by circumstances only partly under their own control. For young residents of neighborhoods, these effects may be even more complex:

> Neighborhoods are not static features of a child's life; instead, neighborhoods change over time as children move through different periods of development, providing unique risks and opportunities at each stage. It follows that neighborhoods have the potential to alter developmental trajectories, and that their influence may be lagged or cumulative. (Sampson, Sharkey, and Raudenbush 2008:851–52)

Researchers have considered the effects of neighborhoods on a range of outcomes for children and adolescents (see Harding, Gennetian, Winship et al. 2011 for a review). The ensuing debates have not been settled by first-rate observational data analysis (e.g., Harding 2003; Sharkey and Elwert 2011; Sampson 2012) or by large-scale social experimentation (see Gennetian, Sanbonmatsu, Katz et al. 2012; Kling, Liebman, Katz 2007; Kling, Ludwig, and Katz 2005; Sampson 2008).

The Causal Effects of Father Absence on Child and Adolescent Development

Bridging work on family background effects with work in family demography, social scientists have considered the consequences of family structure for child and adolescent development (see McLanahan 2004, 2009; McLanahan and Percheski 2008; Wu and

Wolfe 2001). The most prominent strand of this research began as an effort to assess the effect of growing up as the child of a single parent (McLanahan and Sandefur 1994), during a period when single parenthood was on the rise and the subject of intense political debate.

Recently, the literature has come to focus more specifically on father absence. In a review of the latest research, McLanahan and her colleagues conclude:

> We find strong evidence that father absence negatively affects children's social-emotional development, particularly by increasing externalizing behavior. These effects may be more pronounced if father absence occurs during early childhood than middle childhood, and they may be more pronounced for boys than for girls. There is weaker evidence of an effect of father absence on children's cognitive ability.
>
> Effects on social-emotional development persist into adolescence, for which we find strong evidence that father absence increases adolescents' risky behavior, such as smoking or early childbearing. The evidence of an effect on adolescent cognitive ability continues to be weaker, but we do find strong and consistent negative effects of father absence on high school graduation. The latter finding suggests that the effects on educational attainment operate by increasing problem behaviors rather than by impairing cognitive ability. (McLanahan, Tach, and Schneider 2013:422)

A rich array of models has been used to generate estimates of these effects, and yet some controversy remains over how substantial these effects are and whether they should instead be attributed to unmeasured environmental characteristics of families, schools, and neighborhoods, including the complexity of events that often co-occur with father absence.

The Causal Effect of Catholic Schooling on Learning

James S. Coleman and his colleagues presented evidence that Catholic schools are more effective than public schools in teaching mathematics and reading to equivalent high school students (see Coleman and Hoffer 1987; Coleman, Hoffer, and Kilgore 1982; Hoffer, Greeley, and Coleman 1985). Their findings were challenged vigorously by other researchers, who argued that public school students and Catholic school students are insufficiently comparable, even after adjustments for family background and measured motivation to learn (see Alexander and Pallas 1983, 1985; Murnane, Newstead, and Olsen 1985; Noell 1982; Willms 1985; see Bryk, Lee, and Holland 1993 for a summary of the debate). Although the challenges were wide ranging, the most compelling argument raised (and that was foreseen by Coleman and his colleagues) was that students who are most likely to benefit from Catholic schooling are more likely to enroll in Catholic schools net of all observable characteristics. Thus, self-selection on the causal effect itself may generate a mistakenly large apparent Catholic school effect. If students instead were assigned randomly to Catholic and public schools, both types of schools would be shown to be equally effective on average.

To address the possibility that self-selection dynamics create an illusory Catholic school effect, a later wave of studies then assessed whether or not naturally occurring experiments were available that could be used to more effectively estimate the Catholic school effect. Using a variety of variables that predict Catholic school attendance (e.g., share of the local population that is Catholic) and putting forth arguments for why these variables do not directly determine achievement, Evans and Schwab (1995), Hoxby (1996), and Neal (1997) generated support for Coleman's original conclusions.[17]

More recent research has considered whether the Catholic school effect still exists in the twenty-first century, generating similar results with alternative and more recent data sources (see Carbonaro and Covay 2010; Morgan and Todd 2008; West and Woessmann 2010). Similar recent research demonstrates that the case for a Catholic school effect at the primary and middle school levels is considerably weaker (Hallinan and Kubitschek 2012; Jepsen 2003; Reardon, Cheadle, and Robinson 2009).

The Causal Effect of School Vouchers on Learning

In response to a perceived crisis in public education in the United States, policymakers have introduced publicly funded school choice programs into some metropolitan areas in an effort to increase competition among schools on the assumption that competition will improve school performance and resulting student achievement (see Chubb and Moe 1990; see also Fuller and Elmore 1996). Although these school choice programs differ by school district, the prototypical design is the following. A set number of $3,000 tuition vouchers redeemable at private schools are made available to students resident in the public school district, and all parents are encouraged to apply for one of these vouchers. The vouchers are then randomly assigned among those who apply. Students who receive a voucher remain eligible to enroll in the public school to which their residence status entitles them. But they can choose to enroll in a private school. If they choose to do so, they hand over their $3,000 voucher but may then be required to pay top-up tuition and fees.

The causal effects of interest resulting from these programs are numerous. Typically, evaluators are interested in the achievement differences between those who attend private schools using vouchers and other suitable comparison groups. Most commonly, the comparison group is the group of voucher applicants who lost out in the lottery and ended up in public schools (see Howell and Peterson 2002; Hoxby 2003; Ladd 2002; Neal 2002). And, even though these sorts of comparisons may seem entirely straightforward, the published literature shows that considerable controversy surrounds how best to estimate these effects, especially given the real-world complexity that confronts the implementation of randomization schemes (see Jin, Barnard, and Rubin 2010; Krueger and Zhu 2004; Peterson and Howell 2004).

For this example, other effects are of interest as well. A researcher might wish to know how the achievement of students who applied for vouchers but did not receive them changed in comparison with those who never applied for vouchers in the first place (as this would be crucial for understanding how the self-selecting group of voucher applicants may differ from other public school students). More broadly, a researcher might wish to know the expected achievement gain that would be observed for a public

---

[17]See Cohen-Zada and Elder (2009) for a similar instrumental variable approach that is less supportive of Coleman and colleagues' results for test scores. See also Altonji, Elder, and Taber (2005a, 2005b).

school student who was randomly assigned a voucher irrespective of the application process. This goal would necessitate altering the voucher assignment mechanism, and thus it has not been an object of research. Finally, the market competition justification for creating these school choice policies implies that the achievement differences of primary interest are those among public school students who attend voucher-threatened public schools (i.e., public schools that feel as if they are in competition with private schools but that did not feel as if they were in competition with private schools before the voucher program was introduced).

The Causal Effect of Charter Schools on Learning

Complementing the school voucher example, we will also consider as an additional example the contentious research on charter schooling in the United States. In an excellent book on recent academic and public debates on the effectiveness of charter schools, Henig (2008:2) introduces and defines charter schools in the following way:

> Just a little more than fifteen years since the first charter school opened in Minnesota, there are now nearly 4,000 nationwide, serving an estimated 1.1 million students.... The laws governing charter schools differ – sometimes substantially – from state to state, of course, but some general characteristics have emerged. Charter schools receive public funding on a per-student basis, are often responsible for achieving educational outcomes defined by their government chartering entity, and are subject to at least nominal public oversight. They typically are barred from charging tuition on top of the public per-pupil allocation, but are free to pursue other forms of supplementary support from donors, foundations, or corporate sponsors. Although they must observe certain baseline regulations, such as prohibitions on discrimination and the provision of safe environments, they are exempt from many of the rules and regulations that bind regular public schools to specific standards and procedures. This hybrid status... has made charter schools a special focus of attention and helped draw them into ideological whirlpools that raise the stakes surrounding the research into their actual form and consequences.

At their core, the central research questions in the charter school debate are simple to state (and identical in structure to those of the other two schooling examples presented above): Do students who attend charter schools perform better on standardized tests than they would have performed if they had instead attended regular public schools? Would students who attend regular public schools perform better on standardized tests if they had instead attended charter schools?

The contentious research that has addressed these questions is distinguished in many respects (see Abdulkadiroglu, Angrist, Dynarski et al. 2011; Angrist, Dynarski, Kane et al. 2010; Center for Research on Educational Outcomes 2009; Hoxby, Murarka, and Kang 2009; Tuttle, Gill, Gleason et al. 2013). Not only are some of its combatants leading researchers at the nation's top universities, many of them are unusually ideological (as Henig shows brilliantly in his book). Their scholarly energy and policy advocacy is amplified by the public attention that has been paid to charter schools by the national press, which is related to the support that charter schools have received from celebrity donors and from recent presidential aspirants. At the same time, the research that informs the debate is cutting-edge in the best sense. Careful attention is paid to details of measurement, and the research designs that have been adopted are a healthy mixture of basic comparisons of achievement levels as well as daring attempts to leverage quasi-experimental variation from the ways in which charter school programs are administered (e.g., using lotteried-out students for comparison groups, when such groups exist).

What makes estimating the effects of charter schools complex, perhaps even more so than for the Catholic school effect and the school vouchers effect, is the underlying heterogeneity of the real world. The process by which some students become enrolled in charter schools is only partly observed. At the same time, charter schools differ greatly from each other, such that the effect of charter schooling must surely vary because of quality differences, as well as the match between each student and the unique features of each charter school.

The Causal Effect of Worker Training on Earnings

The United States federal government has supported worker training programs for economically disadvantaged citizens for decades (see LaLonde 1995). Through a series of legislative renewals, these programs have evolved substantially, and program evaluations have become an important area of applied work in labor and public economics. The services provided to trainees differ and include classroom-based vocational education, remedial high school instruction leading to a general equivalency degree, and on-the-job training (or retraining) for those program participants who have substantial prior work experience. The types of individuals served by these programs are heterogeneous, including ex-felons, welfare recipients, and workers displaced from jobs by foreign competition. Accordingly, the causal effects of interest are heterogeneous, varying with individual characteristics and the particular form of training provided.

Even so, some common challenges have emerged across most program evaluations. Ashenfelter (1978) discovered what has become known as "Ashenfelter's dip," concluding after his analysis of training-program data that

> all of the trainee groups suffered unpredicted earnings declines in the year prior to training.... This suggests that simple before and after comparisons of trainee earnings may be seriously misleading evidence. (Ashenfelter 1978:55)

Because trainees tend to have experienced a downward spiral in earnings just before receiving training, the wages of trainees would rise to some degree even in the absence of any training. Ashenfelter and Card (1985) then pursued models of these "mean reversion" dynamics, demonstrating that the size of treatment effect estimates is a function of alternative assumptions about pre-training earnings trajectories. They called for the construction of randomized field trials to improve program evaluation.

LaLonde (1986) then used results from program outcomes for the National Supported Work (NSW) Demonstration, a program from the mid-1970s that randomly assigned subjects to alternative treatment conditions. LaLonde argued that most of the econometric techniques used for similar program evaluations failed to match the experimental estimates generated by the NSW data. Since LaLonde's 1986 paper, economists and statisticians have continued to refine procedures for evaluating both experimental and nonexperimental data from training programs, focusing in detail on how to model the training selection mechanism (see Frumento, Mealli, Pacini, and Rubin 2012; Heckman, LaLonde, and Smith 1999; Heckman and Vytlacil 2005, 2007; Smith and Todd 2005; Zhang, Rubin, and Mealli 2008, 2009).

The Causal Effects of Risky Health Behaviors and Peer Relationships on Obesity and Mortality

Health scientists are increasingly concerned with the worldwide increase in obesity rates (see Swinburn, Sacks, Hall et al. 2011). In the United States, social scientists have devoted considerable energy to understanding the interrelationships between risky health behaviors that are sometimes referred to as modifiable risk factors for mortality (see Cawley and Ruhm 2012; Pampel, Krueger, and Denney 2010). Although declines in most risky behaviors are evident since the 1970s, especially in rates of smoking, the gains to health from these trends have been mitigated by a concomitant increase in obesity over the same time period (see Stewart, Cutler, and Rosen 2009).

Many causal controversies exist in efforts to assess the causes and consequences of obesity (see Cawley 2011), but the most vigorously debated has been the claim that obesity is contagious. In a large study of a networked community over 32 years, Christakis and Fowler conclude:

> Our study suggests that obesity may spread in social networks in a quantifiable and discernable pattern that depends on the nature of social ties. Moreover, social distance appears to be more important than geographic distance within these networks. Although connected persons might share an exposure to common environmental factors, the experience of simultaneous events, or other common features (e.g., genes) that cause them to gain or lose weight simultaneously, our observations suggest an important role for a process involving the induction and person-to-person spread of obesity.... Obesity in alters might influence obesity in egos by diverse psychosocial means, such as changing the ego's norms about the acceptability of being overweight, more directly influencing the ego's behaviors (e.g., affecting food consumption), or both. Other mechanisms are also possible. (Christakis and Fowler 2007:377)

Many other scholars have objected to these claims (e.g., Shalizi and Thomas 2011), and their objections have been addressed by counterarguments and additional analyses (e.g., Christakis and Fowler 2013; VanderWeele 2011b).

The Causal Effect of Voting Technology on Valid Voting and Election Outcomes

For specific causal effects embedded in the larger political participation debates, we could focus on particular decision points – the effect of education on campaign contributions, net of income, and so on. However, the politics literature is appealing in another respect: outcomes in the form of actual votes cast and subsequent election victories. These generate finely articulated counterfactual scenarios.

Although recent research has considered a broad range of voting technology effects (see Card and Moretti 2007; Hanmer, Park, Traugott et al. 2010), the most famous example remains the contested 2000 presidential election in the United States, where considerable attention was focused on the effect of voting technology on the election outcome in Florida. Wand, Shotts, Sekhon et al. (2001) published a refined version of their analysis that spread like wildfire on the Internet in the week following the presidential election. They asserted that

> the butterfly ballot used in Palm Beach County, Florida, in the 2000 pres- idential election caused more than 2,000 Democratic voters to vote by mistake for Reform candidate Pat Buchanan, a number larger than George W. Bush's certified margin of victory in Florida. (Wand et al. 2001:793)

Reflecting on efforts to recount votes undertaken by various media outlets, Wand and his colleagues identify the crucial contribution of their analysis:

> Our analysis answers a counterfactual question about voter intentions that such investigations [by media outlets of votes cast] cannot resolve. The inspections may clarify the number of voters who marked their ballot in support of the various candidates, but the inspections cannot tell us how many voters marked their ballot for a candidate they did not intend to choose. (Wand et al. 2001:804)

Herron and Sekhon (2003) then examined invalid votes that resulted from overvotes (i.e., voting for more than one candidate), arguing that such overvotes further hurt Gore's vote tally in two crucial Florida counties. Finally, Mebane (2004) then consid- ered statewide voting patterns, arguing that if voters' intentions had not been thwarted by technology, Gore would have won the Florida presidential election by 30,000 votes. One particularly interesting feature of this example is that the precise causal effect of voting technology on votes is not of interest, only the extent to which such causal effects aggregate to produce an election outcome inconsistent with the preferences of those who voted (see also Yamamoto 2012 for a related point).

1.4 Observational Data and Random-Sample Surveys

When we discuss methods and examples throughout this book, we will usually assume that the data have been generated by a relatively large random-sample survey of a well- defined population. We will also assume that the proportion and pattern of individuals who are exposed to the cause are fixed in the population by whatever process generates causal exposure.

We rely on the random-sample perspective because we feel it is the most natural framing of these methods for the typical social scientist, even though many of the classic applications and early methodological pieces in this literature do not reference random-sample surveys. For the examples just summarized, the first three have been examined primarily with random-sample survey data, but many of the others have not. Some, such as the worker training example, depart substantially from this sort of setup, as the study subjects for the treatment in that example are a heterogeneous collection of welfare recipients, displaced workers, and others.[18]

Pinning down the exact consequences of the data generation and sampling scheme of each application is important for developing estimates of the expected variability of a causal effect estimate. We will therefore sometimes modify the random-sampling background when discussing what is known about the expected variability of the alternative estimators we will present. Nonetheless, our primary focus in this book is on strategies to estimate parameters that can be interpreted as warranted causal effects, and accordingly we will give far less attention to procedures for estimating the standard errors of these parameter estimates. In fact, as the reader will notice in subsequent chapters, we will often assume that the sample is infinite. This preposterous assumption is useful for presentation purposes because it simplifies matters greatly; we can then assume that sampling error is zero and assert, for example, that the sample mean of an observed variable is equal to the population expectation of that variable. But this assumption also signals a critical note of caution: It is meant to appear preposterous and unreasonable in order to reinforce the point that the consequences of sampling error must always be considered in any empirical analysis.[19] Our assumption is that our readers know how to estimate and utilize standard errors for many analysis situations, and so we will discuss these issues only when additional guidance is needed for the particular estimators presented in this book.

Moreover, we will also assume for our presentation that the variables in the data are measured without error. This perfect measurement assumption is, of course, also entirely unreasonable. But it is commonly invoked in discussions of causality and in many, if not most, other methodological pieces. We will indicate in various places throughout the book when random measurement error is especially problematic for the methods that we present. We leave it as self-evident that nonrandom measurement error can be debilitating for all methods.

---

[18]Partly for this reason, some of the literature (e.g., Imbens 2004) has made careful distinctions between the sample average treatment effect (SATE) and the population average treatment effect (PATE). In this book, we will focus most of our attention on the PATE (and other conditional PATEs). We will generally assume that a well-defined population exists (usually a superpopulation with explicit characteristics) and that the available data are a random sample from this population. However, much of our treatment of these topics could be rewritten without the large random-sample perspective and focusing only on the average treatment effect within the sample in hand. Many articles in this tradition of analysis adopt this alternative starting point (especially those relevant for small-scale studies in epidemiology and biostatistics for which the "sample" is generated in such a way that a formal connection to a well-defined population is impossible). We discuss these issues in substantial detail in Chapter 2, especially in its appendix on alternative population models.

[19]Because we will assume in these cases that the sample is infinite, we must then also assume that the population is infinite. This assumption entails adoption of the superpopulation perspective from statistics (wherein the finite population from which the sample is drawn is regarded as one realization of a stochastic superpopulation). Even so, and as we will explain in Chapter 2, we will not clutter the text of the book by making fine distinctions between the observable finite population and its more encompassing superpopulation.

1.5 Causal Graphs as an Introduction to the Remainder of the Book

In Chapters 2 and 3, we will introduce what we regard as the two main pieces of the counterfactual approach to causal analysis for observational social science – the potential outcome model and the directed graph approach to causal analysis. In Chapter 4, we will present the basic conditioning strategy for the estimation of causal effects, after which we will then explain – in Chapters 5 through 7 – why matching, regression, and weighted regression estimators are complementary implementations of the more general conditioning strategy.

We will then make the transition from "easy" to "hard" instances of causal effect estimation, for which simple conditioning will not suffice because relevant variables that are related to causal exposure are not observed. After presenting the general predicament in Chapter 8, we will then offer Chapters 9 through 11 on instrumental variable techniques, mechanism-based estimation of causal effects, and the usage of over-time data to estimate causal effects. Finally, we will consider in Chapter 12 how to proceed when no estimators are available to offer warranted point estimates of causal effects, considering both the literature on set identification and sensitivity analysis.

In conclusion, in Chapter 13 we will provide a summary of some of the objections that others have developed against the counterfactual model. We will also offer a broad discussion of the complementary modes of causal inquiry that comprise causal effect estimation in observational social science.

In part because our detailed Table of Contents already gives an accurate accounting of the material that we will present in the remaining chapters, we will not provide a set of detailed chapter summaries here. Instead, we will conclude this introductory chapter with three causal diagrams and the causal effect estimation strategies that they suggest. These graphs allow us to foreshadow some of the specific causal effect estimation strategies that we will present later.

Because the remainder of the material in this chapter will be reintroduced and more fully explained later (primarily in Chapters 3, 4, 8, and 10), it can be skipped now without consequence. However, our experience in teaching this material suggests that many readers may benefit from a quick graphical introduction to the basic estimation techniques before considering the details of the counterfactual framework for observational data analysis.

Graphical Representations of Causal Relationships

Judea Pearl (2000, 2009) and others have developed a general set of rules for representing causal relationships with graphs.[20] We will provide a more complete introduction to directed graph representations of causal effects in Chapter 3, and for now we use the most intuitive pieces of Pearl's graphical apparatus with only minimal discussion of technical details.

---

[20]These graphs can be interpreted as the most recent incarnation of the path diagrams developed by Sewall Wright (1925, 1934; see Bollen and Pearl 2013).

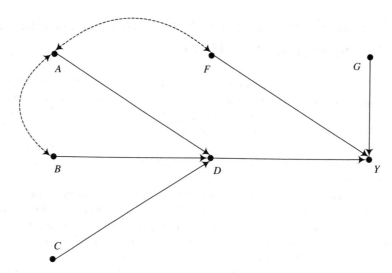

Figure 1.1 A causal graph in which back-door paths from D to Y can be blocked by observable variables and in which C is an instrumental variable for D.

Consider the causal relationships depicted in the graph in Figure 1.1 and suppose that these relationships are derived from a set of theoretical propositions that have achieved consensus in the relevant scholarly community. For this graph, each node represents an observable random variable. Each directed edge (i.e., single-headed arrow) from one node to another signifies that the variable at the origin of the directed edge causes the variable at the terminus of the directed edge. Each curved and dashed bidirected edge (i.e., double-headed arrow) signifies the existence of common unobserved nodes that cause both terminal nodes. Bidirected edges represent common causes only, not mere correlations with unknown sources and not relationships of direct causation between the two variables that they connect.

Now, suppose that the causal variable of primary interest is D and that the causal effect that we wish to estimate is the effect of D on Y. The question to consider is the following: Given the structure of causal relationships represented in the graph, which variables must we observe and then use in a data analysis routine to estimate the size of the causal effect of D on Y?

Three Strategies to Estimate Causal Effects

Although we will consider many strategies for estimating causal effects in this book, we will give our most sustained attention to the following three strategies. First, one can condition on variables (with procedures such as stratification, matching, and regression) to eliminate the noncausal portion of an association between a causal variable and an outcome variable. This strategy is often referred to as conditioning to block all "back-door" paths from the causal variable to the outcome variable, where a back-door path is defined as any path between the causal variable and the outcome variable that begins with an arrow that points to the causal variable. Second, one can use exogenous variation in an appropriate instrumental variable to isolate covariation in the causal and outcome variables. Third, one can establish an exhaustive and isolated mechanism that relates the causal variable to the outcome variable and then calculate the causal effect as it propagates through the mechanism.

Consider the graph in Figure 1.1 and the opportunities it presents to estimate the causal effect of D on Y with the conditioning estimation strategy. First note that there are two back-door paths from D to Y in the graph that generate a supplemental noncausal association between D and Y: (1) $D \leftarrow A \leftarrow\!-\!-\!-\!\rightarrow F \rightarrow Y$ and (2) $D \leftarrow B \leftarrow\!-\!-\!-\!\rightarrow A \leftarrow\!-\!-\!-\!\rightarrow F \rightarrow Y$.[21] Both of these back-door paths can be blocked in order to eliminate the supplemental noncausal association between D and Y by observing and then conditioning on A and B or by observing and then conditioning on F. These two conditioning strategies are general in the sense that they will succeed in producing consistent estimates of the causal effect of D on Y under a variety of conditioning techniques and in the presence of nonlinear effects. They are minimally sufficient in the sense that one can observe and then condition on any subset of the observed variables in $\{A, B, C, F, G\}$ as long as the subset includes either $\{A, B\}$ or $\{F\}$.[22]

Now, consider the second estimation strategy, which is to use an instrumental variable for D to estimate the effect of D on Y. This strategy is completely different from the conditioning strategy just summarized. The goal is not to block back-door paths from the causal variable to the outcome variable but rather to use a localized exogenous shock to both the causal variable and the outcome variable in order to estimate indirectly the relationship between the two. For the graph in Figure 1.1, the variable C is a valid instrument for D because it causes D but does not have an effect on Y except through its effect on D. As a result, one can estimate consistently the causal effect of D on Y by taking the ratio of the relationship between C and Y and between C and D.[23] For this estimation strategy, A, B, F, and G do not need to be observed if the only interest of a researcher is the causal effect of D on Y.

To further consider the differences between these first two strategies, now consider the alternative graph presented in Figure 1.2. There are five possible strategies for estimating the causal effect of D on Y for this graph, and they differ from those

---

[21] As we note later in Chapter 4 when more formally defining back-door paths, the two paths labeled "back-door paths" in the main text here may represent many back-door paths because the bidirected edges may represent more than one common cause of the variables they point to. Even so, the conclusions stated in the main text are unaffected by this possibility because the minimally sufficient conditioning strategies apply to all such additional back-door paths as well.

[22] For the graph in Figure 1.1, one cannot effectively estimate the effect of D on Y by simply conditioning only on A. We explain this more completely in Chapters 3 and 4, where we introduce the concept of a collider variable. The basic idea is that conditioning only on A, which is a collider, creates dependence between B and F within the strata of A. As a result, conditioning only on A does not eliminate the noncausal association between D and Y.

[23] Although all other claims in this section hold for all distributions of the random variables and all types of nonlinearity of causal relationships, one must assume for instrumental variable (IV) estimation what Pearl labels a linearity assumption. What this assumption means depends on the assumed distribution of the variables. It would be satisfied if the causal effect of C on D is linear and the causal effect of D on Y is linear. Both of these would be true, for example, if both C and D were binary variables and Y were an interval-scaled variable, and this is the most common scenario we will consider in this book.

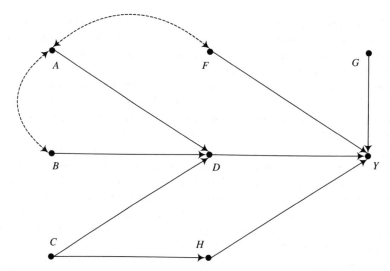

Figure 1.2 A causal graph in which C is no longer an instrumental variable for D.

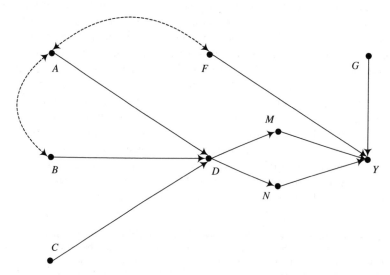

Figure 1.3 A causal diagram in which M and N represent an isolated and exhaustive mechanism for the causal effect of D on Y.

for the set of causal relationships in Figure 1.1 because a third back-door path is now present: $D \leftarrow C \rightarrow H \rightarrow Y$. For the first four strategies, all back-door paths can be blocked by conditioning on $\{A,B,C\}$, $\{A,B,H\}$, $\{F,C\}$, or $\{F,H\}$. For the fifth strategy, the causal effect can be estimated by conditioning on H and then using C as an instrumental variable for D.

Finally, to see how the third mechanistic estimation strategy can be used effectively, consider the alternative graph presented in Figure 1.3. For this graph, four feasible strategies are available as well. The same three strategies proposed for the graph in Figure 1.1 can be used. But, because the variables M and N completely account for the causal effect of D on Y, and because M and N are not determined by anything other than D, the causal effect of D on Y can also be calculated by estimation of the causal effect of D on M and N and then subsequently the causal effects of M and N on Y. And, because this strategy is available, if the goal is to obtain the causal effect of D on Y, then the variables A, B, C, F, and G can be ignored.[24]

In an ideal scenario, all three of these forms of causal effect estimation could be used to obtain estimates, and all three would generate equivalent estimates (subject to the expected variation produced by a finite sample from a population). If a causal effect estimate generated by conditioning on variables that block all back-door paths is similar to a causal effect estimate generated by a valid instrumental variable estimator, then each estimate is bolstered. Better yet, if a mechanism-based strategy then generates a third equivalent estimate, all three causal effect estimates would be even more convincing. And, in this case, an elaborated explanation of how the causal effect comes about is also available, as a researcher could then describe how the causal effect is propagated through the intermediate mechanistic variables M and N.

Implications

The foregoing skeletal presentation of causal effect estimation is, of course, inherently misleading. Rarely does a state of knowledge prevail in a field that allows a researcher to specify causes as cleanly as in the graphs in these figures. Writing down a full graph that represents a consensus position, or a set of graphs that represent alternative positions, can be very difficult, especially if the arguments put forward in alternative pieces of research are open to multiple interpretations. Yet, little progress on estimating causal effects is possible until such graphs are drawn, or at least some framework consistent with them is brought to bear on the questions of central interest.

Beyond introducing the basic estimation strategies, these graphs convey two additional points that are relevant for the material that follows. First, there is often more than one way to estimate a causal effect, and simple rules such as "control for all other causes of the outcome variable" can be poor guides for practice. For example, for Figure 1.1, there are two completely different and plausible conditioning strategies: either condition on F or on A and B. The strategy to "control for all other causes of the outcome variable" is misleading because (1) it suggests that one should condition on G as well, which is unnecessary if all one wants to obtain is the causal effect of D on Y, and (2) it does not suggest that one can estimate the causal effect of D on Y by conditioning on a subset of the variables that cause the causal variable of interest. In this case, one can estimate the causal effect of D on Y without conditioning on any of the other direct causes of Y, but instead by conditioning on the variables that cause D. Even so, this last conditioning strategy should not be taken too far. One

---

[24]Note that, for the graph in Figure 1.3, both M and N must be observed. If, instead, only M were observed, then this mechanistic estimation strategy would not allow for estimation of the full causal effect of D on Y. However, if M and N are isolated from each other, as they are in Figure 1.3, then the portion of the causal effect that passes through M or N can be effectively estimated in the absence of observation of the other. We discuss these issues in detail in Chapter 10.

need not condition on C when also conditioning on both A and B. Not only is this unnecessary (just as for G with the other conditioning strategy), but in doing so one fails to use C in its (possibly) most useful way: as an instrumental variable that can be used to consistently estimate the causal effect of D on Y, ignoring completely A, B, F, and G.

Second, the methods we will present, as we believe is the case with all estimation strategies in the social sciences, are best suited to the targeted estimation of the effects of focal causes. As we will discuss at several points throughout the book, especially in Chapters 6 and 13, the counterfactual approach can also be used to pursue the more ambitious goal of estimating the effects of *all* causes of an outcome of interest. Even so, this more ambitious goal is rarely pursued because of how difficult it is to achieve. The way in which we have presented these graphs is telling on this point. Consider again the question that we posed after introducing Figure 1.1. We asked a simpler version of the following question: Given the structure of causal relationships that relate A, B, C, D, F, G, and Y to each other (represented by presupposed edges that signify causal effects of unknown magnitude), which variables must we observe and then use in a data analysis routine to estimate the size of the causal effect of D on Y? This sort of constrained question (i.e., beginning with the conditional "given" clause) is quite a bit different from seeking to answer the more general questions: What are all of the causes of Y, and how large are their effects relative to each other? The methods that we will present in this book are not irrelevant to this broader question, but they are designed to first answer more targeted questions about the effects of subsets of all causes of an outcome.

The limited nature of the methods that we will present implies two important features of causal effect estimation from the perspective of counterfactual modeling. To offer a precise and defendable causal effect estimate, a well-specified theory is needed to justify assumptions about underlying causal relationships. And, if theory is poorly specified, or divergent theories exist in the relevant scholarly community that support alternative assumptions about underlying causal relationships, then alternative causal effect estimates may be considered valid conditional on the validity of alternative maintained assumptions. We discuss these issues in depth across the chapters of the book, while presenting the framework and the methods that generate estimates that must then be placed in their proper context.

Part II: Counterfactuals, Potential Outcomes, and Causal Graphs

Chapter 2

Counterfactuals and the Potential Outcome Model

In this chapter, we introduce the foundational components of the potential outcome model. We first discuss causal states, the relationship between potential and observed outcome variables, and the usage of the label "counterfactual" to refer to unobserved potential outcomes. We introduce average causal effects and then discuss the assumption of causal effect stability, which is maintained explicitly in most applications that use the potential outcome model. We discuss simple estimation techniques and demonstrate the importance of considering the relationship between the potential outcomes and the process of causal exposure. We conclude by extending our presentation to over-time potential outcome variables for one or more units of analysis, as well as causal variables that take on more than two values.

2.1 Defining the Causal States

The counterfactual framework for observational data analysis presupposes the existence of well-defined causal states to which all members of the population of interest could be exposed.[1] As we will show in the next section, causal effects are then defined based on comparisons of outcomes that would result from exposure to alternative causal states. For a binary cause, the two states are usually labeled treatment and control. When a many-valued cause is analyzed, the convention is to refer to the alternative states as alternative treatments.

Although these labels are simple, the assumed underlying states must be very carefully defined so that the contribution of an empirical analysis based upon them is clear. Some of the relevant issues can only be discussed as we introduce additional

---

[1] We justify the importance of carefully defining the boundaries of the population of interest when presenting average causal effects later in this chapter. We also provide an appendix to this chapter, in which we explain the general superpopulation model that we will adopt when the boundaries of the population can be clearly defined and when we have the good fortune of having a large random sample from the population.

pieces of the full counterfactual framework in this chapter and the next – moving from the definition of individual-level causal effects, through average causal effects, and then to causal graphs and the underlying structural equations that they represent. Nonetheless, some initial clarification of core definitional issues for these causal states is essential.

Fine Articulation. To appreciate the value of finely articulated causal states, consider the examples introduced in Section 1.3. The worker training example is straightforward, and the two states are "entered a training program" (the treatment state) and "did not enter a training program" (the control state). The charter school example is similar. Here, the alternative states are "enrolled in a charter school" (the treatment state) and "enrolled in a regular public school" (the control state, although possibly referred to as an alternative treatment state). One possible complication with these examples is the possibility of inherent differences across training programs, charter schools, and regular public schools. If any such treatment-site heterogeneity exists, then stratified analyses may be necessary, perhaps by regions of the country, size of the program or school, or whatever other dimension suggests that variability of the causal states deserves explicit modeling.[2]

Other examples, at least as executed in the extant research, have causal states that are not finely articulated. Consider the classic political participation line of inquiry. For the relationship between socioeconomic status and political participation, there are many underlying causal effects, such as the effect of having obtained at least a college degree on the frequency of voting in local elections and the effect of having a family income greater than some cutoff value on the amount of money donated to political campaigns. Well-defined causal states exist for these narrow causal effects, but it is not clear at all that well-defined causal states exist for the internally differentiated concept of socioeconomic status, which social scientists have created for their own analytic purposes. It is therefore unsurprising that some of the most recent literature (e.g., Henderson and Chatfield 2011; Kam and Palmer 2008) has specified more finely articulated causal states for this line of research, such as "entered college" (the treatment state) in comparison to "did not enter college" (the control state).

The father absence example is an intermediate case. The original research attempted to estimate the broad effects of single parenthood on the outcomes of children and adolescents (see McLanahan and Sandefur 1994; Wu and Wolfe 2001). The more recent literature has focused on the narrowly defined treatment state of father absence. Even so, much variation remains in both the definition of this treatment state and the relevant comparison (or control) state, as noted in a review piece:

> Studies in this field measured father absence in several ways, which the reader should keep in mind when interpreting and comparing results across studies. Some studies compared children of divorced parents with children of stably married parents; others compared children whose parents married after their child's birth with those parents who never married.... More recently, researchers have started to use even more nuanced categories to

---

[2]For example, Hong and Raudenbush (2006) provide a careful analysis of retention policies in primary education, implementing this type of treatment-site stratification based on the average level of retention in different public schools. We will discuss these types of studies in Section 2.5.

measure family structure – including married biological-parent families, cohabiting biological-parent families, married stepparent families, cohabiting stepparent families, and single parents by divorce and nonmarital birth – reflecting the growing diversity of family forms in society.... We did not identify any studies that used causal methods to study the effects of same-sex unions. (McLanahan et al. 2013:408)

In general, research that takes account of heterogeneity by splitting treatment states into mutually exclusive component states will break new ground if sufficient data are available to estimate the more narrowly defined treatment effects.

Nominal States from Constitutive Features. We take a pragmatic but principled position on the characteristics of causal states, and in this subsection we want to clarify our position for readers who are interested in debates on the nature of causation in philosophy and how those debates are relevant for social science research. Readers who are uninterested in these debates may wish to skim this subsection now and reengage it after reading Chapter 10 on causal mechanisms and Section 13.1 on objections to the counterfactual approach to observational data analysis (pages 438–446). In fact, most scholars who work with counterfactual models in social science research do not take any positions on the issues that we raise in this subsection, and their research shows that much useful work can proceed by taking and using the causal states as measured, without considering the features that give them their capacities to generate effects.

Having offered these warnings, we will now explain why we take the position that each state of each treatment should be regarded as a nominal state with constitutive features (e.g., entities, activities, and relations) that are jointly capable of producing the outcome of interest. Consider the Catholic school example, where the nominal states are "enrolled in a Catholic school" and "enrolled in a public school" and where the outcome is "learning." Each type of school has teachers, classrooms, curricula, administrators, normative environments, affiliated institutions, and networks of peers and parents. The literature on the differences between Catholic schools and public schools suggests that these constitutive features of the schools are interrelated in ways that differ by type of school. Accordingly, while Catholic schools and public schools both produce student learning, the particular ways in which they do so are thought to differ meaningfully across type of school, and in ways that have not been documented comprehensively with available data. Nonetheless, we can still conceive of each student in the population of interest being exposed to each type of school, and we can assume that each student would then experience the learning generated in toto by the joint capacities of the constituent features of each type of school.

Taking this position, while at the same time embracing counterfactual dependence, implies that we see value in mounting causal analysis in the social sciences on top of a foundation that conjoins a metaphysics of causal powers with a metaphysics of counterfactual dependence (see Collins, Hall, and Paul 2004; Mumford and Anjum 2011). The price for such an inclusive pragmatism is an elaborate metaphysics, which most philosophers would likely regard as insufficiently elegant and insufficiently reductive. With reference to Hume's example of billiard balls (Hume 1977[1772]), our position requires that we adopt the following specific (but perhaps painfully elaborate) account of the nature of causation: The cue ball causes the second billiard ball to roll a particular observed distance because billiard balls are spheres *and* because the cue ball was struck by the player's pool cue in such a way that it then struck the second billiard ball at a particular angle and with a particular force. Furthermore, the cue ball would not have caused the second billiard ball to roll the same observed distance if the billiard balls had instead not been spheres or if the cue ball had not been struck by the player's pool cue in the exact same way. Thus, the causal effect of the cue ball on the second billiard ball is a joint product of the spherical feature of the billiard balls as well as the external intervention of the pool player.[3]

For the sorts of social science examples we consider in this book, we will express the effects of causal states using contrasts between observed exposure to one state and what-if counterfactual exposure to another state. However, we will also take the position that any such claims about the effects of exposure to alternative states are incomplete until those claims are accompanied by accounts of the constitutive features of the causal states and how those features are thought to grant the states the power to generate outcomes.[4] The most complete accounts point to evidence that mechanisms exist that are capable of generating the outcomes of interest (and, better yet, that it is reasonable to believe that these mechanisms will be able to explain why exposure to alternative causal states generates differences).

Consider an example where many of these issues are settled. For the Catholic school effect, analysis can proceed within a guiding framework shaped by a rich background literature. The historical events that generated public schools and Catholic schools as coherent institutions suggests that they can be meaningfully compared when studying student achievement because they each aim to produce learning for core academic subjects, even though they each pursue additional distinct goals (see Tyack 1974 for one of the most widely read accounts). In addition, each type of school has a rich set of literature that has examined the mechanisms that generate learning. For Catholic

---

[3]With the goal of reducing the complexity of such an account, philosophers seem inclined to take positions on whether the spherical characteristic of the cue ball (its "causal power") is more fundamental than the striking (i.e., the "counterfactual dependence" induced by the intervention of the player), whether "striking" can be defined in the absence of an intervening pool player, whether a valid explanation can simply be deduced from laws of motion in space and time, whether anything is transferred between the two billiard balls at the moment of impact, and so on. We see no reason to take a position on these matters in this book, and we therefore quite consciously violate rule 4 of Paul and Hall (2013:40), "Thou shalt not be an ontological wimp."

[4]We see little value in placing restrictions on what types of origination accounts are admissible and should be relied upon. For some causal claims, historical narratives are appropriate, to the extent that they focus on especially salient institutional histories while pushing into the background the multitude of specific decisions of all individuals that have given shape to the constitutive features of the alternative states (see Reed 2011 for examples of such "forming" narratives). In other cases, the origins of the states can be explained as contrasting values for built concepts, based on underlying analytic dimensions drawn from the extant social science literature, where these underlying dimensions have been chosen precisely because background evidence exists that they are sources of productive causal power of the nominal causal states of interest (see Goertz 2006 for examples). However, we see one complication for this second type of account, as foreshadowed by our discussion of socioeconomic status above. Causal states drawn from values for a built concept may be real only in the minds of researchers. As a result, explanations based upon them may appear nonsensical to individuals who are purported to be producing the effects of interest. Whether such behind-the-back accounts are to be regarded as powerful or not is almost certainly a domain-specific consideration, which will also vary with the goals of a study.

schools, several complementary narratives exist that provide arguments that suggest why Catholic schools have the capacity to be more effective than public schools. These narratives include those that emphasize the relations embedded in parental network structure alongside an appropriated ideology of the Catholic church (see Coleman and Hoffer 1987), those that emphasize the extent to which Catholic schools are especially responsive to parental feedback and the threat of exit (Chubb and Moe 1990), and those that emphasize the trusting relationships between teachers and administrators that flow from a shared purpose (Bryk, Lee, and Holland 1993).

Of course, the existence of such lower-level claims about the specific mechanistic capacities of features of Catholic schools begs the question: When is it advisable to decompose nominal causal states into component causal states with their own capacities for producing the outcome of interest? We see the answer to this question as subject- and domain-specific. Accordingly, we see little value in making general arguments about when such decomposition is feasible because of the inherent separability of the productive capacities attached to particular constitutive features or is infeasible because of the deeply entangled complementarities among them. And, as we will argue later in Chapter 10, it is generally impossible to take a position on these issues in any given study without first stipulating the causal structure of the mechanisms that are presumed to produce the outcome. Fortunately, as we will demonstrate in the intervening chapters, causal effects defined only by nominal causal states can be sufficiently precise so that their estimation is itself feasible and very much worthwhile.

Local and Reasonable. Consider the literature on socioeconomic status as a fundamental cause of health and mortality, which takes as its defining feature the argument that it is only occasionally useful to identify causal states for the measurable dimensions underneath the fundamental cause of socioeconomic status (see page 19). For these scholars, it is the abundant causal pathways that link socioeconomic status with health and mortality that are most noteworthy because of the robust, total associations that they generate. Isolating a particular causal effect that is attributable to a contrast defined by two clearly defined underlying causal states embedded within socioeconomic status could still be useful, such as for the estimation of a health disparity attributable to a family income difference of $25,000. The claim of this literature is that this narrow exercise could become counterproductive if it detracted from the broader claim of fundamental causality, as would be the case if the analyst were to imply that any such narrow effect is as robust as the total causal effect that socioeconomic status exerts on health and mortality.

A fundamental-cause orientation may be useful in challenging the status quo in research areas that have become too narrowly focused on only a few relevant causal pathways, but widespread adoption of the fundamental-cause orientation to causal analysis would not be productive. In many areas of research, it would not be hard to take collections of narrowly and carefully defined causal contrasts, lump them together into latent constructs, and then assert that, over sufficiently long intervals, the latent construct is a fundamental cause because the mechanisms that are activated by component causes switch on and off over time. Indeed, considering our other examples in Section 1.3, one could argue quite easily that the socioeconomic status of one's parents is a fundamental cause of educational attainment, subsequent labor market earnings, political participation, and fertility decisions. We doubt many scholars would disagree with such broad claims, and most would likely interpret them as consistent with conclusions drawn by scholars working with comparatively coarse data more than six decades ago. More importantly, we think it unlikely that the reassertion of such broad claims would encourage researchers to move in productive new directions. Instead, we see the counterfactual perspective, and the potential outcome model in particular, as enabling the pursuit of a more ambitious goal: the careful delineation of the relevant causal states that lie within any purported fundamental causes and then the estimation of the specific effects generated by contrasts between them. Should there be reason to expect that any such effects vary in time, then their estimation across time demands empirical analysis, not simply the assertion that such effects, by nature of their variability in time, can only be regarded as specialized instantiations of more fundamental causes.

For a related reasonableness concern, consider a specific political participation example. To what extent do restrictions on who can vote determine who wins elections? A highly publicized variant of this question is this: What is the effect on election outcomes of laws that forbid individuals with felony convictions from voting?[5] Uggen and Manza (2002) make the straightforward claim that the 2000 presidential election would have gone in favor of Al Gore if felons and ex-felons had been permitted to vote:

> Although the outcome of the extraordinarily close 2000 presidential election could have been altered by a large number of factors, it would almost certainly have been reversed had voting rights been extended to any category of disenfranchised felons. (Uggen and Manza 2002:792)

Uggen and Manza (2002) then note an important limitation of their conclusion:

> our counterfactual examples rely upon a ceteris paribus assumption – that nothing else about the candidates or election would change save the voting rights of felons and ex-felons. (Uggen and Manza 2002:795)

When thinking about this important qualification, one might surmise that a possible world in which felons had the right to vote would probably also be a world in which the issues (and probably candidates) of the election would be very different. Thus, the most challenging definitional issue here is not who counts as a felon or whether or not an individual is disenfranchised, but rather how well the alternative causal states can be characterized.

A relevant criterion, although necessarily subjective, is whether it "stretches the mind" too much to imagine conceivable alternative worlds in which all else remains the same, except for the instantiation of the alternative causal states. For this particular example, the "too much" criterion was not likely crossed. Scholars in political sociology and criminology supported publication through blind peer review in the discipline's highest prestige journal, the *American Sociological Review*. Reviewers presumably saw this particular line of research as an important contribution to our knowledge on how changing laws to allow felons and ex-felons to vote could have potential effects on

---

[5] Behrens, Uggen, and Manza (2003), Manza and Uggen (2004), and Uggen, Behrens, and Manza (2005) give historical perspective on this question.

election outcomes, and they must have concluded that there was value in understanding such effects in hypothetical isolation from other changes that would also likely co-occur in the real world along with the contemplated legislative changes.

The more general point, however, is that it is important that the "what would have been" nature of the conditionals that define the causal states of interest be carefully considered. When a facile ceteris paribus assumption is invoked to relieve the analyst from having to discuss other contrasts that are nearly certain to occur at the same time, the posited causal states may be open to the charge that they are too improbable or ill-defined to justify the pursuit of a causal analysis based on them.[6]

2.2 Potential Outcomes and Individual-Level Treatment Effects

Given the existence of well-defined causal states, causal inference in the counterfactual tradition proceeds by stipulating the existence of potential outcome random variables that are defined over all individuals in the population of interest. For a binary cause, we will denote potential outcome random variables as Y^1 and Y^0.

We will also adopt the notational convention from statistics in which realized values for random variables are denoted by lowercase letters. Accordingly, y_i^1 is the potential outcome in the treatment state for individual i, and y_i^0 is the potential outcome in the control state for individual i.[7] The individual-level causal effect of the treatment

---

[6]The philosopher Nancy Cartwright (2007a, 2007b) would refer to an analysis that defines potential outcomes (see next section) in terms of ill-conceived causal states as generating "impostor counterfactuals." She stresses the need for full causal models of all interrelated causes of outcomes, so that the effects of causes are not too narrowly assessed. She writes:

> To evaluate counterfactuals ... we need a causal model; and the causal model must contain all the information relevant to the consequent about all the changes presumed in the antecedent. There is no other reasonable method on offer to assess counterfactuals. We may not always produce a model explicitly, but for any grounded evaluation there must be a causal model implicit; and our degree of certainty about our counterfactual judgments can be no higher than our degree of certainty that our causal model is correct. (Cartwright 2007a:193)

We agree with the value of having a causal model, as will become clear in subsequent chapters. However, Cartwright takes this position to an extreme that is counterproductive for practice; see Pearl (2009:362–65).

[7]There is a wide variety of notation in the potential outcome and counterfactuals literature, and we have adopted the notation that we feel is the easiest to grasp. However, we should note that Equation (2.1) and its elements are often written as one of the following alternatives,

$$\Delta_i = Y_{1i} - Y_{0i},$$
$$\delta_i = Y_i^t - Y_i^c,$$
$$\tau_i = y_i(1) - y_i(0),$$

and variants thereof. We use the right-hand superscript to denote the potential treatment state of the corresponding potential outcome variable, but other authors use the right-hand subscript or parenthetical notation. We also use numerical values to refer to the treatment states, but other authors (including us, see Morgan 2001, Winship and Morgan 1999, and Winship and Sobel 2004) use values such as t and c for the treatment and control states, respectively. There is also variation in the usage of uppercase and lowercase letters. We do not claim that everyone will agree that our notation is the easiest to grasp, and it is certainly not as general as, for example, the parenthetic notation. But it

is then defined as

$$\delta_i = y_i^1 - y_i^0. \tag{2.1}$$

Before proceeding, two caveats on this definition of individual-level causal effects should be noted. First, the individual-level causal effect can be defined in ways other than as the linear difference between the two relevant potential outcomes.[8] One obvious possibility is the ratio of one individual-level potential outcome to another, y_i^1/y_i^0. In some research areas, alternative definitions at the individual level may have advantages. The most prominent case is epidemiology, where the goal of estimating risk factors for health outcomes continues to dominate practice and leads to a frequent preference for ratio-based rather than difference-based comparisons. Nonetheless, the overwhelming majority of the literature represents individual-level causal effects as linear differences, as in Equation (2.1).

Second, the individual-level causal effect could be defined as the difference between the expectations of individual-specific random variables, as in $E[Y_i^1] - E[Y_i^0]$, where $E[.]$ is the expectation operator from probability theory (see, for a clear example of this alternative setup, King et al. 1994:76–82). In thinking about individuals self-selecting into alternative treatment states, it can be useful to set up the treatment effects in this way. In many applications, individuals are thought to consider potential outcomes with some recognition of the inherent uncertainty of their beliefs, which may properly reflect true variability in their individual-level potential outcomes. But, with data for which a potential outcome is necessarily observed for any individual as a scalar value (via an observed outcome variable, defined later), this individual-level, random-variable definition is largely redundant. Accordingly, we will denote individual-level potential outcomes as values such as y_i^1 and y_i^0, regarding these as realizations of population-level random variables Y^1 and Y^0 while recognizing, at least implicitly, that they could also be regarded as realizations of individual-specific random variables Y_i^1 and Y_i^0.

2.3   Treatment Groups and Observed Outcomes

For a binary cause with two causal states and associated potential outcome variables Y^1 and Y^0, a corresponding causal exposure variable, D, is specified that takes on two values: D is equal to 1 for members of the population who are exposed to the treatment state and equal to 0 for members of the population who are exposed to the control state. Exposure to the alternative causal states is determined by a particular process, typically an individual's decision to enter one state or another, an outside actor's decision to allocate individuals to one state or another, a planned random allocation carried out by an investigator, or some combination of these alternatives.

By convention, those who are exposed to the treatment state are referred to as the treatment group, whereas those who are exposed to the control state are referred to as the control group. Because D is defined as a population-level random variable (at least

---

does seem to have proven itself in our own classes, offering the right balance between specificity and compactness.

[8]Rubin (2005, figure 1) uses the general notation "v." for "versus" to depict individual-level effects in their most general form.

in most cases in observational data analysis), the treatment group and control group exist in the population as well as the observed data. Throughout this book, we will use this standard terminology, referring to treatment and control groups when discussing those who are exposed to alternative states of a binary cause. If more than two causal states are of interest, then we will shift to the semantics of alternative treatments and corresponding treatment groups, thereby discarding the baseline labels of control state and control group.

Despite our adoption of this convention, we could rewrite all that follows referring to members of the population as what they are – those who are exposed to alternative causal states – and not use the words treatment and control at all. Indeed, we recognize that for some readers the usage of treatment and control language may feel sufficiently heterodox relative to the semantics of the areas in which they work that avoidance of these terms seems prudent. If so, it is perfectly acceptable to adopt parallel language without using the words treatment and control.

When we refer to individuals in the observed treatment and control groups, we will again adopt the notational convention from statistics in which realized values for random variables are denoted by lowercase letters. Accordingly, the random variable D takes on values of $d_i = 1$ for each individual i who is an observed member of the treatment group and $d_i = 0$ for each individual i who is an observed member of the control group.

Given these definitions of Y^1, Y^0, and D (as well as their realizations y_i^1, y_i^0, d_i), we can now define the observed outcome variable Y in terms of them. We can observe values for a variable Y as $y_i = y_i^1$ for individuals with $d_i = 1$ and as $y_i = y_i^0$ for individuals with $d_i = 0$. The observable outcome variable Y is therefore defined as

$$Y = Y^1 \quad \text{if } D = 1,$$
$$Y = Y^0 \quad \text{if } D = 0.$$

This paired definition is often written compactly as

$$Y = DY^1 + (1 - D)Y^0. \tag{2.2}$$

Equation (2.2) implies that one can never observe the potential outcome under the treatment state for those observed in the control state, and one can never observe the potential outcome under the control state for those observed in the treatment state. This impossibility implies that one can never calculate individual-level causal effects.

Holland (1986) describes this challenge as the fundamental problem of causal inference in his widely read introduction to the potential outcome model of counterfactual causality. Table 2.1 depicts the "problem," which one might alternatively refer to as the "fundamental reality of causal analysis." Causal effects are defined by contrasts within rows, which refer to groups of individuals observed in the treatment state or in the control state. However, only the diagonal of the table is observable, thereby rendering impossible the direct calculation of individual-level causal effects merely by means of observation and then subtraction.[9]

---

[9] As Table 2.1 shows, we are more comfortable than some writers in using the label "counterfactual" when discussing potential outcomes. Rubin (2005), for example, avoids the term counterfactual, under the argument that potential outcomes become counterfactual only after treatment assignment has occurred. Thus, no potential outcome is ever ex ante counterfactual. We agree, of course. But, because

Table 2.1 The Fundamental Problem of Causal Inference

Group	Y^1	Y^0
Treatment group ($D = 1$)	Observable as Y	Counterfactual
Control group ($D = 0$)	Counterfactual	Observable as Y

As shown clearly in Equation (2.2), the outcome variable Y, even if we could enumerate all of its individual-level values y_i in the population, reveals only half of the information contained in the underlying potential outcome variables. Individuals contribute outcome information only from the treatment state in which they are observed. This is another way of thinking about Holland's fundamental problem of causal inference. The outcome variables we must analyze – labor market earnings, test scores, and so on – contain only a portion of the information that would allow us to directly calculate causal effects for all individuals.

2.4   The Average Treatment Effect

Because it is typically impossible to calculate individual-level causal effects, we usually focus attention on the estimation of carefully defined aggregate causal effects. When we adopt the linear difference in potential outcomes as the definition of the individual-level causal effect, we typically define aggregate causal effects as averages of these individual-level effects. These average causal effects can de defined for any subset of the population, and throughout this book we will consider many different average effects. In this section, we introduce the broadest possible average effect, which is the average treatment effect (ATE) in the population as a whole.

With $E[.]$ denoting the expectation operator from probability theory, the average treatment effect in the population is

$$E[\delta] = E[Y^1 - Y^0] \tag{2.3}$$
$$= E[Y^1] - E[Y^0].$$

The second line of Equation (2.3) follows from the linearity of the expectation operator: The expectation of a difference is equal to the difference of the two expectations.[10]

For Equation (2.3), the expectation is defined with reference to the population of interest. For the fertility pattern example introduced in Section 1.3, the population would be one or more birth cohorts of women in a particular country. For the election outcome examples, the population would be "all eligible voters" or "all eligible voters in Florida." For other examples, such as the worker training example, the population

---

our focus is on observational data analysis, we find the counterfactual label useful for characterizing potential outcomes that are rendered unobservable ex post to the treatment assignment/selection mechanism.

[10]However, at a deeper level, it also follows from the assumption that the causal effect is defined as a linear difference at the individual level, which allows the application of expectations in this simple way to characterize population-level average effects.

would be "all adults eligible for training," and eligibility would need to be defined carefully. Thus, to define average causal effects and then interpret estimates of them, it is crucial that researchers clearly define the characteristics of the individuals in the assumed population of interest.[11]

Note also that the subscripting on i for the individual-level causal effect, δ_i, has been dropped for Equation (2.3). Even so, the definition of the ATE should not be interpreted to suggest that we now must assume that the treatment effect is constant in the population in any fundamental sense. Rather, we can drop the subscript i in Equation (2.3) because the expected causal effect of a randomly selected individual from the population is equal to the average causal effect across individuals in the population. We will at times throughout this book reintroduce redundant subscripting on i in order to reinforce the inherent individual-level heterogeneity of the potential outcomes and the causal effects they define.[12]

To see all of these pieces put together, consider the Catholic school example. The potential outcome under the treatment, y_i^1, is the what-if achievement outcome of individual i if he or she were enrolled in a Catholic school. The potential outcome under the control, y_i^0, is the what-if achievement outcome of individual i if he or she were enrolled in a public school. Accordingly, the individual-level causal effect, δ_i, is the what-if difference in achievement that could be calculated if we could simultaneously educate individual i in both a Catholic school and a public school. The ATE, $E[\delta]$, is then the average value among all students in the population of these what-if differences in test scores. The ATE is also equal to the expected value of the what-if difference in test scores for a randomly selected student from the population.

An alternative group-level causal effect that we will not consider much in this book is the causal risk ratio,

$$\frac{Pr[Y^1 = 1]}{Pr[Y^0 = 1]}, \tag{2.4}$$

where now the outcomes Y^1 and Y^0 are indicator variables equal to 1 if the outcome of interest is present and 0 if not. This group-level effect is the analog to the individual-level ratio of potential outcomes, y_i^1/y_i^0, noted earlier in Section 2.2. The causal risk ratio is most frequently analyzed in epidemiology and the health sciences, where risk-factor analysis remains dominant and the outcomes are typically onset of a disease or a troubling symptom thereof (see Hernán and Robins 2006a). For our purposes, most outcomes modeled as causal risk ratios can be translated to average treatment effects, interpreting $E[Y^1] - E[Y^0]$ as $Pr(Y^1 = 1) - Pr(Y^0 = 1)$. Expectations of indicator variables are equivalent to probabilities of indicator variables, and an interval metric

---

[11]And, regardless of the characterization of the features of the population, we will assume throughout this book that the population is a realization of an infinite superpopulation. We discuss our decision to adopt this underlying population model in an appendix to this chapter. Although not essential to understanding most of the material in this book, some readers may find it helpful to read that appendix now in order to understand how these definitional issues are typically settled in this literature.

[12]For example, at many times in the book, we will stress that quantities such as the ATE should not be assumed to be equal to the individual-level causal effect for any individual i, which we will express as $\delta_i \neq E[\delta_i] = E[\delta]$ for all i. In words, when individual-level heterogeneity of causal effects is present, individual-level causal effects, δ_i, will not all be equal to the average of these individual-level causal effects, $E[\delta_i]$, which is, by the definition of the expectation operator, equal to $E[\delta]$.

is at least as sensible as a ratio metric for all of the examples we will consider. (The ratio metric might be preferable if we were attempting to make effect comparisons across outcomes with very different base rates, such as the effect of the same treatment on pancreatic cancer and hypertension.)

2.5   The Stable Unit Treatment Value Assumption

In most applications, the potential outcome model retains its tractability through the maintenance of a strong assumption known as the stable unit treatment value assumption or SUTVA (see Rubin 1980b, 1986). In economics, a version of this assumption is sometimes referred to as a no-macro-effect or partial equilibrium assumption (see Garfinkel, Manski, and Michalopoulos 1992, Heckman 2000, 2005, for the history of these ideas, and Manski and Garfinkel 1992 for examples).[13]

SUTVA, as implied by its name, is a basic assumption of causal effect stability that requires that the potential outcomes of individuals be unaffected by changes in the treatment exposures of all other individuals. In the words of Rubin (1986:961), who developed the term,

> SUTVA is simply the a priori assumption that the value of Y for unit u when exposed to treatment t will be the same no matter what mechanism is used to assign treatment t to unit u and no matter what treatments the other units receive.

Consider the idealized example in Table 2.2, in which SUTVA is violated because the treatment effect varies with treatment assignment patterns. For the idealized example, there are three randomly drawn subjects from a population of interest, and the study is designed such that at least one of the three study subjects must receive the treatment and at least one must receive the control. The first column of the table gives the six possible treatment assignment patterns.[14] The first row of Table 2.2 presents all three ways to assign one individual to the treatment and the other two to the control, as well as the potential outcomes for each of the three subjects. Subtraction within the last column shows that the individual-level causal effect is 2 for all three individuals. The second row of Table 2.2 presents all three ways to assign two individuals to the treatment and one to the control. As shown in the last column of the row, the individual-level causal effects implied by the potential outcomes are now 1 instead of 2. Thus, for this idealized example, the underlying causal effects are a function of the treatment assignment patterns, such that the treatment is less effective when more individuals are assigned to it. For SUTVA to hold, the potential outcomes would need to be identical for both rows of the table.

---

[13]SUTVA is a much maligned acronym, and many others use different labels. Manski (2013a:S1), for example, has recently labeled the same assumption the "individualistic treatment response" assumption in order "to mark it as an assumption that restricts the form of treatment response functions."

[14]For this example, assume that the values of y_i^1 and y_i^0 for each individual i are either deterministic potential outcomes or exactly equal to $E[Y_i^1]$ and $E[Y_i^0]$ for each individual i. Also, assume that these three subjects comprise a perfectly representative sample of the population.

Table 2.2 A Hypothetical Example in Which SUTVA Is Violated

Treatment assignment patterns	Potential outcomes	

$$\begin{bmatrix} d_1 = 1 \\ d_2 = 0 \\ d_3 = 0 \end{bmatrix} \text{ or } \begin{bmatrix} d_1 = 0 \\ d_2 = 1 \\ d_3 = 0 \end{bmatrix} \text{ or } \begin{bmatrix} d_1 = 0 \\ d_2 = 0 \\ d_3 = 1 \end{bmatrix}$$

$y_1^1 = 3 \qquad y_1^0 = 1$
$y_2^1 = 3 \qquad y_2^0 = 1$
$y_3^1 = 3 \qquad y_3^0 = 1$

$$\begin{bmatrix} d_1 = 1 \\ d_2 = 1 \\ d_3 = 0 \end{bmatrix} \text{ or } \begin{bmatrix} d_1 = 0 \\ d_2 = 1 \\ d_3 = 1 \end{bmatrix} \text{ or } \begin{bmatrix} d_1 = 1 \\ d_2 = 0 \\ d_3 = 1 \end{bmatrix}$$

$y_1^1 = 2 \qquad y_1^0 = 1$
$y_2^1 = 2 \qquad y_2^0 = 1$
$y_3^1 = 2 \qquad y_3^0 = 1$

This type of treatment effect dilution is only one way in which SUTVA can be violated. More generally, suppose that $\mathbf{d}$ is an $N \times 1$ vector of treatment indicator variables for N individuals (analogous to the treatment assignment vectors in the first column of Table 2.2), and define potential outcomes of each individual as functions across all potential configurations of the elements of vector $\mathbf{d}$. Accordingly, the outcome for individual i under the treatment is $y_i^1(\mathbf{d})$, and the outcome for individual i under the control is $y_i^0(\mathbf{d})$. The treatment effect for each individual i is then

$$\delta_i(\mathbf{d}) = y_i^1(\mathbf{d}) - y_i^0(\mathbf{d}). \tag{2.5}$$

With this more general setup, individual-level treatment effects could be different for every possible pattern of treatment exposure.

SUTVA is what allows us to declare $y_i^1(\mathbf{d}) = y_i^1$ and $y_i^0(\mathbf{d}) = y_i^0$ and, as a result, assert that individual-level causal effects δ_i exist that are independent of the overall configuration of causal exposure. If SUTVA cannot be maintained, then the simplified definition in Equation (2.1) is invalid, and the individual-level treatment effect must be written in its most general form in Equation (2.5), with all ensuing analysis proceeding conditional on alternative vectors $\mathbf{d}$.

Sometimes it is argued that SUTVA is so restrictive that we need an alternative conception of causality for the social sciences. Our position is that SUTVA reveals the limitations of social science data and the perils of immodest causal modeling rather than the limitations of the potential outcome model itself. Rather than consider SUTVA as overly restrictive, researchers should always reflect on the plausibility of SUTVA in each application and use such reflection to motivate a clear discussion of the meaning and scope of all causal effect estimates offered. Such reflection may lead one to determine that only the more general case of the potential outcome framework can be justified, and this may necessitate building the analysis on top of the individual-level treatment effect defined in Equation (2.5) rather than the SUTVA-simplified variant in Equation (2.1). In some cases, however, analysis can proceed assuming SUTVA, as long as all resulting estimates are given restricted interpretations, as we now explain.

Typical SUTVA violations share two interrelated features: (1) influence patterns that result from contact across individuals in social or physical space and (2) dilution/concentration patterns that one can assume would result from changes in the prevalence of the treatment. Neither feature is entirely distinct from the other, and in many cases dilution/concentration effects arise because influence patterns are present. Yet, if the violation can be interpreted as a dilution/concentration pattern, even when generated in part by an underlying influence pattern, then the analyst can proceed by scaling back the asserted relevance of any estimates to situations where the prevalence of the treatment is not substantially different.

For a simple example, consider the worker training example. Here, the plausibility of SUTVA may depend on the particular training program. For small training programs situated in large labor markets, the structure of wage offers to retrained workers may be entirely unaffected by the existence of the training program. However, for a sizable training program in a small labor market, it is possible that the wages on offer to retrained workers would be a function of the way in which the price of labor in the local labor market responds to the movement of trainees in and out of the program (as might be the case in a small company town after the company has just gone out of business and a training program is established). As a result, SUTVA may be reasonable only for a subset of the training sites for which data have been collected.

For an example of where influence patterns are more of a threat to SUTVA, consider the example of the Catholic school effect. For SUTVA to hold, the effectiveness of Catholic schooling cannot be a function of the number (and/or composition) of students who enter the Catholic school sector. For a variety of reasons – endogenous peer effects, capacity constraints, and so on – most school effects researchers would probably expect that the Catholic school effect would change if large numbers of public school students entered the Catholic school sector. As a result, because there are good theoretical reasons to believe that the pattern of effects would change if Catholic school enrollments ballooned, it may be that researchers can estimate the causal effect of Catholic schooling only for those who would typically choose to attend Catholic schools, but also subject to the constraint that the proportion of students educated in Catholic schools remains constant. Accordingly, it may be impossible to determine from any data that could be collected what the Catholic school effect on achievement would be under a new distribution of students across school sectors that would result from a large and effective policy intervention. As a result, the implications of research on the Catholic school effect for research on school voucher programs may be quite limited, and this has not been clearly enough recognized by some (see Howell and Peterson 2002, chapter 6). A similar argument applies to research on charter school effects.

Consider a SUTVA violation for a related example: the evaluation of the effectiveness of mandatory school desegregation plans in the 1970s on the subsequent achievement of black students. Gathering together the results of a decade of research, Crain and Mahard (1983) conducted a meta-analysis of 93 studies of the desegregation effect on achievement. They argued that the evidence suggests an increase of .3 standard deviations in the test scores of black students across all studies.[15] It seems undeniable that SUTVA is violated for this example, as the effect of moving from one school to another must be a function of relative shifts in racial composition across schools. Breaking the analysis into subsets of cities where the compositional shifts were similar could yield conditional average treatment effect estimates that can be more clearly interpreted. In this case, SUTVA would be abandoned in the collection of all desegregation events, but it could then be maintained for some groups (perhaps in cities where the compositional shift was comparatively small).

In general, if SUTVA is maintained but there is some doubt about its validity because dilution or concentration patterns would emerge under shifts in treatment prevalence, then certain types of marginal effect estimates can usually still be defended. The idea here is to state that the estimates of average causal effects hold only for what-if movements of relatively small numbers of individuals from one hypothetical treatment state to another.

If, however, influence patterns are inherent to the causal process of interest, and the SUTVA violation cannot be considered as a type of dilution or concentration, then it will generally not be possible to circumvent the SUTVA violation by proceeding with the same analysis and only offering cautious and conditional interpretations. The most well-developed literature on situations such as these is the literature on the effects of vaccine programs (see Hudgens and Halloran 2008). Here, additional causal effects of interest using the potential outcome framework have been defined, conditional on the overall pattern of treatment assignment:

> The *indirect effect* of a vaccination program or strategy on an individual is the difference between what the outcome is in the individual not being vaccinated in a community with the vaccination program and what the outcome would have been in the individual, again not being vaccinated, but in a comparable community with no vaccination program. It is, then, the effect of the vaccination program on an individual who was not vaccinated. The combined *total effect* in an individual of being vaccinated and the vaccination program in the community is the difference between the outcome in the individual being vaccinated in a community with the vaccination program and what the outcome would be if the individual were not vaccinated and the community did not have the vaccination program. The total effect, then, is the effect of the vaccination program combined with the effect of the person having been vaccinated. The *overall effect* of a vaccination program is the difference in the outcome in an average individual

---

[15] As reviewed by Schofield (1995) and noted in Clotfelter (2004), most scholars now accept that the evidence suggests that black students who were bused to predominantly white schools experienced small positive reading gains but no substantial mathematics gains. Cook and Evans (2000:792) conclude that "it is unlikely that efforts at integrating schools have been an important part of the convergence in academic performance [between whites and blacks], at least since the early 1970s" (see also Armor 1995; Rossell, Armor, and Walberg 2002). Even so, others have argued that the focus on test score gains has obscured some of the true effectiveness of desegregation. In a review of these longer-term effects, Wells and Crain (1994:552) conclude that "interracial contact in elementary and secondary school can help blacks overcome perpetual segregation."

in a community with the vaccination program compared to an average individual in a comparable population with no vaccination program. (Halloran, Longini, and Struchiner 2010:272; italics in the original)

Effectively estimating these types of effects generally requires a nested randomization structure, wherein (1) vaccine programs are randomly assigned to a subset of participating groups and then (2) vaccinations are randomly given to individuals within groups enrolled in vaccine programs. These particular study designs are not possible for most social science applications, but the basic interpretive framework has been adopted to clarify what can be learned from social experiments, in particular, the Moving to Opportunity neighborhood experiment (see Sobel 2006).[16]

Much observational research on social influence patterns proceeds without consideration of these sorts of complications. Consider the contentious literature on whether peer effects have accelerated the obesity epidemic, as presented in Section 1.3 (see page 26). As we noted there, the basic claim of Christakis and Fowler (2007) is that having a friend who becomes obese increases one's own odds of becoming obese. Yet, their full set of claims is substantially more detailed, suggesting that these peer effects travel across network paths of length three before dying out. In particular, one's odds of becoming obese also increase if friends of friends become obese and if friends of friends of friends become obese. The sizes of these three effects diminish with the length of friendship distance.

Now consider whether SUTVA is reasonable for such a schedule of effects across network ties. Holding the social network structure fixed, if obesity increases in the population, then, on average, individuals have more obese friends, more obese friends of friends, and more obese friends of friends of friends. Most theoretical predictions would suggest that the effects on one's own odds of becoming obese that result from having friends of friends of friends who become obese should decline with the proportion of one's own friends who are already obese or who have just become obese.[17] Effects that cascade in these conditional ways, because they are defined across a pattern of interpersonal contact between units, nearly always violate SUTVA.[18]

---

[16]Suitable models for observational data are an active frontier of research (see Hong and Raudenbush 2013; Manski 2013a). Tchetgen Tchetgen and VanderWeele (2010) show that some estimators may be effective for applications with observational data if all relevant patterns of treatment assignment (i.e., **d**) can be attributed to measured treatment-level variables.

[17]This means that, even if the issues raised by critics on the severity of homophily bias are invalid (see VanderWeele 2011b for a convincing case that they have been exaggerated), the pattern of effects only holds under the prevalence of obesity in the data analyzed, which is the pattern of obesity in Framingham, Massachusetts, among adults born in 1948 for whom data was collected between 1971 and 1999 (and for a social network structure elicited by an unconventional name generator). The overall pattern of declining effects may be valid, but the relation of the various lagged regression coefficients offered to well-defined causal effects of general interest may be rather thin.

[18]When we have conveyed this point to network analysis researchers, a common reaction is that the potential outcome model must not, therefore, be suitable for studying causal effects that propagate across networks. The logic of this position eludes us for two reasons. First, the potential outcome model cannot be deemed inappropriate because it makes clear how hard it is to define and estimate the effects that analysts claim that they wish to estimate. Second, the potential outcome model can accommodate SUTVA violations, although not without considerable additional effort. Weihua An (2013) demonstrates the value of counterfactual thinking for modeling peer effects, fully embedded within a social network perspective (see also VanderWeele and An 2013).

2.6 Treatment Assignment and Observational Studies

A researcher who wishes to estimate the effect of a treatment that he or she can control on an outcome of interest typically designs an experiment in which subjects are randomly assigned to alternative treatment and control groups. Other types of experiments are possible, as we described in Chapter 1, but randomized experiments are the most common research design when researchers have control over the assignment of the treatment.

After randomization of the treatment, the experiment is run, and the values of the observed outcome, y_i, are recorded for those in the treatment group and for those in the control group. The mean difference in the observed outcomes across the two groups is then anointed the estimated average causal effect, and discussion (and any ensuing debate) then moves on to the particular features of the experimental protocol and the degree to which the pool of study participants reflects the population of interest for which one would wish to know the average treatment effect.

Consider this randomization research design with reference to the underlying potential outcomes defined earlier. For randomized experiments, the treatment indicator variable D is forced by design to be independent of the potential outcome variables Y^1 and Y^0. (However, for any single experiment with a finite set of subjects, the values of d_i will be related to the values of y_i^1 and y_i^0 because of chance variability.) Knowing whether or not a subject is assigned to the treatment group in a randomized experiment yields no information whatsoever about a subject's what-if outcome under the treatment state, y_i^1, or, equivalently, about a subject's what-if outcome under the control state, y_i^0. Treatment status is therefore independent of the potential outcomes, and the treatment assignment mechanism is said to be ignorable.[19] This independence assumption is usually written as

$$(Y^0, Y^1) \perp\!\!\!\perp D, \tag{2.6}$$

where the symbol $\perp\!\!\!\perp$ denotes independence and where the parentheses enclosing Y^0 and Y^1 stipulate that D must be jointly independent of all functions of the potential outcomes (such as δ). For a properly run randomized experiment, learning the treatment to which a subject has been exposed gives no information whatsoever about the size of the treatment effect.

This way of thinking about randomized experiments and potential outcomes can be confusing to social scientists who work primarily with observational data. The independence relationships represented by Equation (2.6) seem to imply that even a well-designed randomized experiment cannot tell us about the causal effect of the treatment on the outcome of interest. But, of course, this is not so, because Equation (2.6) does not imply that D is independent of Y. Equation (2.6) implies only that in the full population, ex ante to any pattern of treatment assignment, D is independent of Y^0, Y^1, and any causal effects defined from them. Only after a study is undertaken

---

[19] As we will discuss in detail in later chapters, the word "ignorability" has a very specific meaning that is broader than implied in this paragraph. In short, ignorability also holds in the weaker situation in which S is a set of observed variables that characterize treatment assignment patterns and in which $(Y^0, Y^1) \perp\!\!\!\perp D \mid S$. Thus, treatment assignment is ignorable when the potential outcomes are independent of D, conditional on S.

do values for Y emerge, from $Y = DY^1 + (1 - D)Y^0$ in Equation (2.2). If individuals are randomly assigned to both the treatment and the control states, and individual causal effects are nonzero, then Y and D will be dependent because the average value of DY^1 will not be equal to the average value of $(1 - D)Y^0$.

Now consider the additional challenges posed by observational data analysis. These challenges to causal inference are the defining features of an observational study, according to Rosenbaum (2002:vii):

> An *observational study* is an empiric investigation of treatments, policies, or exposures and the effects they cause, but it differs from an experiment in that the investigator cannot control the assignment of treatments to subjects.[20] (Italics in the original)

Observational data analysis in the counterfactual tradition is thus defined by a lack of control over the treatment – and often more narrowly by the infeasibility of randomization designs that allow for the straightforward maintenance of the independence assumption in Equation (2.6). An observational researcher, hoping to estimate a causal effect, begins with observed data in the form of values $\{y_i, d_i\}_i^N$ for an observed outcome variable, Y, and a treatment status variable, D. To determine the causal effect of D on Y, the first step in analysis is to investigate the treatment selection mechanism. Notice the switch in language from assignment to selection. Because observational data analysis is defined as empirical inquiry in which the researcher does not have the capacity to assign individuals to treatments (or, as Rosenbaum states equivalently, to assign treatments to individuals), researchers must instead investigate how individuals are selected into alternative treatment states.

And herein lies the challenge of much scholarship in the social sciences. Although some of the process by which individuals select alternative treatments can be examined empirically, a full accounting of treatment selection is sometimes impossible (e.g., if subjects are motivated to select on the causal effect itself and a researcher does not have a valid measure of the expectations that determine their choices). As much as this challenge may be depressing to a dispassionate policy designer/evaluator, this predicament should not be depressing for social scientists in general. On the contrary, our existential justification rests on the pervasive need to deduce theoretically from a set of basic principles or infer from experience and knowledge of related studies the set of defendable assumptions about the missing components of the treatment selection mechanism. Only through such effort can it be determined whether causal analysis can proceed or whether further data collection and preliminary theoretical analysis are necessary.

2.7 Average Causal Effects and Naive Estimation

The fundamental problem of causal inference requires that we focus on non–individual-level causal effects, maintaining assumptions about treatment assignment and treatment stability that will allow us to give causal interpretations to differences in average

---

[20]Note that Rosenbaum's definition is consistent with the Cox and Reid definition quoted in Chapter 1 (see page 7).

values of observed outcomes. In the remainder of this chapter, we define average treatment effects of varying sorts and then lay out the complications of estimating them. In particular, we consider how average treatment effects vary across those who receive the treatment and those who do not.

2.7.1 Conditional Average Treatment Effects

The unconditional average treatment effect, which is typically labeled the ATE in the counterfactual tradition, was defined in Equation (2.3) as $E[\delta] = E[Y^1 - Y^0]$. This average effect is the most common subject of investigation in the social sciences, and it is the causal effect that is closest to the sorts of effects investigated in the broad foundational examples introduced in Section 1.3.1, such as the effects of family background and mental ability on educational attainment, the effects of educational attainment and mental ability on earnings, and the effects of socioeconomic status on political participation. More narrowly defined average causal effects are of interest as well in virtually all of the other examples introduced in Chapter 1.

Two conditional average treatment effects are of particular interest. The average treatment effect for those who typically take the treatment is

$$E[\delta|D=1] = E[Y^1 - Y^0|D=1] \tag{2.7}$$
$$= E[Y^1|D=1] - E[Y^0|D=1],$$

and the average treatment effect for those who typically do not take the treatment is

$$E[\delta|D=0] = E[Y^1 - Y^0|D=0] \tag{2.8}$$
$$= E[Y^1|D=0] - E[Y^0|D=0],$$

where, as for the ATE in Equation (2.3), the second line of each definition follows from the linearity of the expectation operator. These two conditional average causal effects are often referred to by the acronyms ATT and ATC, which signify the average treatment effect for the treated and the average treatment effect for the controls, respectively.

Consider the examples again. For the Catholic school example, the ATT is the average effect of Catholic schooling on the achievement of those who typically attend Catholic schools rather than across all students who could potentially attend Catholic schools. The difference between the ATE and the ATT can also be understood with reference to individuals. From this perspective, the average treatment effect in Equation (2.3) is the expected what-if difference in achievement that would be observed if we could educate a randomly selected student in both a public school and a Catholic school. In contrast, the ATT in Equation (2.7) is the expected what-if difference in achievement that would be observed if we could educate a randomly selected Catholic school student in both a public school and a Catholic school.

For this example, the ATT is a theoretically important quantity, for if there is no Catholic school effect for Catholic school students, then most reasonable theoretical arguments would maintain that it is unlikely that there would be a Catholic school effect for students who typically attend public schools (at least after adjustments for observable differences between Catholic and public school students). And, if policy interest were focused on whether or not Catholic schooling is beneficial for Catholic school students (and thus whether public support of transportation to Catholic schools is a benevolent government expenditure, etc.), then the Catholic school effect for Catholic school students is the only quantity we would want to estimate. The ATC would be of interest as well if the goal of analysis is ultimately to determine the effect of a potential policy intervention, such as a new school voucher program, designed to move more students out of public schools and into Catholic schools. In fact, an even narrower conditional average treatment effect might be of interest: $E[\delta|D = 0$, CurrentSchool = Struggling], where of course the definition of being currently educated in a struggling school would have to be clearly specified.

The worker training example is similar, in that the subject of first investigation is surely the ATT (as discussed in detail in Heckman et al. 1999). If a cost-benefit analysis of a program is desired, then a comparison of the aggregate net benefits for the treated to the overall costs of the program to the funders is needed. The treatment effect for other potential enrollees in the treatment program could be of interest as well, but this effect is secondary (and may be impossible to estimate for groups of individuals completely unlike those who have enrolled in the program in the past).

The butterfly ballot example is somewhat different. Here, the treatment effect of interest is bound by a narrow question that was shaped by media attention. The investigators were interested only in what actually happened in the 2000 election, and they focused very narrowly on whether the effect of having had a butterfly ballot rather than an optical scan ballot caused some individuals to miscast their votes. And, in fact, they were most interested in narrow subsets of the treated, for whom specific assumptions were more easily asserted and defended (e.g., those who voted for Democrats in all other races on the ballot but who voted for Pat Buchanan or Al Gore for president). In this case, the ATC, and hence the all-encompassing ATE, was of little interest to the investigators (or to the contestants and the media).

As these examples demonstrate, more specific average causal effects (or more general properties of the distribution of causal effects) are often of greater interest than simply the average causal effect in the population. In this book, we will focus mostly on the three types of average causal effects represented by Equations (2.3), (2.7), and (2.8), as well as simple conditional variants of them. But, especially when presenting instrumental variable estimators later and discussing general heterogeneity issues, we will also focus on more narrowly defined causal effects. Heckman (2000), Manski (1995), and Rosenbaum (2002) all give full discussions of the variety of causal effects that may be relevant for different types of applications, such as quantiles of the distribution of individual-level causal effects in subpopulations of interest and the probability that the individual-level causal effect is greater than zero among the treated (see also Heckman, Smith, and Clements 1997).

2.7.2   Naive Estimation of Average Treatment Effects

Suppose again that randomization of the treatment is infeasible and thus that only an observational study is possible. Instead, an autonomous fixed treatment selection regime prevails, where π is the proportion of the population of interest that takes the treatment instead of the control. In this scenario, the value of π is fixed in the population by the behavior of individuals, and it is unknown. Suppose further that we have observed survey data from a relatively large random sample of the population of interest.

Because we are now shifting from the population to data generated from a random sample of the population, we must use appropriate notation to distinguish sample-based quantities from the population-based quantities that we have considered until now. For the sample expectation of a quantity in a sample of size N, we will use a subscript on the expectation operator, as in $E_N[.]$. With this notation, $E_N[d_i]$ is the sample mean of the dummy treatment variable, $E_N[y_i|d_i = 1]$ is the sample mean of the outcome for those observed in the treatment group, and $E_N[y_i|d_i = 0]$ is the sample mean of the outcome for those observed in the control group.[21] The naive estimator is then defined as

$$\hat{\delta}_{\text{NAIVE}} \equiv E_N[y_i|d_i = 1] - E_N[y_i|d_i = 0], \qquad (2.9)$$

which is simply the difference in the sample means of the observed outcome variable Y for the observed treatment and control groups.

In observational studies, the naive estimator rarely yields consistent or unbiased estimates of the ATE because it converges to a contrast, $E[Y|D = 1] - E[Y|D = 0]$, that is not equivalent to (and usually not equal to) any of the average causal effects defined above. To see why, first decompose the ATE in Equation (2.3) as

$$E[\delta] = \{\pi E[Y^1|D = 1] + (1 - \pi)E[Y^1|D = 0]\} \qquad (2.10)$$
$$- \{\pi E[Y^0|D = 1] + (1 - \pi)E[Y^0|D = 0]\}.$$

Equation (2.10) reveals that the ATE is a function of five unknowns: the proportion of the population that is assigned to (or self-selects into) the treatment along with four conditional expectations of the potential outcomes.

With observational data from a random sample of the population and without introducing additional assumptions, we can compute estimates that are consistent and unbiased for only three of the five unknowns on the right-hand side of Equation (2.10). Consider π first, which we have defined as equal to $E[D]$, and which is the fixed proportion of the population that would be assigned to (or would select into) the treatment. The sample-mean estimator, $E_N[d_i]$, is consistent for π, which we write as

$$E_N[d_i] \xrightarrow{p} \pi. \qquad (2.11)$$

Equation (2.11) represents the claim that, as the sample size N increases to infinity, the sample mean of the values for d_i converges to the true value of π, which we assume is a fixed population parameter equal exactly to $E[D]$.[22] Thus, the notation $\xrightarrow{p}$ denotes convergence in probability for a sequence of estimates over a set of samples where the sample size N is increasing to infinity. (Estimators with this property are defined as

---

[21] In other words, the subscript N serves the same basic notational function as an overbar on y_i, as in $\bar{y}_i$. We use this sub-N notation because it allows for greater clarity in aligning sample-level and population-level conditional expectations for subsequent expressions.

[22] Again, see our appendix to this chapter on our assumed superpopulation model. We are implicitly assuming that these sequences are well defined because conditions are such that the law of large numbers is applicable.

"consistent" in the statistical literature on estimation. We can also state that $E_N[d_i]$ is unbiased for π because the expected value of $E_N[d_i]$ over repeated samples of size N from the same population is equal to π as well. However, in this book we focus primarily on the consistency of estimators.)[23]

We can offer similar claims for consistent estimators of two other unknowns in Equation (2.10):

$$E_N[y_i|d_i = 1] \xrightarrow{p} E[Y^1|D = 1], \tag{2.12}$$

$$E_N[y_i|d_i = 0] \xrightarrow{p} E[Y^0|D = 0], \tag{2.13}$$

which indicate that the sample mean of the observed outcome in the treatment group converges to the true average outcome under the treatment state for those in the treatment group (and analogously for the control group and control state).

Unfortunately, however, there is no assumption-free way to compute consistent or unbiased estimates of the two remaining unknowns in Equation (2.10): $E[Y^1|D = 0]$ and $E[Y^0|D = 1]$. These are counterfactual conditional expectations: the average outcome under the treatment for those in the control group and the average outcome under the control for those in the treatment group. Without further assumptions, no estimated quantity based on observed data from a random sample of the population of interest would converge to the true values for these unknown counterfactual conditional expectations. For the Catholic school example, these are the average achievement of public school students if they had instead been educated in Catholic schools and the average achievement of Catholic school students if they had instead been educated in public schools.

2.7.3   The Typical Inconsistency and Bias of the Naive Estimator

In the last section, we concluded that the naive estimator $\hat{\delta}_{\text{NAIVE}}$, which is defined as $E_N[y_i|d_i = 1] - E_N[y_i|d_i = 0]$, converges to a contrast, $E[Y^1|D = 1] - E[Y^0|D = 0]$, that does not necessarily equal the ATE. In this section, we show why this contrast can be uninformative about the causal effect of interest in an observational study by analyzing the typical inconsistency and bias in the naive estimator as an estimator of the ATE.[24] Consider the following rearrangement of the decomposition in

---

[23] Nonetheless, we will often label estimators as "consistent and unbiased" when this is true, even though we will not state the case for unbiasedness. On the one hand, estimators of fixed, finite values in the population that are consistent are necessarily also asymptotically unbiased. On the other hand, some consistent estimators are not unbiased in finite samples (most prominently, for this book, the instrumental variable estimators that we will present in Chapter 9). As with the statistical literature on point estimation, we typically interpret unbiasedness as a desirable property among the consistent estimators that we will present. If an estimator is not consistent (i.e., is inconsistent), then in practice there is little reason to further consider it (as unbiased, etc.), especially when invoking a superpopulation perspective and assuming that a dataset for a large random sample is available.

[24] An important point of this literature is that the inconsistency and bias of an estimator is a function of the target parameter that has been selected for analysis. Because there are many causal effects that can be estimated, general statements about the inconsistency and bias of particular estimators are always conditional on a clear indication of the causal effect of interest.

Equation (2.10):

$$E[Y^1|D=1] - E[Y^0|D=0] = E[\delta] + \{E[Y^0|D=1] - E[Y^0|D=0]\} \qquad (2.14)$$

$$+ (1-\pi)\{E[\delta|D=1] - E[\delta|D=0]\}.$$

The naive estimator converges to the difference on the left-hand side of this equation, and the right-hand side shows that this difference is equal to the true ATE, $E[\delta]$, plus the expectations of two potential sources of inconsistency and bias in the naive estimator.[25] The first source, $\{E[Y^0|D=1] - E[Y^0|D=0]\}$, is a *baseline bias* equal to the difference in the expected outcome in the absence of the treatment between those in the treatment group and those in the control group. The second source, $(1-\pi)\{E[\delta|D=1] - E[\delta|D=0]\}$, is a *differential treatment effect bias* equal to the expected difference in the treatment effect between those in the treatment group and those in the control group (multiplied by the proportion of the population under the fixed treatment selection regime that does not select into the treatment).

To clarify this decomposition, consider a substantive example – the effect of education on an individual's mental ability. Assume that the treatment is college attendance. After administering a test to a group of young adults, we find that individuals who have attended college score higher than individuals who have not attended college. There are three possible reasons that we might observe this finding. First, attending college might make individuals smarter on average. This effect is the ATE, represented by $E[\delta]$ in Equations (2.3) and (2.14). Second, individuals who attend college might have been been smarter in the first place. This source of inconsistency and bias is the baseline difference represented by $E[Y^0|D=1] - E[Y^0|D=0]$. Third, the mental ability of those who attend college may increase more than would the mental ability of those who did not attend college if they had instead attended college. This source of inconsistency and bias is the differential effect of the treatment represented by $E[\delta|D=1] - E[\delta|D=0]$.

To further clarify the last term in the decomposition, consider the alternative hypothetical example depicted in Table 2.3. Suppose, for context, that the potential outcomes are now some form of labor market outcome, and that the treatment is whether or not an individual has obtained a bachelor's degree. Suppose further that 30 percent of the population obtains a bachelor's degree, such that π is equal to .3. As shown on the main diagonal of Table 2.3, the average (or expected) potential outcome under the treatment is 10 for those in the treatment group, and the average (or expected) potential outcome under the control for those in the control group is 5. Now, consider the off-diagonal elements of the table, which represent the counterfactual average potential outcomes. According to these values, those who have bachelor's degrees would have done better in the labor market than those without bachelor's degrees in the counterfactual state in which they did not in fact obtain bachelor's degrees (i.e., on average they would have received 6 instead of 5). Likewise, those who do not obtain bachelor's

---

[25]The referenced rearrangement is simply a matter of algebra. Let $E[\delta] = e$, $E[Y^1|D=1] = a$, $E[Y^1|D=0] = b$, $E[Y^0|D=1] = c$, and $E[Y^0|D=0] = d$ so that Equation (2.10) can be written more compactly as $e = \{\pi a + (1-\pi)b\} - \{\pi c + (1-\pi)d\}$. Rearranging this expression as $a - d = e + a - b - \pi a + \pi b + \pi c - \pi d$ then simplifies to $a - d = e + \{c - d\} + \{(1-\pi)[(a-c) - (b-d)]\}$. Substituting for a, b, c, d, and e then yields Equation (2.14).

Table 2.3 An Example of Inconsistency and
Bias of the Naive Estimator When the ATE Is
the Causal Effect of Interest

Group	$E[Y^1|D]$	$E[Y^0|D]$
Treatment group ($D=1$)	10	6
Control group ($D=0$)	8	5

degrees would not have done as well as those who did obtain bachelor's degrees in the counterfactual state in which they did in fact obtain bachelor's degrees (i.e., on average they would have received 8 rather than 10). Accordingly, the ATT is 4, whereas the ATC is only 3.[26] Finally, if the proportion of the population that has a bachelor's degree is .3, then the ATE is 3.3, which is equal to $.3(10-6)+(1-.3)(8-5)$.

Consider now the inconsistency and bias of the naive estimator. For this example, the naive estimator, as defined in Equation (2.9), would be equal to 5 for an infinite sample (or equal to 5, on average, across repeated samples). Thus, the naive estimator is inconsistent and upwardly biased for the ATE (i.e., yielding 5 rather than 3.3), the ATT (i.e., yielding 5 rather than 4), and the ATC (i.e., yielding 5 rather than 3). Equation (2.14) gives the components of the total expected bias of 1.7 for the naive estimator as an estimate of the ATE. The term $\{E[Y^0|D=1] - E[Y^0|D=0]\}$, which we labeled the expected baseline bias, is $6-5=1$. The term $(1-\pi)\{E[\delta|D=1]-E[\delta|D=0]\}$, which is the expected differential treatment effect bias, is $(1-.3)(4-3)=.7$.[27]

2.7.4 Estimating Causal Effects Under Maintained Assumptions About Potential Outcomes

What assumptions suffice to enable consistent and unbiased estimation of the ATE with the naive estimator? There are two basic classes of assumptions: (1) assumptions about potential outcomes for subsets of the population defined by treatment status and (2) assumptions about the treatment assignment/selection process in relation to the potential outcomes. These two types of assumptions are variants of each other, and each may have a particular advantage in motivating analysis in a particular application.

In this section, we discuss only the first type of assumption, as it suffices for the present examination of the fallibility of the naive estimator. And our point in introducing these assumptions is simply to explain in one final way why the naive estimator will fail in most social science applications to generate a consistent and unbiased estimate of the ATE when randomization of the treatment is infeasible.

---

[26]For the causal effect of education on earnings, there is debate in the recent literature on whether the ATT is larger than the ATC. Cunha and Heckman (2007) and Carneiro, Heckman, and Vytlacil (2011) offer results in support of this pattern, but Brand and Xie (2010) offer results in opposition to it.

[27]In general, the size of this expected differential treatment effect bias declines as more of the population is characterized by the ATT than by the ATC (i.e., as π approaches 1).

Consider the following two assumptions:

$$\text{Assumption 1:} \quad E[Y^1|D=1] = E[Y^1|D=0], \qquad (2.15)$$

$$\text{Assumption 2:} \quad E[Y^0|D=1] = E[Y^0|D=0]. \qquad (2.16)$$

If one asserts these two equalities and then substitutes into Equation (2.10), the number of unknowns is reduced from the original five parameters to the three parameters that we know from Equations (2.11)–(2.13) can be consistently estimated with data generated from a random sample of the population. If both Assumptions 1 and 2 are maintained, then the ATE, ATT, and ATC in Equations (2.3), (2.7), and (2.8), respectively, are all equal. And the naive estimator is consistent and unbiased for all of them.

When would Assumptions 1 and 2 in Equations (2.15) and (2.16) be reasonable? Clearly, if the independence of potential outcomes, as expressed in Equation (2.6), is valid because the treatment has been randomly assigned, then Assumptions 1 and 2 in Equations (2.15) and (2.16) are implied. But, for observational data analysis, for which random assignment is infeasible, these assumptions would rarely be justified.

Consider the Catholic school example. If one were willing to assume that those who choose to attend Catholic schools do so for completely random reasons, then these two assumptions could be asserted. We know from the applied literature that this characterization of treatment selection is false. Nonetheless, one might be able to assert instead a weaker narrative to warrant these two assumptions. One could maintain that students and their parents make enrollment decisions based on tastes for an education with a religious foundation and that this taste is unrelated to the two potential outcomes, such that those with a taste for education with a religious foundation would not be expected to score higher on math and reading tests if educated in Catholic schools rather than public schools. This possibility also seems unlikely, in part because it implies that those with a distaste for education with a religious foundation do not attend Catholic schools. It seems reasonable to assume that these students would perform substantially worse in Catholic schools than the typical students who do attend Catholic schools.

Thus, at least for the Catholic school example, there seems no way to justify the naive estimator as a consistent and unbiased estimator of the ATE. We encourage the reader to consider all of the examples presented in Chapter 1, and we suspect that all will agree that Assumptions 1 and 2 in Equations (2.15) and (2.16) cannot both be sustained for any of them.

Finally, it is important to recognize that assumptions such as these can be evaluated separately. Consider the two relevant cases for Assumptions 1 and 2:

1. If Assumption 1 is true but Assumption 2 is not, then $E[Y^1|D=1] = E[Y^1|D=0]$, whereas $E[Y^0|D=1] \neq E[Y^0|D=0]$. In this case, the naive estimator remains inconsistent and biased for the ATE, but it is now consistent and unbiased for the ATC. This result is true because of the same sort of substitution we noted earlier. We know that the naive estimator $E_N[y_i|d_i=1] - E_N[y_i|d_i=0]$ converges to $E[Y^1|D=1] - E[Y^0|D=0]$. If Assumption 1 is true, then one can

substitute $E[Y^1|D=0]$ for $E[Y^1|D=1]$. Then, one can state that the naive estimator converges to the contrast $E[Y^1|D=0] - E[Y^0|D=0]$ when Assumption 1 is true. This contrast is defined in Equation (2.8) as the ATC.

2. If Assumption 2 is true but Assumption 1 is not, then $E[Y^0|D=1] = E[Y^0|D=0]$, whereas $E[Y^1|D=1] \neq E[Y^1|D=0]$. The opposite result to the prior case follows. One can substitute $E[Y^0|D=1]$ for $E[Y^0|D=0]$ in the contrast $E[Y^1|D=1] - E[Y^0|D=0]$. Then, one can state that the naive estimator converges to the contrast $E[Y^1|D=1] - E[Y^0|D=1]$ when Assumption 2 is true. This contrast is defined in Equation (2.7) as the ATT.

Considering the validity of Assumptions 1 and 2 separately shows that the naive estimator may be inconsistent and biased for the ATE and yet may be consistent and unbiased for either the ATT or the ATC. These possibilities can be important in practice. For some applications, it may be the case that we have good theoretical reason to believe that (1) Assumption 2 is valid because those in the treatment group would, on average, do no better or no worse in the counterfactual control state than those in the control group, and (2) Assumption 1 is invalid because those in the control group would not do nearly as well in the counterfactual treatment state as those in the treatment group. Or, stated more simply, we may have good theoretical reason to believe that the treatment is more effective for the treatment group than it would be for the control group. Under this scenario, the naive estimator will deliver a consistent and unbiased estimate of the ATT, even though it is still inconsistent and biased for both the ATC and the unconditional ATE.

Now, return to the case in which neither Assumption 1 nor Assumption 2 is true. If the naive estimator is therefore inconsistent and biased for the typical average causal effects of interest, what can be done? The first recourse is to attempt to partition the sample into subgroups within which assumptions such as Assumptions 1 and/or 2 can be defended. This strategy amounts to conditioning on one or more variables that identify such strata and then asserting that the naive estimator is consistent and unbiased within these strata for one of the average treatment effects. One can then average estimates from these strata in a reasonable way to generate the average causal effect estimate of interest. We will explain this strategy in great detail in subsequent chapters. Next, we introduce over-time potential outcomes and then extend the framework to many-valued causes.

2.8 Over-Time Potential Outcomes and Causal Effects

Having shown in the last section that the cross-sectional naive estimator will rarely deliver consistent and unbiased estimates of average causal effects of interest when analyzing observational data, it is natural to then wonder whether observing individuals across time and then estimating similar unconditional differences may be more promising. We will take the position in this book that the power of over-time observation is considerable but that it is also too often oversold and misunderstood. In this section, we lay out the basic potential outcome model when observations occur in more than one time period, moving from the case of a single individual or unit to multiple individuals or units. We reserve our full treatment of the strengths and weaknesses of alternative estimators using repeated observations for Chapter 11.

2.8.1 A Single Unit Over Time

Consider the analysis of a single unit, observed during time intervals indexed by a discrete counter t that increases from 1 to T. The outcome variable is Y_t, which has observed values $\{y_1, y_2, y_3, \ldots, y_T\}$. Suppose that we have a two-state causal variable, D_t, that is equal to 1 if the treatment is in place during a time period t and is equal to 0 otherwise.

Because we are considering only one unit of analysis – possibly an individual but more likely a school, organization, city, state, country, or other aggregate unit – we do not have either a control group or a treatment group. Instead, we have a single unit that is exposed to the treatment state and the control state at different points in time. The fundamental problem/reality of causal inference now is that we cannot observe the same unit at the same time in both the treatment state and the control state.

For an analysis of a single unit, it only makes sense to consider designs where we have at least some pretreatment data and where the unit under consideration spends at least one time period in the treatment state. In particular, we will label the time period in which the treatment is initiated as t^*, and our restriction to situations in which pretreatment data are available requires that $1 < t^* \leq T$. We will allow the treatment to persist for one or more time periods, from t^* through $t^* + k$, where $k \geq 0$. Once the treatment ends, following $t^* + k$, we will not allow the treatment to be reintroduced before the full observation window terminates at T.

We can set up the potential outcome model in the following way to capture the basic features of before-and-after designs for a single unit of analysis:

1. Before the treatment is introduced (for $t < t^*$):[28]

$$D_t = 0$$
$$Y_t = Y_t^0$$

2. While the treatment is in place (from t^* through $t^* + k$):

$$D_t = 1$$
$$Y_t = Y_t^1$$
$$Y_t^0 \text{ exists but is counterfactual}$$

---

[28] Although in theory counterfactual values Y_t^1 exist in pretreatment time periods, these values are not typically considered. If one were interested in asking what the treatment effect would have been if the treatment had been introduced in an earlier time period, then these counterfactual values would need to be introduced into the analysis.

3. After the treatment ends (for time periods $t > (t^* + k)$):[29]

$$D_t = 0$$
$$Y_t = Y_t^1$$

Y_t^0 exists but is counterfactual.

For a single unit, the causal effect of the treatment is

$$\delta_t = Y_t^1 - Y_t^0, \tag{2.17}$$

and these effects may exist in more than one time period t, depending on the duration of the treatment and whether the treatment is assumed to be a reversible treatment state or a permanent change that cannot be undone. Studies are often unclear on maintained assumptions such as these, as well as on the distinctions between time periods of types 2 and 3. Our setup is very general and can accommodate many alternative types of studies with only minor modifications, including those for which time periods of types 2 or 3 are unobserved.[30] Because such assumptions and design features will always be application-specific, we offer a worked example next.

The Year of the Fire Horse

For a concrete example that reveals the possible power of over-time analysis for a single unit, consider a variant of the demography example on the determinants of fertility introduced in Section 1.3 (see page 17). In addition to the individual-level effects of family background and other life course events on fertility decisions, the causal effects of religion, values, and more general cultural beliefs have been of long-standing interest as well (see Mayer and Marx 1957; Westoff and Jones 1979; Hayford and Morgan 2008; Thornton, Binstock, Yount et al. 2012).

Suppose that the unit of analysis is the birth rate in a single country, estimated from aggregate census data and vital statistics. The example we will consider is presented in Figure 2.1, which displays birth rates in Japan between 1951 and 1980. Following a post-war baby boom, birth rates in Japan were comparatively stable from the late 1950s through the early 1960s. However, in 1966, the birth rate fell precipitously, after which it rebounded in 1967 and then stabilized. From the 1970s onward, Japan's birth rate then resumed its decline, as its population aged and it continued with its demographic transition to a low mortality and low fertility country, as discussed in general in Chapter 1.[31]

---

[29]Below we will consider an example where $k \leq (T - t^*)$, but we do not mean to imply that the treatment cannot remain in place after the observation window ends at T. In fact, we place no upper bound on values for k. If $(t^* + k) > T$, then none of the time periods of type 3 are observed.

[30]In some applications, the treatment is stipulated to occur between observation intervals. In these cases, time periods of type 2 are assumed to be unobserved. Typically, in this case D_t is assumed to be equal to 1 for at least the first time period of type 3 in order to indicate that the treatment was initiated in the unobserved time periods of type 2. Others studies imply that $t^* + k = T$, such that the treatment is present through the full posttreatment observation window. In these cases, time periods of type 3 are unobserved.

[31]For a comprehensive consideration of the post-1973 "baby bust" in Japan, see Retherford and Ogawa (2006).

One could ask many causal questions about the trend in Japan's birth rate in Figure 2.1, but the natural first question is, What caused the dramatic decline in the birth rate in 1966? The demographic consensus is the following. Every 60 years, two cycles within the Asian zodiac calendar – one over twelve animals and one over five elements – generate a year of the "fire horse." A folk belief exists that families who give birth to babies designated as fire horses will suffer untold miseries, particularly so if the baby is a girl. Enough couples supposedly held this belief in the years around 1966 that they adjusted their fertility behavior accordingly (see Hodge and Ogawa 1991).

In their discussion of causal analysis in demography, Ní Bhrolcháin and Dyson (2007:8) consider this example and write that "demographers naturally interpret this event in a causal way, without worrying about the formalities of causal inference." Although we agree that the assertion of a fire-horse causal effect on Japanese birth rates in 1966 does not require a formal treatment to convince most demographers, we will nonetheless use this example to demonstrate an over-time analysis of a causal effect for a single unit of analysis.

The first issue to consider is measurement of the outcome. The outcome for Figure 2.1 is known as the crude birth rate, which is the number of live births per 1,000 persons alive in the same year. A more refined outcome could be constructed,

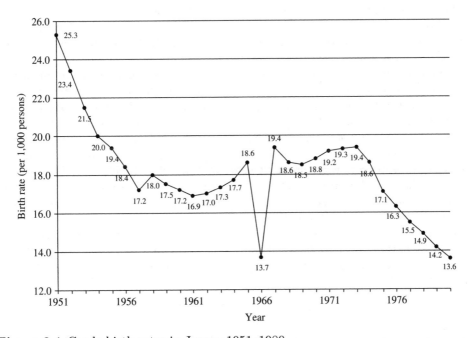

Figure 2.1 Crude birth rates in Japan, 1951–1980.
Source: Table 2-24, Live Births by Sex and Sex Ratio of Live Births, Bureau of Statistics of Japan. Accessed at http://www.stat.go.jp/data/chouki/zuhyou/02-24.xls on March 11, 2013.

standardizing for cohort sizes of women of childbearing age. An analogous drop in standardized birth rates would still be present for 1966.[32]

What are the causal states that generate the seemingly obvious causal effect? At the individual level, the causal states could be quite specific ("believing that having a fire-horse baby is less desirable than having a baby in another year") or considerably more broad ("believing in the relevance of zodiac calendars"). For our purposes here, the particular individual-level causal states do not matter because the causal states are at the national level for this analysis. The states are "fire-horse year in the zodiac calendar" and "not," and they are aligned, not with observed treatment and control groups, but with the observed years, $D_{1966} = 1$ versus $D_{t \neq 1966} = 0$. In the year 1966, the treatment generates two corresponding potential outcome variables, Y^1_{1966} and Y^0_{1966}, which define the effect of interest for the birth rate in Japan in 1966:

$$\delta_{1966} = y^1_{1966} - y^0_{1966}.$$

We can estimate this effect as

$$\hat{\delta}_{1966} = y_{1966} - \hat{y}^0_{1966}$$
$$= 13.7 - \hat{y}^0_{1966}.$$

The estimate $\hat{\delta}_{1966}$ is the observed birth rate in 1966, 13.7, minus an estimate of the counterfactual birth rate, $\hat{y}^0_{1966}$, that would have been observed if 1966 had not been the year of the fire horse.

What is a reasonable estimated value for the true counterfactual outcome, y^0_{1966}? How about 18.6, which is the birth rate for 1965? Ní Bhrolcháin and Dyson (2007:30) reason that "some births that might have taken place in 1966 were transferred – either in fact, or via the year of occurrence reported by the parents at the time of registration – to the adjacent years." Such a possibility is evident in 1965, where the birth rate appears elevated in comparison to prior years. In fact, the birth rate appears to be slightly higher in 1964 as well, relative to the prevailing trend in the early 1960s. What about the value for 1967, 19.4? Again, the birth rate appears to be higher here too, and possibly again in 1968. Following Ní Bhrolcháin and Dyson's reasoning (and also Hodge and Ogawa 1991), these higher rates in 1964, 1965, 1967, and 1968 could be present because parents were especially eager to have children before or after the year of the fire horse and adjusted their behavior accordingly. Given the uncertainties of conception, at least some of these parents started their avoidance early or failed to conceive a child until after the year of the fire horse.

In our experience discussing this example with other researchers, most will settle on the average of the two values for 1963 and 1969, $(17.3 + 18.5)/2 = 17.9$, as a reasonable estimate of the counterfactual value, y^0_{1966}. The result of such a choice is an estimated causal effect,

$$\hat{\delta}_{1966} = 13.7 - 17.9 = -4.2,$$

---

[32] At least some of the overall downward trend would disappear because the trend in Figure 2.1 is produced in part by the aging of the population (which itself is a function of the return to peace following World War II and continuing declines in mortality). Hodge and Ogawa (1991) document these trends and consider alternative adjustments for other demographic trajectories.

that implies a decline in the crude birth rate of 23 percent. At the individual level, and ignoring rates of twins and so on, this estimate suggests that nearly 1 out of 4 mothers who would have given birth in 1966 did not do so because 1966 was the year of the fire horse. Notice also that, in selecting the average of the birth rates in 1963 and 1969 as the most reasonable estimate of the counterfactual value, one is thereby assuming positive fire-horse-year effects in 1964, 1965, 1967, and 1968, which were non–fire-horse years. As a result, in order to estimate the overall effect of the year of the fire horse on the population structure of Japan, as determined by the year-by-year evolution of what is known as the total fertility rate, one would need to model and then appropriately combine five year-specific effects, δ_{1964} through δ_{1968}, based on additional corresponding treatment states for the specific years.

Now consider what has been learned and what has not. Visual inspection of Figure 2.1 is probably convincing on its own that something causal happened in 1966 that altered birth rates in Japan. However, the specific explanation that is accepted in demography is based on substantive knowledge of Asian zodiac calendars, as well as cultural beliefs based upon them that were sufficiently widespread in 1966. The over-time design did not generate this knowledge, even though it provided the incentive to uncover it.

Has anything deeper been learned from this effect? Certainly the effect supports the more general conclusion that cultural beliefs have been a cause of fertility decisions in Japan in the past. This conclusion may then further bolster the overall perspective in empirical demography that fertility decisions across the world are likely shaped by cultural beliefs and cannot be reduced to a cold rational calculus of the direct costs and psychic benefits of producing offspring.

With a little extra work, we have been able to generate the reasonable estimate that the effect in 1966 was a reduction of the birth rate of 23 percent, and we also took the position that there were very likely positive near-fire-horse-year effects in the two years on either side of 1966. Notwithstanding these successes, our analysis yields no information on which types of couples changed their fertility behavior. The relevance of the zodiac calendar surely varies across couples in nonrandom ways, and such patterns would be useful to know in order to develop additional perspective on the broader consequences of the effect. We have also not learned how many women, or which women, had fewer children or instead simply had children who were one or two years younger or older than they would have been if their children had been born in 1966.

We will return to this type of example in Chapter 11, where we will consider the deeper modeling issues that accompany the estimation of these types of effects. Often referred to as interrupted time series (ITS) designs, there are formal time series analysis procedures for generating best estimates of counterfactual values, and these may be easier to defend than our ad hoc choice of 17.9 as the most reasonable estimate of the crucial counterfactual value that determines the size of the casual effect. As we will discuss in detail in Chapter 11, the main weakness of these designs is their rarity. In the observational social sciences, few examples are as clear-cut as the year of the fire horse. More commonly, the causal shocks to outcomes unfolding over time are less dramatic, and, as a result, they are more difficult to separate from underlying trends.

Notice also that we have chosen in this section an aggregate unit – a country – as our example of the analysis of a causal effect for a "single unit." In part, this decision reflects our position on what can be learned by studying individuals in isolation. Surely there are genuine individual-level causal effects, some of which can be discerned from examining the lives of particular individuals over time. The challenge is what in general can be learned from documenting such apparent effects, and whether analyses can be strengthened by considering individuals in groups, especially in representative samples. This is the focus of our next section, where we introduce the potential outcome model for many units over time.

2.8.2   Many Units Over Time

Suppose now that we have a collection of individuals observed at multiple points in time. Generally referred to as panel data or longitudinal data, outcomes and causes now vary over both individuals and time. Consider two examples. First, researchers who study the charter school effect often have access to samples of students observed over multiple grades in both charter schools and regular public schools. Second, researchers who study the effects of father absence that results from divorce typically have access to data from random samples of families. These data often include measures of child development outcomes before and after the divorce that triggers father absence.

We will now extend the potential outcome framework to consider the estimation of causal effects with such over-time data on samples of individuals. In earlier sections of this chapter, we dropped subscripting on i for brevity when discussing causal effects. For this section, in which we must deal now with some quantities that vary only over individuals, others that vary only over time, and others that vary over both, we subscript with i for individuals and with t for time. For a two-state cause, the potential outcome variables are Y_{it}^1, Y_{it}^0, and the observed variables are D_{it} and Y_{it}.[33] As in the last section, we will allow the observation window to run from $t = 1$ to T. Unlike the last section, we will not utilize notation for a focal time interval, t^*, in which the treatment is introduced. Here, we want to preserve the possibility that the treatment is introduced at different times for different individuals.

We will also now distinguish between two different treatment indicator variables. D_{it} is a time-varying variable that indicates whether individual i receives the treatment in time period t. In contrast, D_i^* is a time-constant variable that indicates whether individual i ever receives the treatment at any point in time during the observation window of the study (i.e., in any time period from $t = 1$ to T). D_{it} is best thought of as a *treatment exposure indicator*, and D_i^* is best thought of as a *treatment group indicator*.

The setup for the potential outcome model with panel data follows directly from these definitions. For members of the control group, $Y_{it} = Y_{it}^0$ for all time periods t. For members of the treatment group, $Y_{it} = Y_{it}^0$ before treatment exposure, and $Y_{it} = Y_{it}^1$

---

[33]In some cases, the subscripting is redundant. For example, in Sections 2.4 and 2.7, we represented the causal effect as δ, recognizing that this effect can vary over individuals. In this section, we will represent the individual-level causal effect always as δ_i, so that it is clear that in this form we are assuming that it does not vary with time. For a time-varying causal effect, we would instead need to subscript it as δ_{it}.

after treatment exposure begins. Altogether, Y_{it} is defined with reference to D_{it}, such that

$$Y_{it} = D_{it}Y_{it}^1 + (1 - D_{it})Y_{it}^0, \tag{2.18}$$

where D_{it} remains equal to 0 over time for all members of the control group ($D_i^* = 0$) but varies between 0 and 1 for all members of the treatment group ($D_i^* = 1$).

Consider a concrete application of this notation using both the treatment exposure indicator and the treatment group indicator. If no one is exposed to the treatment in time period $t = 1$ or $t = 2$, then $D_{i1} = 0$, $D_{i2} = 0$, $Y_{i1} = Y_{i1}^0$, and $Y_{i2} = Y_{i2}^0$ for those in the control group ($D_i^* = 0$) and for those in the treatment group ($D_i^* = 1$). But, if the treatment is then introduced to all members of the treatment group in time period $t = 3$, then the values of D_{it} and Y_{it} diverge across the treatment and control groups in time period 3. Now, for those in the control group ($D_i^* = 0$), $D_{i3} = 0$ and $Y_{i3} = Y_{i3}^0$. But, for those in the treatment group ($D_i^* = 1$), $D_{i3} = 1$ and $Y_{i3} = Y_{i3}^1$.

The distinction between D_{it} and D_i^* reveals the potential value of panel data. For a cross-sectional study in which all observation occurs in a single time period, $D_{it} = D_i^*$. However, a panel dataset over multiple time periods allows a researcher to consider how treatment exposure (D_{it}) can be separated from treatment group membership (D_i^*) and then exploit this difference to estimate the causal effect. Consider two types of analysis.

First, when individuals receive the treatment at different times or do not receive the treatment at all, it is possible to observe how Y_{it}^0 changes over time for some individuals after others have received the treatment. As a result, it may be possible to make reasonable predictions about how Y_{it}^0 would have evolved over time for individuals in the treatment group during the posttreatment time period. If predictions of these counterfactual trajectories are reasonable, inferences about the causal effect of the treatment, δ_{it}, in time period t, can be made by comparison of the observed Y_{it} in time period t for those who are treated ($D_i^* = 1$) with predictions of their corresponding counterfactual values of Y_{it}^0 in time period t. The crux of the matter, of course, is how to use observed values of $Y_{it} = Y_{it}^0$ in time period t among those in the control group ($D_i^* = 0$) to make reasonable predictions about posttreatment counterfactual values of Y_{it}^0 for those in the treatment group ($D_i^* = 1$).

Second, in situations where it is reasonable to assume that $E[Y_{it}^0]$ is evolving in parallel fashion across the treatment and control groups and that treatment assignment is unrelated to the values of the outcome prior to the treatment, it is possible to estimate the ATT by offering a model that (1) calculates the average difference in the outcome in the treatment group between the pretreatment and posttreatment time periods and (2) subtracts from this difference the average difference in the outcome for the control group between the same two time periods.[34] In this case, the most important issue to consider is whether it is reasonable to assume parallel trajectories and that treatment assignment is unrelated to pretreatment values of the outcome. The latter assumption would be unreasonable if individuals with comparatively low or comparatively high values of the pretreatment outcome, net of other determinants

---

[34]Moreover, if it is reasonable to assume that self-selection on the causal effect is absent, then the ATT is equal to the ATE and ATC. A consistent and unbiased estimate of the ATT from a difference-based estimator is then also consistent and unbiased for both the ATE and ATC.

of the outcome, select into the treatment, either under the assumption that they are especially suited to the treatment because of their recent strong performance or that they are especially in need of the treatment because of their recent weak performance.

We will discuss a variety of panel data estimation strategies in Chapter 11, but we want to foreshadow two basic conclusions here to temper the optimism that many may feel after considering our prior two paragraphs. First, panel data estimators based only on posttreatment observations do not usually improve on the cross-sectional estimators we will present in this book. In our experience, analysts are often too sanguine about the clarifying power of observing the evolution of outcome variables for those who are always observed under exposure to the treatment of interest (e.g., students always enrolled in Catholic high schools relative to students always enrolled in public high schools, with no data on either group available prior to high school). Second, one needs data from multiple pretreatment time periods and/or very well-developed theory to justify required assumptions before the gains to panel data are clear, especially given the other ancillary patterns, such as panel attrition, that must also often be modeled.

2.9   The Potential Outcome Model for Many-Valued Treatments

So far in this chapter, we have focused our presentation of the potential outcome model on binary causal variables, conceptualized as dichotomous variables that indicate whether individuals are observed in treatment and control states. As we show in this section, the counterfactual framework can be used to analyze causal variables with more than two categories.

Consider the more general setup, in which we replace the two-valued causal exposure variable, D, and the two potential outcomes Y^1 and Y^0 with

1. a set of J treatment states,

2. a corresponding set of J causal exposure dummy variables, $\{Dj\}_{j=1}^{J}$, and

3. a corresponding set of J potential outcome random variables, $\{Y^{Dj}\}_{j=1}^{J}$.

Each individual receives only one treatment, which we denote Dj'. Accordingly, the observed outcome variable for individual i, y_i, is then equal to $y_i^{Dj'}$. For the other $J-1$ treatments, the potential outcomes of individual i exist in theory as $J-1$ other potential outcomes y_i^{Dj} for $j \neq j'$, but they are counterfactual.

Consider the fundamental problem of causal inference for many-valued treatments presented in Table 2.4 (which is simply an expansion of Table 2.1 to many-valued treatments). Groups exposed to alternative treatments are represented by rows with, for example, those who take treatment $D2$ in the second row. For a binary treatment, we showed earlier that the observed variable Y contains exactly half of the information contained in the underlying potential outcome random variables. In general, for a treatment with J values, Table 2.4 shows that the observed outcome variable Y contains only $1/J$ of the total amount of information contained in the underlying

Table 2.4 The Fundamental Problem of Causal Inference for Many-Valued Treatments

Group	Y^{D1}	Y^{D2}	$\cdots$	Y^{DJ}
Takes $D1$	Observable as Y	Counterfactual	$\cdots$	Counterfactual
Takes $D2$	Counterfactual	Observable as Y	$\cdots$	Counterfactual
$\vdots$	$\vdots$	$\vdots$	$\ddots$	$\vdots$
Takes DJ	Counterfactual	Counterfactual	$\cdots$	Observable as Y

potential outcome random variables. Thus, the proportion of unknown and inherently unobservable information increases as the number of treatment values, J, increases.

For an experimentalist, this decline in the relative amount of information in Y is relatively unproblematic. Consider an example in which a researcher wishes to know the relative effectiveness of three pain relievers for curing headaches. The four treatments are "Take nothing," "Take aspirin," "Take ibuprofen," and "Take acetaminophen." Suppose that the researcher rules out an observational study, in part because individuals have constrained choices (i.e., pregnant women may take acetaminophen but may not take ibuprofen; many individuals take a daily aspirin for general health reasons). Instead, she gains access to a large pool of subjects not currently taking any medication and not prevented from taking any of the three medicines.[35] She divides the pool randomly into four groups, and the drug trial is run. Assuming all individuals follow the experimental protocol, at the end of the data-collection period the researcher calculates the mean length and severity of headaches for each of the four groups.

Even though three quarters of the cells in a 4×4 observability table analogous to Table 2.4 are counterfactual, she can easily estimate the relative effectiveness of each of the drugs in comparison with each other and in comparison with the take-nothing control group. Subject to random error, contrasts such as $E_N[y_i|\text{Take aspirin}] - E_N[y_i|\text{Take ibuprofen}]$ reveal all of the average treatment effects of interest. The experimental design allows her to ignore the counterfactual cells in the observability table by assumption. In other words, she can assume that the average counterfactual value of Y^{Aspirin} for those who took nothing, took ibuprofen, and took acetaminophen (i.e., $E[Y^{\text{Aspirin}}|\text{Take nothing}]$, $E[Y^{\text{Aspirin}}|\text{Take ibuprofen}]$, and $E[Y^{\text{Aspirin}}|\text{Take acetaminophen}]$) can all be assumed to be equal to the average observable value of Y for those who take the treatment aspirin, $E[Y|\text{Take aspirin}]$. She can therefore compare sample analogs of the expectations in the cells of the diagonal of the observability table, and she does not have to build contrasts within its rows. Accordingly, for this type of example, comparing the effects of multiple treatments with each other is no more complicated than the bivariate case, except insofar as one nonetheless has more treatments to assign and resulting causal effect estimates to calculate.

---

[35]Note that, in selecting this group, she has adopted a definition of the population of interest that does not include those who (1) take one of these pain relievers regularly for another reason and (2) do not have a reason to refuse to take one of the pain relievers. We will discuss the importance of considering such groups of "always takers" and "never takers" when we present instrumental variable estimators in Chapter 9.

Table 2.5 The Observability Table for Estimating How Education Increases Earnings

Education	Y^{HS}	Y^{AA}	Y^{BA}	Y^{MA}
Obtains HS	Observable as Y	Counterfactual	Counterfactual	Counterfactual
Obtains AA	Counterfactual	Observable as Y	Counterfactual	Counterfactual
Obtains BA	Counterfactual	Counterfactual	Observable as Y	Counterfactual
Obtains MA	Counterfactual	Counterfactual	Counterfactual	Observable as Y

Now consider a variant on the education-earnings example. Suppose that a researcher hopes to estimate the causal effect of different educational degrees on labor market earnings, and further that only four degrees are under consideration: a high school diploma (HS), an associate's degree (AA), a bachelor's degree (BA), and a master's degree (MA). For this problem, we therefore have four dummy treatment variables corresponding to each of the treatment states: HS, AA, BA, and MA. Table 2.5 has the same structure as Table 2.4. Unlike the pain reliever example, random assignment to the four treatments is impossible. Consider the most important causal effect of interest for policy purposes, $E[Y^{BA} - Y^{HS}]$, which is the average effect of obtaining a bachelor's degree instead of a high school diploma.

Suppose that an analyst has survey data on a set of middle-aged individuals for whom earnings at the most recent job and highest educational degree are recorded. To estimate this effect without asserting any further assumptions, the researcher would need to be able to consistently estimate population-level analogs to the expectations of all of the cells of Table 2.5 in columns 1 and 3, including six counterfactual cells off of the diagonal of the table. The goal would be to formulate consistent estimates of $E[Y^{BA} - Y^{HS}]$ for all four groups of differentially educated adults. To obtain a consistent estimate of $E[Y^{BA} - Y^{HS}]$, the researcher would need to be able to consistently estimate $E[Y^{BA} - Y^{HS}|HS = 1]$, $E[Y^{BA} - Y^{HS}|AA = 1]$, $E[Y^{BA} - Y^{HS}|BA = 1]$, and $E[Y^{BA} - Y^{HS}|MA = 1]$, after which these estimates would be averaged across the distribution of educational attainment. Notice that this requires the consistent estimation of some doubly counterfactual contrasts, such as the effect on earnings of shifting from a high school diploma to a bachelor's degree for those who are observed with a master's degree. The researcher might boldly assert that the wages of all high school graduates are, on average, equal to what all individuals would obtain in the labor market if they instead had high school diplomas. But this is very likely to be a mistaken assumption if it is the case that those who carry on to higher levels of education would have been judged more productive workers by employers even if they had not attained more than high school diplomas.

As this example shows, a many-valued treatment creates substantial additional burden on an analyst when randomization is infeasible. For any two-treatment comparison, one must find some way to estimate a corresponding $2(J-1)$ counterfactual conditional expectations, because treatment contrasts exist for individuals in the population whose observed treatments place them far from the diagonal of the observability table.

If estimating all of these counterfactual average outcomes is impossible, analysis can still proceed in a more limited fashion. One might simply define the parameter of interest very narrowly, such as the average causal effect of a bachelor's degree only for those who typically attain high school diplomas: $E[Y^{\mathrm{BA}} - Y^{\mathrm{HS}}|\mathrm{HS} = 1]$. In this case, the causal effect of attaining a bachelor's degree for those who typically attain degrees other than a high school diploma are of no interest for the analyst.

Alternatively, there may be reasonable assumptions that one can invoke to simplify the complications of estimating all possible counterfactual expectations. For this example, many theories of the relationship between education and earnings suggest that, for each individual i, $y_i^{\mathrm{HS}} \leq y_i^{\mathrm{AA}} \leq y_i^{\mathrm{BA}} \leq y_i^{\mathrm{MA}}$. In other words, earnings never decrease as one obtains a higher educational degree. Asserting this assumption (i.e., taking a theoretical position that implies it) may allow one to ignore some cells of the observability table that are farthest from the direct comparison one hopes to estimate. We will discuss these sorts of assumptions in Chapter 12.

Aside from the expansion of the number of causal states, potential outcomes, and treatment effects, all other features of the potential outcome model remain essentially the same. SUTVA is typically still maintained, and, if it is unreasonable, then more general methods must again be used to model treatment effects that may vary with patterns of treatment assignment. Modeling treatment selection remains the same, even though the added complexity of having to model movement into and out of multiple potential treatment states can be taxing. And the same sources of inconsistency and bias in standard estimators must be considered, only here again the complexity can be considerable when there are multiple states beneath each contrast of interest.

To avoid all of this complexity, one temptation is to assume that treatment effects are linear additive in an ordered set of treatment states. For the effect of education on earnings, a researcher might instead choose to move forward under the assumption that the effect of education on earnings is linear additive in the years of education attained. For this example, the empirical literature has demonstrated that this is a particularly poor idea. For the years in which educational degrees are typically conferred, individuals appear to receive an extra boost in earnings. When discussing the estimation of treatment effects using linear regression for many-valued treatments in Section 6.6.1, we will consider a piece by Angrist and Krueger (1999) that shows very clearly how far off the mark these methods can be when motivated by unreasonable linearity and additivity assumptions.

2.10 Conclusions

In this chapter, we have introduced the main components of the potential outcome model, which is a foundational piece of the counterfactual model of causality for observational research. We defined individual-level causal effects as the what-if differences in potential outcomes that would result from being exposed to alternative causal states. We then presented the assumption of causal effect stability – the stable unit treatment value assumption – that is frequently relied on when estimating effects defined by potential outcomes. We defined average causal effects at the population level, considered how ineffective the simple mean-difference estimator is for estimating average causal effects with observational data, and concluded with extensions of the potential outcome model for effects observed over time and for effects defined across many values of the cause. In the next chapter, we introduce the directed graph approach to causal analysis, which we see as the second foundational piece of the counterfactual model of causality for observational research.

2.11 Appendix to Chapter 2: Population and Data Generation Models

In the counterfactual tradition, no single agreed-on way to define the population exists. In a recent piece, for example, Rubin (2005:323) introduces the primary elements of the potential outcome model without taking any particular position on the nature of the population, writing that "'summary' causal effects can also be defined at the level of collections of units, such as the mean unit-level causal effect for all units." As a result, a variety of possible population-based (and "collection"-based) definitions of potential outcomes, treatment assignment patterns, and observed outcomes can be used. In this appendix, we explain the choice of population model that we will use throughout the book (and implicitly, unless otherwise specified).

Because we introduce populations, samples, and convergence claims in this chapter, we have placed this appendix here. Nonetheless, because we have not yet introduced models of causal exposure, some of the fine points in the following discussion may well appear confusing (notably, how "nature" performs randomized experiments behind our backs). For readers who wish to have a full understanding of the implicit super-population model we will adopt, we recommend a quick reading of this appendix now and then a second more careful reading after completing Chapters 5 through 7.

Our Implicit Superpopulation Model. The most expedient population and data generation model to adopt is one in which the population is regarded as a realization of an infinite superpopulation. This setup is the standard perspective in mathematical statistics, in which random variables are assumed to exist with fixed moments for an uncountable and unspecified universe of events. For example, a coin can be flipped an infinite number of times, but it is always a Bernoulli distributed random variable for which the expectation of a fair coin is equal to .5 for both heads and tails. For this example, the universe of events is infinite because the coin can be flipped forever.

Many presentations of the potential outcome framework adopt this basic setup, following Rubin (1977) and Rosenbaum and Rubin (1983b, 1985a). For a binary cause, potential outcomes Y^1 and Y^0 are implicitly assumed to have expectations $E[Y^1]$ and $E[Y^0]$ in an infinite superpopulation. Individual realizations of Y^1 and Y^0 are then denoted y_i^1 and y_i^0. These realizations are usually regarded as fixed characteristics of each individual i.

This perspective is tantamount to assuming a population machine that spawns individuals forever (i.e., the analog to a coin that can be flipped forever). Each individual is born as a set of random draws from the distributions of Y^1, Y^0, and additional variables collectively denoted by S. These realized values y^1, y^0, and s are then given individual identifiers i, which then become y_i^1, y_i^0, and s_i.

The challenge of causal inference is that nature also performs randomized experiments in the superpopulation. In particular, nature randomizes a causal variable D within strata defined by the values of S and then sets the value of Y as y_i equal to y_i^1 or y_i^0, depending on the treatment state that is assigned to each individual. If nature assigns an individual to the state $D = 1$, nature then sets y_i equal to y_i^1. If nature assigns an individual to the state $D = 0$, nature then sets y_i equal to y_i^0. The differential probability of being assigned to $D = 1$ instead of $D = 0$ may be a function in S, depending on the experiment that nature has decided to conduct (see Chapters 4 and 5). Most important, nature then deceives us by throwing away y_i^1 and y_i^0 and giving us only y_i.

In our examples, a researcher with good fortune obtains data from a random sample of size N from a population, which is in the form of a dataset $\{y_i, d_i, s_i\}_{i=1}^N$. The sample that generates these data is drawn from a finite population that is itself only one realization of a theoretical superpopulation. Based on this set-up, the joint probability distribution in the sample $\Pr_N(Y, D, S)$ must converge in probability to the true joint probability distribution in the superpopulation $\Pr(Y, D, S)$ as the sample size approaches infinity. The main task for analysis is to model the relationship between D and S that nature has generated in order use observed data on Y to estimate causal effects defined by Y^1 and Y^0. [Many researchers do not have such good fortune and instead must analyze a dataset with measures of only a subset of the variables in S, which we will typically label X. These researchers have access to a dataset $\{y_i, d_i, x_i\}_{i=1}^N$ and model $\Pr_N(Y, D, X)$, which does not converge to $\Pr(Y, D, S)$.]

Because of its expediency, we will usually write with this superpopulation model in the background, even though the notions of infinite superpopulations and sequences of sample sizes approaching infinity are manifestly unrealistic. We leave the population and data generation model largely in the background in the main text, so as not to distract the reader from the central goals of our book.

Alternative Perspectives. There are two main alternative models of the population that we could adopt. The first, which is consistent with the most common starting point of the survey sampling literature (e.g., Kish 1965), is one in which the finite population is recognized as such but treated as so large that it is convenient to regard it as infinite. Here, values of a sample statistic (such as a sample mean) are said to equal population values in expectation, but now the expectation is taken over repeated samples from the population (see Thompson 2002 for an up-to-date accounting of this perspective). Were we to adopt this perspective, rather than our superpopulation model, much of what we write would be the same. However, this perspective tends to restrict attention to large survey populations (such as all members of a country's population older than 18) and makes it cumbersome to discuss some of the estimators we will consider (e.g., in Chapter 5, where we will sometimes define causal effects only across the common support of some random variables, thereby necessitating a redefinition of the target population).

The second alternative is almost certainly much less familiar to many empirical social scientists but is a common approach within the counterfactual causality literature. It is used often when no clearly defined population exists from which the data can be said to be a random sample (such as when a collection of data of some form is available and an analyst wishes to estimate the causal effect for those appearing in the data). In this situation, a dataset exists as a collection of individuals, and the observed individuals are assumed to have fixed potential outcomes y_i^1 and y_i^0. The fixed potential outcomes have average values for those in the study, but these average values are not typically defined with reference to a population-level expectation. Instead, analysis proceeds by comparison of the average values of y_i for those in the treatment and control groups with all other possible average values that could have emerged under all possible permutations of treatment assignment. This perspective then leads to a form of randomization inference, which has connections to exact statistical tests of null hypotheses most commonly associated with Fisher (1935). As Rosenbaum (2002) shows, many of the results we present in this book can be expressed in this framework (see also Rubin 1990, 1991). But the combinatoric apparatus required for doing so can be cumbersome (and often requires constraints, such as homogeneity of treatment effects, that are too restrictive). Nonetheless, because the randomization inference perspective has some distinct advantages in some situations, we will refer to it at several points throughout the book. And we strongly recommend that readers consult Rosenbaum (2002, 2010) if the data under consideration arise from a sample that has no straightforward and systematic connection to a well-defined population. In this case, sample average treatment effects may be the only well-defined causal effects, and, if so, then the randomization inference tradition is a clear choice.

Chapter 3

Causal Graphs

In his 2009 book titled *Causality: Models, Reasoning, and Inference*, Judea Pearl lays out a powerful and extensive graphical theory of causality.[1] Pearl's work provides a language and a framework for thinking about causality that differs from the potential outcome model presented in Chapter 2. Beyond the alternative terminology and notation, Pearl (2009, section 7.3) shows that the fundamental concepts underlying the potential outcome perspective and his causal graph perspective are equivalent, primarily because they both encode counterfactual causal states to define causality. Yet, each framework has value in elucidating different features of causal analysis, and we will explain these differences in this and subsequent chapters, aiming to convince the reader that these are complementary perspectives on the same fundamental issues.

Even though we have shown in the last chapter that the potential outcome model is simple and has great conceptual value, Pearl has shown that graphs nonetheless provide a direct and powerful way of thinking about full causal systems and the strategies that can be used to estimate the effects within them. Some of the advantage of the causal graph framework is precisely that it permits suppression of what could be a dizzying amount of notation to reference all patterns of potential outcomes for a system of causal relationships. In this sense, Pearl's perspective is a reaffirmation of the utility of graphical models in general, and its appeal to us is similar to the appeal of traditional path diagrams in an earlier era of social science research. Indeed, to readers familiar with path models, the directed graphs that we will present in this chapter will look familiar. There are, however, important and subtle differences between traditional path diagrams and Pearl's usage of directed graphs, which we will explain.

---

[1] Our references throughout will be to the 2009 second edition of Pearl's original 2000 book. Pearl recognizes that his influential work was developed in dialogue with many others (see Pearl 2009:104–5). In focusing heavily on his development of causal graphs in this book, we regrettably do not give enough credit to the active group of researchers who have participated in developing graphical representations of causality. We recommend that interested readers turn to this broader literature to learn the interconnections between the many complementary perspectives that exist, starting with pieces such as Berzuini, Dawid, and Bernardinelli (2012); Dawid (2002); Dechter, Geffner, and Halpern (2010); Elwert (2013); Glymour, Scheines, and Spirtes (2001); Koller and Friedman (2009); Lauritzen (1996); and Robins and Richardson (2010). For readers from philosophy who wish to know why we do not use neuron diagrams at all, we take the same basic position as Hitchcock (2007).

For our purposes in this book, Pearl's work is important for three different reasons. First, directed graphs encode causal relationships that are completely nonparametric and fully interactive, and as a result when considering feasible analysis strategies it is usually unnecessary to specify the nature of the functional dependence of an outcome Y on the variables that cause it. A graph that includes the causal effects $X \to Y$ and $W \to Y$ simply implies that X and W both cause Y, without specifying whether their effects are linear, quadratic, interactive, or any other highly nonlinear function in the values of both X and W. This generality allows for a model of causality without side assumptions about functional form, such as assumptions of linear additivity. Second, directed graphs show clearly the critical importance of what Pearl labels *collider variables*, which are endogenous variables that must be treated with caution in many research scenarios. Finally, Pearl uses directed graphs to develop transparent and clear justifications for the three basic methods for estimating causal effects that we will feature in this book: conditioning on variables to eliminate noncausal associations by blocking all relevant back-door paths from the causal variable, conditioning on variables that allow for estimation by a mechanism, and using an instrumental variable that is an exogenous shock to the causal variable in order to consistently estimate its effect.

In this chapter, we provide the foundations of the directed graph approach to causal analysis and discuss the relationships between directed graphs and the potential outcome model. In the course of this presentation, we provide a brief introduction to conditioning techniques, but we will hold off on the full presentation of conditioning, as well as Pearl's back-door criterion for conditioning strategies, until Chapter 4.

3.1   Identification

To set the stage for our introduction to directed graph representations of causal relationships, it is helpful to define the concept of identification. In Chapter 2, we defined causal effects as contrasts between well-defined potential outcomes and then proceeded to consider some of the conditions under which consistent and unbiased estimators of average causal effects are available. A complementary approach, and the one which motivates the usage of the directed graphs that we will present in this chapter, is to perform an *identification analysis*. Here, the challenges to inference that arise from the finite nature of the available sample are held aside while the analyst considers whether a causal effect could be computed if data on the full population were instead available.

In his 1995 book *Identification Problems in the Social Sciences*, Manski writes,

> it is useful to separate the inferential problem into statistical and identification components. Studies of identification seek to characterize the conclusions that could be drawn if one could use the sampling process to obtain an unlimited number of observations. (Manski 1995:4)

He continues,

> Empirical research must, of course, contend with statistical issues as well as with identification problems. Nevertheless, the two types of inferential difficulties are sufficiently distinct for it to be fruitful to study them

separately. The study of identification logically comes first. Negative identification findings imply that statistical inference is fruitless: it makes no sense to try to use a sample of finite size to infer something that could not be learned even if a sample of infinite size were available. Positive identification findings imply that one should go on to study the feasibility of statistical inference. (Manski 1995:5)

The two most crucial ingredients for an identification analysis are these:

1. The set of assumptions about causal relationships that the analyst is willing to assert based on theory and past research, including assumptions about relationships between variables that have not been observed but that are related to both the cause and the outcome of interest.

2. The pattern of information that one can assume would be contained in the joint distribution of the variables in the observed dataset *if* all members of the population had been included in the sample that generated the dataset.

As we will begin to explain in this chapter, causal graphs can represent these ingredients effectively and efficiently and are therefore valuable tools for conducting identification analyses.[2] We will use the concept of identification frequently in this book, and we will expand upon this brief introduction as we introduce additional strategies for analysis.

3.2 Basic Elements of Causal Graphs

3.2.1 Nodes, Edges, Paths, and Cycles

The primary goal when drawing a causal system as a directed graph is to represent explicitly all causes of the outcome of interest, based on past empirical research and assumptions grounded in theory. As we discussed in Section 1.5, each node of a graph represents a random variable and is labeled by a letter, such as A, B, or C. Nodes that are represented by a solid circle • are observed random variables, whereas nodes that are represented by a hollow circle ○ are unobserved random variables.

Causal effects are represented by directed edges → (i.e., single-headed arrows), such that an edge from one node to another signifies that the variable at the origin of the

---

[2]Nonetheless, an identification analysis can be conducted, and typically is within the econometric tradition, without utilizing directed graphs. Consider our discussion of the naive estimator in Chapter 2. The equalities across the expected values of potential outcomes that we stated as Assumptions 1 and 2 in Equations (2.15) and (2.16) are *identification assumptions*. Maintenance of these particular assumptions would allow an analyst to assert that the naive estimator in Equation (2.9) is consistent and unbiased for the true average treatment effect in Equation (2.3), as explained by the decompositions offered there. As such, the average treatment effect is "identified" or is "identifiable" when these assumptions can be maintained, even though an estimate from a finite sample may, because of sampling error, depart substantially from the true average treatment effect in the population. The value of causal graphs, as we will show in this chapter and the next, is that they allow for an efficient representation of full systems of causal relationships, which can be helpful for determining whether identification assumptions are reasonable.

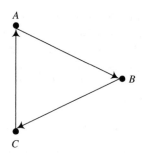

Figure 3.1 A directed graph that includes a cycle.

directed edge causes the variable at the terminus.[3] These "directed" edges are what give graphs composed of nodes and single-headed arrows the general label of "directed graphs."

A *path* is any sequence of edges pointing in any direction that connects one variable to another. A *directed path* is a path in which all edges point in the same direction. A variable is a *descendant* of another variable if it can be reached by a directed path. All kinship terms are then duly appropriated. Most importantly, for directed paths of length one, as in $A \to B$, the variable A is the *parent* while the variable B is the *child*.

In this book, we will consider only a subset of directed graphs known as *directed acyclic graphs* or DAGs. For these graphs, no directed paths emanating from a causal variable also terminate at the same causal variable. In other words, no variable can be its own descendant. Figure 3.1 presents a graph that includes a directed path that forms a cycle, and as a result it is not a DAG (even though it is a directed graph because it includes only directed edges). Unlike some graphical models from the past, the prohibition of cycles in DAGs rules out representations of simultaneous causation and feedback loops.[4] All of our statements about graphs from this point onward assume that only acyclic graphs are under consideration.

Under some circumstances it is useful to use a curved and dashed bidirected edge (as in Figures 1.1–1.3) as a shorthand device to indicate that two variables are mutually dependent on one or more unobserved common causes. In this shorthand, the two graphs presented in Figures 3.2(a) and (b) are equivalent. Such shorthand can be helpful in suppressing a complex set of background causal relationships that are irrelevant

---

[3]In Pearl's framework, each random variable is assumed to have an implicit probability distribution net of the causal effects represented by the directed edges that point to it. This position is equivalent to assuming that background causes of each variable exist that are independent of the causes explicitly represented in the graph by directed edges. We will discuss this assumption in more detail in Section 3.3.2, where we introduce the structural equations that correspond to directed graphs.

[4]As shown in White and Chalak (2009), a broader framework that accommodates cycles is possible (a position acknowledged by Pearl and colleagues for some time, and which has its origins in the interest in reconciling recursive and nonrecursive models since the 1960s). However, the additional details of the broader setup can be daunting, and we recommend that interested readers first carefully consider how much their research questions really do require the full specification of feedback loops that generate cycles. In our experience, most such purportedly necessary loops result from a misplaced unwillingness to consider more tractable empirical research questions that can be confined to shorter spans of analytic time. We do not mean to imply, however, that theory should not attend to such feedback loops, only that most empirical projects can benefit from recursion pragmatism.

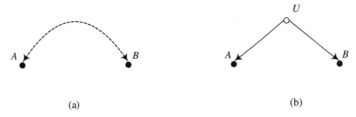

(a)   (b)

Figure 3.2 Two representations of the joint dependence of A and B on unobserved common causes.

to the empirical analysis at hand. Nonetheless, these bidirected edges should not be interpreted in any way other than as we have just stated. They are not indicators of mere associations or correlations between the variables that they connect, and they do not signify that either of the two variables has a direct cause on the other one. Rather, they represent an unspecified set of unobserved common causes of the two variables that they connect.

3.2.2   Causal Graphs for Three Variables

Figure 3.3 presents the three basic patterns of causal relationships that would be observed for any three variables that are connected to each other by only two directed edges: a chain of mediation, a fork of mutual dependence, and an inverted fork of mutual causation. Pearl's analysis of the first two types of relationship is conventional. For the graph in panel (a), A affects B through A's causal effect on C and C's causal effect on B. This type of a causal chain renders the variables A and B unconditionally associated. For the graph in panel (b), A and B are both caused by C. Here, A and B are also unconditionally associated, but now it is because they mutually depend on C.[5]

For the third graph in panel (c), A and B are again connected by a path through C. But now A and B are both causes of C. Pearl labels C a *collider variable*. Formally, a variable is a collider along a particular path if it has two arrows pointing directly at it. Figuratively, the causal effects of A and B "collide" with each other at C. Collider variables are common in social science applications: Any endogenous variable that has two or more causes is a collider along some path.

A path that is connected by a collider variable does not generate an unconditional association between the variables that cause the collider variable. For the mutual causation graph in panel (c) of Figure 3.3, the path between A and B through C does not generate an unconditional association between A and B. As a result, if nothing is known about the value that C takes on, then knowing the value that A takes on yields no information about the value that B takes on. Pearl's language is quite helpful here.

---

[5]The unconditional associations between A and B for both graphs mean that knowing the value that A takes on gives one some information on the likely value that B takes on. This unconditional association between A and B, however, is completely indirect, as neither A nor B has a direct causal effect on each other.

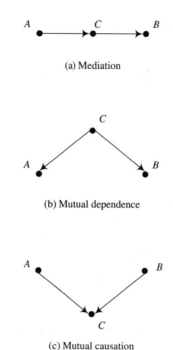

(a) Mediation

(b) Mutual dependence

(c) Mutual causation

Figure 3.3 Basic patterns of causal relationships for three variables.

The path $A \rightarrow C \leftarrow B$ does not generate an association between A and B because the collider variable C "blocks" the possible causal effects of A and B on each other.

Even though collider variables do not generate unconditional associations between the variables that determine them, we will show in the next chapter that incautious handling of colliders can create conditional dependence that can sabotage a causal analysis. The importance of considering collider variables is a key insight of Pearl's framework, and it is closely related to the familiar concerns of selecting on the dependent variable and conditioning on an endogenous variable (see Elwert and Winship 2014). Before turning to these issues in detail in Chapter 4, we first need to continue our presentation of the basic features of the directed graph approach to causal analysis.

3.2.3   A First Look at Confounding and Conditioning

The most common concern of a researcher seeking to estimate a causal effect with observational data is that a causal variable D and an outcome variable Y are determined, in part, by a third variable, C. This common but simple scenario is represented by the two graphs in Figure 3.4, where for panel (a) the variable C is observed and for panel (b) it is not.

For both graphs in Figure 3.4, the total association between D and Y is composed of two pieces: (1) the genuine causal effect of D on Y, represented by $D \rightarrow Y$, and (2) the common dependence of D and Y on C, represented by both $C \rightarrow D$ and $C \rightarrow Y$.

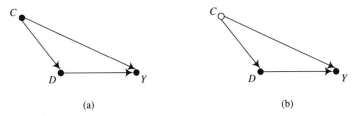

Figure 3.4 Two graphs in which the causal effect of D on Y is confounded by C.

In such cases, it is often said that the causal effect of D on Y is "confounded" by C or, even more simply, that C is a "confounder." Regardless of the label given to C, the causal effects $C \rightarrow D$ and $C \rightarrow Y$ render the total association between D and Y unequal to the causal effect $D \rightarrow Y$.

The frequency of basic confounding has established subgroup analysis as perhaps the most common modeling strategy to prosecute causal questions in social science research. Whether referred to as subclassification, stratification, tabular decomposition, or simply adjustment for a third variable, the data are analyzed after "conditioning" on membership in groups defined by values of the confounder variable. Usage of the word "conditioning" is a reference to the " | " operator, already used extensively in Chapter 2, to define conditional expectations (see, in particular, Section 2.7). From a graphical perspective, this modeling strategy is analogous to disconnecting the conditioning variable from all other variables that it points to in the original graph, rewriting the graph without the edges from the conditioning variable for each value for the conditioning variable, analyzing the data that apply to each of these graphs separately, and then combining the results across graphs to form an overall estimate.

For Figure 3.4(a), but not for Figure 3.4(b), consistent and unbiased estimators of the causal effect of D on Y are available (with a large enough dataset generated by a suitably random sample) because the analyst can condition on the observed variable C and eliminate the portion of the association between D and Y that is generated by their common dependence on C. We will explain this claim more formally and more generally in Chapter 4, where we introduce Pearl's back-door criterion for the identification of a causal effect.

For now, consider the essential operational data analysis routine. For Figure 3.4(a), the effect of D on Y can be estimated by conditioning on C in two steps: (1) calculate the association between D and Y for each subgroup with C equal to c and then (2) average these c-specific associations over the distribution of the values c that the variable C takes on in the sample (which we assume is, again, a large random sample from the population of interest). The resulting weighted average is a consistent and unbiased estimator of the causal effect of D on Y in Pearl's framework, which would be labeled the average treatment effect (ATE) in the potential outcome model introduced in Chapter 2. For Figure 3.4(b), no such data analysis routine is feasible because the analyst has no observed variable C with which to begin.

To make this example more concrete, and to begin to build connections to the examples utilized for Chapter 2, consider the graph in Figure 3.5, where we revisit

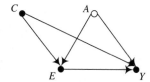

Figure 3.5 A causal graph in which the effect of education (E) on earnings (Y) is confounded by observed variables (C) and by unobserved ability (A).

the research question on ability bias in estimates of the earning returns attributable to completed years of education. Here, education, E, is a cause of earnings, Y, but this causal effect is confounded by two sources: (1) observed variables, C, such as demographic characteristics and family background, and (2) an unobserved variable for ability, A.[6] For Figure 3.5, we have moved one step beyond the analysis of the naive estimator in Chapter 2 (see Tables 2.3 and 2.5), because now we are considering the more typical scenario in which the analyst includes in the directed graph some observed variables in C that past research has established almost certainly have causal effects on both years of completed education and subsequent labor market earnings. The ability bias literature asserts that, even if we were to condition on a rich parameterization of all values of the variables in C, we still would not be able to eliminate the confounding generated by A.

Notice also that we have not asserted that the education variable, E, in Figure 3.5 is a two-valued cause, as was the case for the discussion of Table 2.3. Unless specified otherwise, graphs of this type do not place restrictions on the numbers of values taken on by any of their variables. This particular graph, for example, would apply to either the two-valued education variable discussed in Section 2.7.3 or the four-valued education variable introduced in Section 2.9.

3.3 Graphs and Structural Equations

Having introduced the basic elements of directed graphs, the next step is to introduce the structural equations that lie beneath them. As noted in the introduction to this chapter, directed graphs encode causal relationships that are completely nonparametric and fully interactive. This generality allows for a model of causality without assumptions about functional form, which is a major advantage over traditional path diagrams. An appreciation for this advantage requires first understanding the constraints imposed by the parametric assumptions that were common in the equations associated with these path diagrams and why many researchers had good reason to object to them.

---

[6]In the labor economics literature, it would generally be assumed that both C and A depend on a common unobserved cause U that generates an unconditional association between C and A. We leave out this dependence in this diagram for simplicity. Its inclusion would not change the fact that the effect of D on Y is not identified.

3.3.1 The Appeal and Critique of Traditional Path Diagrams

To explain both the appeal and subsequent critique of traditional linear additive path models, we will use the charter schools example introduced in Section 1.3.2.[7] Consider the path diagram presented in Figure 3.6, and suppose that we have data on all sixth graders in a large metropolitan area. For this path diagram, Y is a standardized test taken at the end of the sixth grade, and D indicates whether or not a student attended a charter school for the past year. The variable P represents an omnibus parental background measure that captures differences in economic standing and other basic dimensions of resources that predict both charter school attendance and school performance. The variable N is neighborhood of residence, and we assume that there are meaningful differences in the extent to which neighborhood environments are conducive to engagement with schooling. Thus, D is the cause of primary interest, P represents individual determinants of D that also have direct causes on the outcome Y, and N is a measure of the social context in which the effect of D on Y occurs.[8]

The path diagram presented in Figure 3.6 is associated with two implicit structural equations for the two endogenous variables:

$$D = a_D + b_P P + e_D, \tag{3.1}$$

$$Y = a_Y + b_D D + b_P P + b_N N + e_Y. \tag{3.2}$$

These structural equations are linear and additive in the variables P, D, and N, and each equation has terms, e_D and e_Y, that are represented in the path diagram as all other determinants of D and Y other than P, D, and N.[9] The structure of the path diagram and its equations imply that the proper empirical regression specification for Y is the same as Equation (3.2). Under the implicit assumption that e_Y is uncorrelated with D, P, and N, the path-model interpretation of least squares estimates of the coefficients b_D, b_P, and b_N is that they are consistent and unbiased estimates of the genuine causal effects of P, D, and N on Y.[10]

How would such a path diagram have been presented and then discussed in a typical research methods class in the 1970s (assuming that the charter school research question was under discussion)? Following an introduction to graphical representations of causal relationships via path diagrams, at least one student would invariably ask the instructor:

---

[7]This section draws on material previously published in Morgan and Winship (2012).

[8]Most path models assumed the existence of unexplained correlations between all "predetermined" or "exogenous" variables, which are P and N for Figure 3.6. In many path diagrams, curved double-headed arrows would be drawn to represent such correlations. To avoid confusion with our usage of bidirected edges throughout this book, we have not added such a double-headed arrow to Figure 3.6.

[9]The standard approach in the early days of path modeling in the social sciences would have been to assume that e_D is uncorrelated with P and that e_Y is uncorrelated with P, D, and N. More complete approaches, as exemplified by Duncan (1975), would not necessarily have maintained such assumptions.

[10]The estimated effects are presumed to be constant across individuals, as specified in Equation (3.2). However, most analysts regarded the estimates as simple average effects across individuals. We will return to this issue in detail when we discuss how regression models do not in general deliver simple average effect estimates (see Chapter 6) and when we then introduce explicit heterogeneity into causal graphs (see Chapter 8).

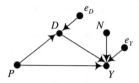

Figure 3.6 A traditional linear additive path diagram for the effects of parental background (P), charter schools (D), and neighborhoods (N) on test scores (Y).

Can the effect of D on Y vary across P? That seems reasonable, since it would seem that the effect of a charter school would depend on family background. Parents with college degrees probably help their kids get more out of school. Actually, now that I think about it, since N captures neighborhood characteristics, don't we think that there are better schools in some neighborhoods? In fact, charter schools are more likely to be established in areas with troubled neighborhood-based schools. And neighborhoods with weaker schools also tend to have stronger deviant subcultures with gangs and such. So the effect of charter schooling probably also depends on the neighborhood in which one lives. How do we represent such variation in effects in the path model?[11]

In response, an instructor would typically explain that one can think of such effects as supplemental arrows from a variable into the middle of another arrow in the path diagram, such that the variable itself modifies the arrow under discussion. Yet, since these sorts of arrows are not formally justified in traditional path diagrams, the instructor would almost surely have then recommended a shift toward a more complex regression specification, such as

$$Y = a_Y + b_D D + b_P P + b_{D \times P}(D \times P) + b_N N + b_{D \times N}(D \times N) + e_Y. \qquad (3.3)$$

In this case, the path diagram ceases to represent an underlying set of structural causal relationships and is instead best interpreted as only a simplified reflection of a more specific regression model. After all, the interaction between the effects of D and P on Y (as well as the interaction between the effects of D and N on Y) can be estimated with little trouble. One need only calculate effects of interest, for example, by plugging in values for $\hat{b}_D D + \hat{b}_P P + \hat{b}_{D \times P}(D \times P)$, after producing standard regression output from estimation of Equation (3.3). The differences then produced can be imbued with causal interpretations based on the same justification as for Equation (3.2), assuming that no other variables that are common causes of P and Y, D and Y, or N and Y have been mistakenly omitted from Equation (3.3).

We see two related outcomes of the rise and then demise of traditional linear path models conceived and estimated in this fashion in the social sciences. First, when it became clear that there was no agreed upon way to represent within path diagrams

---

[11] Were this exchange occurring in the substance of the day, a path model from the status attainment tradition would be the focus of the exchange. The outcome variable Y would be career success, and D would be education. All of the same interactions noted for the charter school case would then apply in this case, although based on different narratives of causation.

the interactions that could be specified in regression equations to capture variability and context, path diagrams came to seem much less useful.[12] Researchers interested in such variability and context may have continued to draw path diagrams on yellow pads in their offices, but rarely did their drawings turn up in published articles.[13] Estimation and reporting became a word and number affair, often with too much of each.

Second, and far more troubling, many scholars apparently chose to retain a linear additive orientation, even while no longer using path diagrams. For some, empirical research could be fruitfully advanced by ignoring the genuine interactive nonlinearity of the real world, in pursuit of a first-approximation, linear pragmatism. This stance might have been an acceptable form of pragmatism if the approximation spirit had carried over to model interpretation. Too frequently it did not, and many causal assertions can be found in the literature based on linear additive models that are overly reductionist.

Overall, the traditional path-modeling literature, and then the more general "age of regression" that we described in Section 1.2.2, opened up quantitative research to the claims of critics that too many practitioners had fallen prey to the belief that linear regression modeling reveals strong causal laws in which variability and context play minor roles. A particularly cogent presentation of this criticism is Abbott's oft-cited "Transcending General Linear Reality" (Abbott 1988). Although its straw-man style is irksome to methodologists who knew of these problems all along, and who urged better practice, it was a reasonable critique of much practice at the time.[14]

3.3.2   The Shift to Nonparametric Structural Equations

For the same substantive example depicted in the path diagram in Figure 3.6, consider now its representation by a directed graph in the new causal graph tradition. Two variants are offered in Figure 3.7. The standard representation that is depicted in panel (a) is then shown again under "magnification" in panel (b).[15] For the latter, each variable is seen, under close examination, to have its own structural "error" or "disturbance" term: e_P, e_D, e_N, and e_Y. Analogous terms are implicitly present in all directed graphs, but they are typically suppressed in their standard representation, as in panel (a), for visual simplicity.

---

[12]We do not mean to imply that methodologists have not proposed solutions. Bollen (1995) offers the elegant proposal of including functions of other variables as separate entities in diagrams and using sawtooth arrows, $\rightsquigarrow$, in order to represent functional assignment relations. For example, an interactive effect of D and P on Y can be represented altogether by three paths $D \to Y$, $P \to Y$, and $(D \times P) \to Y$. The entity $(D \times P)$ is not a new source of variation with an exogenous component but rather is a deterministic function defined in D and P. This functional dependence is signified in the diagram by including both $D \rightsquigarrow (D \times P)$ and $P \rightsquigarrow (D \times P)$.

[13]Freese and Kevern (2013:27) label such causal doodling as "arrow salad."

[14]For similar critiques in sociology in the same time period, see also Lieberson (1985) and Ragin (1987). Abbott (2001), Lieberson and Lynn (2002), and Ragin (2008) offer updates on these earlier critiques. Leamer (1983) is the analog in economics, although written in a completely different style and with alternative suggested remedies.

[15]See Pearl (2009:339) for usage of the word "magnification" to reveal the disturbance/error terms that are implicit in all directed graphs.

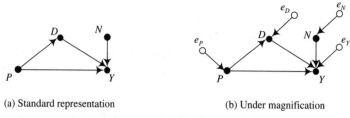

(a) Standard representation   (b) Under magnification

Figure 3.7 Equivalent directed graph representations of the effects of parental background (P), charter schools (D), and neighborhoods (N) on test scores (Y).

What are these terms, e_P, e_D, e_N, and e_Y? Pearl (2009:27) states that such terms "represent errors (or 'disturbances') due to omitted factors" and are always assumed to be independent of each other and of all other variables in the graph. For Figure 3.7(b), the terms e_P, e_D, e_N, and e_Y represent all causes of P, D, N, and Y, respectively, that can be regarded as "idiosyncratic" causes of each variable. They are assumed to be independent of each other and of P, D, N, and Y. As such, they can be suppressed in the standard representation of a directed graph, as in Figure 3.7(a).[16]

In general, all directed graphs in the new causal graph tradition must be drawn in sufficient detail so that any such disturbance terms can be pushed into the background. Pearl cautions,

> The disturbance terms represent independent background factors that the investigator chooses not to include in the analysis. If any of these factors is judged to be influencing two or more variables (thus violating the independence assumption), then that factor must enter the analysis as an unmeasured (or latent) variable and be represented in the causal graph as a hollow node. (Pearl 2009:68)

In other words, one must be able to assume that these structural error terms are mutually independent of each other and of all of the other variables in the graph, such that no pair is mutually dependent on a common third variable that has been mistakenly omitted from the graph. If this assumption is dubious, then one must redraw the graph including hollow nodes for any mistakenly omitted unobserved common causes. These unobserved variables must then be given directed edges that point to the variables that they cause. The new error terms for the re-drawn graph can then be redefined so that they can be pushed into the background (but still rendered visible under magnification).

---

[16]There is considerable debate over the ontological status of these idiosyncratic causes. Their existence implies to some scholars that causality is fundamentally probabilistic. By our reading, Pearl would maintain, for example, that the variables embedded in e_P are simply implicit structural causes of P. Under this interpretation, causality can still be considered a structural, deterministic relation. Freedman (2010, esp. chapter 15) discusses some of the drawbacks for statistical inference of assuming determinism of this form. Although convincing to some degree, his critique does not alter the utility of these sorts of causal graphs for clarifying when causal effects are identified, which is Pearl's primary contribution.

In fact, for the graphs in Figure 3.7, the existing literature on the charter school effect suggests that these graphs are incomplete. Most importantly, it is almost certainly the case that the terms e_D and e_Y in Figure 3.7(b) are mutually dependent on a common cause that has been mistakenly omitted from the graph, analogous to the unobserved ability confounder, A, in Figure 3.5. We will therefore extend this graph in Chapter 8, bringing it more into line with assumptions that analysts have been willing to maintain in existing empirical research. For now, we proceed as if the graphs in Figure 3.7 represent the true causal model, accepted hypothetically (but counterfactually!) by a clear consensus of researchers who have studied the charter school effect.

Mindful of this specific definition of the disturbance/error term in directed graphs, and supposing for now that the two causal graphs in Figure 3.7 represent a valid and accepted causal model for the charter school effect, the corresponding structural equations can now be introduced. Unlike path diagrams in traditional form, where the structural equations are linear and additive and correspond only to the the endogenous variables in the diagram (e.g., D and Y in Figure 3.6), for the new directed graph tradition the structural equations are written for all variables and in unrestricted form (i.e., possibly nonlinear and interactive, as explained below). The two graphs in Figure 3.7 have the same set of structural equations:

$$P = f_P(e_P), \tag{3.4}$$
$$D = f_D(P, e_D), \tag{3.5}$$
$$N = f_N(e_N), \tag{3.6}$$
$$Y = f_Y(P, D, N, e_Y). \tag{3.7}$$

Reading from left to right in the graphs and top to bottom in these equations, P is generated as an unspecified function, $f_P(.)$, with e_P as its sole argument. The next three equations then represent analogous unrestricted functions with inputs that represent all causes of the variables on the left-hand sides of the equations. The last is the most elaborate, in that Y is a function of the three observed variables P, D, and N, as well as e_Y, all of which transmit their effects on Y via the function $f_Y(.)$.

When we say that $f_P(.)$, $f_D(.)$, $f_N(.)$, or $f_Y(.)$ are unrestricted and have no assumed functional form, we mean that a value is produced for the outcome variable of each equation for every combination of the values in the corresponding function on the right-hand sides of these equations. For example, Y takes on a distinct value for each combination of values, $P = p$, $D = d$, and $N = n$ (typically then with the assumption that values of e_Y are drawn at random from some common distribution that is presumed to have finite moments; see Freedman 2010, chapter 15 for discussion of alternative approaches).[17]

An implication of this flexibility deserves particular emphasis, and preexisting knowledge of traditional path models and their implicit linear additive structural

---

[17]Restrictions are not typically placed on how the drawn value of the disturbance/error term is then transmitted by $f(.)$ to the outcome. However, because these terms are independent (by definition) of all other inputs in $f(.)$, their realization in the outcome variable does not typically require any assumption, at least when identification is the focus.

equations can hinder full recognition of its importance. Consider the nonparametric structural equation for Y in Equation (3.7). *All* interactions between the effects of P, D, and N on Y are implicitly permitted by the lack of restrictions placed on $f_Y(.)$. Importantly, this property of the structural equations means that the causal graphs in Figure 3.7 are consistent with all such interactions because the directed edges only signify inclusion in the functions such as $f_Y(.)$. No new arrows, nor any notation of any kind, are needed to represent interactions for more specific parameterizations where, for example, the effect of D on Y varies with the level of P or N.

As a result, even though it may feel natural to want to "see" a specific arrow present in the causal graph to represent an interaction effect that corresponds to a cross-product term in a regression equation, one must learn to suppress such a desire. The key point, in considering an analysis that utilizes causal graphs, is to drop regression models from one's mind when thinking about identification issues. Instead, if one must use a data analytic machine to conceptualize how to perform an appropriate empirical analysis of the puzzle under consideration, one should default to simple conditioning, as introduced in the last section.

As we will explain in far greater detail in the next part of the book, if the graphs in Figure 3.7 were the true causal system that generates the effects of charter schools on test scores, the effect of D on Y would be confounded by the observed parental background variables in P. In addition, the effect of D on Y may vary across the contexts defined by neighborhoods. With a dataset of sufficient size, the analyst can estimate the effect of D on Y for every combination of the values in P and N, adopting a conditioning strategy for identification and estimation. These within-P-and-within-N differences are not confounded and represent average causal effects for students within strata defined by P and N (again, under the assumption that the directed graph is a complete representation of the true causal system). To obtain average causal effects for larger groups of individuals, such as those living within a particular type of neighborhood defined by N, the analyst can combine the strata-specific estimates by forming a weighted average where the weights are proportional to the sample sizes of all strata that correspond to the type of neighborhood chosen.

In this way, causal analysis guided by the specification of directed graphs is an inherently flexible enterprise. Losing sight of the lack of functional form assumed for causal analysis with unrestricted structural equations may still lead one to fail to transcend "general linear reality." If so, the fault lies with the analyst, not the graph or the possibly highly nonlinear structural equations that it represents.

3.4   Causal Graphs and the Potential Outcome Model

What are the connections between the directed graph approach to causal analysis and the potential outcome model introduced in Chapter 2? In short, they are intimately related but very distinct frameworks for considering the same issues, each of which offers unique insight in particular situations. We will return to this basic point repeatedly throughout this book. In this section, we begin to explain their complementary value and then offer a brief examination of the most important formal connection between the two frameworks: how each encodes (potentially counterfactual) causal states.

3.4.1 Complementary Value

We already have shown through our presentation in Chapter 2 that the potential outcome model can be understood and utilized without reference to causal graphs. Individual-level potential outcomes allow one to think independently about the observed data as well as what data would have been observed if individuals had experienced alternative causal states. From these simple pieces, the model allows for transparent definitions of causal effects and encourages the analyst to consider individual-level heterogeneity as a first principle.

We also demonstrated in Chapter 2 how potential outcome random variables enable clear definitions of conditional average causal effects and provide ways to usefully decompose sources of inconsistency and bias in estimators – into unaccounted-for baseline differences between individuals and the differential responsiveness of individuals to the cause of interest. Many other advantages of thinking through causal analysis with potential outcomes will be demonstrated in the remaining chapters of this book.

However, we also began to show in Chapter 2 that the transparency of the potential outcome model begins to cloud over when more than two causal states are considered and when important assumptions, such as the stable unit treatment value assumption (SUTVA) (see Section 2.5), are asserted without considerable reflection. The latter is a specific instance of the complications of maintaining the full notational framework of the potential outcome model when many other causal variables of interest begin to enter the scientific problem.

In this chapter, we have shown that the basic elements of causal graphs do not include potential outcome random variables, such Y^1 and Y^0. The remaining chapters of the book will demonstrate how useful causal graphs can be for empirical research. As we will begin to show in the next chapter, graphs offer a disciplined framework for expressing causal assumptions for entire systems of causal relationships. In many cases, their economy of presentation cannot be matched by potential-outcome-based presentations of the same material, even though equivalent expressions are available. This advantage is especially apparent when the analyst must consider the many causal pathways that may be present in the real applications of observational research. Yet, as we will then also detail later, the price paid for such economy of presentation is then that individuals, and individual-level causal effects, which are so usefully revealed in the potential outcome model, can be covered over by causal graphs, even if the problem is not the causal graphs per se but the shallow interpretations that analysts can too easily attach to them.

Having stated our position on the useful complementarity of these two frameworks, we now lay out the most important point of connection between them. In the next section, we show how each framework encodes causal states in analogous fashion, thereby defining causal effects using counterfactuals.

3.4.2   Potential Outcomes and the $do(.)$ Operator

In Chapter 2, we introduced potential outcomes after first discussing the need to clearly define the states of the causal variables of primary interest. We then moved directly to the definition of potential outcomes defined by the instantiation of particular well-defined causal states, focusing for the most part on the canonical potential outcomes Y^1 and Y^0 for a two-valued cause. Only thereafter did we back out a definition of the corresponding observed variable Y, and only by way of Equation (2.2), $Y = DY^1 + (1 - D)Y^0$.

When we then introduced directed graphs in this chapter, we skipped right over causal states and potential outcomes. We moved straight from the representation of observed random variables A, B, and C to discussions of causal effects invoking the same observed variables utilized earlier in Chapter 2, the observed variables D and Y. Corresponding potential outcomes, Y^1 and Y^0, were not incorporated into our presentation. To demonstrate the connections between potential outcomes and directed graphs, we need to introduce how causal states are represented in Pearl's variant of causal analysis.

Pearl introduces causal states in a different way, using the semantics of an ideal experimental intervention and what he labels the $do(.)$ operator. Recall the basic structure of the directed graph in Figure 3.4(a). For this graph, the causal effect of D on Y is represented by $D \rightarrow Y$, and the directed edge indicates that D is an input in the function, $f_Y(.)$. The $do(.)$ operator, which we present in this section, is what provides the bridge to quantities such as the ATE, $E[\delta] = E[Y^1 - Y^0]$, defined earlier in Equation (2.3).

For Figure 3.4(a), consider the case where D takes on two values, 0 and 1. For Pearl, there are two regimes by which D takes on values of 0 or 1: pre-intervention and under-intervention. In the pre-intervention regime, the value that D takes on for any given unit in the population is determined by the structural equation $D = f_D(C, e_D)$. In the under-intervention regime, the value that D takes on is set by what Pearl sometimes calls an "ideal experiment" (e.g., Pearl 2009:358) and at other times calls an "atomic intervention" (e.g., Pearl 2009:70). Notationally, this intervention is $do(D = 1)$ or $do(D = 0)$.[18]

For Pearl, all causal quantities are defined by under-intervention distributions, not pre-intervention distributions (Pearl 2009, definitions 3.2.1 and 7.1.2-5). For D and Y in Figure 3.4(a), the two probability distributions that define causal effects are $Pr[Y|do(D = 1)]$ and $Pr[Y|do(D = 0)]$, not $Pr[Y|D = 1]$ and $Pr[Y|D = 0]$. In particular, the average causal effect is $E[Y|do(D = 1)] - E[Y|do(D = 0)]$, under the assumption that

---

[18] As we will explain in more detail in the appendix to this chapter, Pearl typically also assumes "modularity," which is the assumption that an intervention on a variable can be carried out without simultaneously altering anything else about the causal relationships encoded in the graph. The atomic intervention is assumed not to generate other interventions on other variables or to open up new causal pathways inconsistent with the structure of the pre-intervention graph. Although modularity is typically assumed, compound interventions are easily accommodated. More difficult, but not impossible, are cases where the assumed intervention generates initially unforeseen counterfactuals. In this case, the pre-intervention causal graph must be redrawn to represent all patterns of unfolding counterfactuals that may follow from prior events.

the individual-level causal effect is defined by the individual-level difference induced by the hypothetical intervention, $[y_i|do(d_i = 1)] - [y_i|do(d_i = 0)])$.[19]

Under this setup, observable associational quantities, based on $Pr[Y|D]$, do not necessarily equal causal quantities, based on $Pr[Y|do(D)]$. Most importantly, for graphs such as Figure 3.4(a), the associational difference $E[Y|D = 1] - E[Y|D = 0]$ does not equal the average causal effect defined by the atomic intervention on D, $E[Y|do(D = 1)] - E[Y|do(D = 0)]$. The confounding from C generates additional dependence between D and Y that enters into the average associational difference, $E[Y|D = 1] - E[Y|D = 0]$, but not the average causal difference, $E[Y|do(D = 1)] - E[Y|do(D = 0)]$. The reason that C does not enter the under-intervention difference is that D is determined by a hypothetical ideal experiment in this regime. In contrast, in the pre-intervention regime, D is determined by the structural equation $D = f_D(C, e_D)$ that has C as one of its inputs.

Consider now the connection with potential outcomes. The $do(.)$ operator is the exact analog to the superscripts given to potential outcomes in order to designate the underlying causal states that define them. In particular, Pearl states that the $do(.)$ operator "is a mathematical device that helps us specify explicitly and formally what is held constant, and what is free to vary" (Pearl 2009:358). The semantics that accompany the $do(.)$ operator – "ideal experiment" and "atomic intervention" – are Pearl's chosen way to express the idea that all units in the population could be assigned to the causal states in which they are observed (the "factual" ones) or to the causal states in which they are not observed (the "counterfactual" ones) and that causal effects are defined by differences attributable to movement between these alternative states. In this sense, $E[Y^1] - E[Y^0]$ is equivalent to $E[Y|do(D = 1)] - E[Y|do(D = 0)]$ for the graph in Figure 3.4(a), differing only in how the causal states are signified.[20]

Even though the $do(.)$ operator is a crucial piece of Pearl's variant of the directed graph approach to causal analysis, we introduced and discussed causal graphs in this chapter without any reference to it. In fact, we asserted that a directed edge signifies a causal effect, and that representations such as $A \rightarrow B$ are equivalent to the assumption that A is a cause of B. Yet, only here, at the end of this chapter, do we note that it is the $do(.)$ operator that defines the causal effects that are signified by these directed edges.

As long as one maintains that it is the $do(.)$ operator that defines causal effects, not associational contrasts such as $E[Y|D = 1] - E[Y|D = 0]$, the $do(.)$ operator does not need to be represented in the causal graph in any explicit way. The primary purpose of the graph is to encode the full set of causal relationships that one assumes characterize a causal effect of interest, so that those causal relationships can be considered in both the pre-intervention regime and the under-intervention regime. The $do(.)$ operator is

---

[19]Other causal quantities, based on comparisons of $Pr[Y|do(D = 1)]$ and $Pr[Y|do(D = 0)]$ could also be defined, analogous to choosing something other than the simple difference to define individual-level causal effects in the potential outcome model or selecting something other than the comparison of population expectations of individual-level potential outcomes. In fact, Pearl prefers to avoid selecting any particular comparison, emphasizing that all such comparisons are less general than focusing on the full under-intervention distribution, $Pr[Y|do(D)]$.

[20]When we introduced our chosen notation for the potential outcomes, we also noted that even for the potential outcome model there is a wide variety of notation adopted to signify causal states (see footnote 7 on page 43).

therefore a piece of the underlying structure of the graphical approach, such as the error terms of nonparametric structural equations, which can be brought into the foreground when necessary to explain an identification result.[21]

And, even though we have not done so, the $do(.)$ operator can be represented in graphical fashion. For each causal graph of the type presented in this chapter, two types of graphs can be drawn to show the associated under-intervention regime. For "mutilated" graphs, the directed edges that point to the causal variable of interest are deleted, leaving a causal variable with no parents because it is set by the atomic intervention. For "augmented" graphs, the original pre-intervention graph is drawn with an additional "forcing" variable that represents the atomic intervention.

For readers who wish to have an introduction to these additional types of graphs, as well as a slightly more formal presentation of the $do(.)$ operator (and its associated modularity condition within an overall definition of a Markovian causal graph), the appendix to this chapter offers an introduction to the primary literature where complete explanations can be found. The appendix also explains why potential outcome variables are rarely depicted in directed graphs. Readers who are uninterested in these details can skip the appendix and will be able to understand all of the material that follows.

3.5   Conclusions

In this chapter, we have introduced the directed graph approach to causal analysis. We first introduced the basic elements of causal graphs and then presented the canonical causal graph for a confounded causal effect. We explained the nonparametric nature of these graphs, as represented by the structural equations that assume no functional form for the effects of causes on outcomes. We concluded by noting the equivalence between causal effects defined by directed edges in a causal graph and causal effects defined by potential outcomes, demonstrating that the equivalence lies in their common invocation of what-if causal states grounded in counterfactual reasoning.

Along the way, we have noted that one goal of writing down a causal graph is to represent the set of causal relationships implied by past research and maintained theories. In all remaining parts of this book, we will enrich our discussion of this first goal, when, for example, we discuss whether simple graphs – such as Figure 3.7(a) for the charter school effect – are sufficiently rich to adequately represent the causal relationships that generate effects in real applications.

Another goal of writing down a causal graph is to assess the feasibility of alternative estimation strategies in light of the data that have been observed. Following up on this second goal, in the next part of the book we offer a full presentation of the rationale for conditioning as a causal effect estimation strategy. We then present three related methods for enacting conditioning estimators – matching, regression, and weighted regression – in three separate chapters.

---

[21]This is similar to the implementation of particular estimation strategies that are motivated by the potential outcome model. In this case, the researcher analyzes data using observed variables Y and D. Potential outcomes, Y^1 and Y^0, are brought out, typically at the beginning of an analysis, to define the causal effects of interest and the assumptions that will be maintained to estimate them.

In subsequent parts of the book, we then further demonstrate how directed graphs can be used to represent heterogeneity and selection on unobserved variables in order to consider how one should analyze causal effects when conditioning on observed variables does not identify the causal effect. We will consider mechanistic and instrumental variable approaches and conclude with estimators that use both pretreatment and posttreatment observations on the outcome variable.

3.6   Appendix to Chapter 3: Graphs, Interventions, and Potential Outcomes

In this appendix, we first explain the intervention foundation of Pearl's variant of causal graph methodology, and we then explain why potential outcomes are not commonly depicted as outcome variables in causal graphs. This appendix is written for curious readers who wish to have an introduction to formal details before consulting the primary literature for more complete explanations.

Graphs That Represent Atomic Interventions. As noted in Section 3.4.2, the key linkage between the potential outcome model and the directed graph approach to causal analysis is Pearl's concept of an atomic intervention, as represented by the $do(.)$ operator. Although the $do(.)$ operator is not visible in the standard representation of a causal graph, additional related graphs can be offered to demonstrate the connection more explicitly.

The graph in Figure 3.8(a) is known as an "augmented" graph because it is the pre-intervention causal graph from Figure 3.4(a), augmented with a representation of the atomic intervention on D.[22] The augmentation takes the form of a special "forcing" variable, F_D, which is placed within □ to denote its special status as an assumed outside force that can produce a hypothetical atomic intervention. Accordingly, this forcing variable takes on three values in this case: $do(D=0)$, $do(D=1)$, and *idle*.

Augmented graphs have accompanying structural equations that represent both the pre-intervention and the under-intervention regimes for the setting of variables subject to the atomic intervention (see Pearl 2009, section 3.2.2). For this graph, the structural equation for Figure 3.4(a), $D = f_D(C, \varepsilon_D)$, is replaced with

$$D = Int[f_D(.), C, \varepsilon_D],$$

where the $Int[.]$ function is defined so that it reduces to 1 if $F_D = do(D=1)$, to 0 if $F_D = do(D=0)$, and to $f_D(C, \varepsilon_D)$ if $F_D = idle$. In other words, the pre-intervention structural equation $f_D(C, \varepsilon_D)$ that generates D becomes the value of F_D when the forcing variable is *idle* because the hypothetical atomic intervention has not been enacted.

Figure 3.8(b) then shows how to think about the graph in Figure 3.8(a) when $F_D = do(D=0)$ and $F_D = do(D=1)$. The two graphs in Figure 3.8(b) are known as the *mutilated* graphs under an atomic intervention. The mutilation refers to the removal of all edges pointing to the variable that is intervened upon. In this case, the

---

[22]Little would be gained by adding forcing variables for either C or Y, since the former has no causes other than ε_C and the latter causes nothing else specified in the graph. Even so, no formal rules prevent the inclusion of both F_C and F_Y in a fully augmented graph.

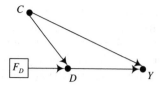

(a) Augmented casual graph with a "forcing"
variable that represents an intervention

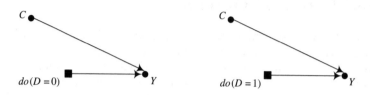

(b) "Mutilated" graphs that demonstrate the
$do(.)$ operator for the two values of D

Figure 3.8 Two alternative representations of assumed interventions in causal graphs
where the effect of D on Y is confounded by C.

observed variable for D, represented in Figure 3.4(a) by •, is replaced by ■'s in two
separate graphs that correspond to the two values of $do(D=0)$ and $do(D=1)$ that
are determined by the intervention. Because D is no longer determined by C, having
been set by a hypothetical atomic intervention, there is no directed edge from C to D
in either graph in Figure 3.8(b).

Now, compare the standard representation of the canonical confounding graph in
Figure 3.4(a) with its augmented variant in Figure 3.8(a). The latter shows the intervention explicitly, and the former leaves it implicit. Dawid (2010:75) claims that Pearl
once regarded the inclusion of forcing variables as crucial components of causal graphs.
Dawid argues that "he [Pearl], and most of those following him, have [recently] been
using only the implicit version, in which the intervention variables F_V are not explicitly
included in the diagram, but (to comply with the Pearlian interpretation) the DAG is
nevertheless to be interpreted as if they were." Dawid regards the suppression of intervention variables as a "retrograde move" that entails a "loss of transparency."[23] For
simple graphs, Dawid is surely correct that forcing variables do increase transparency.
For more complex diagrams, forcing variables can be a visual distraction. Accordingly,
in this book, we will generally follow Pearl and not offer such representations. As we
explain in the next section, however, Pearl builds very precise definitions of causal
graphs, which, when kept in mind while interpreting a causal graph in its standard
representation, leave little doubt about the crucial role of atomic interventions and the
$do(.)$ operator.

---

[23]Dawid prefers a more general influence diagram for causal graphs, resting on top of the decision
theoretic approach he has long championed (see Dawid 2002, 2012).

Criteria for Causal Graphs. For the canonical confounding graph in Figure 3.4(a), only three observable variables are present: C, D, and Y. No reference to the $do(.)$ operator that defines causal effects is visible in the graph. Yet, we wrote in this chapter that this directed graph is also a "causal graph." The careful reader may have noticed that we have presented other directed graphs in this chapter from which we have withheld the label "causal graph." We now explain.

Pearl has specific requirements for when a directed graph can be anointed a causal graph (Pearl 2009, definitions 1.3.1-3, 2.2.2, and 7.1.1), based on whether the candidate graph and its implied joint probability distribution satisfy "Markovian" conditions (Pearl 2009, theorem 1.4.1). These criteria are difficult to convey in direct form without introducing all aspects of Pearl's approach. We offer the following simplified criteria and strongly encourage readers to consult Pearl (2009) for their more complete and original expression.[24] Accordingly, and at the risk of too much oversimplification, a directed graph can be considered a causal graph by Pearl's definitions if

1. All variables in the graph, $\{V\}$, are observed (other than those variables implicitly included in the error terms, $\{e_V\}$, that are only revealed under magnification).

2. All variables in the graph, $\{V\}$, have error terms, $\{e_V\}$, that are independent of all variables in the graph, $\{V\}$, and that are mutually independent of each other.

3. It is reasonable to assume that each variable in the graph, V, can be subjected to a hypothetical intervention, $do(V = v)$, that

   (a) replaces the pre-intervention probability distribution of V, $Pr(V)$, with a single intervention value v,

   (b) removes the directed edges in the graph that terminate at V, and

   (c) changes nothing else in the graph (even though the setting of V to v propagates through the probability distributions of the descendants of V via the directed paths that remain in the graph).

Criteria 1 and 2 should be clear to those familiar with path models; they are stronger versions of the standard linear path model identifying assumptions that "all causal variables are uncorrelated with the error terms of all endogenous variables" and "all error terms on endogenous variables are uncorrelated with each other." Criterion 3 is typically considered to be unique to Pearl's framework, but Pearl himself makes the case that versions of criterion 3 were essential to early motivations of path-modeling techniques and were subsequently forgotten by their inheritors (see Bollen and Pearl 2013).[25]

---

[24]We will not, for example, discuss the basic requirement that the graph, and its associated structural equations, fully determines the joint probability distribution across all variables depicted in the graph. Using our simplified notation, this criterion is "All variables in the graph, $\{V\}$, have probability distributions, $\{Pr(V)\}$, that are generated by associated structural equations $\{f_V(.)\}$ that are functions only in (a) the variables that point directly to them in the graph (i.e, their "parents") and (b) their own error terms, e_V."

[25]In the broader literature on causal graphs beyond Pearl's work, criterion 3 is required only for the subset of the variables in the graph that are immediately relevant for identification of the focal causal effect (see Dawid 2002; Glymour et al. 2001; Robins and Richardson 2010).

Before addressing criterion 3, we should clarify one aspect of criteria 1 and 2. Figure 3.4(a) can be anointed a causal graph because its meets criteria 1 and 2 and because we have implicitly assumed up until now that criterion 3 is met as well. If criterion 1 is not met, but criteria 2 and 3 are, then the directed graph is "semi-Markovian" and typically thought of as if it is a causal graph (because what is observable and what is unobservable is subject to change).

Now consider criterion 3. The lead in – "It is reasonable to assume ..." – is our own writing, but it is consistent with Pearl's presentation of the same material. The key idea is that, by placing such a structure on the graph through a series of finely articulated definitions that are adopted as assumptions, causal effects can be rigorously defined. The payoff to adopting such structure is twofold. First, a directed graph that is a causal graph does not require that forcing variables be displayed to represent the atomic interventions that define all of the causal effects in the graph. Second, if a directed graph is a causal graph that satisfies criteria 1, 2, and 3, then the observed data can be manipulated to deliver estimated causal contrasts that are equal to true causal contrasts defined by the application of the $do(.)$ operator separately to all variables in the graph.[26] (Weaker variants of this implication are available, and should be obvious. Even if the full causal graph does not meet these criteria, it is often possible to identify specific causal effects while other effects in the graph remain unidentified.)

We will demonstrate more fully the second implication in the next chapter, where we introduce the back-door criterion for sufficient conditioning to identify a causal effect.[27] To get a handle on this implication now, it may be helpful to consider it in reverse for the simplest estimators we have considered so far in this book. Whenever conditioning estimators, or simple naive estimators, can be used to recover all causal contrasts in a graph from observed associational contrasts, the directed graph is a causal graph. For example, for Figure 3.4(a) a conditioning estimator (described in brief in Section 3.2.3) can be used to generate a consistent estimate of the average causal effect of D on Y, defined as $E[Y|do(D=1)] - E[Y|do(D=0)]$.[28] This result is true because the graph in Figure 3.4(a) is a causal graph: All variables are observed; the error terms that are viewable under magnification are defined to be independent of all else in the graph; and we have no reason to believe that it is unreasonable to assume that criterion 3 applies (because we have no reason to believe that intervening on D would alter C, etc.).

For another example that also serves as a substantive bridge to the final section of this appendix, recall the graph in Figure 3.5 that includes an unobserved confounder,

---

[26]Written in this form, this implication holds only for an infinite sample, so as to render sampling error unable to destroy the "equal to" within it. The estimates are therefore best interpreted as consistent estimates of the true causal effects.

[27]More generally, Pearl (2009, section 3.4) presents three rules that can be applied to candidate directed graphs and their associated structural equations to assess whether all effects in the graph can be identified. He characterizes such assessment as the application of "do calculus" with the goal of reducing do-defined probability distributions to equivalent probability distributions based only on observable variables (or, more specifically, the joint probability distribution of all variables, as structured by the pre-intervention regime encoded in the graph).

[28]In addition, because the effects of C on both D and Y are unconfounded, assuming criterion 2 holds, the naive estimator (or variants of it if C has more than two values) can be used to consistently estimate these two effects.

A, for the effect of education, E, on earnings, Y. Assuming criteria 2 and 3 are met, this graph is still not fully Markovian, and hence not a causal graph, because criterion 1 is not met. Accordingly, no conditioning estimator can be undertaken with the observed data to generate values that would correspond to $E[Y|do(E=e)]$ for all values e of E. Conditioning only on the observed confounder C does not eliminate the confounding by the unobserved variable A. However, for Pearl, this graph would be semi-Markovian because observation of A would then render it a fully Markovian causal graph, for which an effective conditioning estimator would then become available. Thus, if we can assume that criteria 2 and 3 are met, then we can state that Figure 3.5 would be a causal graph if A were observed.

The Absence of Potential Outcome Variables from Causal Graphs. Imagine a graph that included arrows between D and Y^0 and between D and Y^1, suggesting purported causal effects such as $D \rightarrow Y^0$ and $D \rightarrow Y^1$. Such causal relationships are inherently nonsensical because Y^1 and Y^0 are defined in relation to each other, as they jointly determine the observed variable Y in interaction with the causal states indexed by D.

Recall the definition offered in Equation (2.2):

$$Y = DY^1 + (1-D)Y^0. \tag{3.8}$$

One trivial way to explain why causal effects such as $D \rightarrow Y^1$ are nonsensical is to take Equation (3.8) and rewrite Y^1 in individual realizations as

$$y_i^1 = \frac{y_i - (1-d_i)y_i^0}{d_i}. \tag{3.9}$$

If we think of the supposed causal effect $D \rightarrow Y^1$ as being the theoretical difference in y_i^1 that would result from the action of switching d_i from 0 to 1, we can alternatively substitute 0 and 1 into Equation (3.9) and generate the result that the supposed causal effect of d_i on y_i^1 is the difference between "undefined" and y_i. Doing the same operation for y_i^0 delivers a supposed causal effect of d_i on y_i^0 as the difference between y_i and "undefined."

In other words, it does not make any sense to seek the causal effect of D on Y^1 or of D on Y^0, even though we have already offered examples where, empirically and theoretically, there may be good reason to believe that within the population of interest there may be an association between values of D and Y^1 and/or between D and Y^0 because of the ways individuals enter into the observed causal states (see Section 2.7). From a causal graph perspective, if such associations exist, they are produced by causal relations that connect D and Y but that do not travel through $D \rightarrow Y$. These are exactly the sorts of relationships that we earlier argued, with reference to Equation (2.14), generate inconsistency and bias in the the naive estimator. Individuals with $d_i = 1$ may have higher values, on average, for y_i^0 and/or y_i^1 than those with $d_i = 0$. If so, such average differences must arise because D and Y both mutually depend on a third variable, such as C in Figure 3.4(a).[29]

---

[29] Pearl (2009:342) would state such dependence as "$\{Y^0, Y^1\}$ represents the sum total of all exogenous variables, latent as well as observed, which can influence Y through paths that avoid D" (with notation changes from the original to match our example). The only point which differs from our

Can one represent the same basic insight using causal graphs? Pearl, in his own writing, has largely chosen not to do so (but see Pearl 2009, section 11.3.2).[30] The broader literature in causal graph methodology shows that it is possible to express the same ideas using directed graphs, although these are not graphs that are to be used in the same way as other graphs in this book. (A recent usage similar to ours in this section can be found in de Luna, Waernbaum, and Richardson (2011). See their citations to precursors.)

Recall Figure 3.5, which was used to bridge our first presentation of conditioning in causal graphs with our earlier presentation of the ability bias example used to introduce the potential outcome model in Chapter 2. In that graph, conditioning on C will not generate a consistent and unbiased estimate of the effect of education, E, on earnings, Y, because of the presence of the unobserved ability confounder, A. However, in Chapter 2 we discussed such confounding by ability instead as inconsistency and bias in the naive estimator. In Section 2.7.3, we defined two types of inconsistency and bias, aided by the potential outcome definitions. Baseline bias in the naive estimator was present when $E[Y^0|D=1] \neq E[Y^0|D=0]$, and treatment effect bias was present when $E[\delta|D=1] \neq E[\delta|D=0]$ (which would be present whenever $E[Y^1|D=1] \neq E[Y^1|D=0]$, assuming no highly unlikely canceling produced by the joint distribution of Y^0 and Y^1).

Figure 3.9 presents pairs of directed graphs that express the mutual dependence of potential outcomes and causal variables on exogenous confounders in three different scenarios. Notice first that none of the graphs in Figure 3.9 include either nonsensical causal effects, such as $E \rightarrow Y^0$ or $E \rightarrow Y^1$, or the observed outcome variable earnings, Y. Instead, these graphs are drawn solely to represent the mutual dependence of the potential outcomes for earnings, Y^0 and Y^1, and the focal causal variable, E, on observed confounders, C, and the unobserved ability confounder, A.

Recall that for the representation in Figure 3.5, it was not necessary to stipulate that E took on any particular values because the graph holds under a many-valued version of E. For simplicity of representation, and to match our earlier discussion in Section 2.7.3, we will now consider the case where E takes on only two values: 0 for those who do not obtain a bachelor's degree and 1 for those who do.

For Figure 3.9(a), confounding by C generates noncausal associations between E and both Y^0 and Y^1. Baseline confounding by ability, A, generates a noncausal association between only education and Y^0. This is the case, noted earlier in Section 2.7.3, where those who have college degrees would have higher earnings than those without

---

explanation here is that Pearl uses the $\{.,.\}$ notation to emphasize, as in ignorability, that the exogenous variables can structure a function defined in Y^1 and Y^0, such as the difference between these two. We make the same point in the graphs that follow.

[30] Although Pearl has not devoted a great deal of attention to developing graphical models that include potential outcomes, he and his colleagues have devoted considerable attention to the representation of counterfactuals in causal graphs. Shpitser and Pearl (2007) develop a powerful framework for counterfactual graphs, following on Pearl's earlier work on twinned network graphs; see Pearl (2000:213–14) and citations therein to earlier work; Shpitser (2012a, 2012b) offers particularly clear expositions of these graphs. Counterfactual graphs are beyond the scope of this book, in part because they require a more thorough understanding of Pearl's *do* calculus and full structural causal models than we have space to provide and than is necessary to understand the relevance of his framework for most forms of observational research in the social sciences.

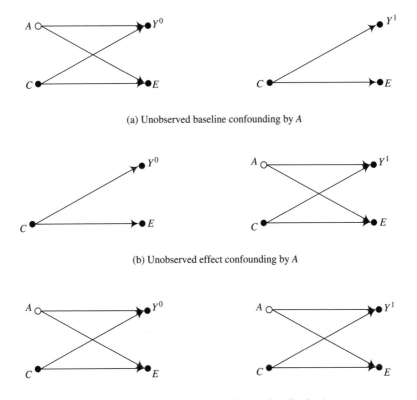

(a) Unobserved baseline confounding by A

(b) Unobserved effect confounding by A

(c) Unobserved baseline and effect confounding by A

Figure 3.9 Alternative graphs for the joint dependence of a two-valued causal variable for education (E) and potential outcomes for earnings (Y^0 and Y^1) on observed confounders (C) and on an unobserved confounder for ability (A).

college degrees in the counterfactual state in which they did not in fact obtain college degrees (even after adjustment for C). In other words, being "smarter" pays off in the labor market, even if one does not carry on to earn a bachelor's degree.

In contrast, for Figure 3.9(b), confounding by ability generates a noncausal association between education and Y^1 but not between education and Y^0. This is a form of treatment effect confounding, wherein those who do not obtain college degrees would not have earnings as high as those who do obtain college degrees in the counterfactual state in which they did in fact obtain college degrees (again, after adjustments for C). Here, being smarter helps one get more of an earnings payoff from obtaining a bachelor's degree, even though, contrary to Figure 3.9(a), being smarter does not lead to higher earnings in the absence of a college degree.

For Figure 3.9(c), both types of confounding by ability are present, as in the pattern presented earlier in Section 2.7.3. In this case, however, there is little advantage in using the paired graph representation instead of the single directed graph in Figure 3.5. For Figures 3.9(a), (b), and (c), the separate graphs for Y^0 and Y^1 allow for distinct patterns of confounding, and such clarity may be useful in some situations.

Part III: Estimating Causal Effects by Conditioning on Observed Variables to Block Back-Door Paths

Chapter 4

Models of Causal Exposure and Identification Criteria for Conditioning Estimators

In this chapter, we present the basic conditioning strategy for the estimation of causal effects. We first provide an account of the two basic implementations of conditioning – balancing the determinants of the cause of interest and adjusting for other causes of the outcome – using the language of "back-door paths." After explaining the unique role that collider variables play in systems of causal relationships, we present what has become known as the *back-door criterion* for sufficient conditioning to identify a causal effect. To bring the back-door criterion into alignment with related guidance based on the potential outcome model, we then present models of causal exposure, introducing the treatment assignment and treatment selection literature from statistics and econometrics. We conclude with a discussion of the identification and estimation of conditional average causal effects by conditioning.

4.1 Conditioning and Directed Graphs

In Section 1.5, we introduced the three most common approaches for the estimation of causal effects, using language from the directed graph literature: (1) conditioning on variables that block all back-door paths from the causal variable to the outcome variable, (2) using exogenous variation in an appropriate instrumental variable to isolate covariation in the causal variable and the outcome variable, and (3) establishing the exhaustive and isolated mechanism that intercepts the effect of the causal variable on the outcome variable and then calculating the causal effect as it propagates through the mechanism. In this chapter, we consider the first of these strategies, which motivates the basic matching, regression, and weighted regression techniques that we will present in Chapters 5, 6, and 7.

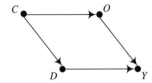

Figure 4.1 A graph in which the causal effect of D on Y is confounded by the back-door path $D \leftarrow C \rightarrow O \rightarrow Y$.

In Chapter 3, we explained the motivation for and execution of very simple conditioning estimators (see Section 3.2.3). In this chapter, we provide a more complete explanation of when, why, and how conditioning estimators will succeed in delivering estimates that can be given causal interpretations. We first reintroduce conditioning in a way that will reorient our perspective away from the "confounder variable" perspective of Chapter 3 to the more complete "back-door path" perspective that we will use in the remainder of this book.

4.1.1   From Confounders to Back-Door Paths

Consider Figure 4.1, which is an elaboration of the canonical confounding graph presented earlier in Figure 3.4(a). For this figure, the intermediate observed variable, O, expands the single-edge causal effect $C \rightarrow Y$ in Figure 3.4(a) to the directed path $C \rightarrow O \rightarrow Y$ in Figure 4.1. We noted in our prior discussion of Figure 3.4(a) that conditioning on C would allow us to generate a consistent and unbiased estimate of the causal effect of D on Y. This result holds for Figure 4.1 as well. However, for Figure 4.1, one could instead condition on O and achieve the same result. In fact, one could condition on both C and O as a third alternative.

The key goal of a conditioning strategy is not to adjust for any particular confounder but rather to remove the portion of the total association between D and Y that is noncausal. For Figure 4.1, the strategy to adjust for C is often referred to as "balancing the determinants of treatment assignment," and it is the standard motivation for matching estimators of causal effects. The alternative strategy to adjust for O is often referred to as "adjusting for all other causes of the outcome," and it is the standard motivation for regression estimators of causal effects.

Pearl characterizes both strategies in a novel way, using the language of back-door paths. As defined in Section 3.2.1, a *path* is any sequence of edges pointing in any direction that connects one variable to another. We now formally define a particular type of path that we have invoked informally already: A *back-door path* is a path between any causally ordered sequence of two variables that begins with a directed edge that points to the first variable.[1] For the directed graph in Figure 4.1, two paths connect D and Y: $D \leftarrow C \rightarrow O \rightarrow Y$ and $D \rightarrow Y$. The path $D \leftarrow C \rightarrow O \rightarrow Y$ is a

---

[1] "Causally ordered" means that the first variable causes the second variable by a directed path of some length. For the example discussed in this paragraph, D and Y are causally ordered because D causes Y by $D \rightarrow Y$. C and Y are also causally ordered because C causes Y by $C \rightarrow O \rightarrow Y$. D and O are *not* causally ordered because the only two paths that connect them ($D \leftarrow C \rightarrow O$ and $D \rightarrow Y \leftarrow O$) are not directed paths. Both of these paths have edges pointing in two directions. Recall also that we

back-door path because it begins with a directed edge pointing to D. Likewise, the path $D \to Y$ is not a back-door path because it does not begin with a directed edge pointing to D.[2]

In Pearl's language, the observed association between D and Y does not identify the causal effect of D on Y in Figure 4.1 because the total association between D and Y is an unknown composite of the true causal effect $D \to Y$ and a noncausal association between D and Y that is generated by the back-door path $D \leftarrow C \to O \to Y$. Fortunately, conditioning on C, O, or both C and O will block the back-door path, leaving within-stratum associations between D and Y that can be given causal interpretations (and can also be suitably averaged to obtain consistent and unbiased estimates of average causal effects of various types). The remainder of this chapter explains this result, as well as many others that are more complex.

4.1.2 Conditioning and Collider Variables

As a technique for estimating causal effects, conditioning is a very powerful and very general strategy. But, it is a much more complicated procedure in general than is suggested by our discussion of the simple directed graphs in Figures 3.4(a) and 4.1. Many of the complications arise when collider variables are present, and Pearl has explained systematically how to resolve these complications.

Recall that a variable is a collider along a particular path if it has two edges pointing directly to it, such as C in the path $A \to C \leftarrow B$ in Figure 3.3(c). For conditioning estimators of causal effects, collider variables must be handled with caution. Conditioning on a collider variable that lies along a back-door path does not help to block the back-door path but instead creates new associations.

The reasoning here is not intuitive, but it can be conveyed by a simple example with the mutual causation graph in Figure 3.3(c). Suppose that the population of interest is a set of applicants to a particular selective college and that C indicates whether applicants are admitted or rejected (i.e., $C = 1$ for admitted applicants and $C = 0$ for rejected applicants). Admissions decisions at this hypothetical college are determined entirely by two characteristics of students that are known to be independent within the population of applicants: SAT scores and a general rating of motivation based on an interview. These two factors are represented by A and B in Figure 3.3(c). Even though SAT scores and motivation are unrelated among applicants in general, they are not unrelated when the population is divided into admitted and rejected applicants. Among admitted applicants, the most motivated students will have lower than average SAT scores, and the least motivated students will have higher than average SAT scores. Thus, the college's sorting of applicants generates a pool of admitted students within which SAT scores and motivation are negatively related.[3]

---

stipulated in Chapter 3 that we will only consider acyclic graphs in this book. Without cycles, causal order is easily discerned by inspecting the directed paths in the graph.

[2]Recall that a directed path is a path in which all edges point in the same direction. Because all back-door paths have edges pointing in two directions, back-door paths are not directed paths. However, they can contain directed paths, such as the directed path $C \to O \to Y$ that is embedded in the back-door path $D \leftarrow C \to O \to Y$.

[3]A negative correlation will emerge for rejected students as well if (1) SAT scores and motivation have similarly shaped distributions and (2) both contribute equally to admissions decisions. As these

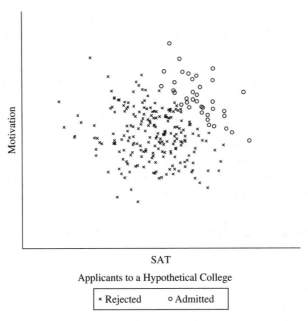

Applicants to a Hypothetical College

| × Rejected | o Admitted |

Figure 4.2 Simulation of conditional dependence within values of a collider variable.

This example is depicted in Figure 4.2 for 250 simulated applicants to this hypothetical college. For this set of applicants, SAT and motivation have a very small positive correlation of .035.[4] Offers of admission are then determined by the sum of SAT and motivation and granted to the top 15 percent of applicants (as shown in the upper right-hand portion of Figure 4.2).[5] Among admitted applicants, the correlation between SAT and motivation is −.641, whereas among rejected applicants the correlation between SAT and motivation is −.232. Thus, within values of the collider (the admissions decision), SAT and motivation are negatively related.

As Pearl documents comprehensively with a wide range of hypothetical examples, this is a very general feature of causal relationships and is present in many real-world applications. Elwert and Winship (2014) present many examples from the social science literature. In the next section, we show that care must be taken when attempting to estimate a causal effect by conditioning because conditioning on a collider variable can spoil an analysis.

---

conditions are altered, other patterns can emerge for rejected students, such as if admissions decisions are a nonlinear function of SAT and motivation.

[4]The values for SAT and motivation are 250 independent draws from standard normal variables. The draws result in an SAT variable with mean of .007 and a standard deviation of 1.01 as well as a motivation variable with mean of −.053 and a standard deviation of 1.02. Although the correlation between SAT and motivation is a small positive value for this simulation, we could drive the correlation arbitrarily close to 0 by increasing the number of applicants for the simulation.

[5]Admission is offered to the 37 of 250 students (14.8 percent) whose sum of SAT and motivation is greater than or equal to 1.5.

4.2 The Back-Door Criterion

With his language of back-door paths, colliders, and descendants, Pearl has developed what he labels the *back-door criterion* for determining whether or not conditioning on a given set of observed variables will identify the causal effect of interest.[6] The overall goal of a conditioning strategy guided by the back-door criterion is to block all paths that generate noncausal associations between the causal variable and the outcome variable without inadvertently blocking any of the paths that generate the causal effect itself. In practice, a conditioning strategy that utilizes the back-door criterion is implemented in two steps:

Step 1: Write down the back-door paths from the causal variable to the outcome variable, determine which ones are unblocked, and then search for a candidate conditioning set of observed variables that will block all unblocked back-door paths.

Step 2: If a candidate conditioning set is found that blocks all back-door paths, inspect the patterns of descent in the graph in order to verify that the variables in the candidate conditioning set do not block or otherwise adjust away any portion of the causal effect of interest.

This two-step procedure is justified by the following reasoning, which constitutes Pearl's back-door criterion:

Back-Door Criterion

If one or more back-door paths connect the causal variable to the outcome variable, the causal effect is identified by conditioning on a set of variables Z if

Condition 1. All back-door paths between the causal variable and the outcome variable are blocked after conditioning on Z, which will always be the case if each back-door path

(a) contains a chain of mediation $A \rightarrow C \rightarrow B$, where the middle variable C is in Z, or

(b) contains a fork of mutual dependence $A \leftarrow C \rightarrow B$, where the middle variable C is in Z, or

(c) contains an inverted fork of mutual causation $A \rightarrow C \leftarrow B$, where the middle variable C and all of C's descendants are *not* in Z;

---

[6]The back-door criterion is meant to be used only when the causal effect of interest is specified as a component of a graph that is a Markovian causal model, or would be if all variables represented in the graph were observed; see Pearl's causal Markov condition for the existence of a causal model (Pearl 2009, section 1.4.2, theorem 1.4.1), as well as our discussion in the appendix to Chapter 3. This requirement can be weakened when the graph is only a locally Markovian causal model, as long as the underspecified causal relations are irrelevant to an evaluation of the back-door criterion for the particular causal effect under consideration. All of the examples we utilize in this book meet these conditions.

and

Condition 2. No variables in Z are descendants of the causal variable that lie on (or descend from other variables that lie on) any of the directed paths that begin at the causal variable and reach the outcome variable.[7]

To explain the back-door criterion, we will first consider Conditions 1(a), (b), and (c), using examples where Condition 2 is met by default because the only descendant of the causal variable D in the graph is the outcome variable Y (and where we assume that the analyst does not consider conditioning on Y itself).

Conditions 1(a) and 1(b) of the back-door criterion should be clear as stated. Return one last time to the simple example in Figure 4.1. Here, there is a single back-door path, $D \leftarrow C \rightarrow O \rightarrow Y$, which includes within it both a fork of mutual dependence ($D \leftarrow C \rightarrow O$) and a chain of mediation ($C \rightarrow O \rightarrow Y$). By the back-door criterion, conditioning on C blocks the path $D \leftarrow C \rightarrow O \rightarrow Y$ because C is the middle variable in a fork of mutual dependence. Likewise, conditioning on O blocks the path $D \leftarrow C \rightarrow O \rightarrow Y$ because O is the middle variable in a chain of mediation. As a result, the candidate conditioning set meets Pearl's back-door criterion if Z is C, O, or both C and O.

Condition 1(c), however, is quite different than Conditions 1(a) and 1(b) and is not intuitive. It states instead that the set of candidate conditioning variables Z *cannot* include collider variables that lie on back-door paths.[8] Consider the following example. A common but poorly justified practice in the social sciences is to salvage a regression model from suspected omitted-variable bias by adjusting for an endogenous variable that can be represented as a proxy for the omitted variable that is unobserved. In many cases, this strategy will fail because the endogenous variable is usually a collider.

Suppose that an analyst is confronted with a directed graph similar to the one in Figure 3.4(b), in which the causal effect of D on Y is confounded by an unobserved variable, such as C. When in this situation, researchers often argue that the effects of the unobserved confounder can be decomposed in principle into a lagged process, using a prior variable for the outcome, Y_{t-1}, and two separate unobserved variables, U and V, as in Figure 4.3. For this graph, there are two back-door paths from D to Y:

1. $D \leftarrow V \rightarrow Y_{t-1} \rightarrow Y$ and

2. $D \leftarrow V \rightarrow Y_{t-1} \leftarrow U \rightarrow Y$.

---

[7]This representation of the back-door criterion is a combination of Pearl's definition of *d-separation* (Pearl 2009:16–17), his original specification of the back-door criterion (Pearl 2009:79), and the generalization of the back-door criterion that was developed and labeled the "adjustment criterion" by Shpitser, VanderWeele, and Robins (2010). In an appendix to this chapter, we clarify our specification of Condition 2, as incorporated from the adjustment criterion. In brief, for Pearl's original back-door criterion, Condition 2 requires more simply (but overly strongly) that no variables in Z can be descendants of the causal variable.

[8]Because the "or" in the Conditions 1(a), (b), and (c) of the back-door criterion is inclusive, one can condition on colliders and still satisfy the back-door criterion if the back-door paths along which the colliders lie are otherwise blocked because Z satisfies Condition 1(a) or Condition 1(b) with respect to another variable on the same back-door path.

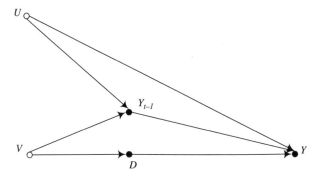

Figure 4.3 A causal diagram in which Y_{t-1} is a collider along a back-door path.

The lagged outcome variable Y_{t-1} lies on both of these back-door paths, but Y_{t-1} does not satisfy the back-door criterion. Notice first that Y_{t-1} blocks the first back-door path because, for this path, Y_{t-1} is the middle variable in a chain of mediation, $V \rightarrow Y_{t-1} \rightarrow Y$. But, for the second path, $D \leftarrow V \rightarrow Y_{t-1} \leftarrow U \rightarrow Y$, Y_{t-1} is a collider because it is the middle variable in an inverted fork of mutual causation, $V \rightarrow Y_{t-1} \leftarrow U$. Accordingly, conditioning on Y_{t-1} would eliminate part of the back-door association between D and Y because Y_{t-1} blocks the first back-door path $D \leftarrow V \rightarrow Y_{t-1} \rightarrow Y$. But, at the same time, conditioning on Y_{t-1} would create a new back-door association between D and Y because conditioning on Y_{t-1} unblocks the second back-door path $D \leftarrow V \rightarrow Y_{t-1} \leftarrow U \rightarrow Y$.

How can conditioning on a collider unblock a back-door path? To see the answer to this question, recall the discussion of conditioning in reference to Figure 3.3(c) and then as demonstrated in Figure 4.2. There, with the example of SAT and motivation effects on a hypothetical admissions decision to a college, we explained why conditioning on a collider variable induces an association between those variables that the collider is dependent on. That point applies here as well, when the causal effect of D on Y in Figure 4.3 is considered. Conditioning on a collider that lies along a back-door path unblocks the back-door path in the sense that it creates an association between D and Y within at least one of the subgroups enumerated by the collider.

Consider the slightly more complex example that is presented in Figure 4.4 (which is similar to Figure 1.1, except that the bidirected edges that signified unspecified and unobserved common causes have been replaced with two specific unobserved variables, U and V). Suppose, again, that we wish to estimate the causal effect of D on Y. For this directed graph, there are two back-door paths between D and Y:

1. $D \leftarrow A \leftarrow V \rightarrow F \rightarrow Y$ and

2. $D \leftarrow B \leftarrow U \rightarrow A \leftarrow V \rightarrow F \rightarrow Y$.

Notice that A is a collider variable in the second back-door path but not in the first back-door path. As a result, the first back-door path generates a noncausal association between D and Y, but the second back-door path does not. We therefore want to block the first path without unblocking the second path.

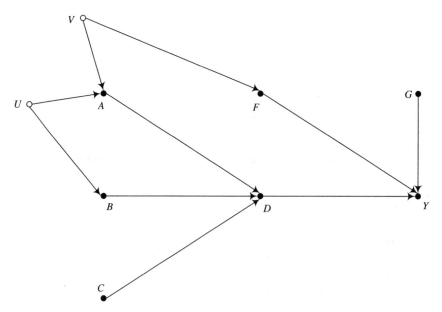

Figure 4.4 A causal diagram in which A is a collider on a back-door path.

For this example, there are two entirely different and effective conditioning strategies available that will identify the causal effect (numbers 1 and 3 in the following list)
and a third one that may appear to work but that will fail (number 2 in the following
list):

1. F is the middle variable in a chain of mediation, $V \rightarrow F \rightarrow Y$, for both back-door
   paths. As a result, F satisfies the back-door criterion, and conditioning on F
   identifies the causal effect of D on Y.

2. A is a middle variable in a chain of mediation, $D \leftarrow A \leftarrow V$, for the first back-door
   path. However, A is a collider variable for the second back-door path because it
   is the middle variable in a fork of mutual causation, $U \rightarrow A \leftarrow V$. As a result,
   A alone does not satisfy the back-door criterion. Conditioning on A does not
   identify the causal effect of D on Y, even though A lies along both back-door
   paths. Conditioning on A would unblock the second back-door path and thereby
   create a new, noncausal, back-door association between D and Y.

3. A is a middle variable in a chain of mediation, $D \leftarrow A \leftarrow V$, for the first back-
   door path. Likewise, B is a middle variable in a chain of mediation, $D \leftarrow B \leftarrow U$,
   for the second back-door path. Thus, even though A blocks only the first back-
   door path (and, in fact, conditioning on it unblocks the second back-door path),
   conditioning on B blocks the second back-door path. As a result, A and B
   together (but not alone) satisfy the back-door criterion, and conditioning on
   them together identifies the causal effect of D on Y.

In sum, for this example the causal effect can be identified by conditioning in one of two minimally sufficient ways: either condition on F or condition on both A and B.[9]

Now, we need to consider the complications introduced by descendants, as stipulated in both Condition 1(c) and Condition 2. Notice that Condition 1(c) states that neither C nor the descendants of C can be in Z. In words, conditioning on a collider *or* the descendant of a collider that lies on a back-door path will unblock the back-door path. Consider an extension of our prior analysis of Figure 4.3. For the graph in Figure 4.5, there are again two back-door paths from D to Y:

1. $D \leftarrow V \rightarrow Y_{t-2} \rightarrow Y_{t-1} \rightarrow Y$ and

2. $D \leftarrow V \rightarrow Y_{t-2} \leftarrow U \rightarrow Y$.

The first path does not contain any colliders and therefore confounds the causal effect of D on Y. The second back-door path, however, is blocked without any conditioning because Y_{t-2} is a collider on it. One might think, therefore, that conditioning on Y_{t-1} will block the first path without unblocking the second path. Condition 1(c) rules out this possibility because Y_{t-1} is a descendant of Y_{t-2}, and the latter is a collider on an already blocked path. The reasoning here is straightforward. The descendant Y_{t-1} is simply a noisy version of Y_{t-2} because its structural equation is

$$Y_{t-1} = f_{Y_{t-1}}(Y_{t-2}, e_{Y_{t-1}}),$$

where the error term, $e_{Y_{t-1}}$, is (as usual) assumed to be independent of all else in the graph. As a result, conditioning on Y_{t-1} has the same consequences for the second back-door path as conditioning on Y_{t-2}.[10]

Having explained Conditions 1(a), (b), and (c) of the back-door criterion, we can now consider Condition 2, which states that none of the variables in the possible conditioning set Z can be descendants of the causal variable that block the causal effect of interest by lying on or descending from any of the directed paths that begin at the causal variable and reach the outcome variable. Recall that a directed path is a path in which all edges point in the same direction. In all prior examples discussed in this section, the only descendant of D has been Y, and thus Condition 2 has been met by default (under the assumption that the analyst has not considered conditioning on the outcome variable Y itself).

We now consider two examples where additional descendants of D are present. Figure 4.6 presents a graph that elaborates Figure 4.1, where now the total effect of D on Y is separated into an indirect pathway through a mediating variable, $D \rightarrow N \rightarrow Y$, and a remaining direct effect, $D \rightarrow Y$. We will discuss models with such mediating

---

[9]One can of course condition in three additional ways that also satisfy the back-door criterion: F and A, F and B, and F, A, and B. These conditioning sets include unnecessary and redundant conditioning.

[10]Hernán, Hernandez-Diaz, and Robins (2004) offer an excellent discussion of examples in epidemiology for which such descendants of colliders are a primary focus. For their types of examples, the outcome is "death," the collider on the back-door path (or elsewhere in the causal graph) is "getting sick enough to be admitted to a hospital for treatment," and the variable that is conditioned on is "in a hospital." Conditioning on "in a hospital" (by undertaking a study of hospital patients) induces associations between the determinants of sickness that can spoil standard analyses. Elwert (2013) and Elwert and Winship (2014) cover many of the same issues from a social science perspective. In an appendix to this chapter, we also provide additional discussion and examples.

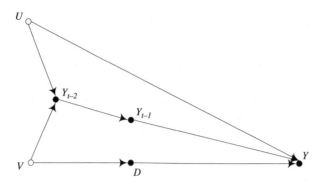

Figure 4.5 A causal diagram in which Y_{t-2} is a collider on a back-door path and Y_{t-1} is its descendant.

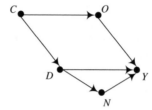

Figure 4.6 A confounded causal effect expressed as an indirect effect and a net direct effect.

mechanisms in substantial detail in Chapter 10, and for now we will use this graph only to introduce a discussion of Condition 2 of the back-door criterion.

As we have noted in this section and elsewhere (and, assuming now, without loss of generality that D is a two-valued treatment), conditioning on either C or O identifies the total average causal effect of D on Y, which we defined in Section 3.4 as $E[Y|do(D=1)] - E[Y|do(D=0)]$. The back-door criterion states that if, in addition to C and/or O, we also conditioned on N, the resulting estimate would no longer identify the casual effect of D on Y. The conditioning set of C, O, and N violates Condition 2 of the back-door criterion because N is a descendant of D that lies on a directed path, $D \to N \to Y$, that reaches Y. As a result, the back-door criterion indicates that the analyst should not condition on N.

We suspect that this conclusion would be clear to readers without consulting the back-door criterion (by reasoning that adjustment for N would rob the total causal effect of D on Y of some of its magnitude, leaving only a partial direct effect that is not equal to $E[Y|do(D=1)] - E[Y|do(D=0)]$).[11] Such reasoning is correct, and Condition 2 of the back-door criterion is, in part, a formalization of this intuition.

---

[11]In this section, we are considering only how to apply the back-door criterion when the goal is to estimate the total effect of the cause D on the outcome Y. If, instead, the analyst is interested in estimating the direct effect of D on Y, net of the indirect effect of D on Y through N, then adjustment for N is necessary. VanderWeele (in press) provides a comprehensive treatment of the identification and estimation of direct effects. We will return to these issues in Chapter 10, where we will consider how to identify causal effects using generative mechanisms.

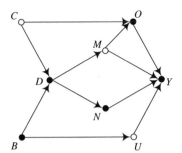

Figure 4.7 A graph where the effect of D on Y is not identified by conditioning on O and B because O is a descendant of D.

However, Condition 2 is more finely articulated than this intuition alone. It invalidates conditioning sets that adjust away any portion of the causal effect of interest, not just those portions that are carried by variables that mediate the causal effect. Just as conditioning on the descendant of a collider has the same consequences as conditioning on the collider itself, conditioning on a descendant of a variable that lies on a directed path from the causal variable to the outcome variable has the same consequences for identification results as conditioning directly on the variable from which it descends.[12]

Consider Figure 4.7, where we are again interested in the effect of D on Y and where we have now specified two mediating variables, M and N, that completely characterize the total effect of D on Y. For this graph, three back-door paths connect D to Y:

1. $D \leftarrow C \rightarrow O \rightarrow Y$,

2. $D \leftarrow C \rightarrow O \leftarrow M \rightarrow Y$, and

3. $D \leftarrow B \rightarrow U \rightarrow Y$.

In addition, the variables C, M, and U are unobserved, and thus they are not available to use in a conditioning estimator.

Suppose that one decides to condition on both O and B. First, note that these two variables do not satisfy Conditions 1(a), (b), and (c) of the back-door criterion.

---

[12]Notice further that Condition 2 applies to descendants of the cause that lie on or descend from variables on directed paths that begin at the causal variable and "reach the outcome variable" rather than "end at the outcome variable." The word "reach" has been chosen to allow for Condition 2 to invalidate conditioning sets that include descendants of the cause that are also descendants of the outcome. Suppose that an additional observed variable W is added to Figure 4.6 along with a directed edge from Y to W, as in $Y \rightarrow W$. The new variable W is a descendant of both D and Y via the two directed paths $D \rightarrow Y \rightarrow W$ and $D \rightarrow N \rightarrow Y \rightarrow W$ that begin at D and reach Y (and, in this case, carry on to W). There is, of course, no reason to condition on W in an attempt to estimate the causal effect of D and Y because W does not lie on a back-door path from D to Y that confounds the causal effect of interest. If an analyst does do so, by adding W to a conditioning set that includes C and/or O, then Conditions 1(a), (b), and (c) of the back-door criterion are met but Condition 2 is not. The variable W is simply a noisy version of Y, and the causal effect of D on Y is therefore embedded within it. Conditioning on W will, in fact, mistakenly suggest that D has no casual effect on Y because within strata defined by W, D and Y are independent (assuming no measurement error and an infinite sample that enables fully nonparametric estimation).

Conditioning on O will block path 1, and conditioning on B will block path 3. However, conditioning on O will unblock path 2 because it is a collider on an already blocked back-door path.

What we want to stress now is that the candidate conditioning set of O and B also does not satisfy Condition 2 of the back-door criterion. O is a descendant of D on a directed path, $D \rightarrow M \rightarrow O \rightarrow Y$, that reaches Y. Furthermore, O is a descendant of M, via $M \rightarrow O$, and M is a variable that lies on another directed path, $D \rightarrow M \rightarrow Y$, that begins at D and reaches Y.

Notice also that the graph in Figure 4.7 is realistic and not contrived. If a researcher follows the (sometimes wrong) conditioning strategy of "adjusting for all other causes of the outcome," and the researcher has not drawn the full causal graph in Figure 4.7, then the researcher would almost certainly condition on O (even if the researcher wisely decides not to condition on the clearly endogenous observed variable N but does decide to condition on the second-order cause B in place of the unobserved direct cause U). In practice, other causes of an outcome that are observed are hard to resist conditioning on, even though many of them are descendants of the cause along directed paths that reach the outcome and thereby violate Condition 2 of the back-door criterion. If a researcher does not observe a mechanistic variable like M, the analyst may not recognize the endogeneity of O with respect to D (especially if the analyst comes to believe that the only mechanistic variable has been observed as N and that the remaining effect of D on Y is an unmediated direct effect).

Condition 2 of the back-door criterion is a general requirement for sufficient conditioning sets, and it is easy to apply. One need only examine each candidate conditioning variable to determine whether it lies on (or descends from a variable that lies on) the directed paths that begin at the causal variable and reach the outcome variable.

Although Condition 2 is easy to apply, we discuss additional examples in an appendix to this chapter that are more challenging to explain. In particular, we reconsider the graphs just presented, building toward a full reexamination of the graph in Figure 4.7, which is even more complex than our presentation here reveals. In the appendix, we show that a full assessment of the consequences of conditioning on descendants of the cause is enabled by drawing the graphs under magnification so that it is easier to recognize all such descendants as colliders. Although these more complex cases are interesting to examine, so as to understand the full range of identification challenges that graphical models help to explain, one need not fully understand them or absorb them to effectively adhere to conditioning strategies that are warranted by the back-door criterion.

The two key points of this section are the following. First, conditioning on variables that lie on back-door paths can be an effective strategy to identify a causal effect. If all back-door paths between the causal variable and the outcome variable are blocked after the conditioning is enacted, then back-door paths do not contribute to the association between the causal variable and the outcome variable. However, it must be kept in mind that conditioning on a collider (or a descendant of a collider) has the opposite effect. Any such conditioning unblocks already blocked back-door paths. And thus, when a conditioning strategy is evaluated, each back-door path must be assessed carefully because a variable can be a collider on one back-door path but not a collider on another.

Second, if a set of conditioning variables blocks all back-door paths, the analyst must then verify that no variables within the conditioning set block the causal effect of interest or otherwise mistakenly adjust it away. If none of the candidate conditioning variables are also descendants of the cause that lie on or descend from directed paths that begin at the causal variable and reach the outcome variable, then the causal effect is identified and a conditioning estimator is consistent and unbiased for the average causal effect.

Pearl's back-door criterion for evaluating conditioning strategies is a generalization (and therefore a unification) of various traditions for how to solve problems that are frequently attributed to omitted-variable bias. From our perspective, Pearl's framework is particularly helpful in two respects. First, it shows clearly that researchers do not need to condition on all omitted direct causes of an outcome variable in order to solve an omitted-variable bias problem. This claim is not new, of course, but Pearl's back-door criterion shows clearly why researchers need to condition on only a minimally sufficient set of variables that (a) renders all back-door paths blocked and (b) does not block the causal effect itself. Second, Pearl's framework shows how to think clearly about the appropriateness of conditioning on endogenous variables. Writing down each back-door path and then determining whether or not each endogenous variable is a collider along any of the back-door paths is a much simpler way to begin to consider the full complications of a conditioning strategy than other approaches.[13]

In the next section, we consider models of causal exposure that have been used in the counterfactual tradition, starting first with the statistics literature and carrying on to the econometrics literature. We will show that the assumptions often introduced in these two traditions to justify conditioning estimation strategies – namely, ignorability and selection on the observables – have close connections to the back-door criterion presented in this section.

---

[13]The back-door criterion is not the only available graphical guide for the selection of conditioning variables. Elwert (2013:256–61) offers a crisp summary of the alternatives. Among these, the *adjustment criterion* is the most general and is guaranteed to select all possible sufficient conditioning sets. Nonetheless, we make the case in the appendix to this chapter that the version of the back-door criterion that we present in the main text of this chapter has advantages relative to the completeness of the adjustment criterion. Among the other criteria, the *disjunctive cause criterion* of VanderWeele and Shpitser (2011) is also particularly useful because it can serve as a guide to conditioning when the analyst is unwilling to commit to a set of assumptions that enable a full directed graph to be drawn for the generation of the outcome. The disjunctive cause criterion instructs the analyst to adjust for all variables that are causes of the treatment or the outcome, but not those variables that are causes neither of the treatment nor of the outcome. The analyst therefore does not need to construct a directed graph, only take a position on whether the candidate conditioning variables should be assumed to be causes of the treatment, causes of the outcome, or neither. Of course, the price to be paid for the simplicity of the disjunctive cause criterion is that it will not necessarily identify the causal effect (because one cannot know about all sources of confounding if one cannot draw the full directed graph for the generation of the outcome) and can suggest redundant conditioning (because one does not need to condition on variables that lie on back-door paths that are already blocked by colliders, variables that lie on back-door paths that are already blocked by conditioning on other variables, or variables that do not lie on any back-door paths). Nonetheless, the disjunctive cause criterion will prevent the analyst from inducing noncausal associations between the treatment and outcome that would result from conditioning on the relevant colliders that already block back-door paths.

4.3   Models of Causal Exposure and Point Identification Based on the Potential Outcome Model

With this general presentation of the conditioning strategy in mind, return to the familiar case of a binary cause D and an observed outcome variable Y. As discussed in Chapter 2, for the potential outcome model we consider Y to have been generated by a switching process between two potential outcome variables, as in $Y = DY^1 + (1-D)Y^0$, where the causal variable D is the switch. To model variation in Y and relate it to the individual-level causal effects defined by the potential outcome variables Y^1 and Y^0, a model for the variation in D must be adopted. This is typically known in the statistics literature as "modeling the treatment assignment mechanism" and in the econometrics literature as "modeling the treatment selection mechanism."

In this section, we first consider the notation and language developed by statisticians, and we then turn to the alternative notation and language developed by econometricians. Although both sets of ideas are equivalent, they each have some distinct conceptual advantages. In showing both, we hope to deepen the understanding of each.

4.3.1   Treatment Assignment Modeling in Statistics

The statistics literature on modeling the treatment assignment mechanism is an outgrowth of experimental methodology and the implementation of randomization research designs. Accordingly, we begin by considering a randomized experiment for which the phrase "treatment assignment" remains entirely appropriate.

As discussed in Chapter 2, if treatment assignment is completely randomized by design, then the treatment indicator variable D is completely independent of the potential outcomes Y^0 and Y^1 as well as any function of them, such as the distribution of δ; see the earlier discussion of Equation (2.6). In this case, the treatment assignment mechanism is known because it is set by the researcher who undertakes the randomization. If the researcher wants treatment and control groups of approximately the same size, then $\Pr[D=1]$ is set to .5. Individual realized values of D for those in the study, denoted d_i generically, are then equal to 1 or 0 and are determined by the flip of a fair coin (or by a computer that runs Bernoulli trials for the random variable D with .5 as the probability).

To facilitate the transition to designs for observational research, consider a slightly more elaborate design where study subjects are stratified first by gender and then assigned with disproportionate probability to the treatment group if female. In this case, the treatment assignment protocol would instead be represented by two conditional probabilities, such as

$$\Pr[D=1|\text{Gender} = \text{Female}] = .7, \tag{4.1}$$

$$\Pr[D=1|\text{Gender} = \text{Male}] = .5. \tag{4.2}$$

These conditional probabilities are typically referred to as "propensity scores" in the literature because they indicate the propensity that an individual with specific characteristics will be observed in the treatment group. Although labeled propensity scores in the literature, they are nonetheless nothing more than conditional probabilities that lie within an interval bounded by 0 and 1. For this example, the propensity score is .7 for female subjects and .5 for male subjects. The general point is that for randomized experiments, the propensity scores are known to the researcher.

In contrast, a researcher with observational data does not possess a priori knowledge of the exact propensity scores that apply to all individuals. However, the researcher may know all of the characteristics of individuals that systematically determine their propensity scores, even if the researcher does not know the specific values of the propensity scores.[14] In this case, treatment assignment patterns are represented by a general conditional probability distribution,

$$\Pr[D=1|S], \tag{4.3}$$

where S now denotes all variables that systematically determine all treatment assignment patterns. Complete observation of S then allows a researcher to assert that treatment assignment is "ignorable" and then consistently estimate the average treatment effect (ATE), as we now explain.

The general idea here is that, within strata defined by S, the remaining variation in the treatment D is completely random and hence the process that generates this remaining variation is labeled "ignorable." The core of the concept of ignorability is the independence assumption that was introduced in Equation (2.6),

$$(Y^0, Y^1) \perp\!\!\!\perp D,$$

where the symbol $\perp\!\!\!\perp$ denotes independence. As defined by Rubin (1978), ignorability of treatment assignment holds when the potential outcomes are independent of the treatment dummy indicator variable, as in this case all variation in D is completely random. Ignorability also holds in the weaker case where

$$(Y^0, Y^1) \perp\!\!\!\perp D \mid S \tag{4.4}$$
$$\text{and}$$
$$\text{all variables in } S \text{ are observed.}$$

In words, the treatment assignment mechanism is ignorable when the potential outcomes (and any function of them, such as δ) are independent of the treatment variable, D, within strata defined by all combinations of values on all variables, S, that systematically determine all treatment assignment patterns. If some components of S are unobserved, the conditional independence condition in Equation (4.4) may still hold, but treatment assignment cannot be considered ignorable with respect to the observed data. In this case, treatment assignment must be regarded as nonignorable, even if it is known that it would be ignorable if all variables in S had instead been observed.[15]

---

[14]This would be the situation for the randomized experiment represented by Equations (4.1) and (4.2) if the experimentalist knew that the probability of being assigned to the treatment differed by gender (and only by gender) but had forgotten the values of .7 and .5.

[15]Rosenbaum and Rubin (1983a) defined strong ignorability to develop the matching literature, which we will discuss later. To Rubin's ignorability assumption, Rosenbaum and Rubin (1983a) required for strong ignorability that each subject have a nonzero probability of being assigned to both the treatment and the control groups. Despite these clear definitions, the term ignorability is

In practice, in order to assert that treatment assignment is ignorable for an observational study, a researcher would

1. determine from related studies and supportable assumptions grounded in theory what the components of S are,

2. measure each of the component variables in S, and

3. collect enough data to be able to consistently estimate outcome differences on the observed variable Y within strata defined by S.

This third step can be weakened if the data are merely sparse, as we will discuss when presenting models based on estimate propensity scores in Chapters 5 and 7. The key point is that a researcher does not need to know the exact propensity scores (i.e., what $\Pr[D = 1|S = s]$ is equal to for all s), only that the systematic features of treatment assignment can be exhaustively accounted for by the data in hand on the characteristics of individuals. The naive estimator can then be calculated within strata defined by values of the variables in S, and a weighted average of these stratified estimates can be formed as a consistent estimate of the ATE.[16]

Consider the Catholic school example. It is well known that students whose parents self-identify as Catholic are more likely to be enrolled in Catholic schools than students whose parents self-identify as non-Catholic. Suppose that parents' religious identity is the only characteristic of students that systematically determines whether they attend Catholic schools instead of public schools. In this case, a researcher can consistently estimate the ATE by collecting data on test scores, students' school sector attendance, and parent's religious identification. A researcher would then estimate the effect of Catholic schooling separately by using the naive estimator within groups of students defined by parents' religious identification and then take a weighted average of these estimates based on the proportion of the population of interest whose parents self-identify as Catholic and as non-Catholic. In the words of the prior section, the researcher can generate consistent and unbiased estimates of the ATE by conditioning on parents' religious identification.

Assumptions of ignorability have a close connection to the back-door criterion of Pearl, given the shared centrality of the conditioning operation. Even so, it is important to recognize some differences. Suppose that we are confronted with the graph in Figure 4.8(a), which includes the causal effect $D \to Y$ but also the bidirected edge $D \leftarrow\!-\!-\!-\!-\! \to Y$. The most common solution is to build an explicit causal model that represents the variables that generate the bidirected edge between D and Y in Figure 4.8(a). The simplest such model is presented in Figure 4.8(b), where $D \leftarrow\!-\!-\!-\!-\! \to Y$ has been replaced with the back-door path $D \leftarrow S \to Y$. If S is observed, then conditioning on S will solve the causal inference problem, according to the back-door criterion

---

often defined in different ways in the literature. We suspect that this varied history of usage explains why Rosenbaum (2002) rarely uses the term in his monograph on observational data analysis, even though he is generally credited, along with Rubin, with developing the ignorability semantics in this literature. And it also explains why some of the most recent econometrics literature uses the words unconfoundedness and exogeneity for the same set of independence and conditional-independence assumptions (see Imbens 2004).

[16] Again, we assume that measurement error does not exist, which requires that step 2 be undertaken without error.

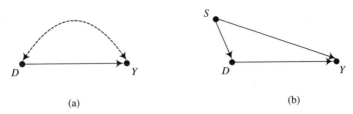

Figure 4.8 Causal diagrams in which treatment assignment is (a) nonignorable and (b) ignorable.

presented in the previous section. In the statistics literature, the same result is explained by noting that treatment assignment is ignorable when S is observed, which is then written as Equation (4.4).

When identification by back-door conditioning is feasible with the observed data, then treatment selection is ignorable with respect to the observed data. However, we do not mean to imply with this statement that the assertion of a valid ignorability assumption will lead the researcher to condition on the exact same variables suggested by the back-door criterion. Recall that the back-door criterion guides the researcher in selecting minimally sufficient conditioning sets. Ignorability can be asserted with respect to these sets or with respect to broader sets of conditioning variables. For an example, let S in Figure 4.8(b) be a set of variables. Suppose that, upon reflection, a researcher decides that one of the variables in S, S', has a direct effect on D but no direct effect on Y. Accordingly, it is appropriate to separate S' from the other variables in S and assert that S' is a cause of D that causes Y only indirectly through D.[17] In this case, S' does not generate any confounding because S' does not lie on a back-door path between D and Y. The minimally sufficient conditioning sets that are judged admissible by the back-door criterion would not include S', even though treatment assignment is ignorable with respect to S' and the remaining variables in S. Thus, when asserting ignorability with respect to the observed data, the researcher may decide to condition on S', even though such conditioning is unnecessary. After all, S' does determine treatment assignment and, hence, does structure the true propensity score. And it is possible that the strong assumption that S' does not have a direct effect on Y is incorrect.

We will begin to discuss specific techniques for conditioning estimators in Chapter 5, where we first present matching estimators of causal effects. But the immediate complications of undertaking a conditioning analysis strategy for the Catholic school example should be clear. How do we determine all of the factors that systematically determine whether a student enrolls in a Catholic school instead of a public school? And can we obtain measures of all of these factors? Attendance at a Catholic school is determined by more than just parents' religious self-identification, and some of these determinants are likely unmeasured. If this is the case, then the treatment assignment mechanism is likely to be nonignorable, as treatment selection is then a function of unobserved characteristics of students that generate confounding. These

---

[17] S' is an instrumental variable, as we will explain in full detail in Chapter 9.

issues are best approached by first drawing a directed graph. It may be that some of the unobserved determinants of treatment assignment do not generate confounding because (a) they do not lie on back-door paths, (b) they lie on back-door paths that are already blocked by colliders, or (c) they lie on back-door paths that can be blocked by conditioning on other variables that lie on the same back-door paths. If so, the researcher may be able to assert an ignorability assumption with respect to only a subset of the variables we have designated as S for this section. However, in most cases, the opposite challenge will dominate: The directed graph will show clearly that only a subset of the variables in S that generate confounding are observed, and the confounding that they generate cannot be eliminated by conditioning with the observed data. Treatment assignment is then nonignorable with respect to the observed data.

4.3.2 Treatment Selection Modeling in Econometrics

The econometrics literature also has a long tradition of analyzing causal effects, and this literature may be more familiar to social scientists. Whereas concepts such as ignorability are used somewhat infrequently in the social sciences, the language of selection bias is commonly used throughout the social sciences. This usage is due, in large part, to the energy that economists have devoted to exploring the complications of self-selection bias.

The selection-bias literature in econometrics is vast, but the most relevant piece that we focus on here is James Heckman's specification of the random-coefficient model for the treatment effects of training programs, which he attributes, despite the difference in substance, to Roy (1951). The clearest specification of this model was presented in a series of papers that Heckman wrote with Richard Robb (see Heckman and Robb 1985, 1986, 1989), but Heckman worked out many of these ideas in the 1970s. Using the notation we have adopted in this book, take Equation (2.2),

$$Y = DY^1 + (1 - D)Y^0,$$

and then rearrange and relabel terms as follows:

$$Y = Y^0 + (Y^1 - Y^0)D \tag{4.5}$$
$$= Y^0 + \delta D$$
$$= \mu^0 + \delta D + v^0,$$

where $\mu^0 \equiv E[Y^0]$ and $v^0 \equiv Y^0 - E[Y^0]$. The standard outcome model from the econometrics of treatment evaluation simply reexpresses Equation (4.5) so that potential variability of δ across individuals in the treatment and control groups is relegated to the error term, as in

$$Y = \mu^0 + (\mu^1 - \mu^0)D + \{v^0 + D(v^1 - v^0)\}, \tag{4.6}$$

where $\mu^1 \equiv E[Y^1]$, $v^1 \equiv Y^1 - E[Y^1]$, and all else is as defined for Equation (4.5).[18] Note that, in evolving from Equation (2.2) to Equation (4.6), the definition of the

---

[18]The original notation is a bit different, but the ideas are the same. Without much usage of the language of potential outcomes, Heckman and Robb (1985, section 1.4) offered the following setup for the random coefficient model of treatment effects to analyze posttreatment earnings differences for a fictitious worker training example. For each individual i, the earnings of individual i if trained are

observed outcome variable Y has taken on the look and feel of a regression model.[19] The first μ^0 term is akin to an intercept, even though it is defined as $E[Y^0]$. The term $(\mu^1 - \mu^0)$ that precedes the first appearance of D is akin to a coefficient on the primary causal variable of interest D, even though $(\mu^1 - \mu^0)$ is defined as the true ATE, $E[\delta]$. Finally, the term in braces, $\{v^0 + D(v^1 - v^0)\}$, is akin to an error term, even though it represents both heterogeneity of the baseline no-treatment potential outcome and of the causal effect, δ, and even though it includes within it the observed variable D.[20]

Heckman and Robb use the specification of the treatment evaluation problem in Equation (4.6), and many others similar to it, to demonstrate all of the major problems created by selection bias in program evaluation contexts when simple regression estimators are used. Heckman and Robb show why a regression of Y on D does not in general identify the ATE, in this case $(\mu^1 - \mu^0)$, when D is correlated with the population-level variant of the error term in braces in Equation (4.6), as would be the case when the size of the individual-level treatment effect, in this case $(\mu^1 - \mu^0) + \{v_i^0 + d_i(v_i^1 - v_i^0)\}$, differs among those who select the treatment and those who do not.

The standard regression strategy that prevailed in the literature at the time was to include additional variables in a regression model of the form of Equation (4.6), hoping

$$y_i^1 = \beta^1 + U_i^1,$$

and the earnings of individual i in the absence of training are

$$y_i^0 = \beta^0 + U_i^0,$$

(where we have suppressed subscripting on t for time from the original presentation and also shifted the treatment state descriptors from subscript to superscript position). With observed training status represented by a binary variable, d_i, Heckman and Robb then substitute the right-hand sides of these equations into the definition of the observed outcome in Equation (2.2) and rearrange terms to obtain

$$y_i = \beta^0 + (\beta^1 - \beta^0)d_i + U_i^0 + (U_i^1 - U_i^0)d_i,$$

which they then collapse into

$$y_i = \beta^0 + \bar{\alpha}d_i + \{U_i^0 + \varepsilon_i d_i\},$$

where $\bar{\alpha} \equiv \beta^1 - \beta^0$ and $\varepsilon_i \equiv U_i^1 - U_i^0$ (see Heckman and Robb 1985, equation 1.13). As a result, $\bar{\alpha}$ is the ATE, which we defined as $E[\delta]$ in Equation (2.3), and ε_i is the individual-level departure of δ_i from the ATE, $E[\delta]$. Although the notation in this last equation differs from the notation in Equation (4.6), the two equations are equivalent. Heckman and Vytlacil (2005, 2007) give a fully nonparametric version of this treatment selection framework, which we draw on below.

[19]Sometimes, Equation (4.6) is written as

$$Y = \mu^0 + [(\mu^1 - \mu^0) + (v^1 - v^0)]D + v^0$$

in order to preserve its random-coefficient interpretation. This alternative representation is nothing other than a more fully articulated version of Equation (4.5).

[20]Statisticians sometimes object to the specification of "error terms" because, among other things, they are said to represent a hidden assumption of linearity. In this case, however, the specification of this error term is nothing other than an expression of the definition of the individual-level causal effect as the linear difference between y_i^1 and y_i^0.

to break the correlation between D and the error term.[21] Heckman and Robb show that this strategy is generally ineffective with the data available on worker training programs because (1) some individuals are thought to enter the programs based on anticipation of the treatment effect itself and (2) none of the available data sources have measures of such anticipation. We will return to this case in detail in Chapter 6, where we discuss regression models.

To explain these complications, Heckman and Robb explore how effectively the dependence between D and the error term in Equation (4.6) can be broken. They proceed by proposing that treatment selection be modeled by specifying a latent continuous variable $\tilde{D}$ as

$$\tilde{D} = Z\phi + U, \tag{4.7}$$

where Z represents all observed variables that determine treatment selection, ϕ is a coefficient (or a vector of coefficients if Z includes more than one variable), and U represents both systematic unobserved determinants of treatment selection and completely random idiosyncratic determinants of treatment selection. The latent continuous variable $\tilde{D}$ in Equation (4.7) is then related to the treatment selection dummy, D, by

$$D = 1 \quad \text{if } \tilde{D} \geq 0,$$
$$D = 0 \quad \text{if } \tilde{D} < 0,$$

where the threshold 0 is arbitrary because the term U has no inherent metric (because it is composed of unobserved and possibly unknown variables).

To see the connection between this econometric specification and the one from the statistics literature introduced in the last section, first recall that statisticians typically specify the treatment selection mechanism as the general conditional probability distribution $\Pr[D = 1|S]$, where S is a vector of all systematic observed determinants of treatment selection.[22] This is shown in the graph in Figure 4.8(b). The corresponding causal diagram for the econometric selection equation is presented in two different graphs in Figure 4.9, as there are two scenarios corresponding to whether or not all elements of S have been observed as Z.

For the case in which Z in Equation (4.7) is equivalent to the set of variables in S in Equation (4.3), treatment selection is ignorable, as defined in Equation (4.4), because conditioning on Z is exactly equivalent to conditioning on S. In the econometric tradition, this situation would not, however, be referred to as a case for which treatment assignment/selection is ignorable. Rather, treatment selection would be characterized as "selection on the observables" because all systematic determinants of treatment selection are included in the observed treatment selection variables Z. This phrase is widely used by social scientists because it conveys the essential content

---

[21] Barnow, Cain, and Goldberger (1980:52) noted that "the most common approach" is to "simply assume away the selection bias after a diligent attempt to include a large number of variables" in the regression equation.

[22] When more specific, the basic model is usually a Bernoulli trial, in which $\Pr[D = 1|S = s]$ gives the specific probability of drawing a 1 and the complement of drawing a 0 for individuals with S equal to s.

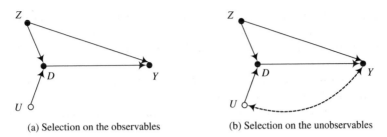

(a) Selection on the observables   (b) Selection on the unobservables

Figure 4.9 Causal diagrams for the terminology from econometric modeling of treatment selection.

of the ignorability assumption: All systematic determinants of treatment selection have been observed.[23]

The scenario of selection on the observables is depicted in Figure 4.9(a). The variable S in Figure 4.8(b) is simply relabeled Z, and there are no back-door paths from D to Y other than the one that is blocked by Z. The remaining idiosyncratic random variation in D is attributed in the econometric tradition to a variable U, which is presented in Figure 4.9(a) as a cause of D that is conditionally independent of both Z and Y. This error term U represents nothing other than completely idiosyncratic determinants of treatment selection. It could therefore be suppressed in Figure 4.9(a), which would render this graph the same as the one in Figure 4.8(b).[24]

Now consider the case in which the observed treatment selection variables in Z are only a subset of the variables in S. In this particular case, some components of S enter into the treatment selection latent variable $\tilde{D}$ through the error term, U, of Equation (4.7). In this case, treatment selection is nonignorable. In the words of econometricians, "selection is on the unobservables" (or, more completely, "selection is on the observables Z and the unobservables U"). The scenario of selection on the unobservables is depicted in Figure 4.9(b), where there are now back-door paths from D to Y represented by $D \leftarrow U \dashleftarrow\dashrightarrow Y$. Conditioning on Z for this graph does not block all back-door paths.

In spite of differences in language and notation, there is little that differentiates the statistics and econometrics models of treatment selection, especially now that the outcome equations used by economists are often completely general nonparametric versions of Equation (4.6) (see Heckman and Vytlacil 2005, which we will discuss in a few different places later in the book, such as Chapters 9 and 12). For now, the key point is that both the statistics and econometric specifications consider the treatment indicator variable, D, to be determined by a set of systematic treatment selection variables

---

[23]And, as with ignorability, the variables in Z are not necessarily equivalent to those in the minimally sufficient conditioning sets suggested by the back-door criterion. The admissible sets according to the back-door criterion may be subsets of the variables in Z or even variables not in Z that are proximate determinants of Y and that, when conditioned on, block all back-door paths generated by the variables in Z.

[24]The latent variable specification in the econometric tradition can be made equivalent to almost all particular specifications of the statement $\Pr[D = 1|S]$ in the statistics tradition by the choice of an explicit probability distribution for U. The full nonparametric equivalence in the causal graph tradition would be $D = f_D(S, e_D) = f_D(Z, U)$.

in S. When all of these variables are observed, the treatment selection mechanism is ignorable and selection is on the observables only. When some of the variables in S are unobserved, the treatment selection mechanism is nonignorable and selection is on the unobservables.

Finally, the qualifications we noted in the prior section about the differences between ignorability assumptions and the back-door criterion apply in analogous fashion to assumptions of selection on the observables. Minimally sufficient conditioning sets suggested by the back-door criterion may be subsets of the variables in Z or entirely different variables that, when conditioned on, eliminate the confounding generated by Z. Variables of the latter type would typically be proximate determinants of Y that intercept the effects of the variables in Z on Y, such as the variable F in Figure 4.4.

4.3.3   Point Identification of Conditional Average Treatment Effects by Conditioning

At the beginning of this chapter, we indicated that we would implicitly focus our presentation of directed graphs and identification issues on the estimation of the unconditional ATE. This narrow focus is entirely consistent with the graphical tradition, in which parameters such as the average treatment effect for the treated (ATT) in Equation (2.7) and the average treatment effect for the controls (ATC) in Equation (2.8) are given considerably less attention than in the potential outcome modeling tradition in both statistics and econometrics. Some comments on the connections may be helpful at this point to foreshadow some of the specific material on causal effect heterogeneity that we will present in the next three chapters.

Identification When the Unconditional ATE Is Identified

If one can identify and consistently estimate the unconditional ATE with conditioning techniques, then one can usually estimate some of the conditional average treatment effects that may be of interest as well. As we will show in the next three chapters, consistent estimates of conditional average treatment effects can usually be formed by specification of alternative weighted averages of the average treatment effects for subgroups defined by values of the conditioning variables. Thus, calculating average effects other than the unconditional ATE may be no more complicated than simply adding one step to the more general conditioning strategy we have presented in this chapter.

Consider again the graph presented in Figure 4.8(b). The back-door path from D to Y is blocked by S. As a result, a consistent estimate of the ATE in Equation (2.3) can be obtained by conditioning on S. But, in addition, consistent estimates of the ATT in Equation (2.7) and the ATC in Equation (2.8) can be obtained by properly weighting conditional differences in the observed values of Y. In particular, first calculate the sample analogs to the differences $E[Y|D=1, S=s] - E[Y|D=0, S=s]$ for all values s of S. Then, weight these differences by the conditional distributions $\Pr[S|D=1]$ and $\Pr[S|D=0]$ to calculate the ATT and the ATC, respectively.

Identification When the Unconditional ATE Is Not Identified

If selection is on the unobservables, conditioning strategies will typically fail to identify unconditional ATEs. Nonetheless, weaker assumptions may still allow for the identification and subsequent estimation by conditioning of various conditional average treatment effects. We will present these specific weaker assumptions in the course of explaining matching and regression techniques in the next three chapters, but for now we give a brief overview of the identification issues in relation to the graphical models presented in this chapter. (See also the discussion in Section 2.7.4 of similar issues with regard to the inconsistency and bias of the naive estimator.)

Suppose, for example, that the graph in Figure 4.9(b) now obtains, and hence that a back-door path from D to Y exists via unobserved determinants of the cause, U. In this case, conditioning on Z will not identify the unconditional ATE. Nonetheless, conditioning on Z may still identify a conditional average treatment effect of interest, as narrower effects can be identified if weaker assumptions can be maintained even though unblocked back-door paths may still exist between D and Y.

Consider a case for which partial ignorability holds, such that $Y^0 \perp\!\!\!\perp D|S$ is true but $(Y^0, Y^1) \perp\!\!\!\perp D \mid S$ is not. Here, conditioning on S generates a consistent estimate of the ATT even though S does not block the back-door path from D to Y. The opposite is, of course, also true. If partial ignorability holds in the other direction, such that $Y^1 \perp\!\!\!\perp D|S$ holds but $(Y^0, Y^1) \perp\!\!\!\perp D|S$ does not, then the ATC can be estimated consistently.[25]

Consider the first case, in which only $Y^0 \perp\!\!\!\perp D \mid S$ holds. Even after conditioning on S, a back-door path remains between D and Y because Y^1 still differs systematically between those in the treatment and control groups and Y is determined in part by Y^1; see Equation (2.2). Nonetheless, if, after conditioning on S, the outcome under the no-treatment-state, Y^0, is independent of exposure to the treatment, then the ATT can be estimated consistently. The average values of Y, conditional on S, can be used to consistently estimate the average what-if values for the treated if they were instead in the control state. This type of partial ignorability is akin to Assumption 2 in Equation (2.16), except that it is conditional on S. We will give a full explanation of the utility of such assumptions when discussing matching estimates of the ATT and the ATC in the next chapter.

Graphs Do Not Clearly Reveal the Identification Possibilities for the ATT and ATC When the ATE Is Not Also Identified

In the prior section, we noted in our presentations of ignorability and selection on the observables that graphs help guide researchers toward minimally sufficient conditioning sets that may differ from the conditioning sets suggested by the statistics and econometrics literature. The back-door criterion is especially helpful in this regard because of its targeted focus on back-door paths that generate noncausal associations between the causal variable and the outcome variable. However, it must also be stated, as implied by this section, that directed graphs will not clearly reveal effective analysis

---

[25] And, as we will show in the next chapter, the required assumptions are even simpler because the entire distributions of Y^0 and Y^1 need not be conditionally independent of D. As long as the stable unit treatment value assumption (SUTVA) holds, only mean independence must be maintained.

strategies to identify either the ATT or the ATC in situations where the ATE cannot also be identified by conditioning.

The overall implication of this point is that researchers should learn all three frameworks for approaching causal identification challenges. Graphs help immensely in selecting conditioning sets when the target parameter is the ATE. When the ATE is not identified by any feasible conditioning sets, the potential outcome model – either as deployed in statistics or econometrics – can still guide the researcher to identification strategies for narrower conditional average effects, most commonly the ATT or the ATC. Directed graphs remain a useful tool in these situations, as they help to organize one's thinking about the full system of causal relationships that are relevant. But, the identification strategy that is then selected to estimate either the ATT or the ATC is likely to emerge from thinking through the possibilities from within the potential outcome model.

4.4   Conditioning to Balance and Conditioning to Adjust

When presenting Pearl's back-door criterion for determining a sufficient set of conditioning variables, we noted that for some applications more than one set of conditioning variables is sufficient. In this section, we return to this point as a bridge to the following three chapters that present both matching and regression implementations of conditioning. Although we will show that matching and regression can be considered variants of each other, here we point to the different ways in which they are usually invoked in applied research. Matching is most often considered a technique to balance the determinants of the causal variable, and regression is most often considered a technique to adjust for other causes of the outcome.

To frame this discussion, consider first the origins of the balancing approach in the randomized experiment tradition. Here, the most familiar approach is a randomized experiment that ensures that treatment status is unassociated with all observed and unobserved variables that determine the outcome (although only in expectation). When treatment status is unassociated with an observed set of variables W, the data are balanced with respect to W. More formally, the data are balanced if

$$\Pr[W|D=1] = \Pr[W|D=0], \tag{4.8}$$

which requires that the probability distribution of W be the same within the treatment and control groups.

Now consider the graph presented in Figure 4.10. Back-door paths are present from D to Y, represented by $D \leftarrow S \leftarrow\!-\!-\!-\!-\!\rightarrow X \rightarrow Y$, where S is the complete set of variables that are direct causes of treatment assignment/selection, X is the complete set of variables other than D that are direct causes of Y, and the bidirected edge between S and X signifies that they are mutually caused by some set of common unobserved causes.[26]

---

[26]For this example, we could have motivated the same set of conclusions with other types of causal graphs. The same basic conclusions would hold even if X and S include several variables within them in which some members of X cause D directly and some members of S cause Y directly. In other

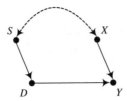

Figure 4.10 A causal diagram in which sufficient conditioning can be performed with respect to S or X.

Because neither S nor X is a collider, all back-door paths in the graph can be blocked by conditioning on either S or X (and we write "paths" because there may be many back-door paths through the bidirected edge between S and X). Conditioning on S is considered a balancing conditioning strategy, whereas conditioning on X is considered an adjustment-for-other-causes conditioning strategy. If one observes and then conditions on S, the variables in S and D are no longer associated within the subgroups defined by the conditioning. The treatment and control groups are thereby balanced with respect to the distribution of S. Alternatively, if one conditions on X, the resulting subgroup differences in Y across D within X can be attributed to D alone. In this case, the goal is not to balance X but rather to partial out its effects on Y in order to isolate the net effect of D on Y.

The distinction between balancing and adjustment for other causes is somewhat artificial (see Hansen 2008). For the graph in Figure 4.10, balancing X identifies the causal effect. Thus, it is technically valid to say that one can identify a causal effect by balancing a sufficient set of other causes of Y. Nonetheless, the graph in Figure 4.10 demonstrates why the distinction is important. The ultimate set of systematic causes that generates the relationship between S and X is unobserved, as it often is in many applied research situations. Because one cannot condition on these unobserved variables, one must condition on either S or X in order to identify the causal effect. These two alternatives may be quite different in their practical implementation.[27]

Should one balance the determinants of a cause, or should one adjust for other causes of the outcome? The answer to this question is situation specific, and it depends on the quality of our knowledge and measurement of the determinants of D and Y.

---

words, all that we need to make the distinction between balancing and adjustment for other direct causes is two sets of variables that are related to each other, with at least one variable in one set that causes D but not Y and at least one variable in the other set that causes Y but not D.

[27]Note also that the ingredients utilized to estimate the ATE (as well as the ATT and the ATC) will differ based on the particular conditioning routine, and this will allow alternative expressions of underlying heterogeneity. If S is observed, then conditional average treatment effects can be calculated for those who are subject to the cause for different reasons, based on the values of S that determine D. If X is observed, then conditional average treatment effects can be calculated holding other causes of Y at chosen values of X. Each of these sets of conditional average treatment effects has its own appeal, with the relative appeal of each depending on the application. In the potential outcome tradition, average treatment effects conditional on S would likely be of more interest than average treatment effects conditional on X. But for those who are accustomed to working within an all-cause regression tradition, then average treatments effects conditional on X might be more appealing.

One answer is that the researcher should do both.[28] Nonetheless, there is a specific advantage of balancing that may tip the scales in its favor if both strategies are feasible: Balancing diminishes the inferential problems that can be induced by data-driven specification searches, as we will explain in Chapter 6.[29]

4.5   Conclusions

In this chapter, we have used the causal graphs introduced in Chapter 3 to explain the rationale for conditioning estimators of causal effects, focusing on the back-door criterion for selecting sufficient sets of conditioning variables that identify the ATE. We then introduced models of treatment assignment and treatment selection from statistics and econometrics that are used to assert similar claims about conditioning estimators, focusing also on the ATT and ATC.

   In the next three chapters, we present details and connections between the three main types of conditioning estimation strategies utilized in the social sciences: matching, regression, and weighted regression. We show how they typically succeed when selection is on the observables and fail when selection is on the unobservables. We lay out the specific assumptions that allow for the identification of unconditional average treatment effects, as well as the weaker assumptions that allow for the identification of narrower conditional average treatment effects, such as the ATT and ATC. In later chapters, we then present additional methods for identifying and estimating causal effects when conditioning methods do not suffice because crucial variables on back-door paths are unobserved.

4.6   Appendix to Chapter 4: The Back-Door and Adjustment Criteria, Descendants, and Colliders Under Magnification

In order to properly utilize the back-door criterion when evaluating alternative conditioning sets, an analyst does not need to understand all details of all scenarios in which its violation will prevent a conditioning estimator from generating consistent and unbiased estimates of causal effects of interest. However, a deeper examination of additional scenarios provides insight into common methodological challenges, while also demonstrating the powerful contribution that graphical methods are likely to make to social science research in the coming decades.

   In this appendix, we offer additional examples where colliders and descendants must be carefully considered in order to understand the fine points of why the back-door criterion warrants causal inference. We show how Condition 2 of the back-door

---

[28]As we discuss in later chapters, many scholars have argued for conditioning on both S and X. Robins, for example, argues for this option as a double protection strategy that offers two chances to effectively block the back-door paths between D and Y (see Robins and Rotnitzky 2001).

[29]We phrase this guidance with "may" because it must be evaluated alongside another concern. Balancing all variables that determine D, including those that do not generate confounding, is unnecessary and inefficient. As a result, it may render conditioning infeasible in datasets of the size typically available to social scientists.

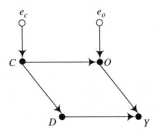

Figure 4.11 A causal graph with a confounded causal effect and where the variables along the back-door path are viewed under magnification.

criterion, which appears only to prevent the analyst from inadvertently conditioning on variables that adjust away the causal effect of interest, also plays a role in blocking hidden as-if back-door paths that also result from conditioning on endogenous variables. We show that this result can only be seen when endogenous variables are viewed under magnification, so that it becomes clear that these variables are also colliders. We conclude with an examination of the connections between Pearl's original back-door criterion, its generalization as the adjustment criterion of Shpitser, VanderWeele, and Robins (2010), and our blended alternative, which retains the targeted spirit and inherent practicality of Pearl's original back-door criterion while also incorporating some of the insight furnished by the admirably broad adjustment criterion.

Like the appendix to Chapter 3, nothing in this appendix is necessary for understanding the next three chapters that follow. We offer it for curious readers who crave a deeper understanding of the back-door criterion and who wish to dive into the original literature after reading this book. We also offer it to explain to the community of causal graph methodologists why we have chosen the version of the back-door criterion that we have.

The Back-Door Criterion Explained with Causal Graphs Viewed Under Selective Magnification. The first step in fully considering why violations of the back-door criterion prevent conditioning estimators from delivering consistent and unbiased estimates is to repeat some of the material from the main body of the chapter, after redrawing the causal graphs under magnification. As we introduced in Section 3.3.2 using Figure 3.7, under magnification the unobserved structural error terms of the associated structural equations are brought into view. For Figure 4.11, we have redrawn Figure 4.1 under what we will call "selective magnification." In this case, we magnify the nodes of all variables that we might consider conditioning on, leaving the error terms on the causal variable and the outcome variable hidden as in the standard representation for visual simplicity. Accordingly, Figure 4.11 shows two additional causal effects, $e_C \to C$ and $e_O \to O$, that were only implicit in Figure 4.1.

Now, reconsider our explanation of the back-door criterion for this graph. We wrote earlier that conditioning on C and/or O would identify the causal effect. As a transition to the additional examples in this appendix, we will now present the same explanation considering how the inclusion of $e_C \to C$ and $e_O \to O$ in the graph does not change this claim.

In the directed graph tradition, C is a "root" variable because it has no causes other than those embedded in its structural error term. Root variables do not need to be considered in any other way when viewed under magnification because their structural error terms are nothing more than the source of the variation observed for these variables. Accordingly, it is still the case that conditioning on C blocks the sole back-door path $D \leftarrow C \rightarrow O \rightarrow Y$ because C is the middle variable of a fork of mutual dependence that lies within the path. Because C is also not a descendant of D, it satisfies the back-door criterion.

For O, an additional complication should be clear. Under magnification, it is revealed that O is a collider variable because $C \rightarrow O$ is now accompanied by $e_O \rightarrow O$. In general, all variables in a causal graph that are not root variables will appear as colliders when viewed under magnification. We have urged caution when handling colliders, and yet only now do we reveal that all variables such as O are, in fact, colliders as well. Accordingly, conditioning on O will induce an association between C and e_O. Fortunately, this conditioning does not invalidate the back-door criterion by unblocking an already blocked back-door path, as we will now explain.

Recall our prior discussions of colliders in Sections 3.2.2 and 4.1.2. We have said that for a blocked path, $A \rightarrow C \leftarrow B$, where C is the collider, conditioning on C opens up this path. We now introduce a piece of notation to convey this result: a conditioning "button," $\odot$, that can be deployed to show the genesis of a new induced association, $\cdots \odot \cdots$.[30] For example, for the blocked path $A \rightarrow C \leftarrow B$, conditioning on C generates a new association, $A \cdots \odot \cdots B$, where the conditioning action itself, $\odot$, generates the association between A and B. For our hypothetical college admissions example in 4.1.2, the notation represents the following action: A button is pushed for "admissions decisions," and the population of applicants is then divided into those admitted and those rejected. Within both groups, A and B are now associated, as shown in Figure 4.2. The only way to eliminate the induced association is to undo the conditioning that generates it (i.e., release the conditioning button).

Return to Figure 4.11 so that we can demonstrate this new notation when considering the conclusions suggested by an evaluation of the back-door criterion. Suppose again that O is our candidate conditioning variable, and we know that conditioning on O will block the back-door path $D \leftarrow C \rightarrow O \rightarrow Y$. When seen under magnification, it should be clear that now conditioning on O also generates a new association, $C \cdots \odot \cdots e_O$. In fact, conditioning on O also generates a second association, $e_C \cdots \odot \cdots e_O$.[31] Pearl (2009:339) explains, for examples such as this one, that these induced associations create two as-if back-door paths:

1. $D \leftarrow C \cdots \odot \cdots e_O \rightarrow O \rightarrow Y$ and

2. $D \leftarrow C \leftarrow e_C \cdots \odot \cdots e_O \rightarrow O \rightarrow Y$.

---

[30]Pearl and other authors often use alternative notation to convey the same associations, most commonly A—B. We prefer our more active "button," which can also be directly extended to actual operator status, as in $\cdots \odot (.) \cdots$. If one wanted to show conditioning on more than one variable, we would have $\cdots \odot (C, O) \cdots$, etc.

[31]Conditioning on a collider, or a descendant of a collider, induces associations between all ancestors of the collider on separate directed paths that reach the collider.

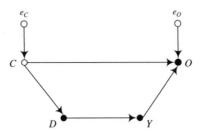

Figure 4.12 A diagram where the causal effect of D on Y is not confounded and where the observed variable O on the back-door path is a descendant of both D and Y.

Fortunately, both of these as-if back-door paths are also blocked when O is conditioned on because O is the middle variable in a chain of mediation, $e_O \to O \to Y$, for both of them. Thus, O satisfies the back-door criterion because O is not a descendant of D and because conditioning on O eliminates all back-door associations between D and Y, including two associations created by as-if back-door paths that are only revealed under magnification.

Of course, we already knew that O satisfied the back-door criterion from our consideration of Figure 4.1. Viewing the graph under selective magnification has forced us to reckon with the unobserved structural error terms, and little insight has been gained. We will now consider two additional examples where consideration of the structural error terms leads to additional insight that will convince the reader of how simple and effective the back-door criterion is.

Consider Figure 4.12, where again we have four variables, and where all variables but D and Y are displayed under magnification. The only back-door path between D and Y, which is $D \leftarrow C \to O \leftarrow Y$, is blocked by the collider O. Although it may seem awkward to refer to $D \leftarrow C \to O \leftarrow Y$ as a back-door path between D and Y (because it does not end with $\to Y$), it satisfies our definition because it is a path between two causally ordered variables D and Y that begins with $D \leftarrow$. As a result, it is still a back-door path, even though it does not terminate with $\to Y$ (unlike all other back-door paths displayed so far in our examples).[32] Because the sole back-door path between D and Y does not generate a noncausal association between D and Y, there is no need to adjust for any variables in order to generate a consistent and unbiased estimate of the causal effect of D on Y.

In fact, all that one can do is make the situation worse. Suppose that an analyst believes mistakenly that the path $D \leftarrow C \to O \leftarrow Y$ is an unblocked back-door path that must be blocked in order to generate a consistent and unbiased estimate of the effect of D on Y. Because C is unobserved, suppose that the analyst decides to condition on O instead, under the rationale that it is the only observed variable that lies on the back-door path.

---

[32] We could bring this graph in line with prior ones by replacing $Y \to O$ with $O \leftarrow U \to Y$. The explanation would then change a bit because O would no longer be a descendant of D via a directed path, but the analysis with respect to Condition 1 of the back-door criterion would be the same.

The variable O violates both conditions of the back-door criterion. Consider Condition 1 first. Because O is a collider variable, conditioning on O generates many new associations, including $C \cdots \odot \cdots Y$, $e_C \cdots \odot \cdots Y$, $C \cdots \odot \cdots e_O$, and $e_O \cdots \odot \cdots Y$. These induced associations generate three as-if back-door paths:

1. $D \leftarrow C \cdots \odot \cdots Y$,

2. $D \leftarrow C \leftarrow e_C \cdots \odot \cdots Y$, and

3. $D \leftarrow C \cdots \odot \cdots e_O \cdots \odot \cdots Y$.

Because C is unobserved, these as-if back-door paths remain unblocked after conditioning on O, and therefore O does not satisfy Condition 1 of the back-door criterion.

Now consider Condition 2, and notice that O lies on a directed path, $D \to Y \to O$, that begins at D and reaches Y. As we noted earlier, we chose to specify Condition 2 of the back-door criterion with the words "reaches the outcome variable" rather than "ends at at the outcome variable" in order to capture cases such as this one. In this case, O lies on the directed path that represents the causal effect of interest, even though O does not mediate the casual effect itself. The intuition of Condition 2 of the back-door criterion should nonetheless still be clear. Because the causal effect of D on Y is fully embedded within the variation in O, adjusting for O would explain away the causal effect itself.[33]

Overall, then, conditioning on O in this graph would make the situation considerably worse. No unblocked back-door paths needed to be blocked in the first place, and conditioning on O would generate as-if back-door paths that induce new noncausal associations between D and Y while at the same time robbing the causal effect of (possibly all) of its magnitude.

To now transition to a more complicated case, pause to consider how our language in this appendix differs from the language used in the main body of the chapter. Up until now, we have expressed similar results to those for Figure 4.12 on the perils of conditioning on colliders using more brief language. For this graph, the simpler language would be the following:

> Conditioning on the collider variable O unblocks the already blocked back-door path, $D \leftarrow C \to O \leftarrow Y$.

This explanatory syntax is concise and correct. Even so, it should now be clear that a more laborious way to understand this result is the following:

> Conditioning on the variable O that is a collider on the back-door path $D \leftarrow C \to O \leftarrow Y$ creates additional as-if back-door paths such as $D \leftarrow C \cdots \odot \cdots Y$, $D \leftarrow C \leftarrow e_C \cdots \odot \cdots Y$, and $D \leftarrow C \cdots \odot \cdots e_O \cdots \odot \cdots Y$. These as-if back-door paths are unblocked when conditioning on O.

---

[33]In an infinite sample, wherein one could stratify the data on all values of O, D and Y would be independent within the strata of O. In a finite sample, possibly necessitating the use of a different type of conditioning estimator, the conditional association between D and Y might not equal zero but would still not be a consistent estimate of the causal effect.

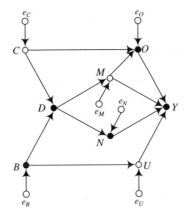

Figure 4.13 A graph where the effect of D on Y is not identified by conditioning on O and B because O is a descendant of D.

For the graphs considered in the main body of this chapter, this more tortured language would not have been helpful and therefore was not used. However, for this particular graph, the more tortured language may offer a clearer explanation. The brief explanation does not make clear why conditioning on O generates a back-door association by unblocking $D \leftarrow C \rightarrow O \leftarrow Y$, given that this back-door path does not end with $\rightarrow Y$. The more tortured language, which uses the conditioning button, does provide the proper imagery because the as-if back-door path, $D \leftarrow C \cdots \odot \cdots Y$, does not end with $\leftarrow Y$.

For an additional example, which draws many of these issues together, consider Figure 4.13. This graph is a redrawn version Figure 4.7, but now with all variables other than D and Y displayed under magnification. As noted earlier for Figure 4.7, there are three back-door paths from D to Y:

1. $D \leftarrow C \rightarrow O \rightarrow Y$,

2. $D \leftarrow C \rightarrow O \leftarrow M \rightarrow Y$, and

3. $D \leftarrow B \rightarrow U \rightarrow Y$.

Again, suppose that one decides to condition on both O and B. Using the conditioning button to reveal all of the associations generated by the conditioning action itself would generate many new as-if back-door paths. Consider just two of these. The induced association $C \cdots \odot \cdots M$ creates the as-if back-door path $D \leftarrow C \cdots \odot \cdots M \rightarrow Y$ that remains unblocked after conditioning on O and B. This as-if back-door path is similar to those considered for Figure 4.12. But now consider how the induced association $D \cdots \odot \cdots e_M$ creates the as-if back-door path $D \cdots \odot \cdots e_M \rightarrow M \rightarrow Y$. This path is also unblocked after conditioning on O and B. But, most importantly, this as-if back-door path only comes to the foreground when the causal graph is viewed under magnification. Except under magnification, it is all too easy to fail to recognize that M is a collider and that conditioning on a descendant of M will induce an association between D and e_M. For all but the most experienced causal graph practitioners, this as-if back-door path would be hidden by the standard representation of the graph.

As a result, the conditioning set of O and B does not satisfy Conditions 1(a), (b), and (c) of the back-door criterion. As we used this graph to show in the main body of the chapter, this graph also does not satisfy Condition 2 of the back-door criterion either. This result does not need to be explained in any different fashion when the graph is viewed under selective magnification. Overall, the adjusted effect of D on Y, which results from conditioning on O and B, would steal some of the total effect of D on Y. Yet, as we show here, the conditioning action also generates new back-door associations, one through the unobserved mediating variable M. This last as-if back-door path is completely hidden from view in the standard representation of a causal graph. Fortunately, all such hidden as-if back-door paths that emerge under conditioning will never creep into an empirical analysis as long as the back-door criterion is properly evaluated.

And this is the key point of this last example: Condition 2 of the back-door criterion must be heeded for two reasons. First, as was also presented in the main body of the chapter, conditioning on descendants of the cause that lie on (or descend from) directed paths that begin at D and that reach Y will always mistakenly adjust away some of the causal effect the analyst is interested in estimating. Second, as we have now shown in this appendix, conditioning on the descendants of the cause (such as O) that are also descendants of unobserved colliders that are are themselves descendants of the cause (such as M) will always generate unblocked back-door associations that spoil the analysis. Fortunately, if one uses the back-door criterion to select conditioning sets, then the threat of such hidden unblocked back-door associations can be avoided.

The Back-Door and Adjustment Criteria Considered. As we noted earlier, our version of the back-door criterion incorporates insight gained from the more recent development of the *adjustment criterion* by Shpitser et al. (2010). Figure 4.14 presents a graph that reveals the differences between Pearl's original back-door criterion and the more recent generalization that is the adjustment criterion. We will use this graph to explain why we have modified Pearl's original back-door criterion to a small degree but also why we have not adopted the adjustment criterion as a whole.

As we have specified it in the main body of this chapter, the back-door criterion is evaluated in the following two steps. The first step is to write down the back-door paths from D to Y, determine which ones are unblocked, and then search for a conditioning set that will block all unblocked back-door paths. For Figure 4.14, the two back-door paths are

1. $D \leftarrow A \rightarrow B \leftarrow C \rightarrow Y$ and

2. $D \leftarrow E \rightarrow Y$.

Path 1 is already blocked by the collider B, and so only path 2 must be blocked by conditioning. Here, the solution is simple: Condition on E because it is the middle variable of a fork of mutual dependence.

The second step is to verify that the variables in the candidate conditioning set are not descendants of the causal variable that lie along, or descend from, all directed paths from the causal variable that reach the outcome variable. For Figure 4.14, it

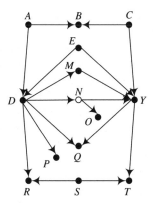

Figure 4.14 A directed graph that reveals the differences between the back-door criterion and the adjustment criterion.

is easy to see that E is a root variable and as such is not a descendant of any other variable in the graph.

However, to understand the differences between the back-door criterion in its original form and the adjustment criterion, we need to consider this step in more detail. For more elaborate causal graphs, in order to evaluate Condition 2 of the back-door criterion one needs to ascend the family tree by looking "upstream" through the arrows that point to each candidate conditioning variable (i.e., from the parent, to the grandparent, to the great grandparent, and so on) to determine whether the causal variable is a direct ancestor of any of the candidate conditioning variables. If the causal variable is not found to be an ancestor of any of the candidate conditioning variables in the set under consideration, then by definition none of the candidate conditioning variables lies on or descends from directed paths that begin at the causal variable and that reach the outcome variable. In this case, a conditioning estimator that uses the variables under consideration will generate consistent and unbiased estimates of the causal effect of interest.

If the causal variable is discovered to be an ancestor of any of the candidate conditioning variables in the set under consideration, then for our version of the back-door criterion (which incorporates insight from the adjustment criterion), one has to determine whether the directed paths from the causal variable that establish the descent of the conditioning variable(s) also reach the outcome variable. If not, then one can still condition on any such variables and generate consistent and unbiased estimates. If, instead, any of the candidate conditioning variables are descendants of the causal variable *and* lie on or descend from directed paths that begin at the causal variable and reach the outcome variable, then the conditioning set will block or adjust away some of the causal effect. In this case, the conditioning set will not generate consistent and unbiased estimates of the causal effect of interest.

To now transition to a parallel consideration of the adjustment criterion, and why we have not adopted it in whole, return to the case of Figure 4.14 and consider how the back-door criterion might be utilized in practice by an experienced analyst. As we noted above in this appendix, conditioning on E alone satisfies the back-door criterion. Notice, however, that, in addition to E, one could also condition on A, on C, on A and B, on B and C, or on A, B, and C. The back-door criterion does not rule out these possibilities, but it does not encourage the analyst to consider A and C as members of the candidate conditioning set because the back-door path on which these variables lie is already blocked in the absence of any conditioning by B. Instead, the back-door criterion guides the analyst to the essential information: (1) path 1 is already blocked by B and can be left alone; (2) path 2 must be blocked by conditioning on E.

Nonetheless, an experienced analyst might reason that, although one should not condition only on B, one could condition on B as long as either A or C is also conditioned on. The experienced analyst might then choose to offer two estimates, one that conditions only on E and one that conditions on B and C as well (perhaps because a fair critic has a theory that says that the $B \leftarrow C$ in Figure 4.14 should instead be $B \rightarrow C$). The experienced analyst might reach this position even though she regards conditioning on B and C as unnecessary and inefficient, given that she truly does believe that her theoretical assumptions represented by the original graph are beyond reproach. Nevertheless, she might reason that it is prudent to show a fair critic that even if his alternative assumptions are correct, the main results of the ensuing empirical analysis remain the same.

Once one begins to allow supplementary but unnecessary conditioning variables into the conditioning set, a question arises as to whether the back-door criterion covers all cases. This is the inspiration for the adjustment criterion, developed by Shpitser et al. (2010). It is designed to allow the analyst to identify all permissible conditioning sets, including those that include unnecessary and redundant conditioning, as we now explain.

The adjustment criterion also has two steps. For the first step, the analyst considers all paths in the graph that begin at the causal variable and then categorizes all paths into (1) "causal paths," which are defined as all directed paths from the causal variable to the outcome variable, and (2) "noncausal paths," which are defined as all paths from the causal variable that are not causal paths. For Figure 4.14, one would enumerate the following paths, categorizing them as follows:

1. $D \leftarrow A \rightarrow B \leftarrow C \rightarrow Y$ (noncausal),

2. $D \leftarrow E \rightarrow Y$ (noncausal),

3. $D \rightarrow M \rightarrow Y$ (causal),

4. $D \rightarrow N \rightarrow Y$ (causal),

5. $D \rightarrow N \rightarrow O$ (noncausal),

6. $D \rightarrow P$ (noncausal),

7. $D \rightarrow Q \leftarrow Y$ (noncausal),

8. $D \rightarrow R \leftarrow S \rightarrow T \leftarrow Y$ (noncausal).

Notice that paths 1 and 2 are back-door paths for the back-door criterion but, for the adjustment criterion, are classified as members of the more encompassing set of noncausal paths. The first step of the adjustment criterion then requires that the analyst search for variables that, when conditioned on, will ensure that all noncausal paths between D and Y are blocked. This is analogous to the back-door criterion, but now all noncausal paths, not just back-door paths, must be considered. For Figure 4.14, this part of the first step requires the analyst to search for all permissible conditioning sets that will block the noncausal paths 1, 2, 5, 6, 7, and 8. For the case of Figure 4.14, more sets than we wish to enumerate would be selected. In brief, all sets would include E and exclude Q. Yet, these sets would include many combinations of additional unnecessary conditioning variables, including appropriate combinations of A, B, C, N, O, P, R, S, and T, such as the maximal blocking set $\{A, B, C, E, N, O, P, R, S, T\}$.

For the second step of the adjustment criterion, one then implements the same second step as the version of the back-door criterion we have specified in the main body of the chapter. For Figure 4.14, this step requires that the candidate conditioning sets not include the variables M, N, or O. Presumably M would not have been considered as a conditioning variable in the first place because it is not a variable that lies on any of the noncausal paths that the analyst is attempting to block. Nonetheless, the analyst must also strike all conditioning sets that include N and O, and this conclusion has to be discerned from the graph, not the list of the paths that does not reveal full patterns of descent. (This is also the case for the back-door criterion.)

The advantage of the adjustment criteria is that it will reveal all permissible conditioning sets, such as the maximal permissible conditioning set $\{A, B, C, E, P, R, S, T\}$, assuming that the analyst devotes the necessary energy to delineating all patterns of permissible conditioning. The disadvantage is that, for graphs such as the one in Figure 4.14, the adjustment criterion is laborious and is likely to lead one to condition on more variables than is necessary to identify the causal effect.

We have therefore chosen to stick with the back-door criterion in the main body of the chapter, with only one substantial modification to Pearl's original specification in deference to the completeness of the adjustment criterion. As we noted earlier in our footnote on Condition 2 of our version of the back-door criterion, Pearl's original back-door criterion requires more simply (but overly strongly) that no variables in the conditioning set Z can be descendants of the causal variable. For our version of the back-door criterion, we weaken Pearl's original no-descendants condition to allow conditioning on variables, such as P in Figure 4.14, that are descendants of the cause D but that do not lie on or descend from directed paths that begin at D and reach Y. We doubt an analyst would think to condition on P when evaluating the back-door criterion for this graph (because the back-door criterion focuses attention more narrowly on back-door paths, rather than all noncausal paths). Still, this modification of the back-door criterion allows it to identify more of the permissible conditioning sets that are identified by the more laborious but admirably complete adjustment criterion.

Chapter 5

Matching Estimators of Causal Effects

The rise of the counterfactual model to prominence has increased the popularity of data analysis routines that are most clearly useful for estimating the effects of causes. The matching estimators that we will review and explain in this chapter[1] are perhaps the best example of a classic technique that has reemerged in the past three decades as a promising procedure for estimating causal effects.[2] Matching represents an intuitive method for addressing causal questions, primarily because it pushes the analyst to confront the process of causal exposure as well as the limitations of available data. Accordingly, among social scientists who adopt a counterfactual perspective, matching methods are fast becoming an indispensable technique for prosecuting causal questions, even though they usually prove to be the beginning rather than the end of causal analysis on any particular topic.

We begin with a brief discussion of the past use of matching methods. Then, we present the fundamental concepts underlying matching, including stratification of the data, weighting to achieve balance, and propensity scores. Thereafter, we discuss how matching is usually undertaken in practice, including an overview of various matching algorithms.

In the course of presentation, we will offer four hypothetical examples that demonstrate some of the essential claims of the matching literature, progressing from idealized examples of stratification and weighting to the implementation of alternative matching algorithms on simulated data for which the treatment effects of interest are known by construction. As we offer these examples, we add real-world complexity in order to demonstrate how such complexity can overwhelm the power of the techniques. In

---

[1]This chapter has its origins in Morgan and Harding (2006), and David Harding's contributions are gratefully acknowledged.

[2]Matching techniques can be motivated as estimators without invoking causality. Just as with regression modeling, which we discuss in detail in Chapters 6 and 7, matching can be used to adjust the data in search of a meaningful descriptive fit to the data in hand. Given the nature of this book, we will focus on matching as an estimator of causal effects. We will, however, discuss the descriptive motivation for regression estimators in Chapter 6.

the next two chapters, we build on the same examples when presenting regression and weighted regression estimators. The overall goals of Chapters 5, 6, and 7 are to explain the interconnections between the most prominent techniques for implementing conditioning estimators that are justified by the back-door criterion presented in Chapter 4, but that are also typically motivated in the social science literature by assumptions of ignorability of treatment assignment or selection on the observables.

5.1 Origins of and Motivations for Matching

Matching techniques have origins in experimental work from the first half of the twentieth century. Relatively sophisticated discussions of matching as a research design can be found in early methodological texts in the social sciences (e.g., Greenwood 1945) and also in attempts to adjudicate between competing explanatory accounts in applied demography (Freedman and Hawley 1949). This early work continued in sociology (e.g., Althauser and Rubin 1970, 1971; Yinger, Ikeda, and Laycock 1967) right up to the key foundational literature in statistics (Rubin 1973a, 1973b, 1976, 1977, 1979, 1980a) that provided the conceptual foundation for the new wave of matching techniques that we will present in this chapter.

In the early 1980s, matching techniques, as we conceive of them now, were advanced in a set of papers by Rosenbaum and Rubin (1983a, 1984, 1985a, 1985b) that offered solutions to a variety of practical problems that had limited matching techniques to very simple applications in the past.[3] Variants of these new techniques found some use immediately in sociology (Berk and Newton 1985; Berk, Newton, and Berk 1986; Hoffer, Greeley, and Coleman 1985), continuing with work by Smith (1997). Since the late 1990s, economists and political scientists have contributed to the development of new matching techniques (e.g., Heckman et al. 1999; Heckman, Ichimura, Smith, and Todd 1998; Heckman, Ichimura, and Todd 1997, 1998 in economics and Ho, Imai, King, and Stuart 2007 and Diamond and Sekhon 2013 in political science). Given the growth of this literature, and the applications that are accumulating, we expect that matching will complement other types of modeling in the social sciences with greater frequency in the future.

In the methodological literature, matching is usually introduced in one of two ways: (1) as a method to form quasi-experimental contrasts by sampling comparable treatment and control cases from among two larger pools of such cases or (2) as a nonparametric method of adjustment for treatment assignment patterns when it is feared that ostensibly simple parametric regression estimators cannot be trusted.

For the first motivation, the archetypical example is an observational biomedical study in which a researcher is called on to assess what can be learned about a particular treatment. The investigator is given access to two sets of data, one for individuals who have been treated and one for individuals who have not. Each dataset includes a measurement of current symptoms, Y, and a set of characteristics of individuals, as a vector of variables X, that are drawn from demographic profiles and health histories.

---

[3]See Rubin (2006) for a compendium. See Guo and Fraser (2010), Sekhon (2009), and Stuart (2010) for reviews that connect the early literature to the current state of practice, but with different points of emphasis than we offer in this chapter.

Typically, the treatment cases are not drawn from a population by means of any known sampling scheme. Instead, they emerge as a result of the distribution of initial symptoms, patterns of access to the health clinic, and then decisions to take the treatment. The control cases, however, may represent a subsample of health histories from some known dataset. Often, the treatment is scarce, and the control dataset is much larger than the treatment dataset.

In this scenario, matching is a method of strategic subsampling from among treated and control cases. The investigator selects a nontreated control case for each treated case based on the characteristics observed as x_i. All treated cases and matched control cases are retained, and all nonmatched control cases are discarded. Differences in the observed y_i are then calculated for treated and matched cases, with the average difference serving as the treatment effect estimate for the group of individuals given the treatment.[4]

The second motivation has no archetypical substantive example, as it is similar in form to any attempt to use regression to estimate causal effects with survey data. Suppose, for a general example, that an investigator is interested in the causal effect of an observed dummy variable, D, on an observed outcome, Y. For this example, it is assumed that a simple bivariate regression, $Y = \alpha + \delta D + \varepsilon$, will yield an estimated coefficient $\hat{\delta}$ that is an inconsistent and biased estimate of the causal effect of interest because the causal variable D is associated with variables included in the error term, ε. For a particular example, if D is the receipt of a college degree and Y is a measure of economic success, then the estimate of interest is the causal effect of obtaining a college degree on subsequent economic success. However, family background variables are present in ε that are correlated with D, and this relationship produces omitted-variable bias for a college-degree coefficient estimated from a bivariate ordinary least squares (OLS) regression of Y on D.

In comparison with the biomedical example just presented, this motivation differs in two ways: (1) in most applications of this type, the data represent a random sample from a well-defined population and (2) the common practice in the applied literature is to use regression to estimate effects. For the education example, a set of family background variables in X is assumed to predict both D and Y. The standard regression solution is to estimate an expanded regression equation: $Y = \alpha + \delta D + X\beta + \varepsilon^*$. With this strategy (which we will discuss in detail in the next chapter), the goal is to estimate simultaneously the causal effects of X and D on the outcome, Y.

In contrast, a matching estimator nonparametrically balances the variables in X across D solely in the service of obtaining the best possible estimate of the causal effect of D on Y. The most popular technique is to estimate the probability of D for each individual i as a function of X and then to select for further analysis only matched sets of treatment and control cases that contain individuals with equivalent values for these predicted probabilities. This procedure results in a subsampling of cases, comparable with the matching procedure described for the biomedical example, but for a single

---

[4]A virtue of matching, as developed in this tradition, is cost-effectiveness for prospective studies. If the goal of a study is to measure the evolution of a causal effect over time by measuring symptoms at several points in the future, then discarding nontreated cases unlike any treated cases can cut expenses without substantially affecting the quality of causal inferences that a study can yield. See Stuart and Ialongo (2010).

dimension that is a function of the variables in X. In essence, the matching procedure throws away information from the joint distribution of X and Y that is unrelated to variation in the treatment variable D until the remaining distribution of X is equivalent for both the treatment and control cases. When this equivalence is achieved, the data are said to be balanced with respect to X.[5] Under specific assumptions, the remaining differences in the observed outcome between the treatment and matched control cases can then be regarded as attributable solely to the effect of the treatment.[6]

At most points in the remainder of this chapter, we will adopt this second scenario because research designs in which data are drawn from random-sample surveys are much more common in the social sciences.[7] Thus, we will assume that the data in hand were generated by a relatively large random-sample survey (in some cases an infinite sample to entirely remove sampling error from consideration), in which the proportion and pattern of individuals who are exposed to the cause are fixed in the population by whatever process generates causal exposure.

5.2 Matching as Conditioning via Stratification

In this section we introduce matching estimators in idealized research conditions, drawing connections with the broad perspective on conditioning introduced in Chapter 4. Thereafter, we proceed to a discussion of matching in more realistic scenarios, which is where we explain the developments of matching techniques that have been achieved in the past three decades.

5.2.1 Estimating Causal Effects by Stratification

Suppose that those who take the treatment and those who do not are very much unlike each other, and yet the ways in which they differ are captured exhaustively by a set of observed treatment assignment/selection variables S. For the language we will adopt in this chapter, knowledge and observation of S allow for a "perfect stratification"of the data. By "perfect,"we mean precisely that individuals within groups defined by values on the variables in S are entirely indistinguishable from each other in all ways except for (1) observed treatment status and (2) differences in the potential outcomes that are independent of treatment status. Under such a perfect stratification of the data, even though we would not be able to assert Assumptions 1 and 2 in Equations

---

[5]As we will discuss later, in many applications balance can be hard to achieve without some subsampling from among the treatment cases. In this case, the causal parameter that is identified is narrower even than the ATT (and is usually a type of marginal treatment effect pinned to the common support of treatment and control cases).

[6]A third motivation, which is due to Ho et al. (2007; see also Iacus and King 2012), has now emerged. Matching can be used as a data preprocessor that prepares a dataset for further causal modeling with a parametric model. We discuss this perspective along with others when we introduce particular matching techniques that are currently in use in the applied literature, including the "coarsened exact matching" of Iacus, King, and Porro (2011, 2012a).

[7]See our earlier discussion in Section 1.4 of this random-sample setup.

(2.15) and (2.16), we would be able to assert conditional variants of those assumptions:

$$\text{Assumption 1-S: } E[Y^1|D=1,S] = E[Y^1|D=0,S], \tag{5.1}$$

$$\text{Assumption 2-S: } E[Y^0|D=1,S] = E[Y^0|D=0,S]. \tag{5.2}$$

These assumptions would suffice to enable consistent and unbiased estimation of the average treatment effect (ATE) because the treatment can be considered randomly assigned within groups defined by values on the variables in S.

When in this situation, researchers often assert that the naive estimator in Equation (2.9) is subject to bias (either generic omitted-variable bias or individually generated selection bias). But, because a perfect stratification of the data can be formulated, treatment assignment is ignorable – see the earlier discussion of Equation (4.3) – or treatment selection is on the observable variables only – see the earlier discussion of Equation (4.7). This is a bit imprecise, however, because Assumptions 1-S and 2-S are implied by ignorability and selection on the observables (assuming S is observed). For ignorability and selection on the observables to hold more generally, the full distributions of Y^1 and Y^0 (and any functions of them) must be independent of D conditional on S; see the discussion of Equation (4.4). Thus Assumptions 1-S and 2-S are weaker than assumptions of ignorability and selection on the observables, but they are sufficient to identify the three average causal effects of primary interest.

Recall the directed graph in Figure 4.8(b), where S lies on the only back-door path from D to Y. As discussed there, conditioning on S allows for consistent and unbiased estimation of the unconditional ATE, as well as the average treatment effect for the treated (ATT) and the average treatment effect for the controls (ATC). Although we gave a conceptual discussion in Chapter 4 of why conditioning works in this scenario, we will now explain more specifically with a demonstration. First note why everything works out so cleanly when a set of perfect stratifying variables is available. If Assumption 1-S is valid, then

$$\begin{aligned} E[\delta|D=0,S] &= E[Y^1 - Y^0|D=0,S] \\ &= E[Y^1|D=0,S] - E[Y^0|D=0,S] \\ &= E[Y^1|D=1,S] - E[Y^0|D=0,S] \\ &= E[Y|D=1,S] - E[Y|D=0,S]. \end{aligned} \tag{5.3}$$

If Assumption 2-S is valid, then

$$\begin{aligned} E[\delta|D=1,S] &= E[Y^1 - Y^0|D=1,S] \\ &= E[Y^1|D=1,S] - E[Y^0|D=1,S] \\ &= E[Y^1|D=1,S] - E[Y^0|D=0,S] \\ &= E[Y|D=1,S] - E[Y|D=0,S]. \end{aligned} \tag{5.4}$$

The last line of Equation (5.3) is identical to the last line of Equation (5.4), and neither line includes counterfactual conditional expectations. Accordingly, one can consistently estimate the difference in the last line of Equation (5.3) and the last line of Equation (5.4) for each value of S. To then form consistent estimates of alternative average treatment effects, one simply averages the stratified estimates over the distribution of S, as we show in the following demonstration.

Matching Demonstration 1

Consider a completely hypothetical example in which Assumptions 1 and 2 in Equations (2.15) and (2.16) cannot be asserted because positive selection ensures that those who are observed in the treatment group are more likely to benefit from the treatment than those who are not. But assume that a three-category perfect stratifying variable S is available that allows one to assert Assumptions 1-S and 2-S in Equations (5.1) and (5.2). Moreover, suppose for simplicity of exposition that our sample is infinite so that sampling error is zero. In this case, we can assume that the sample moments in our data equal the population moments (i.e., $E_N[y_i|d_i = 1] = E[Y|D = 1]$ and so on).

If it is helpful, think of Y as a measure of an individual's economic success at age 40, D as an indicator of receipt of a college degree, and S as a mixed family-background and preparedness-for-college variable that completely accounts for the pattern of self-selection into college that is relevant for lifetime economic success. Note, however, that no one has ever discovered such a variable as S for this particular causal effect.

Suppose that, for our infinite sample, the sample mean of the outcome for those observed in the treatment group is 10.2, whereas the sample mean of the outcome for those observed in the control group is 4.4. In other words, we have data that yield $E_N[y_i|d_i = 1] = 10.2$ and $E_N[y_i|d_i = 0] = 4.4$, and for which the naive estimator would yield a value of 5.8 (i.e., $10.2 - 4.4$).

Consider, now, an underlying set of potential outcome variables and treatment assignment patterns that could give rise to a naive estimate of 5.8. Table 5.1 presents the joint probability distribution of the treatment variable D and the stratifying variable S in its first panel as well as expectations, conditional on S, of the potential outcomes under the treatment and control states. The joint distribution in the first panel shows that individuals with S equal to 1 are more likely to be observed in the control group, individuals with S equal to 2 are equally likely to be observed in the control group and the treatment group, and individuals with S equal to 3 are more likely to be observed in the treatment group.

As shown in the second panel of Table 5.1, the average potential outcomes conditional on S and D imply that the average causal effect is 2 for those with S equal to 1 or S equal to 2, but 4 for those with S equal to 3 (see the last column). Moreover, as shown in the last row of the table, where the potential outcomes are averaged over the within-D distribution of S, $E[Y|D = 0] = 4.4$ and $E[Y|D = 1] = 10.2$, matching the initial setup of the example based on a naive estimate of 5.8 from an infinite sample.

Table 5.2 shows what can be calculated from the data, assuming that S offers a perfect stratification of the data. The first panel presents the sample expectations of the observed outcome variable conditional on D and S. The second panel of Table 5.2 presents corresponding sample estimates of the conditional probabilities of S given D.

The existence of a perfect stratification (and the supposed availability of data from an infinite sample) ensures that the estimated conditional expectations in the first panel of Table 5.2 equal the population-level conditional expectations of the second panel of Table 5.1. When stratifying by S, the average observed outcome for those in the control/treatment group with a particular value of S is equal to the average potential outcome under the control/treatment state for those with a particular value

Table 5.1 The Joint Probability Distribution and Conditional Population Expectations for Matching Demonstration 1

	Joint probability distribution of S and D		
	$D=0$	$D=1$	
$S=1$	$\Pr[S=1, D=0] = .36$	$\Pr[S=1, D=1] = .08$	$\Pr[S=1] = .44$
$S=2$	$\Pr[S=2, D=0] = .12$	$\Pr[S=2, D=1] = .12$	$\Pr[S=2] = .24$
$S=3$	$\Pr[S=3, D=0] = .12$	$\Pr[S=3, D=1] = .2$	$\Pr[S=3] = .32$
	$\Pr[D=0] = .6$	$\Pr[D=1] = .4$	

	Potential outcomes		
	Under the control state	Under the treatment state	
$S=1$	$E[Y^0|S=1] = 2$	$E[Y^1|S=1] = 4$	$E[Y^1-Y^0|S=1] = 2$
$S=2$	$E[Y^0|S=2] = 6$	$E[Y^1|S=2] = 8$	$E[Y^1-Y^0|S=2] = 2$
$S=3$	$E[Y^0|S=3] = 10$	$E[Y^1|S=3] = 14$	$E[Y^1-Y^0|S=3] = 4$

$$E[Y^0|D=0] \qquad\qquad E[Y^1|D=1]$$
$$= \tfrac{.36}{.6}(2) + \tfrac{.12}{.6}(6) \qquad = \tfrac{.08}{.4}(4) + \tfrac{.12}{.4}(8)$$
$$+ \tfrac{.12}{.6}(10) \qquad\qquad + \tfrac{.2}{.4}(14)$$
$$= 4.4 \qquad\qquad\qquad = 10.2$$

Table 5.2 Estimated Conditional Expectations and Probabilities for Matching Demonstration 1

	Estimated mean observed outcome conditional on s_i and d_i	
	Control group	Treatment group
$s_i=1$	$E_N[y_i|s_i=1, d_i=0] = 2$	$E_N[y_i|s_i=1, d_i=1] = 4$
$s_i=2$	$E_N[y_i|s_i=2, d_i=0] = 6$	$E_N[y_i|s_i=2, d_i=1] = 8$
$s_i=3$	$E_N[y_i|s_i=3, d_i=0] = 10$	$E_N[y_i|s_i=3, d_i=1] = 14$

	Estimated probability of S conditional on D	
$s_i=1$	$\Pr_N[s_i=1|d_i=0] = .6$	$\Pr_N[s_i=1|d_i=1] = .2$
$s_i=2$	$\Pr_N[s_i=2|d_i=0] = .2$	$\Pr_N[s_i=2|d_i=1] = .3$
$s_i=3$	$\Pr_N[s_i=3|d_i=0] = .2$	$\Pr_N[s_i=3|d_i=1] = .5$

of S. Conversely, if S were not a perfect stratifying variable, then the sample means in the first panel of Table 5.2 would not equal the expectations of the potential outcomes in the second panel of Table 5.1. The sample means would be based on heterogeneous groups of individuals who differ systematically within the strata defined by S in ways that are related with individual-level treatment effects.

If S offers a perfect stratification of the data, then one can estimate from the numbers in the cells of the two panels of Table 5.2 both the ATT as

$$(4-2)(.2) + (8-6)(.3) + (14-10)(.5) = 3$$

and the ATC as

$$(4-2)(.6) + (8-6)(.2) + (14-10)(.2) = 2.4.$$

Finally, if one calculates the appropriate marginal distributions of S and D (using sample analogs for the marginal distribution from the first panel of Table 5.1), one can estimate the unconditional ATE either as

$$(4-2)(.44) + (8-6)(.24) + (14-10)(.32) = 2.64$$

or as

$$3(.4) + 2.4(.6) = 2.64.$$

Thus, for this hypothetical example, the naive estimator would be inconsistent and (asymptotically) upwardly biased for the ATT, ATC, and the ATE. But, by appropriately weighting stratified estimates of the treatment effect, one can obtain consistent and unbiased estimates of all three of these average treatment effects.

In general, if a stratifying variable S completely accounts for all systematic differences between those who take the treatment and those who do not, then conditional-on-S estimators yield consistent and unbiased estimates of the average treatment effect conditional on a particular value s of S:

$$\{E_N[y_i|d_i = 1, s_i = s] - E_N[y_i|d_i = 0, s_i = s]\} \tag{5.5}$$
$$\xrightarrow{p} E[Y^1 - Y^0|S = s] = E[\delta|S = s].$$

Weighted sums of these stratified estimates can then be taken, such as for the unconditional ATE:

$$\sum_s \{E_N[y_i|d_i = 1, s_i = s] - E_N[y_i|d_i = 0, s_i = s]\} \Pr{}_N[s_i = s] \xrightarrow{p} E[\delta]. \tag{5.6}$$

Substituting into this last expression the distributions of S conditional on the two possible values of D (i.e., $\Pr{}_N[s_i = s|d_i = 1]$ or $\Pr{}_N[s_i = s|d_i = 0]$), one can obtain consistent and unbiased estimates of the ATT and ATC.

The key to using stratification to solve the causal inference problem for all three causal effects of primary interest is twofold: finding the stratifying variable and then obtaining the marginal probability distribution $\Pr[S]$ as well as the conditional probability distribution $\Pr[S|D]$. Once these steps are accomplished, obtaining consistent and unbiased estimates of the within-strata treatment effects is straightforward. Thereafter, estimates of other average treatment effects can be formed by taking appropriate weighted averages of the stratified estimates.

This simple example shows all of the basic principles of matching estimators that we will present in greater detail in the remainder of this chapter. Treatment and control subjects are matched together in the sense that they are grouped together into strata. Then, an average difference between the outcomes of treatment and control subjects is estimated, based on a weighting of the strata (and thus the individuals within them) by a common distribution.

5.2.2   Overlap Conditions for Estimation of the ATE

Suppose again that a perfect stratification of the data is available, such that within values of a stratifying variable S individuals are indistinguishable from each other as defined in the last section. But now suppose that there is a stratum of the population (and hence of the observed data) in which no member of the stratum ever receives the treatment. In this case, the ATE is ill-defined, and the analyst will only be able to generate a consistent and unbiased estimate of the ATT, as we show in the following demonstration.[8]

Matching Demonstration 2

For the example depicted in Tables 5.3 and 5.4, S again offers a perfect stratification of the data. The setup of these two tables is exactly equivalent to that of the prior Tables 5.1 and 5.2 for Matching Demonstration 1. We again assume that the data are generated by a random sample of a well-defined population, and for simplicity of exposition that the sample is infinite. The major difference is evident in the joint distribution of S and D presented in the first panel of Table 5.3. As shown in the first cell of the second column, no individual in the population with S equal to 1 would ever be observed in the treatment group of a dataset of any size because the joint probability of S equal to 1 and D equal to 1 is zero. Corresponding to this structural zero in the joint distribution of S and D, the second panel of Table 5.3 shows that there is no corresponding conditional expectation of the potential outcome under the treatment state for those with $S = 1$. And, thus, as shown in the last column of the second panel, no average causal effect is presented for individuals with $S = 1$ because this particular average causal effect is ill-defined.[9]

Adopting the same framing as for the college-degree example used in Matching Demonstration 1, this hypothetical example asserts that there is a subpopulation of individuals from such disadvantaged backgrounds that no individuals with $S = 1$ have ever graduated from college. For this group of individuals, we assume in this example that there is simply no justification for using the wages of those from more advantaged social backgrounds to extrapolate to the what-if wages of the most disadvantaged individuals if they had somehow overcome the obstacles that prevented them from obtaining college degrees.

---

[8]In this section, we focus on the lack of overlap that may exist in a population (or superpopulation). For now, we ignore the lack of overlap that can emerge in observed data solely because of the finite size of a dataset. We turn to these issues in the next section, where we discuss solutions to sparseness.

[9]By "ill-defined," we mean the following. No information about $E[Y|S = 1, D = 1]$ or $E[Y^1|S = 1, D = 1]$ exists in the population (and, as a result, the data will never give us a value for $E_N[y_i|s_i = 1, d_i = 1]$ because no individuals in the data will ever have both $s_i = 1$ and $d_i = 1$).

Table 5.3 The Joint Probability Distribution and Conditional Population Expectations for Matching Demonstration 2

	Joint probability distribution of S and D		
	$D=0$	$D=1$	
$S=1$	$\Pr[S=1, D=0] = .4$	$\Pr[S=1, D=1] = 0$	$\Pr[S=1] = .4$
$S=2$	$\Pr[S=2, D=0] = .1$	$\Pr[S=2, D=1] = .13$	$\Pr[S=2] = .23$
$S=3$	$\Pr[S=3, D=0] = .1$	$\Pr[S=3, D=1] = .27$	$\Pr[S=3] = .37$
	$\Pr[D=0] = .6$	$\Pr[D=1] = .4$	

Potential outcomes

	Under the control state	Under the treatment state	
$S=1$	$E[Y^0|S=1] = 2$		
$S=2$	$E[Y^0|S=2] = 6$	$E[Y^1|S=2] = 8$	$E[Y^1-Y^0|S=2] = 2$
$S=3$	$E[Y^0|S=3] = 10$	$E[Y^1|S=3] = 14$	$E[Y^1-Y^0|S=3] = 4$

$$E[Y^0|D=0] \qquad E[Y^1|D=1]$$
$$= \tfrac{.4}{.6}(2) + \tfrac{.1}{.6}(6) \qquad = \tfrac{.13}{.4}(8) + \tfrac{.27}{.4}(14)$$
$$+ \tfrac{.1}{.6}(10)$$
$$= 4 \qquad\qquad = 12.05$$

Table 5.4 Estimated Conditional Expectations and Probabilities for Matching Demonstration 2

	Estimated mean observed outcome conditional on s_i and d_i	
	Control group	Treatment group
$s_i=1$	$E_N[y_i|s_i=1, d_i=0] = 2$	
$s_i=2$	$E_N[y_i|s_i=2, d_i=0] = 6$	$E_N[y_i|s_i=2, d_i=1] = 8$
$s_i=3$	$E_N[y_i|s_i=3, d_i=0] = 10$	$E_N[y_i|s_i=3, d_i=1] = 14$

	Estimated probability of S conditional on D	
$s_i=1$	$\Pr_N[s_i=1|d_i=0] = .667$	$\Pr_N[s_i=1|d_i=1] = 0$
$s_i=2$	$\Pr_N[s_i=2|d_i=0] = .167$	$\Pr_N[s_i=2|d_i=1] = .325$
$s_i=3$	$\Pr_N[s_i=3|d_i=0] = .167$	$\Pr_N[s_i=3|d_i=1] = .675$

Table 5.4 shows what can be estimated consistently for this example. Even though S offers a perfect stratification of the data, the fact that $\Pr[S=1, D=1] = 0$ prevents the analyst from using the data for the stratum with $S=1$ to estimate a stratum-level causal effect. No value exists for $E_N[y_i|s_i=1, d_i=1]$.

Fortunately, the analyst can consistently estimate the average effect of the treatment separately for the two strata with $S=2$ and $S=3$. And, because all members of the treatment group belong to these two strata, the analyst can therefore consistently estimate the ATT as

$$(8 - 6)(.325) + (14 - 10)(.675) = 3.35.$$

Still, no consistent and unbiased estimates of the ATC or the ATE are available.[10]

Are examples such as this one ever found in practice? For an example that is more realistic than the causal effect of a college degree on economic success, consider the evaluation of a generic program in which there is an eligibility rule. The benefits of enrolling in the program for those who are ineligible cannot be estimated from the data, even though, if some of those individuals were enrolled in the program, they would likely be affected by the treatment in some way. Developing such estimates would require going well beyond the data, introducing assumptions that allow for extrapolation off of the joint distribution of S and D.

More generally, even in best-case data availability scenarios where the sample size is infinite, it may not be possible to consistently estimate all average causal effects of theoretical or practical interest because the distribution of the treatment across all segments of the population is incomplete. However, at other times, the data may appear to suggest that no causal inference is possible for some group of individuals even though the problem is simply a small sample size. There is a clever solution to sparseness of data for this latter type of situation, which we discuss in the next section.

5.3   Matching as Weighting

As shown in the last section, if all of the variables in S have been observed such that a perfect stratification of the data would be possible with an infinitely large random sample from the population, then a consistent and unbiased estimator is available in theory for each of the average causal effects of interest defined in Equations (2.3), (2.7), and (2.8) as the ATE, ATT, and ATC, respectively. Unfortunately, in many (if not most) datasets of finite size, it may not be possible to use the simple estimation methods of the last section to generate consistent and unbiased estimates. Treatment and control cases may be missing at random within some of the strata defined by S, such that some strata contain only treatment or only control cases. In this situation, some within-stratum causal effect estimates cannot be calculated with the available data. We now introduce a set of weighting estimators that rely on estimated propensity scores to solve the sparseness problems that afflict samples of finite size.

---

[10]The naive estimate can be calculated for this example, and it would equal 8.05 for an infinite sample because $[8(.325) + 14(.675)] - [2(.667) + 6(.167) + 10(.167)]$ is equal to 8.05. See the last row of Table 5.3 for the population analogs to the two pieces of the naive estimator. This means that, without determining the lack of overlap by stratifying the data, an incautious analyst might offer the naive estimate and then discuss its relationship to the ATE, which is itself a target parameter that is ill-defined.

5.3.1 The Utility of Known Propensity Scores

An estimated propensity score is the estimated probability of taking the treatment as a function of variables that predict treatment assignment. Before the attraction of estimated propensity scores is explained, there is value in understanding why known propensity scores would be useful in an idealized context such as a perfect stratification of the data. (See also our prior discussion of propensity scores in Section 4.3.1.)

Within a perfect stratification, the true propensity score is nothing other than the within-stratum probability of receiving the treatment, or $\Pr[D = 1|S]$. For the hypothetical example in Matching Demonstration 1, the propensity scores are

$$\Pr[D = 1|S = 1] = \frac{.08}{.44} = .182,$$

$$\Pr[D = 1|S = 2] = \frac{.12}{.24} = .5,$$

$$\Pr[D = 1|S = 3] = \frac{.2}{.32} = .625.$$

Why is the propensity score useful? As shown earlier for Matching Demonstration 1, if a perfect stratification of the data is available, then the final ingredient for calculating estimates of the ATT and ATC is the conditional distribution $\Pr[S|D]$. One can recover $\Pr[S|D]$ from the propensity scores by applying Bayes' rule using the marginal distributions of D and S. For example, for the first stratum,

$$\Pr[S = 1|D = 1] = \frac{\Pr[D = 1|S = 1]\Pr[S = 1]}{\Pr[D = 1]} = \frac{(.182)(.44)}{(.4)} = .2.$$

Thus, the true propensity scores encode all of the necessary information about the joint dependence of S and D that is needed to estimate and then combine conditional-on-S treatment effect estimates into estimates of the ATT and the ATC. Known propensity scores are thus useful for unpacking the inherent heterogeneity of causal effects and then averaging over such heterogeneity to calculate average treatment effects.

Of course, known propensity scores are almost never available to researchers working with observational rather than experimental data. Thus, the literature on matching more often recognizes the utility of propensity scores for addressing an entirely different concern: solving comparison problems created by the sparseness of data in any finite sample. These methods rely on estimated propensity scores, as we discuss next.

5.3.2 Weighting with Estimated Propensity Scores to Address Sparseness

Suppose again that a perfect stratification of the data exists and is known. In particular, Assumptions 1-S and 2-S in Equations (5.1) and (5.2) are valid, and S is observed. But, suppose now that (1) there are multiple variables in S, (2) some of the variables in S take on many values, and (3) the true propensity score is greater than 0 and less than 1 for every stratum defined by S. In this scenario, there may be many strata in the available data from a finite sample in which no treatment and/or no control cases are observed, even though the true propensity score is between 0 and 1 for every stratum in the population (i.e., every population-level stratum includes individuals in both the treatment and control states).

Can average treatment effects be consistently estimated in this scenario? Rosenbaum and Rubin (1983a) answer this question affirmatively. The essential points of their argument are the following (see the original article for a formal proof): First, the sparseness that results from the finiteness of a sample is random, conditional on the joint distribution of S and D. As a result, within each stratum for a perfect stratification of the data, the probability of having a zero cell in the treatment or the control state is solely a function of the propensity score. Because such sparseness is conditionally random, strata with identical propensity scores (i.e., different combinations of values for the variables in S but the same within-stratum probability of treatment) can be combined into more coarse strata. Over repeated samples from the same population, zero cells would emerge with equal frequency across all strata within these coarse propensity-score-defined strata.

Because sparseness emerges in this predictable fashion, stratifying on the propensity score itself (rather than more finely on all values of the variables in S) solves the sparseness problem because the propensity score can be treated as a single stratifying variable. In fact, as we show in the next hypothetical example, one can obtain consistent estimates of treatment effects by weighting the individual-level data by an appropriately chosen function of the estimated propensity score, without ever having to compute any stratum-specific causal effect estimates.

But how does one obtain the propensity scores for data from a random sample of the population of interest? Rosenbaum and Rubin (1983a) argue that, if one has observed the variables in S, then the propensity score can be estimated using standard methods, such as logit modeling. That is, one can estimate the propensity score, assuming a logistic distribution,

$$\Pr[D=1|S] = \frac{\exp(S\phi)}{1+\exp(S\phi)}, \tag{5.7}$$

and invoke maximum likelihood to estimate a vector of coefficients $\hat{\phi}$. One can then stratify on the index of the estimated propensity score, $e(s_i) = s_i\hat{\phi}$, or appropriately weight the data, and all of the results established for known propensity scores then obtain.[11] Consider the following hypothetical example, in which weighting is performed only with respect to the estimated propensity score, resulting in consistent and unbiased estimates of average treatment effects even though sparseness problems are severe.

---

[11] As Rosenbaum (1987) later clarified (see also Rubin and Thomas 1996), the estimated propensity scores do a better job of balancing the observed variables in S than the true propensity scores would in any actual application, because the estimated propensity scores correct for the chance imbalances in S that characterize any finite sample. This insight has led to a growing literature that seeks to balance the observed variables in S by various computationally intensive but powerful nonparametric techniques (e.g., Diamond and Sekhon 2013; Lee, Lessler, and Stuart 2009; McCaffrey, Ridgeway, and Morral 2004). We discuss this literature later, and for now we use only parametric models for the estimation of propensity scores, as they dominate the foundational literature on matching.

Matching Demonstration 3

Consider the following Monte Carlo simulation, which is an expanded version of the hypothetical example in Matching Demonstration 1 in two respects. First, for this example, there are two stratifying variables, A and B, each of which has 100 separate values. As for Matching Demonstration 1, these two variables represent a perfect stratification of the data and, as such, represent all of the variables in the set of perfect stratifying variables, defined earlier as S. Second, to demonstrate the properties of alternative estimators, this example utilizes 50,000 samples of data, each of which is a random realization of the same set of definitions for the constructed variables and the stipulated joint distributions between them.

Generation of the 50,000 Datasets. For the simulation, we gave the variables A and B values of .01, .02, .03, and upward to 1. We then cross-classified the two variables to form a 100×100 grid and stipulated a propensity score, as displayed in Figure 5.1, that is a positive, nonlinear function in both A and B.[12] We then populated the resulting 20,000 constructed cells (100×100 for the $A \times B$ grid multiplied by the two values of D) using a Poisson random-number generator with the relevant propensity score as the Poisson parameter for the 10,000 cells for the treatment group and one minus the propensity score as the Poisson parameter for the 10,000 cells for the control group. This sampling scheme generates (on average across simulated datasets) the equivalent of 10,000 sample members, assigned to the treatment instead of the control as a function of the probabilities plotted in Figure 5.1.

Across the 50,000 simulated datasets, on average 7,728 of the 10,000 possible combinations of values for both A and B had no individuals assigned to the treatment, and 4,813 had no individuals assigned to the control. No matter the actual realized pattern of sparseness for each simulated dataset, all of the 50,000 datasets are afflicted, such that a perfect stratification on all values for the variables A and B would result in many strata within each dataset for which only treatment or control cases are present.

To define treatment effects for each dataset, two potential outcomes were defined as linear functions of individual values for A and B:

$$y_i^0 = 100 + 3a_i + 2b_i + v_i^0, \tag{5.8}$$
$$y_i^1 = 102 + 6a_i + 4b_i + v_i^1, \tag{5.9}$$

where both v_i^0 and v_i^1 are independent random draws from a normal distribution with expectation 0 and standard deviation of 5. Then, as in Equation (2.2), individuals from the treatment group were given an observed y_i equal to their simulated y_i^1, and individuals from the control group were given an observed y_i equal to their simulated y_i^0.

With this setup, the simulation makes available 50,000 datasets for which the individual treatment effects can be calculated exactly (because true values of y_i^1 and

---

[12]The parameterization of the propensity score is a constrained tensor product spline regression for the index function of a logit. See Ruppert, Wand, and Carroll (2003) for examples of such parameterizations. Here, $S\phi$ in Equation (5.7) is equal to $-2 + 3(A) - 3(A - .1) + 2(A - .3) - 2(A - .5) + 4(A - .7) - 4(A - .9) + 1(B) - 1(B - .1) + 2(B - .7) - 2(B - .9) + 3(A - .5)(B - .5) - 3(A - .7)(B - .7)$.

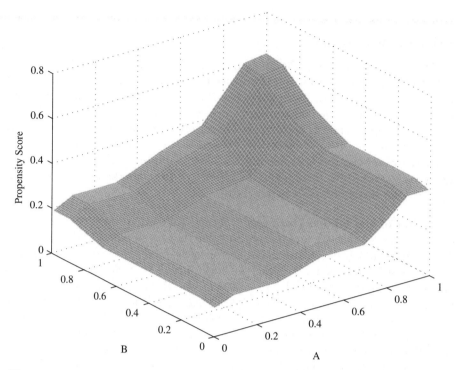

Figure 5.1 The propensity score specification for Matching Demonstration 3.

y_i^0 are available for all simulated individuals). As a result, the true ATE, ATT, and ATC are known for each simulated dataset, and these known average effects can serve as baselines against which alternative estimators that use data only on y_i, d_i, a_i, and b_i can be compared.

The first row of Table 5.5 presents true Monte Carlo means and standard deviations of the three average treatments effects, calculated across the 50,000 simulated datasets. The mean of the true ATE across all datasets is 4.525, whereas the means of the true ATT and ATC are 4.892 and 4.395, respectively. Similar to the hypothetical example in Matching Demonstration 1, this example represents a form of positive selection, in which those who are most likely to be in the treatment group are also those most likely to benefit from the treatment. Accordingly, the ATT is larger than the ATC.

Methods for Treatment Effect Estimation. The last three rows of Table 5.5 present results for three propensity-score-based weighting estimators. For the estimates in the second row, it is (incorrectly) assumed that the propensity score can be estimated consistently with a logit model with linear terms for A and B (i.e., assuming that, for Equation (5.7), a logit with $S\phi$ specified as $\alpha + \phi_A A + \phi_B B$ will yield consistent estimates of the propensity score surface plotted in Figure 5.1). After the logit model was estimated for each of the 50,000 datasets with the wrong specification, the

Table 5.5 Monte Carlo Means and Standard
Deviations of True and Estimated Treatment Effects
for Matching Demonstration 3

	ATE	ATT	ATC
True treatment effects	4.525	4.892	4.395
	(.071)	(.139)	(.083)
Propensity-score-based weighting estimators:			
Misspecified propensity score estimates	4.456	4.913	4.293
	(.122)	(.119)	(.128)
Perfectly specified propensity score estimates	4.526	4.892	4.396
	(.120)	(.127)	(.125)
True propensity scores	4.527	4.892	4.396
	(.127)	(.127)	(.132)

estimated propensity score for each individual was then calculated,

$$\hat{p}_i = \frac{\exp(\hat{\alpha} + \hat{\phi}_A a_i + \hat{\phi}_B b_i)}{1 + \exp(\hat{\alpha} + \hat{\phi}_A a_i + \hat{\phi}_B b_i)}, \tag{5.10}$$

along with the estimated odds of the propensity of being assigned to the treatment

$$\hat{r}_i = \frac{\hat{p}_i}{1 - \hat{p}_i}, \tag{5.11}$$

where $\hat{p}_i$ is as constructed in Equation (5.10).

To estimate the ATT, we then implemented a weighting estimator by calculating the average outcome for the treated and subtracting from this average outcome a counterfactual average outcome using weighted data on those from the control group:

$$\hat{\delta}_{\text{ATT,weight}} \equiv \left(\frac{1}{n^1} \sum_{i:d_i=1} y_i \right) - \left(\frac{\sum_{i:d_i=0} \hat{r}_i y_i}{\sum_{i:d_i=0} \hat{r}_i} \right), \tag{5.12}$$

where n^1 is the number of individuals in the treatment group and $\hat{r}_i$ is the estimated odds for each individual i of being in the treatment group instead of in the control group, as constructed in Equations (5.10) and (5.11). The weighting operation in the second term gives more weight to control group individuals equivalent to those in the treatment group; see Rosenbaum (1987, 2002); see also Imbens (2004) and Hainmueller (2012).[13] To estimate the ATC, we then implemented a weighting estimator that is

---

[13] As we will describe in Chapter 7 when discussing the connections between matching and regression, the weighting estimator in Equation (5.12) can be written as a weighted regression estimator.

the mirror image of the one in Equation (5.12):

$$\hat{\delta}_{\text{ATC,weight}} \equiv \left(\frac{\sum\limits_{i:d_i=1} y_i/\hat{r}_i}{\sum\limits_{i:d_i=1} 1/\hat{r}_i} \right) - \left(\frac{1}{n^0} \sum\limits_{i:d_i=0} y_i \right), \qquad (5.13)$$

where n^0 is the number of individuals in the control group. Finally, the corresponding estimator of the unconditional ATE is

$$\hat{\delta}_{\text{ATE,weight}} \equiv \left(\frac{1}{n} \sum_i d_i \right) \left(\hat{\delta}_{\text{ATT,weight}} \right) + \left[\left(1 - \frac{1}{n} \sum_i d_i \right) \right] \left(\hat{\delta}_{\text{ATC,weight}} \right), \qquad (5.14)$$

where $\hat{\delta}_{\text{ATT,weight}}$ and $\hat{\delta}_{\text{ATC,weight}}$ are as defined in Equations (5.12) and (5.13), respectively.

The same basic weighting scheme is implemented for the third row of Table 5.5, but the estimated propensity scores utilized to define the estimated odds of treatment, $\hat{r}_i$, are instead based on results from a flawlessly estimated propensity score equation (i.e., one that uses the exact same specification that was fed to the random-number generator that assigned individuals to the treatment; see footnote on page 153 for the specification). Finally, for the last row of Table 5.5, the same weighting scheme is implemented, but, in this case, the estimated odds of treatment, $\hat{r}_i$, are replaced with the true odds of treatment, r_i, as calculated with reference to the exact function that generated the propensity score for Figure 5.1.

Monte Carlo Results. On average across all $50,000$ simulated datasets, the naive estimator yields a value of 5.388, which is substantially larger than all three of the average values for the true ATE, ATT, and ATC presented in the first row of Table 5.5. The reason is simple. The two variables A and B mutually cause both D and Y (in a structure analogous to Figure 3.5). The two back-door paths, $D \leftarrow A \rightarrow Y$ and $D \leftarrow B \rightarrow Y$, generate noncausal associations between D and Y. These paths must be blocked, and this is the motivation for the weighting estimators.

The second row of the table presents three estimates from the weighting estimators in Equations (5.12), (5.13), and (5.14), using weights based on the misspecified logit described above. These estimates are closer to the true average values presented in the first row (and much closer than the average value of the naive estimate), but the misspecification of the propensity-score-estimating equation appears to generate systematic bias in the estimates, suggesting that they are unlikely to be consistent estimates. The third row of the table presents another three weighting estimates, using a perfect specification of the propensity-score-estimating equation, and now the estimates appear to be consistent and unbiased for the ATE, ATT, and ATC. Finally, the last row presents weighting estimates that utilize the true propensity scores, which we know by construction are consistent and asymptotically unbiased (but, as shown by Rosenbaum 1987, more variable than those based on the flawlessly estimated propensity score; see also Hahn 1998; Hirano, Imbens, and Ridder 2003; Rosenbaum 2002). The last two rows demonstrate the most important claim of the literature: If one can obtain consistent estimates of the true propensity scores, one can solve the problems created by sparseness of data.

This example shows the potential power of propensity-score-based modeling. If treatment assignment can be modeled perfectly, one can solve the sparseness problems that afflict finite datasets. At the same time, this simulation also develops two important qualifications of this potential power. First, this solution only holds in expectation over repeated samples (or in the limit as the sample size increases to infinity). For any single dataset, any resulting point estimate of a treatment effect will differ from the true target parameter to some degree because of sampling variability.

Second, without a perfect specification of the propensity-score-estimating equation, one cannot rest assured that consistent and unbiased estimates can be obtained. Because propensity scores achieve their success by "undoing" the treatment assignment patterns, analogous to weighting a stratified sample so that it is representative of the population, systematically incorrect estimated propensity scores can generate systematically incorrect weighting schemes that yield inconsistent and biased estimates of treatment effects.

There are two common sources of inconsistency and bias that can be considered separately. As discussed at length in Chapter 4, if the conditioning set leaves one or more back-door paths unblocked, then Assumption 1-S and/or Assumption 2-S in Equations (5.1) and (5.2) are/is invalid. We avoided this problem in Matching Demonstration 3 because we used A and B to estimate the propensity scores, and we know by construction that S is defined as the set $\{A, B\}$. Had we mistakenly used only A or only B, then we would not have conditioned fully on S, thereby leaving a back-door path unblocked. We will have much more to say about this source of inconsistency and bias in the remainder of this chapter.

The second source of inconsistency and bias is misspecification of the equation that estimates the propensity scores, and this is especially important to consider for the sort of propensity-score-based weighting estimators utilized for Matching Demonstration 3. For the results reported in the second row of Table 5.5, we included both A and B in the propensity-score estimating equation. But, we did not do so while also choosing a flexible enough parameterization of A and B that would allow the data to generate a sufficiently accurate set of estimated propensity scores (which, in expectation, would match the shape of the surface in Figure 5.1, when plotted in three dimensions). As a result, the estimated effects in the Monte Carlo simulation were systematically biased when the weights based on these estimated propensity scores were used. Only when the correct specification was used to generate the weights were we able to generate unbiased estimates of the ATE, ATT, and ATC.

These possible weaknesses aside, one concluding question should be answered: In what sense are the individual-level weighting estimators of the hypothetical example in Matching Demonstration 3 equivalent to matching estimators? For the hypothetical examples in Matching Demonstrations 1 and 2, we explained how stratification estimators have a straightforward connection to matching. The strata that are formed represent matched sets, and a weighting procedure is then used to average stratified treatment effect estimates in order to obtain the average treatment effects of interest. The propensity-score weighting estimators presented in this section have a less straightforward connection. Here, the data are, in effect, stratified coarsely by the estimation of the propensity score, and then the weighting is performed directly across individuals instead of across the strata. This individual-level weighting is made necessary by sparseness, since some of the fine strata for which propensity scores are estimated necessarily contain only treatment or control cases, thereby preventing the direct calculation of stratified treatment effect estimates.

5.4   Matching as a Data Analysis Algorithm

Algorithmic matching estimators differ primarily in (1) the number of matched cases designated for each to-be-matched target case and (2) how multiple matched cases are weighted if more than one is utilized for each target case. In this section, we describe the four main types of matching estimators as well as recent extensions to them.

Heckman, Ichimura, and Todd (1997, 1998) and Smith and Todd (2005) outline a general framework for representing alternative matching estimators, and we follow their lead. With our variant of their notation, all matching estimators of the ATT can be expressed as some variation of

$$\hat{\delta}_{\text{ATT,match}} = \frac{1}{n^1} \sum_i \left[(y_i | d_i = 1) - \sum_j \omega_{i,j}(y_j | d_j = 0) \right], \qquad (5.15)$$

where n^1 is the number of treatment cases, i is the index over treatment cases, j is the index over control cases, and $\omega_{i,j}$ represents a set of scaled weights that measure the distance between each control case and the target treatment case. In Equation (5.15), the weights are entirely unspecified. For the ATC, an opposite expression is available, with alternative weights $\omega_{j,i}$ instead attached to the treatment cases

$$\hat{\delta}_{\text{ATC,match}} = \frac{1}{n^0} \sum_j \left[(y_j | d_j = 0) - \sum_i \omega_{j,i}(y_i | d_i = 1) \right], \qquad (5.16)$$

and where n^0 is the number of control cases.

Alternative matching estimators can be represented as different procedures for deriving the weights represented by $\omega_{i,j}$ and $\omega_{j,i}$ in these two expressions. As we will describe next, the weights can take on many values, indeed as many as $n^1 \times n^0$ different values, because alternative weights can be used when constructing the counterfactual value for each target case. The difference in the propensity score between the target case and each potential matched case is the most common distance measure used to construct weights. Other measures of distance are available, including the estimated odds of the propensity score, the index of an estimated logit, probit, or other parametric binary outcome model, and the Mahalanobis metric.[14]

Before describing the four main types of matching algorithms, and their extensions, we note three important points. First, for simplicity of presentation, in the remainder of this section we will focus on matching estimators of the ATT. Each of the following matching algorithms could be described in reverse, explaining how treatment cases can be matched to control cases in order to construct an estimate of the ATC, relying on Equation (5.16) rather than Equation (5.15). We mention this, in part, because it

---

[14]The Mahalanobis metric is $(S_i - S_j)'\Sigma^{-1}(S_i - S_j)$, where Σ is the covariance matrix of the variables in S (usually calculated for the target cases only). There is a long tradition in this literature of using Mahalanobis matching, sometimes in combination with propensity-score matching.

is sometimes implied in the applied literature that the matching techniques that we are about to summarize are useful for estimating only the ATT. This is false. If (1) all variables in S are known and observed, such that a perfect stratification of the data could be formed with a suitably large dataset because both Assumptions 1-S and 2-S in Equations (5.1) and (5.2) are valid and (2) the ranges of all of the variables in S are the same for the treatment group and the control group, then simple variants of the matching estimators that we will present in this section can be formed that are consistent for both the ATT and ATC (and, as a result, for the ATE as a weighted average).

Second, if it is the case that one only wants to estimate the ATT, one does not need to assume full ignorability of treatment assignment or that both Assumptions 1-S and 2-S in Equations (5.1) and (5.2) are valid. Instead, only Assumption 2-S (i.e., $E[Y^0|D = 1, S] = E[Y^0|D = 0, S]$) must hold. In other words, to estimate the ATT, it is sufficient to assume that, conditional on S, the average level of the outcome under the control for those in the treatment is equal, on average, to the average level of the outcome under the control for those in the control group.[15] This assumption is still rather stringent, in that it asserts that those in the control group do not disproportionately gain from exposure to the control state more than would those in the treatment group if they were instead in the control group. But it is surely weaker than having to assert Assumptions 1-S and 2-S together.[16]

Third, the matching algorithms we summarize next are data analysis procedures that can be used more generally when ignorability, or related assumptions, cannot be assumed to hold because some of the variables in the perfect stratification set S are unobserved. Matching routines are still useful, according to Rosenbaum (2002) and others, as techniques that generate provisional estimates that can then be subjected to further analysis in pursuit of warranted causal inferences.

5.4.1 Basic Variants of Matching Algorithms

Exact Matching

Exact matching for the ATT constructs the counterfactual for each treatment case using the control cases with identical values on all of the variables in S. In the notation of Equation (5.15), exact matching uses weights equal to $1/k_i$ for the matched control cases, where k_i is the number of exact matches identified for each target treatment case i. Weights of 0 are given to all unmatched control cases. If only one exact match is chosen at random from among available exact matches, then $\omega_{i,j}$ is set to 1 for the randomly selected match and to 0 for all other control cases.

---

[15]There is an ignorability variant of this mean-independence assumption: D is independent of Y^0 conditional on S. One would always prefer a study design in which this more encompassing form of independence holds. Resulting causal estimates would then hold under transformations of the potential outcomes. This would be particularly helpful if the directly mapped Y – defined as $DY^1 + (1-D)Y^0$ – is not observed but some monotonic transformation of Y is observed (as could perhaps be generated by a feature of measurement).

[16]And this is again weaker than having to assert an assumption of ignorability of treatment assignment.

If S includes more than one or two variables and/or the sample size of the available data is limited, then exact matching is typically infeasible, since many treatment cases will remain unmatched. As a result, exact matching is rarely used on its own and is instead most commonly used in combination with one of the other matching methods described in the following sections. The analyst performs an exact match on one or two of the variables in S and then utilizes another matching algorithm for the remaining variables in S.

Nearest-Neighbor, Caliper, and Radius Matching

Nearest-neighbor matching for the ATT constructs the counterfactual for each treatment case using the control cases that are closest to the treatment case on a unidimensional distance measure constructed from the variables in S, most commonly an estimated propensity score (see Althauser and Rubin 1970; Cochran and Rubin 1973; Rosenbaum and Rubin 1983a, 1985a, 1985b; Rubin 1973a, 1973b, 1976, 1980a, 1980b). As noted in our discussion of Equation (5.15), other distance metrics are sometimes used.

The traditional algorithm randomly orders the treatment cases and then selects for each treatment case the single control case with the smallest distance on the chosen metric. The algorithm can be run with or without replacement. With replacement, a control case is returned to the pool after a match and can be matched later to another treatment case. Without replacement, a control case is taken out of the pool once it is matched.

One weakness of the traditional algorithm when used without replacement is that the estimate will vary depending on the order in which the the treatment cases are passed to the matching algorithm. Moreover, any single estimate may not be based on the minimum aggregate distance between all treatment cases and their matched control cases. A form of optimal matching, which we discuss below, has been developed to remedy these weaknesses in the traditional algorithm by selecting the best overall match of the data from among all of those that are possible.

An analyst can also match on a fixed number of multiple nearest neighbors for each target treatment case, such as 5 nearest neighbors for each target treatment case. The decision of whether to set the algorithm to select more than one nearest neighbor represents a subtle trade-off. Matching more control cases to each treatment case results in lower expected variance of the treatment effect estimate but also tends to increase bias because the probability of making more poor matches increases with the number of matches.

If only one nearest neighbor is selected for each treatment case, as in the traditional algorithm, then $\omega_{i,j}$ is set equal to 1 for the matched control case. If multiple nearest neighbors are selected, the weights $\omega_{i,j}$ are set equal to $1/k_i$ for each matched nearest neighbor, where k_i is the number of matches selected for each target treatment case i. As with exact matching, the weights are set to 0 for all unmatched control cases.

A danger with nearest-neighbor matching, especially when the algorithm is forced to find a fixed multiple of matches such as 5, is that it may result in some very poor matches for some treatment cases. A version of nearest-neighbor matching, known as caliper matching, is designed to remedy this drawback by restricting matches to a chosen maximum distance, such as .25 standard deviations of the distance metric when calculated only for the treatment cases. This distance restriction then also allows for variable numbers of multiple nearest neighbors (e.g., ≤ 5 nearest neighbors within the caliper for each target treatment case). However, with this type of matching, some treatment cases may not receive matches, since for some of these cases no control cases will fall within their caliper. If this occurs, the resulting effect estimate then applies only to the subset of the treatment cases that have received matches (even if ignorability holds and there is simply sparseness in the data). Because such a data-induced shift of focus in the target parameter may be undesirable, a common strategy is to then use a hybrid approach, where in a second step all treatment cases without any caliper-based matches are then matched to a single nearest neighbor outside of the caliper.

Finally, for radius matching, there is variation in terminology and definitions in the literature. Most commonly, all control cases within a particular distance of each target treatment case are matched, and as a result the "radius" is functionally equivalent to the caliper in caliper matching. And, again, the weights $\omega_{i,j}$ are set equal to $1/k_i$ for the matched nearest neighbors, where k_i is the number of matches selected for each target treatment case i. The difference from caliper matching is simply that all potential matches within the caliper of each treatment case are anointed as matches for the target treatment case, which means that the matching is performed with replacement.[17] As with caliper matching, supplemental single nearest-neighbor matching may be needed if the analyst wishes to keep all treatment cases in the analysis. For some researchers, such additional forced matching is an essential component of radius matching, unlike caliper matching.

Interval Matching

Interval matching (also referred to as subclassification and stratification matching) for the ATT sorts the treatment and control cases into segments of a unidimensional distance metric, usually the estimated propensity score (see Cochran 1968; Rosenbaum and Rubin 1983a, 1984; Rubin 1977). For the traditional estimator of the ATT, the intervals are defined by cutpoints on the distance metric that subdivide the treatment cases into a chosen number of equal-sized subgroups (e.g., the intervals that subdivide 1,000 treatment cases into five groups of 200). More recent variants of interval matching, sometimes referred to as full matching, use an optimization step to first generate intervals of variable size that minimize the within-subgroup average difference in the distance metric.

No matter how defined, for each interval a variant of the matching estimator in Equation (5.15) is then estimated separately. The weights $\omega_{i,j}$ are set to give the same amount of weight to the treatment cases and control cases within each interval. The ATT is then calculated as the mean of the interval-specific treatment effects, weighted by the number of treatment cases in each interval.

---

[17]In contrast, caliper matching can be performed without replacement, although it is most commonly performed with replacement.

Kernel Matching

Kernel matching for the ATT constructs the counterfactual for each treatment case using all control cases but weights each control case based on its distance from the treatment case (see Heckman, Ichimura, and Todd 1997, 1998). The weights represented by $\omega_{i,j}$ in Equation (5.15) are calculated with a kernel function, $G(.)$, that transforms the distance between the selected target treatment case and all control cases in the study. When the estimated propensity score is used to measure the distance, kernel-matching estimators define the weight as

$$\omega_{i,j} = \frac{G(\frac{\hat{p}_j - \hat{p}_i}{a_n})}{\sum_j G(\frac{\hat{p}_j - \hat{p}_i}{a_n})}, \tag{5.17}$$

where a_n is a bandwidth parameter that scales the difference in the estimated propensity scores based on the sample size and $\hat{p}$ is the estimated propensity score.[18] The numerator of this expression yields a transformed distance between each control case and the target treatment case. The denominator is a scaling factor equal to the sum of all the transformed distances across control cases, which is needed so that the sum of $\omega_{i,j}$ is equal to 1 across all control cases when matched to each target treatment case.

Although kernel-matching estimators appear complex, they are a natural extension of nearest-neighbor caliper and radius matching: All control cases are matched to each treatment case but weighted so that those closest to the treatment case are given the greatest weight. Smith and Todd (2005) offer an excellent intuitive discussion of kernel matching along with generalizations to local linear matching (Heckman, Ichimura, Smith, and Todd 1998) and local quadratic matching (Ham, Li, and Reagan 2011).

5.4.2   Recent Matching Routines That Seek Optimal Balance

In the description of the matching algorithms above, we have given no indication of which algorithm will work best. We will defer this question until after we offer a demonstration of a variety of matching techniques below. Instead, in this section we will shift the motivation of matching slightly in order to present the most recent matching estimators that we will discuss in this section.

Consider our order of presentation of the main issues so far in this chapter. We offered Matching Demonstrations 1 and 2 to explain the basic conditioning strategy that underlies matching and to clarify why consistent estimates of the ATT, ATC, and ATE are all available when the set of perfect stratifying variables S is observed. We then turned to Matching Demonstration 3 to explain how estimated propensity scores can address the problems posed by finite sample sizes, under the recognition that a full stratification of the data on all of the variables in S will rarely be feasible if S includes many variables and/or those variables take on many values.

The recent methodological literature on matching assumes that the reader already knows these points and instead moves directly to the consideration of an omnibus

---

[18]Increasing the bandwidth increases bias but lowers variance. Smith and Todd (2005) find that estimates are fairly insensitive to the size of the bandwidth.

performance criterion that motivates most recent matching routines: the capacity of matching estimators to balance the variables that have been matched on. We therefore need to explain the concept of balance in more detail, first in an abstract sense and then in relation to the particular matching estimators considered so far in this chapter. (See also our earlier brief discussion of balance in Section 4.4.)

Consider a case where we have many variables in the perfect stratification set S, where all of these variables are observed, but where we have a finite sample afflicted by rampant sparseness. In such a dataset, many combinations of values on the variables in S will not be present, even though individuals with these patterns of values exist in the population. To capture this point formally, let $\Pr_N[s_i]$ represent the observed joint distribution across the realized values s of all variables in S for a particular sample of size N. For a dataset with substantial sparseness, the joint distribution $\Pr_N[s_i]$ will not in general be equal to the population distribution of S, denoted $\Pr[S]$. Generic sampling variation will lead to over-representation and under-representation of most combinations of values on the variables in S, and many of the rarest combinations in the population will not be present in the sample.

To now consider within-sample balance with respect to a treatment effect parameter, we must consider the observed joint distribution of S conditional on membership in the observed treatment and control groups, $\Pr_N[s_i|d_i = 1]$ and $\Pr_N[s_i|d_i = 0]$, respectively. For all examples considered so far in this chapter, the observed data are imbalanced with respect to treatment effect estimates for D because

$$\Pr_N[s_i|d_i = 1] \neq \Pr_N[s_i|d_i = 0]. \tag{5.18}$$

In words, the joint distributions for the observed versions of the variables in the perfect stratification set S are not the same in the treatment and control groups.

The underlying goal of all matching estimators is to transform the data in such a way that they can be analyzed as if they are balanced with respect to the treatment effect of interest, which will be the case if

$$\Pr_N[s_i|d_i = 1] = \Pr_N[s_i|d_i = 0] \tag{5.19}$$

in the matched dataset. If the ATT is the parameter of interest, then the data will only be balanced when matching has yielded a set of matched control cases that have the same joint distribution for the observed variables in S that the treatment cases have in the original observed dataset. If the ATC and ATE are also parameters of interest, then the data will only be balanced with respect to these additional parameters if the same routine also yields a set of matched treatment cases that have the same joint distribution for the observed variables in S that the control cases have in the original observed dataset.

Consider classic exact matching in a scenario where all treatment cases have one available exact match among the control cases, and where by exact we mean again that the resulting matched pairs have the same values on all of the variables in S. In this case, optimal balance is achieved for estimation of the ATT. In particular, all unmatched cases are tossed out of the dataset, and the remaining treatment and control cases have the exact same observed joint distribution across all variables in S. These data are not, however, balanced with respect to either the ATC or the ATE.[19]

For a closely related example, where the data are balanced with respect to the ATT, ATC, and ATE, consider our hypothetical stratification example in Matching Demonstration 1. For that example, we assumed an infinite sample where the distribution of S (see Table 5.1) differs across the treatment and control groups, such that the population and sample have treatment groups with disproportionately low numbers of individuals with $s_i = 1$ and disproportionately high numbers of individuals with $s_i = 3$. Accordingly, the observed data are imbalanced. However, when the data are analyzed within the strata of S, the average values for the outcome Y are generated from balanced comparisons with respect to S because, within each stratum, all individuals have the same value for S.[20] To generate estimates of the ATT, ATC, and ATE, a common distribution of S is then passed back to the estimator as stratum-level weights. In this sense, a stratification estimator, if available, is a generalization of exact matching that allows one to optimally balance the data for all average treatment effects that can be defined as functions in the underlying strata, including the ATT, ATC, and ATE.

As we noted above, finite datasets generally render exact matching and full stratification infeasible. Indeed, this was the impetus for the development of matching estimators, such as propensity-score matching, that utilize distance metrics for matching that are lower-dimensional than the full joint distribution of the perfect stratification set S. In these cases, perfect balance is generally impossible to achieve, as we now explain.

Consider our hypothetical example in Matching Demonstration 3. For the assumed data there, the distributions of A and B (which jointly represent the perfect stratification set S) differ across the treatment and control groups, such that the 50,000 simulated samples, on average, have treatment groups with disproportionately high values for A and B. Similar to Matching Demonstration 1, individuals are grouped into strata, but now only implicitly by estimation of the propensity score and only coarsely (i.e., not based on the full stratification defined by the full joint distribution of A and B). Individuals are then weighted within these propensity-score-defined strata, although indirectly by individual-level weights, in order to ensure that the *expected* distribution of S is the same within the treatment and control cases that are then used to estimate the ATT, ATC, and ATE.

Notice the inclusion of the word "expected" in the last sentence. Unlike exact matching and our full stratification example in Matching Demonstration 1, the data

---

[19]They would be balanced with respect to the ATC and ATE if the unmatched control cases also had the same observed joint distribution for S as the treatment cases. This would be the case if (1) all treatment cases have two exact matches among the control cases, (2) there are no other control cases that are exact matches, and (3) for each treatment case the exact match that is chosen from among the available two is determined by the toss of fair coin. We implicitly assume that this is not the case for this example because, if it were, then the original dataset would have been balanced in the first place and no matching would have been performed.

[20]In other words, the stratification match allows the unbalanced data to be analyzed in a transformed fashion (i.e., within strata) where the data are balanced. In particular, the probability distribution of S collapses to a single value and becomes degenerate within each stratum, at which point stratum-level effects are calculated. The distribution of S is then reanimated with stratum-level weights that generate average treatments effects weighted by a distribution of S common to the treatment and control cases.

for this example are only balanced in expectation. For any single sample among the 50,000 samples in the simulation, imbalance in the observed joint distribution of A and B will almost certainly exist within some of the coarse strata defined by the estimated propensity scores. Any imbalance that resides within these coarse strata then cumulates to the sample level.[21]

Now, consider the traditional matching algorithms presented in the last section. Because nearest-neighbor, caliper, radius, interval, and kernel matching also utilize a unidimensional distance metric to form matches, they have the same basic features as the weighting estimators for Matching Demonstration 3. For a single dataset, a set of perfect matches on the estimated propensity score, or any other unidimensional metric, does not guarantee balance on the observed joint distribution of S. In fact, if such optimal balance were possible, then exact matching or a full stratification of the data could have been used in the first place. Although the applied matching literature is replete with misleading claims that propensity-score matching is a complete remedy for covariate imbalance in a single sample (and, even more distressing, even for selection on the unobservables!), the core methodological literature is clear on these points (see Rubin 2006).

The most recent literature has focused on developing estimators that optimize within-sample balance, typically by more closely tying the estimation steps to balance assessment. Ongoing research on matching methods can be grouped into four non-mutually exclusive streams: new methods for balance assessment, more flexible estimation of the distance metric on which matching is performed, optimization of the matching step that follows the estimation of the distance metric, and coarsening the stratification variables before matching is performed.

Enhancing Routines for Balance Assessment

Assessing the balance of the variables on which the cases have been matched can be difficult for two reasons. First, if exact matching or full stratification are not possible, then it will not in general be possible to achieve perfect balance. Yet, there are no clear standards, nor any sense that there must be common standards across all applications, in what features of balance can be traded off against others. Surely it is correct that one should attempt to match the mean values of variables that strongly predict both the treatment and the outcome, as opposed to the mean values of variables that only strongly predict one or the other. But, beyond this obvious point, there is no clear consensus on the relative importance of alternative components of overall balance. Second, the use of any hypothesis test of similarity for the features of two distributions has two inherent associated dangers: (a) with small samples, a null hypothesis of no difference may be accepted when in fact the data are far from balanced; (b) with very large samples, almost any difference, however small, is likely to be statistically significant, leading to the possibility that no amount of balance would ever be deemed

---

[21]More formally, the foundational propensity-score literature claims only that the observed data will be balanced after propensity-score matching *on average over repeated sampling from the same population.* For any single finite sample, either Equation (5.18) or (5.19) could be true. And, unfortunately, except for the simplest stratification sets S, it is far more likely that the observed data will remain unbalanced to some degree.

acceptable. As such, hypothesis tests are generally less useful for assessing balance than measures that focus on differences in magnitude.[22]

Against the backdrop of these acknowledged challenges, substantial attention has been devoted to the creation of new methods for balance assessment. The favorite measure of balance, following the sole consensus position just outlined, is that balance should first be assessed based on the standardized difference in sample means,

$$\frac{|E_N[x_i|d_i=1] - E_N[x_i|d_i=0]|}{\sqrt{\frac{1}{2}\text{Var}_N[x_i|d_i=1] + \frac{1}{2}\text{Var}_N[x_i|d_i=0]}}, \tag{5.20}$$

which is calculated twice for every variable X that is matched on: (1) before any matching is performed in order to establish the baseline level of imbalance in the original data and (2) across the matched treatment and control cases to determine how much of the initial imbalance has been removed by matching. Because this measure of balance is a scaled absolute difference in the means of each X across treatment and control cases, variable-specific measures can be compared between alternative Xs. In addition, one can take the mean of these standardized differences across all variables that have been matched on, and this single value then summarizes the quality of the balance in means across all variables.

Equation (5.20) is but the tip of the iceberg among all possible balance measures that one could construct. Most of the matching software in use has improved tremendously in the past decade, offering researchers more ways to examine how much balance has been achieved by particular matching routines, based on analogs to standardized differences for higher moments of the distributions of the matching variables, as well as indices based on differences between full distributions and between quantile-quantile plots (see Austin 2009a; Hansen and Bowers 2008; Iacus et al. 2011; Imai et al. 2008; Sekhon 2007).

With these new tools for balance assessment in hand, an analyst can optimize balance by pursuing two different strategies: reestimating the distance metric on which matching is performed and optimizing the matching algorithm that uses these distances. In some of the most recent matching routines, these two steps are blended, but for clarity we present them next with some artificial separation.

Reestimating the Distance Metric

If the covariates remain substantially imbalanced after a particular matching routine has been performed, the first recourse is to return to the model that estimates the distances between the treatment and control cases. Typically, this involves respecifying the propensity-score-estimating equation progressively in pursuit of improvements in balance (which, of course, can only be measured after the analyst then passes the reestimated distance metric through to the matching algorithm of choice). The analyst can try adding interaction terms, quadratic terms, and other higher-order terms,

---

[22]For a full discussion of these issues, and somewhat different conclusions, see Hansen and Bowers (2008), Heller, Rosenbaum, and Small (2010), Iacus et al. (2011), Imai, King, and Stuart (2008), and Rubin and Thomas (1992, 1996).

guided by the balance indices that are now on offer.[23] Such respecification can be labor intensive and frustrating because alternative respecifications can improve balance in some variables while simultaneously eroding balance in others. In the end, there is no guarantee that through such a respecification loop the analyst will find the specification that will deliver the best possible balance for the single sample under analysis. For this reason, some of the most recent literature has moved toward more flexible ways of estimating the difference metric, attempting to remove the analyst from the initial specification, and then any resulting respecification loop, by harnessing available computational power through data mining protocols. We will discuss this literature next.

A second, not mutually exclusive, option is to assess whether balance improvement may result from the relaxation or alteration of the parametric assumptions one may have implicitly adopted in estimating the unidimensional distance metric. Estimated propensity scores are only one distance metric that can be adopted before matching is performed, and from its origins the matching literature has been especially enamored of one alternative, the Mahalanobis metric, which reduces the data in a different fashion, preserving more of the multidimensional content of the matching variables (but in a way that makes verbal description difficult; see the note on page 158). Even for the propensity score, the preference for a default logit specification is rarely justified in any formal sense and rather is typically adopted because the literature suggests that it has worked well enough in the past.

Recently, multiple articles have tested for the sensitivity of the logit functional form in simulation studies and with real data (e.g., Harder, Stuart, and Anthony 2010; Hill et al. 2011). Our reading of this literature is that logit specifications have performed moderately well. Still, in some head-to-head competitions, nonparametric estimation has performed comparatively well (e.g., Harder et al. 2010), and these results may portend that the future of propensity score estimation lies in regression tree modeling (e.g, McCaffrey et al. 2004).[24]

Such a trajectory may be appealing because a case can be made that the whole business of asserting a parametric model and then searching for a balance-maximizing specification by trying out one after another is the sort of error-prone sausage making that it would be best to avoid. Jasjeet Sekhon's position is common, even if infrequently expressed in writing:

> The literature on matching, particularly the debate over the LaLonde (1986) data, makes clear the need to find algorithms which produce matched datasets with high levels of covariate balance. The fact that so many talented researchers over several years failed to produce a propensity score

---

[23]Because this type of respecification does not involve examining the effects of the matching variables on the outcome of interest, it is not considered one of the pernicious forms of data mining (e.g., where false effects are claimed when variables sneak through specification screens as a function solely of sampling error; see our discussion later in Section 6.6.2).

[24]As we will show in a later demonstration, the main danger to effect estimates is still unblocked back-door paths that exist because some variables in S have either been left out of the estimating equation or unobserved. This position has been reinforced by a related set of evaluations, this time with a "yoked experiment," where mode of data analysis was far less important that the selection of the variables with which to balance or adjust the data (see Cook, Steiner, and Pohl 2009; Steiner, Cook, Shadish, and Clark 2010).

model which had a high degree of covariate balance is a cautionary tale. In situations like these, machine learning can come to the rescue. There is little reason for a human to try all of the multitude of models possible to achieve balance when a computer can do this more systematically and much faster. Even talented and well trained researchers need aid. (Sekhon 2007:8)

From this orientation, Diamond and Sekhon (2013) have proposed a general multivariate matching method that uses a genetic algorithm to search for the match that achieves the best possible balance. Although their algorithm can be used to carry out matching after the estimation of a propensity score, their technique is more general and can almost entirely remove the analyst from having to make any specification choices other than designating the matching variables. Diamond and Sekhon (2013) show that their matching algorithms provide superior balance in both Monte Carlo simulations and a test with genuine data.[25] Imai and Ratkovic (2014) offer a related approach, but it is one that integrates covariate balancing into the estimation of the propensity score.

Optimizing the Matching Algorithm

Suppose that an analyst eschews methods such as those proposed by Diamond and Sekhon (2013) and instead decides to preserve the traditional two-step estimation strategy. Even with a distance metric in hand from an iterative first step that delivers the best observed balance (ascertained by running candidate distance metrics through the intended matching algorithm), there is no guarantee that the chosen matching algorithm will then optimize the balance that the distance metric may be able to deliver.

For example, as we noted above, nearest-neighbor matching without replacement can generate suboptimal matching patterns because of order effects that emerge in the matching process. Although these problems are not large when the pool of available control cases includes many good matches for each treatment case, they can be substantial for even moderate-sized datasets and/or when the number of control cases is comparatively small. Consideration of these effects has led to the more general recognition that any serial matching of pairs that passes over the target cases only once will not necessarily guarantee that the overall average distance on the unidimensional metric within matched pairs is minimized. If this average distance is not minimized, then there is little chance that the underlying joint distribution of the matching variables will be balanced as completely as the distance metric may allow.

To address the weaknesses of nearest-neighbor matching, and other related algorithms, Rosenbaum (1989) began a literature on optimal matching that has now fully matured. The key advances are to (1) consider the closeness of a match pattern across all treatment and control cases using a global metric that is a function in all pairwise differences in the chosen unidimensional distance metric and (2) use a computationally

---

[25] A natural end-state of this orientation is the full nonparametric estimation of the effects, dispensing with the intermediate estimation of a propensity score (see Hill 2011; Hill et al. 2011).

rich algorithm to search through possible matching patterns for the one that delivers the smallest average within-pair distance.

Although optimal matching algorithms vary and allow the user to specify many additional constraints on the match patterns searched (see Hansen 2004; Rosenbaum 2002, 2010; Zubizarreta 2012), optimal matching routines are based on the idea of minimizing the average distance between the estimated propensity scores (or some other unidimensional distance metric) for the cases that are matched together. The most recent literature combines optimal pair matching with attempts to achieve near exact matching on the variables deemed most crucial to balance (see Yang, Small, Silber, and Rosenbaum 2012; Rosenbaum 2012; Zubizarreta 2012).

Direct Coarsening of the Stratification Variables

One way to improve balance is to ask less of the matching variables, while invoking a theoretical justification for doing so. Consider a political participation example. If one seeks to estimate the average effect of receiving a college degree on participation (according to some index of participatory acts; see Section 1.3.1, page 16), theory may suggest that one can block all back-door paths by conditioning on five variables: self-reported racial-ethnic identity, gender, state of residence, household income, and marital status.[26] For datasets of finite size, it may be difficult to balance these five variables if they are measured in full detail. Accordingly, an analyst might consider conditioning on a coarse representation of state of residence (reliable red state, reliable blue state, and swing state), marital status (collapsing widows and widowers with those currently married, but leaving never married and formerly married as separate categories), self-reported race (collapsing Hispanic ethnicity into two categories, Cuban and non-Cuban), and household income (using quintiles of household income rather than the full interval scale). One suspects that researchers have been making such decisions for decades, using reasoned theoretical judgment and background knowledge for doing so.

Iacus et al. (2012a; see also Iacus and King 2012) have refined this approach into a new matching strategy that they label "coarsened exact matching." As suggested by the label, the key idea is to perform only exact matching, but after the matching variables have been coarsened in some principled fashion. They situate this new method within a broader class of "monotone imbalance bounding" (MIB) matching methods that they propose (see Iacus et al. 2011), all of which have the attractive property of allowing the analyst to avoid setbacks in variable-specific balance when respecifying the routine by altering the specifications for other variables.[27]

The primary trade-off of using exact matching, even with coarsened data, is that cases will typically be dropped from both the treatment and control groups in the observed data. Such a narrowing of the dataset shifts the target parameter to something narrower than the population-based ATT, ATC, and ATE that we have focused

---

[26]For a recent debate on these issues for this particular example, see the discussion of alternative matching estimates in Henderson and Chatfield (2011), Kam and Palmer (2008, 2011), and Mayer (2011).

[27]Because the matching is exact, no propensity score is estimated. Rather, here the respecification involves coarsening the matching variables in alternative ways, which may then also shift the target parameter, as we discuss in the main text.

on in this book. Iacus et al. (2012a:5) imply that the most common target parameter of coarsened exact matching is the local sample average treatment effect for the treated (local SATT), which they define as "the treatment effect averaged over only the subset of treated units for which good matches exist among available controls." This is a well-defined parameter in theory, but in practice it moves with each coarsening decision because these decisions recalibrate what counts as a good match.[28]

When are such methods likely to lead to improvements in estimation relative to other matching methods? Iacus et al. (2012a:1) write that "the key goal of matching is to prune observations from the data so that the remaining data have better balance between the treated and control group." As such, coarsened exact matching is a natural endpoint for the data preprocessing perspective on matching methods introduced into the literature by Ho et al. (2007). Overall, coarsened exact matching is likely to work quite well for analysts who can accept that the key goal of matching is to "prune" the data in pursuit of an estimate of a local SATT. And this will often be a very reasonable position when there is no clear correspondence between the data and a well-defined population, such as for non-random collections of individuals who have received a treatment in a biomedical study or for collections of aggregate units that are defined by social processes, such as nation states or congressional districts.

This book reflects our tastes for target parameters anchored in well-defined populations and random samples of them, and our tastes are grounded in the tradition of random-sample survey analysis that dominates sociology and demography. Yet, the overall matching literature, especially the streams that have always seen matching as a type of practical sampling in the service of best estimates of hard-to-estimate average treatment effects, is closer to the motivation of coarsened exact matching than is our treatment of matching in this book.[29]

5.4.3   Which of These Matching Algorithms Works Best?

Very little specific guidance exists in the literature on which of these matching algorithms works best. We have given some indication of the relative strengths and weaknesses in our introduction of each technique, but the strengths and weaknesses of each will necessarily vary in their relevance for alternative applications.

We do not mean to imply that no one advocates for the superiority of particular matching estimators. On the contrary, the strongest advocates for any particular estimators would appear to be those who have had a hand in developing them. Heckman, Ichimura, and Todd (1997, 1998), for example, argue for the advantages of kernel matching. Diamond and Sekhon (2013) prefer genetic matching for similar situations. Rosenbaum (2010) defends optimal matching, and now Iacus et al. (2011) advocate for coarsened exact matching. We could continue.

---

[28] In particular, as coarsening proceeds variable by variable, more of the cases are exactly matched, until the local SATT becomes the SATT. At the same time, the ability to defend the exact matching estimate as unbiased for the SATT declines as coarsening is pursued.

[29] From the perspective of Iacus et al. (2012a), we are willing to accept some model dependence in final estimates, so as to preserve the correspondence between sample and population-based quantities. Our orientation will become clear over the next two chapters, where we endorse modeling strategies, such as weighted regression, that blend matching methods with more traditional forms of regression analysis.

These scholars are very likely correct that the matching estimators they have developed are the best for the applications they have been developed to model. This does not mean that it is easy to choose from among the alternatives, as we showed in the first edition of this book, and as others have since also documented (e.g., Austin 2009b; Harder et al. 2010; Hill et al. 2011). Because there is no clear guidance on which of these matching estimators is "best," we offer a fourth hypothetical example to give a sense of how often alternative matching estimators yield appreciably similar estimates.

Matching Demonstration 4

The debate over the size of the causal effect of Catholic schooling on the test scores of high school students has spanned more than three decades (see Section 1.3.2, page 22). The example we offer here is based on Morgan (2001), although we will use simulated data for which we have defined the potential outcomes and treatment assignment patterns so that we can explore (a) the relative performance of alternative matching estimators and (b) the consequences of conditioning on only a subset of the variables in the set of perfect stratification variables S. Similar to Matching Demonstration 3, we will repeat the simulation for multiple samples, but for many fewer of them for the reasons we explain below.

Generation of the Data. The simulated datasets that we constructed mimic the real dataset from the National Education Longitudinal Study (NELS) analyzed by Morgan (2001). For that application, regression and matching estimators were used to estimate the effect of Catholic schooling on the achievement of high school students in the United States. For our simulation here, we generated datasets of 10,000 individuals with values for baseline variables that resemble closely the joint distribution of the similar variables in Morgan (2001). The variables for respondents include variables for race, region, urbanicity, whether they have their own bedrooms, whether they live with two parents, an ordinal variable for number of siblings, and a continuous variable for socioeconomic status. Departing from Morgan (2001), we also created a cognitive skill variable, assumed to reflect innate and acquired skills in unknown proportions, that we assume was measured just prior to the decision of whether or not to enroll in a Catholic school.[30]

---

[30]To be precise, each sample, with a fixed N of 10,000, was generated using a multinomial distribution with parameters calibrated by a 40-cell cross-tabulation of race (five categories of white, Asian, Hispanic, black, and Native American), region (four categories), and urbanicity (two categories), based on the data in Morgan (2001). Values for socioeconomic status were then drawn from normal distributions with means and standard deviations estimated separately for each of the race-by-region-by-urbanicity cells using the NELS data with respect to socioeconomic status. Thereafter, all other variables were generated from socioeconomic status, using parameter values for suitable random distributions based on auxiliary estimates from the NELS data. Because we relied on standard parametric distributions, and did not build interactions between these additional variables into the simulation routine, the data for these additional variables are smoother than the original NELS data. But, because the sampling cross-tabulation has 40 cells, for which socioeconomic status is then parameterized differently for each, the simulation is initiated as a mixture distribution for socioeconomic status that is itself rather lumpy, given the pronounced differences in socioeconomic status between racial groups and across geographic space in the referent NELS data.

We defined potential outcomes for all $10,000$ individuals of each dataset, assuming that the observed outcome of interest is a standardized test taken at the end of high school. For the potential outcome under the control (i.e., a public school education), we generated what-if test scores as

$$y_i^0 = 100 + 2(Asian_i) - 3(Hispanic_i) - 4(Black_i)$$
$$-5(Native\,American_i) - 1(Urban_i) + .5(Northeast_i)$$
$$+.5(North\,Central_i) - .5(South_i) + .02(Number\,of\,Siblings_i) \quad (5.21)$$
$$+.05(Own\,Bedroom_i) + 1(Two\,Parent\,Household_i)$$
$$+2(Socioeconomic\,Status_i) + 4(Cognitive\,Skills_i) + v_i^0,$$

where the values for v_i^0 are independent random draws from a normal distribution with expectation 0 and standard deviation of 12.[31] We then assumed that the what-if test scores under the treatment (i.e., a Catholic school education) would be equal to the outcome under the control plus a boosted outcome under the treatment that is a function in race, region, and cognitive skills (under the assumption, based on the dominant position in the extant literature, that black and Hispanic respondents from the north, as well as all students with high preexisting cognitive skills, are disproportionately likely to benefit from Catholic secondary schooling). Accordingly, we generated the potential outcomes under the treatment as

$$y_i^1 = y_i^0 + \delta_i' + \delta_i'', \quad (5.22)$$

where the values for δ_i' are independent random draws from a normal distribution with expectation 6 and standard deviation of 0.5. To this common but stochastic individual-level treatment effect, we added a second component with values for δ_i'' that vary systematically over individuals. These values were constructed as draws from a normal distribution with expectation equal to

$$0 + 1(Hispanic_i \times Northeast_i) + .5(Hispanic_i \times North\,Central_i)$$
$$+1.5(Black_i \times Northeast_i) + .75(Black_i \times North\,Central_i) \quad (5.23)$$
$$+.5(Cognitive\,Skills_i)$$

and with a standard deviation of 2, after which we subtracted the mean of all drawn values in the simulated sample (in order to center the draws for δ_i'' on 0). Taken together, the values of δ_i' and δ_i'' for each individual represent two additive components of their individual-level treatment effects, as is clear from a rearrangement of Equation (5.22), with reference to the definition of the individual-level treatment effect in Equation (2.1), as $\delta_i = y_i^1 - y_i^0 = \delta_i' + \delta_i''$.

We then defined the probability of attending a Catholic school using a logistic distribution,

$$\Pr[D_i = 1|S_i] = \frac{\exp(S_i\phi)}{1 + \exp(S_i\phi)}, \quad (5.24)$$

---

[31] Across simulated datasets, the standard deviations of socioeconomic status and cognitive skills varied but were typically close to 1.

where

$$S_i\phi = -4.6 - .69(Asian_i) + .23(Hispanic_i) - .76(Black_i)$$
$$- .46(Native\,American_i) + 2.7(Urban_i) + 1.5(Northeast_i)$$
$$+ 1.3(North\,Central_i) + .35(South_i) - .02(Number\,of\,Siblings_i)$$
$$- .018(Own\,Bedroom_i) + .31(Two\,Parent\,Household_i)$$
$$+ .39(Socioeconomic\,Status_i) + .33(Cognitive\,Skills_i)$$
$$- .032(Socioeconomic\,Status_i^2) - .32(Cognitive\,Skills_i^2)$$
$$- .084(Socioeconomic\,Status_i \times Cognitive\,Skills_i)$$
$$- .37(Two\,Parent\,Household_i \times Black_i)$$
$$+ 1.6(Northeast_i \times Black_i) - .38(North\,Central_i \times Black_i)$$
$$+ .72(South_i \times Black_i) + .23(Two\,Parent\,Household_i \times Hispanic)$$
$$- .74(Northeast_i \times Hispanic_i) - 1.3(North\,Central_i \times Hispanic_i)$$
$$- 1.3(South_i \times Hispanic_i) + .25\delta_i''.$$

(5.25)

The specification for Equation (5.25) is based on the results of Morgan (2001), along with an additional assumed self-selection dynamic in which individuals are slightly more likely to select the treatment as a function of the relative size of the systematic component of the individual-specific shock to their treatment effect, δ_i''.

The probabilities defined by Equation (5.24) were then specified as the parameter for random draws from a binomial distribution, generating a treatment variable d_i for each simulated student. Finally, following Equation (2.2), simulated students in Catholic schools were given observed values for y_i equal to their simulated y_i^1, while all others were given observed values for y_i equal to their simulated y_i^0.

The ATT as the Target Parameter. The presence of the self-selection term, δ_i'', in both Equation (5.22) and Equation (5.25) creates a nearly insurmountable challenge for the estimation of the ATE. We will consider cases such as this one in detail in later chapters, but for now we will explain briefly why the only target parameter that can be estimated with any degree of confidence is the ATT.

Consider the directed graph presented in Figure 5.2, and consider why this graph indicates that the ATE cannot be estimated with the available data. Equations (5.21) and (5.25) imply that all of the variables whose names are written out in this figure are mutual causes of both D and Y. All of these variables are on the right-hand sides of both of these equations, and the nonparametric nature of the directed graph also represents the variety of higher-order and cross-product terms embedded in Equation (5.25). If we observe and then condition on all of these variables, we will be able to block many back-door paths that would otherwise confound the causal effect of D on Y. Unfortunately, there is an additional back-door path, represented by the bidirected edge $D \leftarrow\!\text{-}\text{-}\text{-}\text{-}\!\rightarrow Y$, that will remain unblocked after such conditioning. This is the noncausal back-door association that is generated by the presence of δ_i'' in both Equation (5.22) and Equation (5.25). As with many real-world applications, this simulation assumes that individuals (in this case students, although surely informed by parents and others) select from among alternative treatments based on accurate expectations,

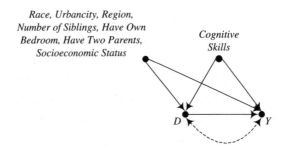

Figure 5.2 The directed graph implied by the underlying data generation model for Matching Demonstration 4.

unavailable as measures to researchers, of their likely gains from alternative treatment regimes. Accordingly, prospective Catholic school students with comparatively high values of δ_i'' are more likely to be in Catholic schools, according to Equation (5.25), and are also more likely to benefit from Catholic schooling, according to Equations (5.22) and (5.23). This pattern is the most common type of selection on the unobservables in the social sciences.

The directed graph reveals clearly why the ATE is not identified, and yet it does not at all suggest that the ATT is, in fact, still identified in cases such as this one. Here, it is helpful to consult the equations again. The set of variables that was used to generate y_i^0 in Equation (5.21) does not include the individual-specific shock, δ_i''. We therefore do not need to observe δ_i'' in order estimate reasonable counterfactual values for y_i^0 among simulated students enrolled in Catholic schools. As discussed earlier – see Equations (2.16) and (5.2) – these are the only counterfactual values that are needed in order to effectively estimate the ATT.[32]

The problem for the ATE is that the ATC is not identified. Because of the way in which the generation of y_i^1 and d_i are entangled by their joint dependence on δ_i'', we cannot estimate reasonable counterfactual values for y_i^1 among those who are enrolled in public schools. Net of all of the other observed variables, the students most likely to benefit from Catholic schooling are in Catholic schools, and they therefore have values for $y_i = y_i^1$ that are too positive, even net of the observables, to enable effective estimation of the counterfactual values of y_i^1 for public school students.

(Although the directed graphs we have introduced so far do not make the identification of the ATT particularly clear when the ATE is not also identified, in Chapter 8 we will offer elaborated directed graphs, which use latent classes and variables for individuals' own expectations, to communicate results such as these. Still, it is often the case that targeted identification results are best conveyed by equations, whether

---

[32]Notice that we did not specify self-selection on the causal effect in both directions. If we had, the ATT would not be identified either. We have built this simulation assuming that public schools serve as a baseline state, out of which students may exit in order to capture additional learning gains that may be available to them from Catholic schooling. The more complicated scenario would be one where Equation 5.21 also included δ_i'' but with a negative coefficient (such that, net of observables, those who are least likely to do well in Catholic schooling are those most likely to do well in public schools).

these are equations furnished by the potential outcome model or based on the nonparametric structural equations that underlie all directed graphs. We will consider these issues in substantial detail in the next part of the book.)

Methods for Treatment Effect Estimation. In Tables 5.6 and 5.7, we offer 19 separate matching estimates of the ATT, where we match the simulated control cases to the simulated treatment cases. These estimates are produced by routines written for R and Stata by others (see the notes to Table 5.6 for citations to the software), and the row labels we have chosen for the tables map onto the terminology that we used to present the matching estimators in Sections 5.4.1 and 5.4.2.

We estimate all matching estimators under two basic scenarios. First, we offer a set of estimates based on an incomplete specification of treatment assignment, where we omit the cognitive skills variable. In addition, for routines that utilize estimated propensity scores, we also omitted the higher-order and cross-product interaction terms present in Equation (5.25). The exclusion of the cognitive skills variable is particularly important because it has correlations of 0.4 with the outcome variable and 0.2 with the treatment variable, on average across the simulated datasets. For the second scenario, which we label the complete specification scenario, we included the cognitive skills variable, and, for those routines that utilize propensity scores, the higher-order and cross-product interaction terms present in Equation (5.25). Both scenarios lack an adjustment for the self-selection dynamic, in which individuals select into the treatment partly as a function of an accurate expectation of their own individual-specific treatment effect. In this sense, the complete specification is only complete enough to deliver a consistent and asymptotically unbiased estimate of the ATT, not of the ATC or the ATE as explained above.

Regarding the specific settings for the alternative matching estimators, we list some relevant information in the row labels for each (i.e., caliper size, kernel specification). Beyond these settings, we generally accepted the default options for all estimators as specified in the relevant software, on the argument that this "horse race" between estimators ought to be fair. As we will note below, each of the estimates we offer could be tailored to this specific application more so than for the analysis that we present, and perhaps some estimators could therefore get closer on average to the target parameter and be shown to outperform all others. We have made the datasets available, on the book's Web site (www.cambridge.org/9781107065079), for any readers who wish to try. We encourage instructors using this book to share these datasets with their students and to use them to try out additional matching estimators beyond those we have presented here, including any matching routines developed after publication of this edition of the book.

We will not provide estimated standard errors in the tables, although we will indicate their range in the text. As we describe following the demonstration (see Section 5.5), there is no common methodology that allows the estimation of standard errors in analogous ways across all estimators, which makes informative comparisons across estimators difficult (in terms of relative efficiency or expected mean-squared error). Nonetheless, the literature agrees that estimators that use more of the data, such as interval matching and kernel matching, have smaller estimated standard errors than

Table 5.6 Matching Estimates of the ATT, Catholic Schooling on Achievement for One Simulated Dataset

Method	Specification of treatment assignment variables:		Number of treatment cases retained for the estimate of the ATT
	Incomplete	Complete	
Nearest-neighbor match:			
1 without replacement (ps2)	7.37	7.03	1052
1 without replacement (MI)	7.55	7.09	1052
1 with replacement (ps2/psc)	7.77	7.17	1052
1 with replacement (MI)	7.96	7.38	1052
1 with replacement and			
caliper = .05 SD (ps2)	7.77	7.15	1051
1 with replacement and			
caliper = .05 SD (MI)	7.27	6.43	1051
5 with replacement (ps2)	7.51	6.42	1052
5 with replacement (MI)	8.06	7.29	1052
5 with replacement and			
caliper = .05 SD (ps2)	7.55	6.37	1051
5 with replacement and			
caliper = .05 SD (MI)	8.00	7.12	1051
Radius match:			
Caliper = .05 SD (ps2)	7.61	6.36	1051
Caliper = .05 SD (psc)	8.23	7.84	1051
Interval match:			
10 fixed blocks (MI)	8.71	8.71	1052
Variable blocks (ps2)	7.50	6.60	1052
Kernel match:			
Epanechnikov (ps2/psc)	7.57	6.58	1052
Gaussian (ps2/psc)	7.70	6.82	1052
Optimal match (MI-opt)	6.84	6.78	1052
Genetic match (MI-gen)	7.80	6.46	1052
Coarsened exact match (cem)	7.54	6.59	1015/973

Notes: The software utilized is denoted "ps2" for Leuven and Sianesi (2012), "MI" for Ho et al. (2011), "psc" for Becker and Ichino (2005), "opt" for Hansen, Fredrickson, and Buckner (2013), "gen" for Sekhon (2013), and "cem" for Iacus et al. (2012b).

those that discard more of the data, such as nearest-neighbor matching. The latter have a claim to lower bias, since bad matches are thrown away, and yet also have greater sampling variability. In this sense, comparisons of matching estimators represent a classic trade-off between bias and variance.

Results. Table 5.6 presents matching estimates of the ATT for a single dataset, as would be the case in nearly all applications of the methodology presented in this book. However, one important difference from the typical scenario is that we know for this simulated dataset that the true ATT is equal to 6.92. And, even though we will not estimate them here, we also know that the true ATE and ATC are equal to 6.00 and 5.90, respectively. By construction, the ATT is larger than the ATC because those who are most likely to benefit from Catholic schooling are more likely to enroll in Catholic schooling, both as a function of observed variables that lie on the back-door paths in Figure 5.2 and because of the self-selection on the individual-level treatment effect itself.

Estimates based on the incomplete specification are reported in the first column of Table 5.6. As expected, the estimates are generally much larger than the true value of the ATT, which is 6.92 (although the optimal match estimate yields a surprisingly close value of 6.84 for this single dataset). Most of the positive bias of these estimates is due to the mistaken exclusion of the variable for cognitive skills from the model for treatment assignment.

Estimates based on the complete specification are then reported in the second column of Table 5.6. On the whole, this second set of estimates is considerably closer to the true ATT, oscillating on either side of the true value of 6.92 (i.e., 10 have negative bias for the ATT, and 9 have positive bias for the ATT). The standard errors for these estimates, which all software routines provide, fall within a range between .43 and .66, although using different estimation techniques. The point values reported in the second column are consistent with what one would expect for variation produced by sampling error alone. However, the variation across the point estimates does not arise from sampling error, since all estimators use the same sample, but rather from the different ways in which the estimators use the data. As such, it is unclear how comforting is the claim that this level of variation is consistent with what would be produced by sampling error alone.

Notice also that the third column reports the number of treatment cases retained for the estimate of the ATT. For estimators that do not impose a nearness criterion (i.e., a caliper or radius), all 1,052 treatment cases were retained for the estimate. Even when quite narrow calipers are selected, only one treatment case is discarded for the relevant estimates for this single dataset. This pattern indicates that these data are therefore well suited for matching estimators that force the utilization of all treatment cases, such as traditional nearest-neighbor matching without a caliper, because there are many suitable control cases on the support of the matching variables for the treatment cases. This will not always be the case for datasets with more extreme patterns of treatment assignment and/or a more limited sample size.

Much more could be said about each pair of estimates, but we will note only a few additional patterns. First, in some cases software programs that utilized the same routine yielded estimates that were quite different. The reasons for these differences are not obvious, but they tend to occur for estimators that process the matching algorithm in ways that could vary or that utilize difference calculations across propensity scores within calipers that can be sensitive to rounding error. Second, although it may be tempting to conclude that the optimal match is superior to all others, we will show

Table 5.7 Bias for Matching Estimates of the ATT, Catholic Schooling on Achievement Across 10 Simulated Datasets

Method	Specification of treatment assignment variables: Incomplete specification		Complete specification	
	Min, Max	Average	Min, Max	Average
Nearest-neighbor match:				
1 without replacement (ps2)	−.37, 2.17	.79	−1.17, 0.68	−.04
1 without replacement (MI)	−.23, 2.08	.75	−1.11, 0.56	.00
1 with replacement (ps2/psc)	−.35, 2.00	.77	−1.18, 0.40	−.13
1 with replacement (MI)	−.25, 2.22	.95	−.70, 0.84	.19
1 with replacement and				
caliper = .05 SD (ps2)	−.32, 2.04	.78	−1.21, 0.35	−.13
1 with replacement and				
caliper = .05 SD (MI)	−.16, 1.89	.98	−.99, 1.16	.07
5 with replacement (ps2)	−.01, 1.77	.79	−.79, 0.83	−.04
5 with replacement (MI)	.43, 2.29	1.17	−.44, 1.34	.50
5 with replacement and				
caliper = .05 SD (ps2)	−.01, 1.71	.77	−.81, 0.75	−.04
5 with replacement and				
caliper = .05 SD (MI)	.15, 2.43	1.19	−.49, 1.18	.39
Radius match:				
Caliper = .05 SD (ps2)	−.18, 2.07	.81	−.82, 1.02	−.08
Caliper = .05 SD (psc)	.36, 2.42	1.17	.10, 2.03	.89
Interval match:				
10 fixed blocks (MI)	.84, 3.02	1.68	.84, 3.02	1.68
Variable blocks (ps2)	.03, 2.18	.88	−.85, 1.12	−.03
Kernel match:				
Epanechnikov (ps2/psc)	.08, 2.25	.89	−.76, 1.25	.01
Gaussian (ps2/psc)	.08, 2.34	1.00	−.59, 1.44	.19
Optimal match (MI-opt)	.45, 2.89	1.28	−.36, 1.53	.54
Genetic match (MI-gen)	.09, 1.66	.96	−.89, 1.55	.23
Coarsened exact match (cem)	.58, 2.66	1.28	−.43, 1.59	.35

Notes: see notes to Table 5.6 for software details.

below that this is partly a chance result for this single sample (even though optimal matching is designed to outperform traditional estimators and can be expected to do so). Third, the interval match with fixed blocks does not improve for the complete specification because the inclusion of the variable for cognitive skills does not have consequences for the ordering of the data, and hence it does not move cases in between the 10 intervals that are used. For the alternative interval matching estimator that uses variable blocks, selected in order to maximize within-interval mean balance on the propensity score, the utilization of the variable for cognitive skills does improve the estimate.

The only estimates that require more detailed explanation are those based on coarsened exact matching. For this estimator, we could have coarsened the data more in order to retain all treatment cases. Or we could have coarsened the data less and retained fewer cases. We chose a strategy that kept most of the treatment cases so that the assumed local SATT is not too far from the target parameter for all other matching estimators, the true ATT.

In particular, for the incomplete specification, we made coarsening decisions that, collectively, defined 868 strata that contained either treatment or control cases, of which 244 had both treatment and control cases present. Overall, 37 of 1,052 treatment cases fell into strata without any control cases and were therefore discarded from the analysis. The specific coarsening decisions were to coarsen the variable for number of siblings into three categories (0, 1 or 2, and 3 or more) and socioeconomic status into five categories as equal-sized quintiles. No other variables were coarsened.

For the complete specification, we then included the variable for cognitive skills, which we also coarsened into quintiles. This additional variable increased to 1,540 the number of strata that contained either treatment or control cases, of which 316 had both treatment and control cases present. Because of the increasingly specific strata, 79 treatment cases fell into strata without any control cases and were discarded from the analysis. In the end, the local SATT for the complete specification is estimated for 973 of the 1,052 treatment cases (or 92.5 percent). It is unclear, therefore, whether it is sensible to compare the resulting estimate to the true ATT for this example, as we do here. Yet, it could also be unfair to compare it to its own implied target parameter, the true local SATT, without similarly doing so for caliper and radius estimators, for example, that also discard some treatment cases.

In sum, when we consider a single sample, and when we know the true value for the ATT, it is clear that omitting a crucial variable, such as the variable for cognitive skills in this application, will prevent matching estimators from effectively estimating the ATT. When the complete specification is adopted, matching appears to perform quite well, although the particular matching estimators give point value estimates that oscillate around the true target parameter.

How general are these results? In one sense, they are very general. We know from a rich existing literature that matching estimators can be an effective tool for estimating causal effects in these situations. And we could, therefore, with suitable computational power and investigator patience, perform a Monte Carlo version of this simulation across 50,000 simulated datasets, as for Matching Demonstration 3.[33] A less ambitious multiple-sample simulation will suffice to reveal the general results.

Table 5.7 presents results from 10 simulated samples, of which one was used for the estimates reported already in Table 5.6.[34] Across all 10 samples, the true values for the ATT vary only slightly, from a low of 6.87 to a high of 6.98. Rather than present

---

[33]Doing so for 50,000 datasets would require substantial time, since some estimators, such as the genetic matching estimator of Sekhon (2013), take far more time to estimate than the simple weighted averages analyzed for Matching Demonstration 3.

[34]In fact, for the estimates reported in Table 5.6, we chose the sample from among these 10 that, on average over all estimators, delivered estimates that were closest to their respective true ATT. In

the specific point estimates of the ATT for all 10 samples (i.e., an additional 9 versions of Table 5.6), we instead offer the minimum, maximum, and average bias across all 10 estimates of the true ATT for each matching estimator.

As shown in the first two columns, for the incomplete specification the average bias of all estimators was positive and substantial, from a low of .75 to a high of 1.68. Moreover, the minimum amount of bias was positive for many estimators, and the maximum amount of bias was never less than 1.66. When the complete specification was used, the average amount of bias was considerably smaller, and the minima and maxima for each estimator were typically negative and positive, respectively.

If we were to repeat this exercise for 50,000 simulated datasets, we would have more confidence that the average values for the bias inform us of how well these matching estimators perform for this particular pattern of simulated data. We do not do so, in part, because this could give a misleading impression that one of these estimators should be more generally preferred, even for applications very much unlike this one. Instead, we hope to have convinced the reader of a more basic point. Even for examples such as this one, where there is considerable overlap between the treatment and control groups, such that even the traditional estimators that have been around for decades can be used with some confidence, the specific point values generated by matching estimators can differ considerably. It is possible that some of the variability of these estimates can be reduced by routine-specific respecifications of the matching variables after assessing the achieved balance of each estimator. It is unlikely that all variability can be eliminated, and therefore researchers who wish to use a matching strategy should make sure that their conclusions hold across a reasonable range of matching estimators.

We have demonstrated three general points with this example. First, the sort of self-selection dynamic built into this example – in which individuals choose Catholic schooling as a function of their expected gains from Catholic schooling – does not prevent one from consistently estimating the ATT (because Assumption 2-S may still hold), even though it makes estimation of both the ATC and ATE impossible (because Assumption 1-S does not hold). If all variables in S other than anticipation of the individual-level causal effects are observed, then the ATT can be estimated consistently.[35]

Second, even when the ATT is formally identified by conditioning on S while maintaining Assumption 2-S, matching estimators cannot compensate for a decision to mistakenly exclude an observed covariate in S. Failure to condition on the variable

---

contrast, Table 5.7 shows that it will often be the case that matching estimators perform substantially worse.

[35] At the same time, this example shows that even our definition of a "perfect stratification" is somewhat underspecified. According to the definition stated above, if self-selection on the causal effect occurs, a perfect stratification is available only if variables that accurately measure anticipation of the causal effect for each individual are also available and duly included in S. Thus, perhaps it would be preferable to refer to three types of perfect stratification: ATC-perfect stratification for which Assumption 1-S is valid (which enables estimation of the ATC), ATT-perfect stratification for which Assumption 2-S is valid (which enables estimation of the average treatment for the treated), and our unconditionally perfect stratification for which both are valid, enabling estimation of the ATT, ATC, and ATE.

for cognitive skills in this example would invalidate Assumption 2-S (in addition to Assumption 1-S that is already invalidated by the self-section on the causal effect itself). The matching routines will still attempt to balance the matching variables, but the resulting estimates will remain inconsistent and biased for the ATT, ATC, and the ATE.

Third, in cases such as this one, where a researcher attempts to estimate the ATT and where there is a large reservoir of suitable control cases to match to the treatment cases, many alternative specific matching estimators can be used. They will all tend to offer slightly different point estimates, and there is no clear guidance for which ones should be preferred. We therefore recommend that, in these situations, many point estimates be offered and that conclusions not be selected that depend on only a subset of all permissible estimates. However, for applications that depart from the features of this example, some of the more recently developed matching estimators may have a clear advantage, particularly optimal matching when the sample size is small, genetic matching when there is very little prior research to help establish a specification for a propensity score, and coarsened exact matching when one wishes to eliminate incomparable cases in a principled way and when one is comfortable narrowing the analysis to a subset of the treatment cases.

5.5 Remaining Practical Issues in Matching Analysis

In this section, we discuss the remaining practical issues that analysts who consider using matching estimators must confront. We first discuss the possibility of using matching to estimate parameters when the ATE, ATT, and ATC are not identified. We then consider the literature on when, why, and how an analyst may want to estimate effects only after restricting the analysis to the range of common support for the matching variables. We then discuss what is known about the sampling variance of alternative matching estimators, and we conclude with some guidance on matching estimators for causal variables with more than two values.

5.5.1 Matching When Treatment Assignment Is Nonignorable

What if neither Assumption 1-S nor Assumption 2-S is valid because we observe only a subset of the variables in S, which we will now denote by X? We can still match on X using the techniques just summarized, as we did for the incomplete specification in the hypothetical example for Matching Demonstration 4. In that example, we showed that if the complete specification is available, one should of course use it. Yet, in practice this will often not be possible, and yet matching can still be used as a data analysis technique with substantial payoff, assuming that the results are properly interpreted.

Consider, for example, the paper of Sekhon (2004), in which a matching algorithm is used to balance various predictors of voting at the county level in an attempt to determine whether or not John Kerry lost votes in the 2004 presidential election campaign because optical scan voting machines were used instead of direct electronic voting machines in many counties (see Section 1.3.2, page 26, on voting technology effects in Florida for the 2000 election). Sekhon shows that it is unlikely that voting technology caused John Kerry to lose votes. In his analysis, ignorability is not asserted in strict form, as it is quite clear that unobserved features of the counties may well have been related to both the distribution of votes and voting technology decisions. Nonetheless, the analysis is convincing because the predictors of treatment assignment are quite rich, and Sekhon is cautious in his interpretations.

When in this position, however, it is important to concentrate on estimating only one type of treatment effect (usually the ATT, although perhaps the ATE). Because a crucial step must be added to the project – assessing the level of inconsistency and bias that may arise from possible nonignorability of treatment – focusing on a very specific treatment effect of primary interest helps to ground a discussion of an estimate's limitations. Then, after using one of the matching estimators of the last section, one should use the data to minimize bias in the estimates and, if possible, proceed thereafter to a sensitivity analysis (which we will discuss later in Chapter 12). We will return to this issue in depth in Chapter 7, where we will present techniques that combine matching and regression estimators, often for examples wherein assumptions of ignorability are knowingly suspect.

5.5.2   Matching Only on the Region of Common Support

Treatment cases may have no counterparts among the control cases in the observed data because they are said to be "off the support" of S, and vice versa for the control cases.[36] Typically, researchers will notice this situation when the distributions of the estimated propensity scores for the treatment and control cases have substantially different minimum and maximum values, and that is the situation we will consider in this section.

In some cases, the lack of observed overlap in estimated propensity scores may reflect generic sparseness because unusual treatment and control cases will often fail to be sampled for finite datasets. In other cases, there is good reason to believe that some of the lack of observed overlap may have emerged from systematic sources, often related to the choice behavior of individuals. In these situations, it is not a sparseness problem. Instead, a more fundamental mismatch between the observed treatment and control cases exists in the population, as in our earlier hypothetical example in Matching Demonstration 2. In still other situations, unlike those that we focus on in this book, the data exist for a collection of units that is either not drawn at random from a well-defined population (as in many biomedical examples) or where the population is not well-defined (as in many social science examples when administrative units, or other socially defined units, are analyzed). In these cases, the treatment and control cases may be off of the support of each other, and matching is invoked precisely to take account of this predicament, using a variant of the data preprocessing perspective proposed by Ho et al. (2007) to define a useful comparison that can move the relevant literature forward.

---

[36]Support is often given slightly different definitions depending on the context, although most definitions are consistent with a statement such as this one: Support is the union of all intervals of a probability distribution that have true nonzero probability mass.

When in any of these situations, applied researchers who use matching techniques to estimate treatment effects may choose to estimate narrower treatment effects. For example, analysis may be confined only to treatment cases whose estimated propensity scores fall between the minimum and maximum estimated propensity scores of the control cases. Resulting estimates are then interpreted as estimates of a treatment effect that is narrower even than the ATT and is typically labeled the common-support ATT (see Heckman, Ichimura, and Todd 1997, 1998; see also Crump, Hotz, Imbens, and Mitnik 2009). Although using the estimated propensity score to find the region of overlap may not capture all dimensions of the common support (as there may be interior spaces in the joint distribution defined by the matching variables), subsequent matching is then expected to finish the job of eliminating incomparable cases.

Sometimes matching on the region of common support helps to clarify and sharpen the contribution of a study. Even if imposing a common-support condition results in throwing away some of the treatment cases, this can be considered an important substantive finding for some applications. Any resulting estimate is a conditional average treatment effect that is informative only about those in the treatment and control groups who are substantially equivalent with respect to observed treatment assignment and selection variables. In some applications, this is precisely the estimate needed, such as when evaluating whether a program should be expanded in size in order to accommodate more treatment cases but without changing eligibility criteria. (We will discuss these marginal treatment effects in Chapter 9.)

The literature on these common-support strategies is now well developed. Heckman, Ichimura, and Todd (1998; see also Smith and Todd 2005) recommend trimming the region of common support to eliminate cases in regions of the common support with extremely low density (and not just with respect to the estimated propensity score but for the full distribution of the matching variables). This involves selecting a minimum density (labeled the "trimming level") that is greater than zero. Heckman and his colleagues have found that estimates are sensitive to the level of trimming in small samples, with greater bias when the trimming level is lower. More recently, Crump et al. (2009) have developed optimal weighting estimators that are more general but designed to achieve the same goals. Coarsened exact matching can be seen as motivated by the same basic strategy, guided directly by the matching variables rather than their unidimensional reduction in the estimated propensity score. The common support ATT may coincide with the local SATT of coarsened exact matching, but the latter will always be at least as narrow if the underlying matching variables are the same at the outset before any coarsening is applied.

5.5.3 The Expected Variance of Matching Estimates

After computing a matching estimate of some form, most researchers naturally desire a measure of its expected variability across samples of the same size from the same population, either to conduct hypothesis tests or to offer an informed posterior distribution for the causal effect that can guide subsequent research.[37] We did not, however, report

---

[37] There is also a related set of randomization inference techniques, built up from consideration of all of the possible permutations of treatment assignment patterns that could theoretically emerge from alternative enactments of the same treatment assignment routine (see Rosenbaum 2002). These

standard errors for the treatment effect estimates reported in Tables 5.6 or 5.7 for the hypothetical example in Matching Demonstration 4 (although we did indicate in the text the range of the estimates that are provided by the relevant software routines).

Although most of the available software routines provide estimated standard errors, many rely on different methodologies for calculating them. Given their lack of agreement, we caution against too strong of a reliance on the standard error estimates produced by any one software routine, at least at present. Much remains to be worked out before commonly accepted standards for calculating standard errors for all types of matching estimators are available (see Abadie and Imbens 2009, 2012; Hill and Reiter 2006; Imbens and Wooldridge 2009). For now, our advice is to report a range of standard errors produced by alternative software routines for corresponding matching estimates.[38]

We recommend caution for the following reasons. In some simple cases, there is widespread agreement on how to properly estimate standard errors for matching estimators. For example, if a perfect stratification of the data can be found, the data can be analyzed as if they are a stratified random sample with the treatment randomly assigned within each stratum. In this case, the variance estimates from stratified sampling apply. But rarely is a perfect stratification available in practice without substantial sparseness in the data at hand. Once stratification is performed with reference to an estimated propensity score, the independence that is assumed within strata for standard error estimates from stratified sampling methodology is no longer present. And, if one adopts a Bayesian perspective, the model uncertainty of the propensity-score-estimating equation must be represented in the posterior (see Abadie and Imbens 2009; An 2010).

Even so, there is now also widespread agreement that convergence results from nonparametric statistics can be used to justify standard error estimates for large samples. A variety of scholars have begun to work out alternative methods for calculating such asymptotic standard errors for matching estimators, after first rewriting matching estimators as forms of nonparametric regression (see Abadie and Imbens 2006, 2011; Hahn 1998; Heckman, Ichimura, and Todd 1998; Hirano et al. 2003; Imbens 2004; Imbens and Wooldridge 2009). For these large-sample approaches, however, it is

---

permutation ideas generate formulas for evaluating specific null hypotheses, which, from our perspective, are largely uncontroversial. They are especially reasonable when the analyst has deep knowledge of a relatively simple treatment assignment regime and has reason to believe that treatment effects are constant in the population. Although Rosenbaum provides large-sample approximations for these permutation-based tests, the connections to the recent econometrics literature that draws on nonparametric convergence results have not yet been established.

[38]Many matching software routines allow one to calculate bootstrapped standard errors. This is presumably because these easy-to-implement methods were once thought to provide a general framework for estimating the standard errors of alternative matching estimators and hence were a fair and quite general way to compare the relative efficiency of alternative matching estimators (see Tu and Zhou 2002). Unfortunately, Abadie and Imbens (2008) show that conventional bootstrapping will not work for nearest neighbor matching and related estimators. In essence, these matching estimators are a type of two-stage sampling in which a set of treatment and control cases are sampled first and then the possible matching cases are subsampled again based on nearness to the target cases. Yet, there is too much dependence between the first and second sampling stages in the single observed sample, such that resampling within the first stage using the observed sample of target cases does not then generate suitable variation among matching cases in the second stage.

generally assumed that matching is performed directly with regard to the variables in S, and the standard errors are appropriate only for large samples in which sparseness is vanishing. Accordingly, the whole idea of using propensity scores to solve rampant sparseness problems is almost entirely dispensed with, and estimated propensity scores then serve merely to clean up whatever chance variability in the distribution of S across treatment and control cases remains in a finite sample.

Given that the literature on methods to estimate standard errors for matching estimates is still developing, and that software developments lag the literature, it seems prudent to (1) report the standard errors offered by the relevant software routines in sufficient detail so that a reader can understand the method adopted but (2) avoid drawing conclusions that depend on accepting any one particular method for calculating standard errors.

5.5.4 Matching Estimators for Many-Valued Causes

Given the prevalence of studies of many-valued causes, it is somewhat odd to place this section under the more general heading of practical issues. But this is appropriate because most of the complications of estimating many-valued treatment effects are essentially practical, even though very challenging in some cases.

Recall the setup for many-valued causes from Section 2.9, where we have a set of J treatment states, a set of J causal exposure dummy variables, $\{Dj\}_{j=1}^{J}$, and a corresponding set of J potential outcome random variables, $\{Y^{Dj}\}_{j=1}^{J}$. The treatment received by each individual is Dj', and the outcome variable for individual i, y_i, is then equal to $y_i^{Dj'}$. For $j \neq j'$, the potential outcomes of individual i exist as $J-1$ counterfactual outcomes y_i^{Dj}.

There are two basic approaches to matching with many-valued treatments (see Rosenbaum 2002, section 10.2.4). The most straightforward and general approach is to form a series of two-way comparisons between the multiple treatments, estimating a separate propensity score for each contrast between each pair of treatments.[39] After the estimated propensity scores are obtained, treatment effect estimates are calculated pairwise between treatments. Care must be taken, however, to match appropriately on the correct estimated propensity scores. The observed outcomes for individuals with equivalent values on alternative propensity scores cannot be meaningfully compared (see Imbens 2000, section 5).

For example, for three treatments with J equal to 1, 2, and 3, one would first estimate three separate propensity scores, corresponding to three contrasts for the three corresponding dummy variables: $D1$ versus $D2$, $D1$ versus $D3$, and $D2$ versus $D3$.

---

[39]Some simplification of the propensity score estimation is possible. Rather than estimate propensity scores separately for each pairwise comparison, one can use multinomial probit and logit models to estimate the set of propensity scores (see Lechner 2002a, 2002b; see also Hirano and Imbens 2004; Imai and van Dyk 2004; Imbens 2000; Zhao, van Dyk, and Imai 2012). One must still, however, extract the right contrasts from such a model in order to obtain an exhaustive set of estimated propensity scores.

One would obtain three estimated propensity scores: $\text{Pr}_N[d1_i = 1 | d1_i = 1 \text{ or } d2_i = 1, s_i]$, $\text{Pr}_N[d1_i = 1 | d1_i = 1 \text{ or } d3_i = 1, s_i]$, and $\text{Pr}_N[d2_i = 1 | d2_i = 1 \text{ or } d3_i = 1, s_i]$. One would then match separately for each of the three contrasts leaving, for example, those with $d3_i = 1$ unused and unmatched when matching on the propensity score for the comparison of treatment 1 versus treatment 2. At no point would one match together individuals with equivalent values for alternative estimated propensity scores. For example, there is no meaningful causal comparison between two individuals, in which for the first individual $d2_i = 1$ and $\text{Pr}_N[d1_i = 1 | d1_i = 1 \text{ or } d2_i = 1, s_i] = .6$ and for the second individual $d3_i = 1$ and $\text{Pr}_N[d1_i = 1 | d1_i = 1 \text{ or } d3_i = 1, s_i] = .6$.

When the number of treatments is of modest size, such as only four or five alternatives, there is much to recommend in this general approach. However, if the number of treatments is considerably larger, then this fully general approach may be infeasible. One might then choose to simply consider only a subset of causal contrasts for analysis, thereby reducing the aim of the causal analysis.

If the number of treatments can be ordered, then a second approach developed by Joffe and Rosenbaum (1999) and implemented in Lu, Zanutto, Hornik, and Rosenbaum (2001) is possible. These models generally go by the name of dose-response models because they are used to estimate the effects of many different dose sizes of the same treatment, often in comparison with a base dosage of 0 that signifies no treatment.

Rather than estimate separate propensity scores for each pairwise comparison, an ordinal probability model is estimated and the propensity score is defined as a single dimension of the predictors of the model (i.e., ignoring the discrete shifts in the odds of increasing from one dosage level to the next that are parameterized by the estimated cut-point parameters for each dosage level). Thereafter, one typically performs a slightly different form of matching in which optimal matched sets are formed by two criteria, which Lu et al. (2001:1249) refer to as "close on covariates; far apart on doses." The idea here is to form optimal contrasts between selected sets of comparable individuals to generate estimates of counterfactually defined responses. The goal is to be able to offer a predicted response to any shift in a dosage level from any k' to k'', where both k' and k'' are between the smallest and largest dosage values observed. Again, however, these methods assume that the treatment values can be ordered, and further that the propensity scores can be smoothed across dose sizes after partialing out piecewise shifts. Even so, these assumptions are no more severe than what is typically invoked implicitly in standard parametric regression modeling approaches to causality, as we discuss in later chapters.

Some work has continued to examine how the rationale for propensity scores can be usefully generalized without necessarily adopting the full structure of dose-response models. Ordered and nonordered multinomial probability models for modeling treatment assignment are again the foundations of these models (see Hirano and Imbens 2004; Imai and van Dyk 2004; and Zhao et al. 2012 for further details). This literature has progressed slowly because analogs to balance checking across many values of a treatment have not been developed, placing these particular approaches outside of the balance-optimizing agenda that has driven the most recent research on matching methods, as discussed in Section 5.4.2.

5.6 Conclusions

We conclude this chapter by discussing the strengths and weaknesses of matching as a method for causal inference from observational data. Some of the advantages of matching methods are not inherent or unique to matching itself but rather are the result of the analytical framework in which most matching analyses are conducted. Matching focuses attention on the heterogeneity of the causal effect, and it suggests clearly why thinking separately about the ATT, ATC, and ATE is crucial. Matching also forces the analyst to examine the alternative distributions of covariates across those exposed to different levels of the causal variable, and it may suggest that for some applications the only estimates worth interpretation are those that are restricted to the region of common support. Matching, as will become clear in the next chapter, is comparatively conservative in a modeling sense because it does not implicitly invoke the interpolation and extrapolation that characterize parametric regression models.

Although these are the advantages of matching, it is important that we not oversell the potential power of the techniques. In much of the applied literature on matching, the propensity score is presented as a single predictive dimension that can be used to balance the distribution of important covariates across treatment and control cases, thereby warranting causal inference. If one does not observe and utilize variables that, in an infinite sample, would yield a perfect stratification, then simply predicting treatment status from a more limited set of observed variables with a logit model and then matching on the estimated propensity score does not solve the causal inference problem. The estimated propensity scores will balance those variables (in expectation) across the treatment and control cases. But the study will remain open to the sort of "hidden bias" explored by Rosenbaum (2002), which is often labeled selection on the unobservables in the social sciences. Matching is thus a statistical method for analyzing available data, which may have some advantages in some situations. The regression estimators that we will present in the next two chapters are complementary, and our understanding of them has been enriched by the matching literature presented in this chapter.

Chapter 6

Regression Estimators of Causal Effects

Regression models are perhaps the most common form of data analysis used to evaluate alternative explanations for outcomes of interest to quantitatively oriented social scientists. In the past 50 years, a remarkable variety of regression models have been developed by statisticians. Accordingly, most major data analysis software packages allow for regression estimation of the relationships between interval and categorical variables, in cross sections and longitudinal panels, and in nested and multilevel patterns. In this chapter, however, we restrict our attention to ordinary least squares (OLS) regression, focusing mostly on the regression of an interval-scaled variable on a binary causal variable. As we will show, the issues are complicated enough for these models, and it is our knowledge of how least squares models work that allows us to explain this complexity. In addition, nearly all of the insight that can be gained from a deep examination of OLS models carries over to more complex regression models because the identification and heterogeneity issues that generate the complexity apply in analogous fashion to all regression-type models.

In this chapter, we present least squares regression from three different perspectives: (1) regression as a descriptive modeling tool, (2) regression as a parametric adjustment technique for estimating causal effects, and (3) regression as a matching estimator of causal effects. We give more attention to the third of these three perspectives on regression than is customary in methodological texts because this perspective allows one to understand the others from a counterfactual perspective. At the end of the chapter, we will draw some of the connections between least squares regression and more general models, and we will discuss the estimation of causal effects for many-valued causes.

6.1 Regression as a Descriptive Tool

Least squares regression can be justified without reference to causality because it can be considered nothing more than a method for obtaining a best-fitting descriptive model under entailed linearity constraints. Goldberger (1991), for example, motivates least squares regression as a technique to estimate a best-fitting linear approximation to a conditional expectation function that may be nonlinear in the population.

Consider this descriptive motivation of regression a bit more formally. If X is a collection of variables that are thought to be associated with Y in some way, then the conditional expectation function of Y, viewed as a function in X, is denoted $E[Y|X]$. Each particular value of the conditional expectation for a specific realization x of X is then denoted $E[Y|X = x]$.

Least squares regression yields a predicted surface $\hat{Y} = X\hat{\beta}$, where $\hat{\beta}$ is a vector of estimated coefficients from the regression of the realized values y_i on x_i. The predicted surface, $X\hat{\beta}$, does not necessarily run through the specific points of the conditional expectation function, even for an infinite sample, because (1) the conditional expectation function may be a nonlinear function in one or more of the variables in X and (2) a regression model can be fit without parameterizing all nonlinearities in X. An estimated regression surface simply represents a best-fitting linear approximation of $E[Y|X]$ under whatever linearity constraints are entailed by the chosen parameterization of the estimated model.[1]

The following demonstration of this usage of regression is simple. Most readers know this material well and can skip ahead to the next section. But, even so, it may be worthwhile to read the demonstration quickly because we will build directly on it when shifting to the consideration of regression as a causal effect estimator.

Regression Demonstration 1

Recall the stratification example presented as Matching Demonstration 1 (see Section 5.2.1, page 145). Suppose that the same data are being analyzed, as generated by the distributions presented in Tables 5.1 and 5.2; features of these distributions are reproduced in Table 6.1 in more compact form. As before, assume that well-defined causal states continue to exist and that S serves as a perfect stratification of the data.[2] Accordingly, the conditional expectations in the last three panels of Table 6.1 are equal as shown.

But, for this demonstration of regression as a descriptive tool, suppose that a cautious researcher does not wish to rush ahead and attempt to estimate the specific underlying causal effect of D on Y, either averaged across all individuals or averaged across particular subsets of the population. Instead, the researcher is cautious and is willing to assert only that the variables S, D, and Y constitute some portion of a larger system of causal relationships. In particular, the researcher is unwilling to assert anything about the existence or nonexistence of other variables that may also lie on the directed paths that reach D and Y. This is tantamount to doubting the claim that S offers a perfect stratification of the data, even though that claim is true by construction for this example.

---

[1] One can fit a large variety of nonlinear surfaces with regression by artful parameterizations of the variables in X, but these surfaces are always generated by a linear combination of a coefficient vector and values on some well-defined coding of the variables in X.

[2] For this section, we will also stipulate that the conditional variances of the potential outcomes are constant across both of the potential outcomes and across levels of S.

Table 6.1 The Joint Probability Distribution and
Conditional Population Expectations for Regression
Demonstration 1

	Joint probability distribution of S and D	
	Control group: $D=0$	Treatment group: $D=1$
$S=1$	$\Pr[S=1,D=0]=.36$	$\Pr[S=1,D=1]=.08$
$S=2$	$\Pr[S=2,D=0]=.12$	$\Pr[S=2,D=1]=.12$
$S=3$	$\Pr[S=3,D=0]=.12$	$\Pr[S=3,D=1]=.2$

	Potential outcomes under the control state	
$S=1$	$E[Y^0|S=1,D=0]=2$	$E[Y^0|S=1,D=1]=2$
$S=2$	$E[Y^0|S=2,D=0]=6$	$E[Y^0|S=2,D=1]=6$
$S=3$	$E[Y^0|S=3,D=0]=10$	$E[Y^0|S=3,D=1]=10$

	Potential outcomes under the treatment state	
$S=1$	$E[Y^1|S=1,D=0]=4$	$E[Y^1|S=1,D=1]=4$
$S=2$	$E[Y^1|S=2,D=0]=8$	$E[Y^1|S=2,D=1]=8$
$S=3$	$E[Y^1|S=3,D=0]=14$	$E[Y^1|S=3,D=1]=14$

	Observed outcomes	
$S=1$	$E[Y|S=1,D=0]=2$	$E[Y|S=1,D=1]=4$
$S=2$	$E[Y|S=2,D=0]=6$	$E[Y|S=2,D=1]=8$
$S=3$	$E[Y|S=3,D=0]=10$	$E[Y|S=3,D=1]=14$

In this situation, suppose that the researcher simply wishes to estimate the best lin-
ear approximation to the conditional expectation $E[Y|D,S]$ and does not wish to then
give a causal interpretation to any of the coefficients that define the linear approx-
imation. The six true values of $E[Y|D,S]$ are given in the last panel of Table 6.1.
Notice that the linearity of $E[Y|D,S]$ in D and S is present only when $S \leq 2$. The
value of 14 for $E[Y|D=1,S=3]$ makes $E[Y|D,S]$ nonlinear in D and S over their full
distributions.

Now consider the predicted surfaces that would result from the estimation of two
alternative least squares regression models with data from a sample of infinite size (to
render sampling error zero). A regression of Y on D and S that treats D as a dummy
variable and S as an interval-scaled variable would yield a predictive surface of

$$\hat{Y} = -2.71 + 2.69(D) + 4.45(S). \tag{6.1}$$

This model constrains the partial association between Y and S to be linear. It represents a sensible predicted regression surface because it is a best-fitting, linear-in-
the-parameters model of the association between Y and the two variables D and S,

where "best" is defined as minimizing the average squared differences between the fitted values and the true values of the conditional expectation function.

For this example, one can offer a better descriptive fit at little interpretive cost by using a more flexible parameterization of S. An alternative regression that treats S as a discrete variable represented in the estimation routine by dummy variables $S2$ and $S3$ (for S equal to 2 and S equal to 3, respectively) would yield a predictive surface of

$$\hat{Y} = 1.86 + 2.75(D) + 3.76(S2) + 8.92(S3). \qquad (6.2)$$

Like the predicted surface for the model in Equation (6.1), this model is also a best linear approximation to the six values of the true conditional expectation $E[Y|D,S]$. The specific estimated values are

$$D = 0, S = 1: \ \hat{Y} = 1.86,$$
$$D = 0, S = 2: \ \hat{Y} = 5.62,$$
$$D = 0, S = 3: \ \hat{Y} = 10.78,$$
$$D = 1, S = 1: \ \hat{Y} = 4.61,$$
$$D = 1, S = 2: \ \hat{Y} = 8.37,$$
$$D = 1, S = 3: \ \hat{Y} = 13.53.$$

In contrast to the model in Equation (6.1), for this model the variable S is given a fully flexible coding. As a result, parameters are fit that uniquely represent all values of S.[3] The predicted change in Y for a shift in S from 1 to 2 is 3.76 (i.e., $5.62 - 1.86 = 3.76$

---

[3]The difference between a model in which a variable is given a fully flexible coding and one in which it is given a more constrained coding is clearer for a simpler conditional expectation function. For $E[Y|S]$, consider the values in the cells of Table 6.1. The three values of $E[Y|S]$ can be obtained from the first and fourth panels of Table 6.1 as follows:

$$E[Y|S=1] = \frac{.36}{(.36+.08)}(2) + \frac{.08}{(.36+.08)}(4) = 2.36,$$

$$E[Y|S=2] = \frac{.12}{(.12+.12)}(6) + \frac{.12}{(.12+.12)}(8) = 7,$$

$$E[Y|S=3] = \frac{.12}{(.12+.2)}(10) + \frac{.2}{(.12+.2)}(14) = 12.5.$$

Notice that these three values of $E[Y|S]$ do not fall on a straight line; the middle value of 7 is closer to 2.36 than it is to 12.5.

For $E[Y|S]$, a least squares regression of Y on S, treating S as an interval-scaled variable, would yield a predictive surface of

$$\hat{Y} = -2.78 + 5.05(S).$$

The three values of this estimated regression surface lie on a straight line -2.27, 7.32, and 12.37 – and they do not match the corresponding true values of 2.36, 7, and 12.5. A regression of Y on S, treating S as a discrete variable with dummy variables $S2$ and $S3$, would yield an alternative predictive surface of

$$\hat{Y} = 2.36 + 4.64(S2) + 10.14(S3).$$

This second model uses a fully flexible coding of S, and each value of the conditional expectation function is a unique function of the parameters in the model (that is, $2.36 = 2.36$, $4.64 + 2.36 = 7$, and $10.14 + 2.36 = 12.5$). Thus, in this case, the regression model would, in a suitably large sample, estimate the three values of $E[Y|S]$ exactly.

and $8.37 - 4.61 = 3.76$), whereas the predicted change in Y for a shift in S from 2 to 3 is 5.16 (i.e., $10.78 - 5.62 = 5.16$ and $13.53 - 8.37 = 5.16$).

Even so, the model in Equation (6.2) constrains the parameter for D to be the same without regard to the value of S. And, because the level of Y depends on the interaction of S and D, specifying more than one parameter for the three values of S does not bring the predicted regression surface into alignment with the six values of $E[Y|D,S]$ presented in the last panel of Table 6.1. Thus, even when S is given a fully flexible coding (and even for an infinitely large sample), the fitted values do not equal the true values of $E[Y|D,S]$.[4] As we discuss later, a model that is saturated fully in both S and D – that is, one that adds two additional parameters for the interactions between D and both $S2$ and $S3$ – would yield predicted values that would exactly match the six true values of $E[Y|D,S]$ in a dataset of sufficient size.

Recall the more general statement of the descriptive motivation of regression analysis presented above, in which the predicted surface $\hat{Y} = X\hat{\beta}$ is estimated for the sole purpose of obtaining a best-fitting linear approximation to the true conditional expectation function $E[Y|X]$. When the purposes of regression are so narrowly restricted, the outcome variable of interest, Y, is not generally thought to be a function of potential outcomes associated with well-defined causal states. Consequently, it would be inappropriate to give a causal interpretation to any of the estimated coefficients in $\hat{\beta}$.

This perspective implies that if one were to add more variables to the predictors, embedding X in a more encompassing set of variables W, then a new set of least squares estimates $\hat{\gamma}$ could be obtained by regressing Y on W. The estimated surface $W\hat{\gamma}$ then represents a best-fitting, linear-in-the-parameters, descriptive fit to a more encompassing conditional expectation function, $E[Y|W]$. Whether one then prefers $W\hat{\gamma}$ to $X\hat{\beta}$ as a description of the variation in Y depends on whether one finds it more useful to approximate $E[Y|W]$ than $E[Y|X]$. The former regression approximation is often referred to as the long regression, with the latter representing the short regression. These labels are aptly chosen, when regression is considered nothing more than a descriptive tool, because there is no inherent reason to prefer a short to a long regression if neither is meant to be interpreted as anything other than a best-fitting linear approximation to its respective true conditional expectation function.

In many applied regression textbooks, the descriptive motivation of regression receives no direct explication. And, in fact, many textbooks state that the only correct specification of a regression model is one that includes all explanatory variables. Goldberger (1991) admonishes such textbook writers, countering their claims:

---

[4] Why would one ever prefer a constrained regression model of this sort? Consider a conditional expectation function, $E[Y|X]$, where Y is earnings and X is years of education (with 21 values from 0 to 20). A fully flexible coding of X would fit 20 dummy variables for the 21 values of X. This would allow the predicted surface to change only modestly between some years (such as between 7 and 8 and between 12 and 13) and more dramatically between other years (such as between 11 and 12 and between 15 and 16). However, one might wish to treat X as an interval-scaled variable, smoothing these increases from year to year by constraining them to a best-fitting line parameterized only by an intercept and a constant slope. This constrained model would not fit the conditional expectation function as closely as the model with 20 dummy variables, but it might be preferred in some situations because it is easier to present and uses fewer degrees of freedom, which could be important if the model is estimated with a small sample.

> An alternative position is less stringent and is free of causal language. Nothing in the CR [classical regression] model itself requires an exhaustive list of explanatory variables, nor any assumption about the direction of causality. (Goldberger 1991:173)

Goldberger is surely correct, but his perspective nonetheless begs an important question on the ultimate utility of descriptively motivated regression. Clearly, if one wishes to know only predicted values of the outcome Y for those not originally studied but whose variables in X are known, then being able to form the surface $X\hat{\beta}$ is a good first step (and perhaps a good last step). And, if one wishes to build a more elaborate regression model, allowing for an additional variable in W or explicitly accounting for multilevel variability by modeling the nested structure of the data, then regression results will be useful if the aim is to generate descriptive reductions of the data. But, if one wishes to know the value of Y that would result for any individual in the population if a variable in X were shifted from a value k to a value k', then regression results may be uninformative.

Many researchers (perhaps a clear majority) who use regression models in their research are very much interested in causal effects. Knowing the interests of their readers, many textbook authors offer presentations of regression that sidestep these issues artfully by, for example, discussing how biased regression coefficients result from the omission of important explanatory variables but without introducing explicit, formal notions of causality into their presentations. Draper and Smith (1998:236), for example, write of the bias that enters into estimated regression coefficients when only a subset of the variables in the "true response relationship" are included in the fitted model. Similarly, Greene (2000:334) writes of the same form of bias that results from estimating coefficients for a subset of the variables from the "correctly specified regression model."[5] And, in his presentation of regression models for social scientists, Stolzenberg (2004:188) equivocates:

> Philosophical arguments about the nature of causation notwithstanding (see Holland, 1986), in most social science uses of regression, the *effect* of an independent variable on a dependent variable is the *rate* at which differences in the independent variable are associated with (or cause) differences or changes in the dependent variable. (italics in the original)

We assume that the readers of our book are interested in causal effect estimators. And thus, although we recognize the classical regression tradition, perhaps best defended by Goldberger (1991) as interpretable merely as a descriptive data reduction tool, we will consider regression as a causal effect estimator in the remaining sections of this chapter. And we further note that, in spite of our reference to Goldberger (1991), in other writing Goldberger has made it absolutely clear that he too was very much interested in the proper usage of regression models to offer warranted causal claims. This is perhaps most clear in work in which he criticized what he regarded as unwarranted causal claims generated by others using regression techniques, such as in his robust critique of Coleman's Catholic schools research (see Goldberger and Cain 1982). We will return

---

[5] There are, of course, other textbooks that do present a more complete perspective, such as Angrist and Pischke (2009), Berk (2004), Freedman (2005), and Gelman and Hill (2007).

to a discussion of the notion of a correct specification of a regression model in the final section of the chapter, where we discuss the connections between theoretical models and regressions as all-cause perfect specifications. Until then, however, we return to the same basic scenario considered in our presentation of matching in Chapter 5: the estimation of a single causal effect that may be confounded by other variables.

6.2 Regression Adjustment as a Strategy to Estimate Causal Effects

In this section, we consider the estimation of causal effects in which least squares regression is used to adjust for variables thought to be related to both the causal variable and the outcome variable. We first consider the textbook treatment of the concept of omitted-variable bias, with which most readers are probably well acquainted. Thereafter, we consider the same set of ideas after specifying the potential outcome variables that the counterfactual tradition assumes lie beneath the observed data.

6.2.1 Regression Models and Omitted-Variable Bias

Suppose that one is interested in estimating the causal effect of a binary variable D on an observed outcome Y. This goal can be motivated as an attempt to obtain a consistent and unbiased estimate of a coefficient δ in a generic bivariate regression equation,

$$Y = \alpha + \delta D + \varepsilon, \tag{6.3}$$

where α is an intercept and ε is a summary random variable that represents all other causes of Y (some of which may be related to the causal variable of interest, D). When Equation (6.3) is used to represent the causal effect of D on Y without any reference to individual-varying potential outcomes, the parameter δ is implicitly cast as an invariant, structural causal effect that applies to all members of the population of interest.[6]

The OLS estimator of this bivariate regression coefficient is then

$$\hat{\delta}_{\text{OLS, bivariate}} \equiv \frac{\text{Cov}_N(y_i, d_i)}{\text{Var}_N(d_i)}, \tag{6.4}$$

where $\text{Cov}_N(.)$ and $\text{Var}_N(.)$ are consistent and unbiased, sample-based estimates from a sample of size N of the population-level covariance and variance of the variables that are their arguments.[7] Because D is a binary variable, $\hat{\delta}_{\text{OLS, bivariate}}$ is exactly equivalent to the naive estimator, $E_N[y_i|d_i = 1] - E_N[y_i|d_i = 0]$, presented in Equation

---

[6]Although this is generally the case, there are of course introductions to regression that explicitly define δ as the mean effect of D on Y across units in the population of interest or, as was noted in the last section, without regard to causality at all.

[7]Notice that we are again focusing on the essential features of the methods. When we present least squares regression estimators in this chapter, we will maintain three implicit assumptions to simplify inference complications in order to focus on identification issues. First, we ignore degree-of-freedom adjustments because we assume that the available sample is very large. To be more precise in justifying unbiasedness for a finite sample, we would want to indicate that the sample variance of D does not equal the population-level variance of D in the absence of such a degree-of-freedom

(2.9) (i.e., the sample mean of y_i for those in the treatment group minus the sample mean of y_i for those in the control group). Our analysis thus follows quite closely the discussion of the naive estimator in Section 2.7.3. The difference is that here we will develop the same basic claims with reference to the relationship between D and ε rather than the general implications of heterogeneity of the causal effect.

Consider first a case in which D is randomly assigned, as when individuals are randomly assigned to the treatment and control groups. In this case, D would be uncorrelated with ε in Equation (6.3), even though there may be a chance correlation between D and ε in any finite set of study subjects.[8] The literature on regression, when presented as a causal effect estimator, maintains that, in this case, (1) the estimator $\hat{\delta}_{\text{OLS, bivariate}}$ is consistent and unbiased for δ in Equation (6.3) and (2) δ can be interpreted as the causal effect of D on Y.

To understand this claim, it is best to consider a counterexample in which D is correlated with ε in the population because D is correlated with other causes of Y that are implicitly embedded in ε. For a familiar example, consider again the effect of education on earnings. Individuals are not randomly assigned to the treatment "completed a bachelor's degree." It is generally thought that those who complete college would be more likely to have had high levels of earnings in the absence of a college education. If this is true, D and the population-level error term ε are correlated because those who have a 1 on D are more likely to have high values rather than low values for ε. For this example, the least squares regression estimator $\hat{\delta}_{\text{OLS, bivariate}}$ in Equation (6.4) would not yield an estimate of δ that can be regarded as consistent or unbiased for the causal effect of D on Y. Instead, $\hat{\delta}_{\text{OLS,bivariate}}$ must be interpreted as inconsistent and upwardly biased. In the substance of the college-degree example, $\hat{\delta}_{\text{OLS, bivariate}}$ would be a poor estimate of the causal effect of a college degree on

---

adjustment, and so on. We merely label $\text{Var}_N(.)$ as signifying such a consistent and unbiased estimate of the population-level-variance of that which is its argument. Thus, $\text{Var}_N(.)$ implicitly includes the proper degree-of-freedom adjustment, which would be $N/(N-1)$ and which would then be multiplied by the average of squared deviations from the sample mean. Second, we will assume that the sampling-error components of the regression error terms have "zero conditional mean"; see Wooldridge (2010, equation 4.3); see also the "strict exogeneity" assumption of Hayashi (2000, equation 1.1.7). Under this assumption, the finite sample bias of the OLS estimator of each coefficient has expectation equal to 0, even though we allow the predictors to be random variables rather than a fixed feature of the design. As will become clear, we will have a lot to say about assumptions regarding regression error terms in this chapter, and for now we invoke this assumption only in the limited sense that it allows us to eliminate finite sample bias from our consideration. Third, our perfect measurement assumption rules out measurement error in predictors, which eliminates attenuation bias in all regression coefficient estimates.

[8] We will frequently refer to D and ε as being uncorrelated for this type of assumption, as this is the semantics that most social scientists seem to use and understand when discussing these issues. Most textbook presentations of regression discuss very specific exogeneity assumptions for D that imply a correlation of 0 between D and ε. Usually, in the social sciences the assumption is defined either by mean independence of D and ε or as an assumed covariance of 0 between D and ε. Both of these imply a correlation between D and ε of 0. In statistics, one often finds a stronger assumption: D and ε must be completely independent of each other. The argument in favor of this stronger assumption, which is convincing to statisticians, is that an inference is strongest when it holds under any transformation of Y (and thus any transformation of ε). When full independence of D and ε holds, mean independence of D and ε, a covariance of 0 between D and ε, and a 0 correlation between D and ε are all implied.

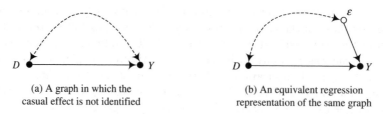

(a) A graph in which the   (b) An equivalent regression
casual effect is not identified   representation of the same graph

Figure 6.1 Graphs for a regression equation of the causal effect of D on Y.

earnings because it would suggest that the effect of obtaining a college degree is larger than it really is.[9]

Figure 6.1(a) presents a graph where D and Y are connected by two types of paths, the direct causal effect $D \to Y$ and an unspecified number of back-door paths represented by $D \dashleftarrow\dashrightarrow Y$. (Recall that bidirected edges $\dashleftarrow\dashrightarrow$ represent an unspecified number of common causes of the two variables that they connect.) For Figure 6.1(a), the causal effect of D on Y is not identified because no observable variables are available to block the back-door paths represented by $D \dashleftarrow\dashrightarrow Y$.

Figure 6.1(b) is the regression analog to the graph in Figure 6.1(a). It contains three edges: $D \to Y$, $\varepsilon \to Y$, and $D \dashleftarrow\dashrightarrow \varepsilon$, where the node for ε is represented by a hollow circle $\circ$ rather than a solid circle $\bullet$ in order to indicate that ε is an unobserved variable. The unblocked back-door paths from D to Y now run through the error term ε, and the dependence represented by the bidirected edge confounds the causal effect of D on Y. Bivariate regression results, when interpreted as warranted causal effect estimates, assume that there are no such unblocked back-door paths from the causal variable to the outcome variable.

For many applications in the social sciences, a correlation between D and ε is conceptualized as a problem of omitted variables. For the example in this section, a bivariate OLS estimate of the effect of a college degree on labor market earnings would be said to be biased because intelligence is unobserved but is correlated with both education and earnings. Its omission from Equation (6.3) leads the estimate of the effect of a college degree on earnings from that equation to be larger than it would have been if a variable for intelligence were instead included in the equation.

This perspective, however, has led to much confusion, especially in cases in which a correlation between D and ε emerges because subjects choose different levels of D based on their expectations about the variability of Y, and hence their own expectations of the causal effect itself. For example, those who attend college may be more likely to benefit from college than those who do not, even independent of the unobserved ability factor. Although this latent form of anticipation can be labeled an omitted

---

[9]Consider for one last time the alternative and permissible descriptive interpretation: The least squares regression estimator $\hat{\delta}_{\text{OLS, bivariate}}$ in Equation (6.4) could be interpreted as consistent and unbiased for δ, where the regression surface generated by the estimation of δ in Equation (6.3) is considered only a descriptively motivated, best linear prediction of the conditional expectation function, $E[Y|D]$ (i.e., where $\hat{\alpha}$ is consistent and unbiased for $E[Y|D=0]$ and $\hat{\alpha}+\hat{\delta}$ is consistent and unbiased for $E[Y|D=1]$). In the substance of the college-degree example, the estimate could be regarded as an estimate of the mean difference between the earnings of those who have obtained a college degree and those who have not, without requiring or warranting any causal interpretation.

variable, it is generally not. Instead, the language of research shifts toward notions such as self-selection bias, and this is less comfortable territory for the typical applied researcher.

To clarify the connections between omitted-variable bias and self-selection bias within a more general presentation, we utilize the potential outcome model in the next section. We break the error term in Equation (6.3) into component pieces defined by underlying potential outcome variables and allow for the more general forms of causal effect heterogeneity that are implicitly ruled out by constant-coefficient models.

6.2.2 Potential Outcomes and Omitted-Variable Bias

Consider the same set of ideas but now use the potential outcome model to define the observed variables. Here, we will build directly on the variant of the potential outcome model presented in Section 4.3.2. From that presentation, recall Equation (4.6), which we reintroduce here as

$$Y = \mu^0 + (\mu^1 - \mu^0)D + \{v^0 + D(v^1 - v^0)\}, \tag{6.5}$$

where $\mu^0 \equiv E[Y^0]$, $\mu^1 \equiv E[Y^1]$, $v^0 \equiv Y^0 - E[Y^0]$, and $v^1 \equiv Y^1 - E[Y^1]$. We could rewrite this equation to bring it into closer alignment with Equation (6.3) by stipulating that $\alpha = \mu^0$, $\delta = (\mu^1 - \mu^0)$, and $\varepsilon = v^0 + D(v^1 - v^0)$. But note that these equalities would redefine what is typically meant by the terms α, δ, and ε in Equation (6.3). The parameters α and δ in Equation (6.3) are usually not considered to be equal to $E[Y^0]$ or $E[\delta]$ for two reasons: (1) models are usually asserted in the regression tradition (e.g., in Draper and Smith 1998) without any reference to underlying causal states tied to potential outcomes and (2) the parameters α and δ are usually implicitly held to be constant structural effects that do not vary over individuals in the population. Similarly, the error term, ε, in Equation (6.3) is almost never separated into two pieces as a function of the definition of potential outcomes and their relationship to D. For these reasons, Equation (6.5) is quite different from the traditional bivariate regression in Equation (6.3), in the sense that it is more finely articulated but also irretrievably tied to a particular formalization of a causal effect that is allowed to vary across individuals.

Suppose that we are interested in estimating the average treatment effect (ATE), denoted $(\mu^1 - \mu^0)$ here. The causal variable D could be correlated with the population-level variant of the error term $v^0 + D(v^1 - v^0)$ in Equation (6.5) in two ways. First, suppose that there is a net baseline difference in the hypothetical no-treatment state that is correlated with membership in the treatment group, but the size of the individual-level treatment effect does not differ on average between those in the treatment group and those in the control group. In this case, v^0 would be correlated with D, generating a correlation between $\{v^0 + D(v^1 - v^0)\}$ and D, even though the $D(v^1 - v^0)$ term in $\{v^0 + D(v^1 - v^0)\}$ would be equal to zero on average because $v^1 - v^0$ does not vary with D. Second, suppose there is a net treatment effect difference that is correlated with membership in the treatment group, but there is no net baseline difference in the absence of treatment. Now, $D(v^1 - v^0)$ would be correlated with D, even though v^0 is not, because the average difference in $v^1 - v^0$ varies across those in the treatment

Table 6.2 Examples of the Two Basic Forms of Bias for Least Squares
Regression

			Differential baseline bias only				
	y_i^1	y_i^0	v_i^1	v_i^0	y_i	d_i	$v_i^0 + d_i(v_i^1 - v_i^0)$
In treatment group	20	10	5	5	20	1	5
In control group	10	0	−5	−5	0	0	−5

			Differential treatment effect bias only				
	y_i^1	y_i^0	v_i^1	v_i^0	y_i	d_i	$v_i^0 + d_i(v_i^1 - v_i^0)$
In treatment group	20	10	2.5	0	20	1	2.5
In control group	15	10	−2.5	0	10	0	0

			Both types of bias				
	y_i^1	y_i^0	v_i^1	v_i^0	y_i	d_i	$v_i^0 + d_i(v_i^1 - v_i^0)$
In treatment group	25	5	5	−2.5	25	1	5
In control group	15	10	−5	2.5	10	0	2.5

group and those in the control group. In either case, an OLS regression of the realized
values of Y on D would yield an inconsistent and biased estimate of $(\mu^1 - \mu^0)$.

It may be helpful to see precisely how these sorts of bias come about with reference
to the potential outcomes of individuals. Table 6.2 presents three simple two-person
examples in which the least squares bivariate regression estimator $\hat{\delta}_{\text{OLS, bivariate}}$ in
Equation (6.4) is biased. Each panel presents the potential outcome values for two
individuals and then the implied observed data and error term in the braces from
Equation (6.5). Assume for convenience that there are only two types of individuals in
the population, both of which are homogeneous with respect to the outcomes under
study and both of which comprise one half of the population. For the three examples
in Table 6.2, we have sampled one of each of these two types of individuals for study.

For the example in the first panel, δ_i is equal to 10 for both individuals. As a
result, the true ATE is also equal to 10. The values of v_i^1 and v_i^0 are deviations of y_i^1
and y_i^0 from $E[Y^1]$ and $E[Y^0]$, respectively. Because $E[Y^1] = 15$, v_i^1 is equal to 5 for
the individual in the treatment group and −5 for the individual in the control group.
Likewise, because $E[Y^0] = 5$, v_i^0 is also equal to 5 for the individual in the treatment
group and −5 for the individual in the control group.

As noted earlier, the bivariate regression estimate of the coefficient on D is equal
to the naive estimator, $E_N[y_i|d_i = 1] - E_N[y_i|d_i = 0]$. Accordingly, a regression of the
values for y_i on d_i would yield a value of 0 for the intercept and a value of 20 for
the coefficient on D. This estimated value of 20 is an upwardly biased estimate for the
true average causal effect because the values of d_i are positively correlated with the
values of the error term $v_i^0 + d_i(v_i^1 - v_i^0)$. In this case, the individual with a value of 1

for d_i has a value of 5 for the error term, whereas the individual with a value of 0 for d_i has a value of -5 for the error term.

For the example in the second panel, the relevant difference between the individual in the treatment group and the individual in the control group is in the value of y_i^1. In this variant, both individuals would have had the same outcome if they were both in the control state, but the individual in the treatment group would benefit relatively more from being in the treatment state. Consequently, the values of d_i are correlated with the values of the error term in the last column because the true treatment effect is larger for the individual in the treatment group than for the individual in the control group. A bivariate regression would yield an estimate of 10 for the ATE, even though the true ATE is only 7.5 in this case.[10]

Finally, in the third panel of the table, both forms of baseline and net treatment effect bias are present, and in opposite directions. In combination, however, they still generate a positive correlation between the values of d_i and the error term in the last column. This pattern results in a bivariate regression estimate of 15, which is upwardly biased for the true ATE of 12.5.

For symmetry, and some additional insight, now consider two additional two-person examples in which regression gives an unbiased estimate of the average causal effect. For the first panel of Table 6.3, the potential outcomes are independent of D, and as a result a bivariate regression of the values y_i on d_i would yield an unbiased estimate of 10 for the true ATE. But the example in the second panel is quite different. Here, the values of v_i^1 and v_i^0 are each correlated with the values of d_i, but they cancel each other out when they jointly constitute the error term in the final column. Thus, a bivariate regression yields an unbiased estimate of 0 for the true ATE of 0. And, yet, with knowledge of the values for y_i^1 and y_i^0, it is clear that these results mask important heterogeneity of the causal effect. Even though the average causal effect is indeed 0, the individual-level causal effects are equal to 10 and -10 for the individuals in the treatment group and control group, respectively. Thus, regression gives the right answer, but it hides the underlying heterogeneity that one would almost certainly wish to know.[11]

Having considered these examples, we are now in a position to answer, with reference to the potential outcome model, the question that so often challenges students when first introduced to regression as a causal effect estimator: What is the error term of a regression equation? Compare the third and fourth columns with the final column in Tables 6.2 and 6.3. The regression error term, $v^0 + D(v^1 - v^0)$, is equal to v^0 for those in the control group and v^1 for those in the treatment group. This can be seen without reference to the examples in the tables. Simply rearrange $v^0 + D(v^1 - v^0)$ as

---

[10]Note, however, that 10 is the true average treatment effect for the treated (ATT), or in this case the treatment effect for the treated for the sole individual in the treatment group. This is the same essential pattern as for Matching Demonstration 4, where the ATT is identified because, in this case, Assumption 2 in Equation (2.16) would hold because $(y_i^0|d_i = 1) = (y_i^0|d_i = 0) = 10$.

[11]Assumptions 1 and 2 in Equations (2.15) and (2.16) are therefore sufficient but not necessary for the naive estimator and bivariate OLS regression estimator to consistently estimate the ATE. We have not mentioned this qualification until now because we do not believe that perfect cancellation occurs in social science applications. This assumption is sometimes labeled "faithfulness" in the counterfactuals literature.

Table 6.3 Two-Person Examples in Which Least Squares Regression Estimates Are Unbiased

| | Independence of (Y^1, Y^0) from D | | | | | | |
	y_i^1	y_i^0	v_i^1	v_i^0	y_i	d_i	$v_i^0 + d_i(v_i^1 - v_i^0)$
In treatment group	20	10	0	0	20	1	0
In control group	20	10	0	0	10	0	0

| | Offsetting dependence of Y^1 and Y^0 on D | | | | | | |
	y_i^1	y_i^0	v_i^1	v_i^0	y_i	d_i	$v_i^0 + d_i(v_i^1 - v_i^0)$
In treatment group	20	10	5	−5	20	1	5
In control group	10	20	−5	5	20	0	5

$Dv^1 + (1 - D)v^0$ and then rewrite Equation (6.5) as

$$Y = \mu^0 + (\mu^1 - \mu^0)D + \{Dv^1 + (1 - D)v^0\}. \qquad (6.6)$$

It should be clear that the error term now appears very much like the definition of Y presented earlier as $DY^1 + (1 - D)Y^0$ in Equation (2.2). Just as Y switches between Y^1 and Y^0 as a function of D, the error term switches between v^1 and v^0 as a function of D. Given that v^1 and v^0 can be interpreted as Y^1 and Y^0 centered around their respective population-level expectations $E[Y^1]$ and $E[Y^0]$, this should not be surprising.

Even so, few presentations of regression characterize the error term of a bivariate regression in this way. Some notable exceptions do exist. The connection is made to the counterfactual tradition by specifying Equation (6.3) as

$$Y = \alpha + \delta D + \varepsilon_{(D)}, \qquad (6.7)$$

where the error term $\varepsilon_{(D)}$ is considered to be an entirely different random variable for each value of D (see Pratt and Schlaifer 1988). Consequently, the error term ε in Equation (6.3) switches between $\varepsilon_{(1)}$ and $\varepsilon_{(0)}$ in Equation (6.7) depending on whether D is equal to 1 or 0.[12]

6.2.3   Regression as Adjustment for Otherwise Omitted Variables

The basic strategy behind regression analysis as an adjustment technique to estimate a causal effect is to add a sufficient set of "control variables" to the bivariate regression in Equation (6.3). The goal is to break a correlation between the treatment variable

---

[12]This is the same approach taken by Freedman (see Berk 2004; Freedman 2005), and he refers to Equation (6.7) as a response schedule. See also the discussion of Sobel (1995). For a continuous variable, Garen (1984) notes that there would be an infinite number of error terms (see his discussion of his equation 10).

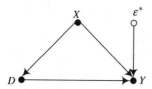

Figure 6.2 A causal graph for a regression equation in which the causal effect of D on Y is identified by conditioning on X.

D and the error term ε, as in

$$Y = \alpha + \delta D + X\beta + \varepsilon^*, \tag{6.8}$$

where X represents one or more variables, β is a coefficient (or a conformable vector of coefficients if X represents more than one variable), ε^* is a residualized version of the original error term ε from Equation (6.3), and all else is as defined for Equation (6.3).

For the multiple regression analog to the least squares bivariate regression estimator $\hat{\delta}_{\text{OLS, bivariate}}$ in Equation (6.4), the observed data values d_i and x_i are embedded in an all-encompassing $\mathbf{Q}$ matrix, which is $N \times K$, where N is the number of respondents and K is the number of variables in X plus 2 (one for the constant and one for the treatment variable D). The OLS estimator for the parameters in Equation (6.8) is then written in matrix form as

$$\hat{\delta}_{\text{OLS, multiple}} \equiv (\mathbf{Q}'\mathbf{Q})^{-1}\mathbf{Q}'\mathbf{y}, \tag{6.9}$$

where $\mathbf{y}$ is an $N \times 1$ vector for the observed outcomes y_i. As all regression textbooks show, there is nothing magical about these least squares computations, even though the matrix representation may appear unfamiliar to some readers. OLS regression is equivalent to the following three-step regression procedure with reference to Equation (6.8) – and without reference to the perhaps overly compact Equation (6.9):

1. Regress y_i on x_i and calculate $y_i^* = y_i - \hat{y}_i$;

2. Regress d_i on x_i and calculate $d_i^* = d_i - \hat{d}_i$;

3. Regress y_i^* on d_i^*.

The regression coefficient on d_i^* yielded by step 3 is the OLS estimate of δ in Equation (6.8), which is typically declared consistent and unbiased for the true value of δ if the correlation between D and ε^* is assumed to be equal to zero. Thus, in this simple example, OLS regression is equivalent to estimating the relationship between residualized versions of Y and D from which their common dependence on other variables in X has been "subtracted out."

Even though the variables in X might be labeled control variables in a regression analysis of a causal effect, this label expresses the intent rather than the outcome of their utilization. The goal of such a regression adjustment strategy is to find variables in X that can be used to redraw the graph in Figure 6.1(b) as the causal graph in Figure 6.2. If this can be done, then one can condition on X in order to consistently estimate the causal effect of D on Y because X blocks the only back-door path between D and Y.

If D is uncorrelated with ε^* (i.e., the error term net of adjustment for X), then least squares regression yields an estimate that is ostensibly freed of the bias generated by the correlation of the treatment D with the error term ε in Equation (6.3). However, even in this case some complications remain when one invokes the potential outcome model.

First, if one assumes that δ is truly constant across individuals (i.e., that $y_i^1 - y_i^0$ is equal to the same constant for all individuals i), then the OLS estimate is consistent and unbiased for δ and for $(\mu^1 - \mu^0)$. If, however, $y_i^1 - y_i^0$ is not constant, then the OLS estimate represents a conditional-variance-weighted estimate of the underlying causal effects of individuals, δ_i, in which the weights are a function of the conditional variance of D (see Angrist 1998 and Angrist and Pischke 2009, as well as our explanation of this result in the next section). Under these conditions, the OLS estimate is consistent and unbiased and for this particular weighted average, which is usually not a causal parameter of primary interest.

Second, note that the residualized error term, ε^*, in Equation (6.8) is not equivalent to either ε from Equation (6.3) or to the multipart error term $\{v^0 + D(v^1 - v^0)\}$ from Equation (6.5). Rather, it is defined by whatever adjustment occurs within Equation (6.8), as represented by the term $X\beta$. Consequently, the residualized error term ε^* cannot be interpreted independently of decisions about how to specify the vector of adjustment variables in X, and this can make it difficult to define when a net covariance between D and ε^* can be assumed to be zero.

We explain these two complications and their important implications in the following sections of this chapter and the next, where we consider a variety of examples that demonstrate the connections between matching and regression estimators of causal effects. Before developing these explanations, however, we conclude this section with two final small-N examples that demonstrate how the regression adjustment strategy does and does not work.

Table 6.4 presents two six-person examples. For both examples, a regression of Y on D yields a biased estimate of the true ATE. And, in fact, both examples yield the same biased estimate because the observed values y_i and d_i are the same for both examples. Moreover, an adjustment variable X is also available for both examples, and its observed values x_i have the same associations with the observed values y_i and d_i for both examples. But the underlying potential outcomes differ substantially between the two examples. These differences render regression adjustment by X effective for only the first example.

For the example in the first panel, a regression of Y on D would yield an estimate of the coefficient for D of 11.67, which is an upwardly biased estimate of the true average causal effect of 10. The bias arises because the correlation between the error term in the last column and the realized values for d_i is not zero but is instead .33.

For the example in the second panel, a regression of Y on D would yield an estimate of the coefficient for D of 11.67 because the values for y_i and d_i are exactly the same as for the example in the first panel. Moreover, this estimate is also upwardly biased because the error term in the last column is positively correlated with the realized

Table 6.4 Two Six-Person Examples in Which Regression Adjustment Is Differentially Effective

	y_i^1	y_i^0	v_i^1	v_i^0	y_i	d_i	x_i	$v_i^0 + d_i(v_i^1 - v_i^0)$
			Regression adjustment with X generates an unbiased estimate for D					
In treatment group	20	10	2.5	2.5	20	1	1	2.5
In treatment group	20	10	2.5	2.5	20	1	1	2.5
In treatment group	15	5	−2.5	−2.5	15	1	0	−2.5
In control group	20	10	2.5	2.5	10	0	1	2.5
In control group	15	5	−2.5	−2.5	5	0	0	−2.5
In control group	15	5	−2.5	−2.5	5	0	0	−2.5
			Regression adjustment with X does not generate an unbiased estimate for D					
In treatment group	20	10	2.83	2.5	20	1	1	2.83
In treatment group	20	10	2.83	2.5	20	1	1	2.83
In treatment group	15	5	−2.17	−2.5	15	1	0	−2.17
In control group	18	10	.83	2.5	10	0	1	2.5
In control group	15	5	−2.17	−2.5	5	0	0	−2.5
In control group	15	5	−2.17	−2.5	5	0	0	−2.5

values of d_i. However, here the patterns are more complex. The underlying potential outcomes are different, and individual-level heterogeneity of the causal effect is now present. One member of the control group has an individual-level treatment effect of only 8, and as a result the true ATE is only 9.67. Consequently, the same bivariate regression coefficient of 11.67 has a larger upward bias in this second example, and the correlation between the values of d_i and the error term in the last column is now .39 rather than .33.

This underlying difference in potential outcomes also has consequences for the capacity of regression adjustment to effectively generate unbiased estimates of the ATE. This is easiest to see by rearranging the rows in Table 6.4 for each of the two examples based on the values of X for each individual, as in Table 6.5. For the first example, the values of d_i are uncorrelated with the error term within subsets of individuals defined by the two values of X. In contrast, for the second example, the values of d_i remain positively correlated with the error term within subsets of individuals defined by the two values of X. Thus, conditioning on X breaks the correlation between D and the error term in the first example but not in the second example. Because the observed data are the same for both examples, this difference is entirely a function of the underlying potential outcomes that generate the data.

Table 6.5 A Rearrangement of the Example in Table 6.4 That Shows How Regression Adjustment Is Differentially Effective

	y_i^1	y_i^0	v_i^1	v_i^0	y_i	d_i	x_i	$v_i^0+d_i(v_i^1-v_i^0)$
	Regression adjustment with X generates an unbiased estimate for D							
For those with $X=1$								
In treatment group	20	10	2.5	2.5	20	1	1	2.5
In treatment group	20	10	2.5	2.5	20	1	1	2.5
In control group	20	10	2.5	2.5	10	0	1	2.5
For those with $X=0$								
In treatment group	15	5	-2.5	-2.5	15	1	0	-2.5
In control group	15	5	-2.5	-2.5	5	0	0	-2.5
In control group	15	5	-2.5	-2.5	5	0	0	-2.5
	y_i^1	y_i^0	v_i^1	v_i^0	y_i	d_i	x_i	$v_i^0+d_i(v_i^1-v_i^0)$
For those with $X=1$								
In treatment group	20	10	2.83	2.5	20	1	1	2.83
In treatment group	20	10	2.83	2.5	20	1	1	2.83
In control group	18	10	.83	2.5	10	0	1	2.5
For those with $X=0$								
In treatment group	15	5	-2.17	-2.5	15	1	0	-2.17
In control group	15	5	-2.17	-2.5	5	0	0	-2.5
In control group	15	5	-2.17	-2.5	5	0	0	-2.5

Section headers in table body: "Regression adjustment with X generates an unbiased estimate for D" (top block), and "Regression adjustment with X does not generate an unbiased estimate for D" (bottom block).

This example demonstrates an important conceptual point. Recall that the basic strategy behind regression analysis as an adjustment technique is to estimate

$$Y = \alpha + \delta D + X\beta + \varepsilon^*,$$

where X represents one or more control variables, β is a coefficient (or a conformable vector of coefficients if X represents more than one variable), and ε^* is a residualized version of the original error term ε from Equation (6.3); see our earlier presentation of Equation (6.8). The literature on regression often states that an estimated coefficient $\hat{\delta}$ from this regression equation is consistent and unbiased for the average causal effect if ε^* is uncorrelated with D. But, because the specific definition of ε^* is conditional on the specification of X, many researchers find this requirement of a zero correlation difficult to interpret and hence difficult to evaluate.

The crux of the idea, however, can be understood without reference to the error term ε^* but rather with reference to the simpler and (as we have argued in Section

6.2.2) more clearly defined error term $v^0 + D(v^1 - v^0)$ from Equation (6.5) or, equivalently, $Dv^1 + (1 - D)v^0$ from Equation (6.6). Regression adjustment by X in Equation (6.8) will yield a consistent and unbiased estimate of the ATE when

1. D is mean independent of (and therefore uncorrelated with) $v^0 + D(v^1 - v^0)$ for each subset of respondents identified by distinct values on the variables in X,

2. the causal effect of D does not vary with X, and

3. a fully flexible parameterization of X is used.

Consider the relationship between this set of conditions and what was described in Section 4.3.1 as an assumption that treatment assignment is ignorable. Switching notation from S to X in Equation (4.4) results in

$$(Y^0, Y^1) \perp\!\!\!\perp D \mid X, \qquad (6.10)$$

where, again, the symbol $\perp\!\!\!\perp$ denotes independence. Now, rewrite the assumption, deviating Y^0 and Y^1 from their population-level expectations:

$$(v^0, v^1) \perp\!\!\!\perp D \mid X. \qquad (6.11)$$

This switch from (Y^0, Y^1) to (v^0, v^1) does not change the assumption, at least insofar as it is relevant here (because we have defined the individual-level causal effect as a linear difference, because the expectation operator is linear, and because $E[Y^0]$ and $E[Y^1]$ do not depend on who is in the treatment state and who is in the control state). Consequently, ignorability of treatment assignment can be defined only with respect to individual-level departures from the true average potential outcomes across all members of the population under the assumptions already introduced.

Given that an assumption of ignorable treatment assignment can be written as Equation (6.11), the connections between this assumption and the set of conditions that we have said justify a regression estimator as consistent and unbiased for the ATE can be explored. Two important points should be noted. First, if treatment assignment is ignorable, as defined in Equation (6.11), then an estimator that conditions fully on all values of X exists that will yield a consistent and unbiased estimate of the ATE (and that can be implemented in practice if the sample size is large enough). For reasons we will explain in the next section, a regression estimator with dummy variables for all but one categories of X will only be the appropriate estimator if, in addition, our number 2 just above also holds: the causal effect of D does not vary with X. Second, ignorability stipulates full independence. Instead, for our set of conditions, v^0 and v^1 – as well as functions of them, such as $v^0 + D(v^1 - v^0)$ – must only be mean independent of D conditional on X, not fully independent of D conditional on X.

Finally, we should note that our three conditions are not the only ones that would establish least squares estimators as consistent and unbiased for the ATE, but they are the most common ones that would apply in most research situations.[13] Our point in

---

[13]For example, the second condition can be dropped if the heterogeneity of the causal effect is modeled as a function of X (i.e., the parameterization is fully saturated in both D and X). In this case, however, regression then becomes a way of enacting a stratification of the data, as for the matching techniques presented in Chapter 5.

laying out these conditions is not to provide a rigid guideline applicable to all types of regression models in all situations but instead to show why the earlier statement that "ε^* must be uncorrelated with D" is insufficiently articulated from a counterfactual perspective.

A larger point of this section, however, is that much of the received wisdom on regression modeling breaks down in the presence of individual-level heterogeneity of a causal effect, as would be present in general when causal effects are defined with reference to underlying potential outcomes tied to well-defined causal states. In the next section, we begin to explain these complications more systematically, starting from the assumption, as in prior chapters, that causal effects are inherently heterogeneous and likely to vary systematically between those in the treatment and control groups. Beginning with the next section and continuing in Chapter 7 on weighted regression, we then present the connections among regression, matching, and stratification, building directly on our presentation of matching in Chapter 5.

6.3   Regression as Conditional-Variance-Weighted Matching

In this section, we return to the demonstrations utilized to explain matching estimators in Chapter 5. Our goal is to show why matching routines and least squares regression yield different results, even though a researcher is attempting to adjust for the same set of variables.

We first show why least squares regression can yield misleading causal effect estimates in the presence of individual-level heterogeneity of causal effects, even if the only variable that needs to be adjusted for is given a fully flexible coding (i.e., when the adjustment variable is parameterized with a dummy variable for each of its values, save one for the reference category).[14] In these cases, least squares estimators implicitly invoke conditional-variance weighting of individual-level causal effects. This weighting scheme generates a conditional-variance-weighted estimate of the average causal effect, which is not an average causal effect that is often of any inherent interest to a researcher.[15] Angrist (1998) provides a more formal explanation of the following results, which is then placed in the context of a larger class of models in Angrist and Krueger (1999) and Angrist and Pischke (2009).

---

[14]When we write of a fully flexible coding of a variable, we are referring to a dummy variable coding of that variable only (i.e., one dummy for a two-category variable, two dummies for a three-category variable, and so on). As we will discuss later, a saturated model entails a fully flexible coding of each variable *as well as all interactions between the dummy variables of each them*. For the models discussed in this section, a saturated model would include interactions between the causal variable D and each dummy variable for all but one of the values of S. For a model with only a fully flexible coding of S, these interactions are left out.

[15]It could be of interest to a researcher who seeks a minimum-variance estimate and who has reason to believe that the inconsistency and bias of the regression estimate is modest. We discuss this point later, but we hope to show that most applied researchers have good reason to want consistent and unbiased estimates rather than minimum mean-squared-error estimates that remain inconsistent and biased.

Regression Demonstration 2

Reconsider Regression Demonstration 1 (see page 189), but now step back from the cautious mindset of the fictitious descriptively oriented researcher. Suppose that a causality-oriented researcher had performed the same exercise and obtained the same results for the regression model reported above in Equation (6.2):

$$\hat{Y} = 1.86 + 2.75(D) + 3.76(S2) + 8.92(S3). \tag{6.12}$$

We know from Matching Demonstration 1 (see page 145), on which Regression Demonstration 1 is based, that for this hypothetical example the ATT is 3, the average treatment effect for the controls (ATC) is 2.4, and the unconditional ATE is 2.64. If the causality-oriented researcher were to declare that the coefficient on D of 2.75 in Equation (6.12) is a good estimate of the causal effect of D on Y, then the researcher would be incautious but not appreciably incorrect. The value of 2.75 is indeed close to the true ATE of 2.64, and we know from the setup of Regression Demonstration 1 that the variable S continues to serve as a perfect stratifying variable, as defined earlier.[16] Thus, if the researcher were to state that the regression model in Equation (6.12) statistically controls for the common effect of S on both D and Y, as in Equation (6.8), where S is specified as the sole element of X but as two dummy variables $S2$ and $S3$, then the researcher is not horribly off the mark. The researcher has offered an adjustment for S and gotten close to the true ATE.

Unfortunately, the closeness of the estimate to the true ATE is not a general feature of this type of a regression estimator. Under this particular specification of the regression equation, the OLS estimator yields precisely the value of 2.75 in an infinite sample as the sum of sample analogs to three terms:

$$\frac{\text{Var}[D|S=1]\Pr[S=1]}{\sum_S \text{Var}[D|S=s]\Pr[S=s]}\left\{E[Y|D=1,S=1]-E[Y|D=0,S=1]\right\} \tag{6.13}$$

$$+\frac{\text{Var}[D|S=2]\Pr[S=2]}{\sum_S \text{Var}[D|S=s]\Pr[S=s]}\left\{E[Y|D=1,S=2]-E[Y|D=0,S=2]\right\}$$

$$+\frac{\text{Var}[D|S=3]\Pr[S=3]}{\sum_S \text{Var}[D|S=s]\Pr[S=s]}\left\{E[Y|D=1,S=3]-E[Y|D=0,S=3]\right\}.$$

These three terms are not as complicated as they may appear. First, note that the differences in the braces on the right-hand side of each term are simply the

---

[16]Moreover, S satisfies the back-door criterion by construction. However, this does not imply that every conditioning estimator will deliver consistent and unbiased estimates of the ATT, ATC, or ATE that could be calculated nonparametrically in a sufficiently large sample. In this case, as we show in this section, the least squares estimator introduces parametric constraints that deliver an alternative average causal effect. The back-door criterion gives a correct result – a causal inference of some type is indeed warranted by conditioning via a regression model on variables that satisfy the back-door criterion – but the causal effect estimate that is produced by the regression estimator is not one of the average causal effects that the analyst typically wants.

stratum-specific differences in the outcomes, which in this case are

$$E[Y|D=1,S=1] - E[Y|D=0,S=1] = 4-2, \tag{6.14}$$

$$E[Y|D=1,S=2] - E[Y|D=0,S=2] = 8-6, \tag{6.15}$$

$$E[Y|D=1,S=3] - E[Y|D=0,S=3] = 14-10. \tag{6.16}$$

The left-hand portion of each term is then just a weight, exactly analogous to the stratum-specific weights that were used for Matching Demonstration 1 to average the stratum-specific causal effect estimates in various ways to obtain consistent and unbiased estimates of the ATE, ATT, and ATC. But, rather than use the marginal distribution of S, $\Pr[S]$, or the two conditional distributions of S, $\Pr[S|D=1]$ and $\Pr[S|D=0]$, a different set of weights is implicitly invoked by the least squares operation. In this case, the weights are composed of three pieces: (1) the variance of the treatment variable within each stratum, $\text{Var}[D|S=s]$, (2) the marginal probability of S for each stratum, $\Pr[S=s]$, and (3) a summation of the product of these two terms across S so that the three weights sum to 1.

Accordingly, the only new piece of this estimator that was not introduced and examined for Matching Demonstration 1 is the conditional variance of the treatment, $\text{Var}[D|S=s]$. Recall that the treatment variable is distributed within each stratum solely as a function of the stratum-specific propensity score, $\Pr[D|S=s]$. Thus, the treatment variable is a Bernoulli distributed random variable within each stratum. As can be found in any handbook of statistics, the variance of a Bernoulli distributed random variable is $p(1-p)$, where p is the Bernoulli probability of success (in this case D equal to 1) instead of failure (in this case D equal to 0). Accordingly, the expected variance of the within-stratum treatment variable D is $(\Pr[D|S=s])\,(1-\Pr[D|S=s])$.

For this example, the conditional variances, $\text{Var}[D|S=s]$, contribute to the numerator of each weight as follows:

$$Var[D|S=1]\Pr[S=1] = \left[\left(\frac{.08}{.08+.36}\right)\left(1-\frac{.08}{.08+.36}\right)\right](.08+.36), \tag{6.17}$$

$$Var[D|S=2]\Pr[S=2] = \left[\left(\frac{.12}{.12+.12}\right)\left(1-\frac{.12}{.12+.12}\right)\right](.12+.12), \tag{6.18}$$

$$Var[D|S=3]\Pr[S=3] = \left[\left(\frac{.2}{.2+.12}\right)\left(1-\frac{.2}{.2+.12}\right)\right](.2+.12). \tag{6.19}$$

The terms in brackets on the right-hand sides of Equations (6.17)–(6.19) are $\text{Var}[D|S=1]$, $\text{Var}[D|S=2]$, and $\text{Var}[D|S=3]$. The terms in the last set of parentheses on the right-hand sides of Equations (6.17)–(6.19) are the marginal probability of S for each stratum, $\Pr[S=1]$, $\Pr[S=2]$, and $\Pr[S=3]$. For example, for the stratum with $S=1$, $\text{Var}[D|S=1] = \left[\left(\frac{.08}{.08+.36}\right)\left(1-\frac{.08}{.08+.36}\right)\right]$ and $\Pr[S=1] = (.08+.36)$. Finally, the denominator of each of the three stratum-specific weights in Equation (6.13) for this example is the sum of Equations (6.17)–(6.19). The denominator is constant across all three weights and scales the weights so that they sum to 1.

With an understanding of the implicit stratum-specific weights of least squares regression, the regression estimator can be seen clearly as an estimator for the ATE but with supplemental conditional-variance weighting. Weighting is indeed performed with respect to the marginal distribution of individuals across strata, as for the ATE in Matching Demonstration 1, but weighting is *also* performed with respect to the conditional variance of the treatment variable across strata as well. Thus, net of the weight given to stratum-specific effects solely as a function of $\Pr[S]$, the conditional-variance terms give more weight to stratum-specific causal effects in strata with propensity scores close to .5 (where $\mathrm{Var}[D|S]$ approaches its maximum of $.5 \times .5$) and less weight to stratum-specific causal effects in strata with propensity scores close to either 0 or 1 (where $\mathrm{Var}[D|S]$ approaches its minimum of 0×1 or 1×0).

Why would the OLS estimator implicitly invoke conditional-variance weighting as a supplement to weighting simply by the marginal distribution of S? OLS is a minimum-variance-based estimator of the parameter of interest. As a result, it gives more weight to stratum-specific effects with the lowest expected variance, and the expected variance of each stratum-specific effect is an inverse function of the stratum-specific variance of the treatment variable D. Thus, if the two pieces of the weighting scheme are not aligned (i.e., the propensity score is close to 0 or 1 for strata that have high total probability mass but close to .5 for strata with low probability mass), then a regression estimator of this form, even under a fully flexible coding of S, can yield estimates that are far from the true ATE even in an infinite sample.

To see the effects that supplemental weighting by the conditional variance of the treatment can have on a regression estimate, we consider alternative joint distributions for S and D, which we then impose on the setup for Matching Demonstration 1 and Regression Demonstration 1. In particular, the values of $E[Y^0|S,D]$, $E[Y^1|S,D]$, and $E[Y|S,D]$ in the final three panels of Table 6.1 again obtain, such that S continues to offer a perfect stratification of the data. Now, however, we assume two different joint distributions of S and D in two variants reported in Table 6.6. For these two alternative joint distributions of S and D, the marginal distribution of S remains the same as for Table 6.1: $\Pr[S=1] = .44$, $\Pr[S=2] = .24$, and $\Pr[S=3] = .32$. As a result, the unconditional ATE is the same for both variants of the joint distribution of S and D depicted in Table 6.6, and it matches the unconditional ATE for the original demonstration represented fully in Table 6.1. In particular, the same distribution of stratum-specific causal effects results in an unconditional ATE of 2.64. The difference represented by each variant of the joint distributions in Table 6.6 is in the propensity score for each stratum of S, which generates an alternative marginal distribution for D and thus alternative true ATTs and ATCs (and, as we will soon see, alternative regression estimates from the same specification).

For Variant I in Table 6.6, those with S equal to 1 or 2 are much less likely to be in the treatment group, and those with S equal to 3 are now only equally likely to be in the treatment group and the control group. As a result, the marginal distribution of D is now different, with $\Pr[D=0] = .76$ and $\Pr[D=1] = .24$. The ATT is now 3.33 whereas the ATC is 2.42. Both of these effects are larger than was the case for Table 6.1 because (1) a greater proportion of those in the control group have $S=3$ (i.e., $\frac{.16}{.76} > \frac{.12}{.6}$), (2) a greater proportion of those in the treatment group have $S=3$ (i.e., $\frac{.16}{.24} > \frac{.2}{.4}$), and (3) those with $S=3$ gain the most from the treatment.

Table 6.6 The Joint Probability Distribution for Two
Variants of the Stratifying and Treatment Variables in
Prior Regression Demonstration 1

	Joint probability distribution of S and D	
	Control group: $D=0$	Treatment group: $D=1$
	Variant I	
$S=1$	$\Pr[S=1, D=0] = .40$	$\Pr[S=1, D=1] = .04$
$S=2$	$\Pr[S=2, D=0] = .20$	$\Pr[S=2, D=1] = .04$
$S=3$	$\Pr[S=3, D=0] = .16$	$\Pr[S=3, D=1] = .16$
	Variant II	
$S=1$	$\Pr[S=1, D=0] = .40$	$\Pr[S=1, D=1] = .04$
$S=2$	$\Pr[S=2, D=0] = .12$	$\Pr[S=2, D=1] = .12$
$S=3$	$\Pr[S=3, D=0] = .03$	$\Pr[S=3, D=1] = .29$

For Variant II, those with S equal to 1 are still very unlikely to be in the treatment group, but those with S equal to 2 are again equally likely to be in the treatment group. In addition, those with S equal to 3 are now very likely to be in the treatment group. As a result, the marginal distribution of D is now different again, with $\Pr[D=0] = .55$ and $\Pr[D=1] = .45$, and the ATT is now 3.29, whereas the ATC is 2.11. Both of these are smaller than for Variant I because a smaller proportion of both the treatment group and the control group have $S=3$.

For these two variants of the joint distribution of S and D, we have examples in which the unconditional ATE is the same as it was for Regression Demonstration 1, but the underlying ATT and ATC differ considerably. Does the reestimation of Equation (6.12) for these variants of the example still generate an estimate for the coefficient on D that is (1) relatively close to the true unconditional ATE and (2) closer to the unconditional ATE than either the ATT or the ATC?

For Variant I, the regression model yields

$$\hat{Y} = 1.90 + 3.07(D) + 3.92(S2) + 8.56(S3) \tag{6.20}$$

for an infinite sample. In this case, the coefficient of 3.07 on D is not particularly close to the unconditional ATE of 2.64, and in fact it is closer to the ATT of 3.33 (although still not particularly close). For Variant II, the regression model yields

$$\hat{Y} = 1.96 + 2.44(D) + 3.82(S2) + 9.45(S3). \tag{6.21}$$

In this case, the coefficient of 2.44 on D is closer to the unconditional ATE of 2.64, but not as close as was the case for Equation (6.12) when applied to the original data specified for Regression Demonstration 1. It is now relatively closer to the ATC, which is 2.11 (although, again, still not particularly close).

For Variant I, the regression estimator is weighted more toward the stratum with $S=3$, for which the propensity score is .5. For this stratum, the causal effect is 4. For

Variant II, the regression estimator is weighted more toward the stratum with $S = 2$, for which the propensity score is .5. And, for this stratum, the causal effect is 2.[17]

What is the implication of these alternative setups of the same basic demonstration? Given that the unconditional ATE is the same for all three joint distributions of S and D, it would be unwise for the incautious researcher to believe that this sort of a regression specification will provide a reliably close estimate to the unconditional ATE, the ATT, or the ATC when there is reason to believe that these three average causal effects differ because of individual-level heterogeneity. The regression estimate will be weighted toward stratum-specific effects for which the propensity score is closest to .5, net of all else.

In general, regression models do not offer consistent or unbiased estimates of the ATE when causal effect heterogeneity is present, even when a fully flexible coding is given to the only necessary adjustment variable(s). Regression estimators with fully flexible codings of the adjustment variables do provide consistent and unbiased estimates of the ATE if either (1) the true propensity score does not differ by strata or (2) the average stratum-specific causal effect does not vary by strata.[18] The first condition would almost never be true (because, if it were, one would not even think to adjust for S because it is already independent of D). And the second condition is probably not true in most applications, because rarely are investigators willing to assert that all consequential heterogeneity of a causal effect has been explicitly modeled.

Instead, for this type of a regression specification, in which all elements of a set of perfect stratifying variables S are given fully flexible codings (i.e., a dummy variable coding for all but one of the possible combinations of the values for the variables in S), the OLS estimator $\hat{\delta}_{\text{OLS, multiple}}$ in Equation (6.9) is equal to

$$\frac{1}{c} \sum_s \text{Var}_N[d_i|s_i = s] \, \Pr{}_N[s_i = s] \{E_N[y_i|d_i = 1, s_i = s] - E_N[y_i|d_i = 0, s_i = s]\} \quad (6.22)$$

in a sample of size N. Here, c is a scaling constant equal to the sum (over all combinations of values s of S) of the terms $\text{Var}_N[d_i|s_i = s]\Pr_N[s_i = s]$.

There are two additional points to emphasize. First, the weighting scheme for stratified estimates in Equation (6.22) applies only when the fully flexible parameterization of S is specified. Under a constrained specification of S – e.g., in which some elements of S are constrained to have linear effects, as in Equation (6.1) – the weighting scheme is more complex. The weights remain a function of the marginal distribution of S and the stratum-specific conditional variance of D, but the specific form of each of these components becomes conditional on the specification of the regression model

---

[17]Recall that, because by construction the marginal distribution of S is the same for all three joint distributions of S and D, the $\Pr[S = s]$ pieces of the weights remain the same for all three alternatives. Thus, the differences between the regression estimates are produced entirely by differences in the $\text{Var}[D|S = s]$ pieces of the weights.

[18]As a by-product of either condition, the ATE must be equal to the ATT and the ATC. Thus, the regression estimator would be consistent and unbiased for both of these as well.

(see section 2.3.1 of Angrist and Krueger 1999). The basic intuition here is that a linear constraint on a variable in S in a regression model entails an implicit assumption that the underlying propensity scores are also linear in the values of S.[19]

Second, regression can make it all too easy to overlook the same sort of fundamental overlap problems that were examined for Matching Demonstration 2 (see page 148). Regression will implicitly drop strata for which the propensity score is either 0 or 1 in the course of forming its weighted average by Equation (6.22). As a result, a researcher who interprets a regression result as a decent estimate of the ATE, but with supplemental conditional-variance weighting, may be entirely wrong. No meaningful average causal effect may exist in the population. This second point is best explained by the following demonstration.

---

[19]For a binary causal variable D, a many-valued variable S that is treated as an interval-scaled variable, and a regression equation

$$\hat{Y} = \hat{\alpha} + \hat{\delta}(D) + \hat{\beta}(S) \; ,$$

the OLS estimator $\hat{\delta}$ is equal to

$$\frac{1}{l} \sum_s \widetilde{\mathrm{Var}}_N[\hat{d}_i|s_i = s] \, \widetilde{\mathrm{Pr}}_N[s_i = s] \left\{ E_N[y_i|d_i = 1, s_i = s] - E_N[y_i|d_i = 0, s_i = s] \right\}$$

in a sample of size N, where l is a scaling constant equal to the sum of $\widetilde{\mathrm{Var}}_N[\hat{d}_i|s_i = s] \, \widetilde{\mathrm{Pr}}_N[s_i = s]$ over all s of S.

The distinction between $\widetilde{\mathrm{Var}}_N[\hat{d}_i|s_i = s] \, \widetilde{\mathrm{Pr}}_N[s_i = s]$ and $\mathrm{Var}_N[d_i|s_i = s] \, \mathrm{Pr}_N[s_i = s]$ in the main text results from a constraint on the propensity score that is implicit in the regression equation. In specifying S as an interval-scaled variable, least squares implicitly assumes that the true propensity score $\mathrm{Pr}[D|S]$ is linear in S. As a result, the first portion of the stratum-specific weight is

$$\widetilde{\mathrm{Var}}_N[\hat{d}_i|s_i = s] \equiv \hat{p}_s(1 - \hat{p}_s),$$

where $\hat{p}_s$ is equal to the predicted stratum-specific propensity score from a linear regression of d_i on s_i: $\hat{p}_s = \hat{\xi} + \hat{\phi}_s s$.

Perhaps somewhat less clear, the term $\widetilde{\mathrm{Pr}}_N[s_i = s]$ is also a function of the constraint on S. $\widetilde{\mathrm{Pr}}_N[s_i = s]$ is not simply the marginal distribution of S in the sample, as $\mathrm{Pr}_N[s_i = s]$ is. Rather, one must use Bayes' rule to determine the implied marginal distribution of S, given the assumed linearity of the propensity score across levels of S. Rearranging

$$\mathrm{Pr}[d_i = 1|s_i] = \frac{\mathrm{Pr}[s_i|d_i = 1]\,\mathrm{Pr}[d_i = 1]}{\mathrm{Pr}[s_i]}$$

as

$$\mathrm{Pr}[s_i] = \frac{\mathrm{Pr}[s_i|d_i = 1]\,\mathrm{Pr}[d_i = 1]}{\mathrm{Pr}[d_i = 1|s_i]},$$

and then substituting $\hat{p}_s$ for $\mathrm{Pr}[d_i = 1|s_i]$, we then find that

$$\widetilde{\mathrm{Pr}}_N[s_i = s] = \frac{\mathrm{Pr}_N[s_i = s|d_i = 1]\,\mathrm{Pr}_N[d_i = 1]}{\hat{p}_s}.$$

The terms $\mathrm{Pr}_N[s_i = s|d_i = 1]$ and $\mathrm{Pr}_N[d_i = 1]$ are, however, unaffected by the linearity constraint on the propensity score. They are simply the true conditional probability of S equal to s given D equal to d as well as the marginal probability of D equal to d for a sample of size N.

Note that, if the true propensity score is linear in S, then the weighting scheme here is equivalent to the one in the main text.

Regression Demonstration 3

Reconsider the hypothetical example presented as Matching Demonstration 2 (see page 148), which is reproduced in Table 6.7. The assumed relationships that generate the hypothetical data for this demonstration are very similar to those we have just considered. However, in this case no individual for whom S is equal to 1 in the population is ever exposed to the treatment because $\Pr[S=1, D=1]=0$ and $\Pr[S=1, D=0]=.4$. As a result, the population, and any sample from it, does not include an individual in the treatment group with $s_i = 1$.[20] Because of this structural zero in the joint distribution of S and D, the three conditional expectations, $E[Y^0|S=1, D=1]$, $E[Y^1|S=1, D=1]$, and $E[Y|S=1, D=1]$, are properly regarded as ill-defined and hence are omitted from the last three panels of Table 6.7.

As shown for Matching Demonstration 2, the naive estimator can still be calculated and will be equal to 8.05 in an infinite sample. Moreover, the ATT can be estimated

Table 6.7 The Joint Probability Distribution and Conditional Population Expectations for Regression Demonstration 3

	Joint probability distribution of S and D	
	Control group: $D=0$	Treatment group: $D=1$
$S=1$	$\Pr[S=1, D=0]=.4$	$\Pr[S=1, D=1]=0$
$S=2$	$\Pr[S=2, D=0]=.1$	$\Pr[S=2, D=1]=.13$
$S=3$	$\Pr[S=3, D=0]=.1$	$\Pr[S=3, D=1]=.27$
	Potential outcomes under the control state	
$S=1$	$E[Y^0|S=1, D=0]=2$	
$S=2$	$E[Y^0|S=2, D=0]=6$	$E[Y^0|S=2, D=1]=6$
$S=3$	$E[Y^0|S=3, D=0]=10$	$E[Y^0|S=3, D=1]=10$
	Potential outcomes under the treatment state	
$S=1$	$E[Y^1|S=1, D=0]=4$	
$S=2$	$E[Y^1|S=2, D=0]=8$	$E[Y^1|S=2, D=1]=8$
$S=3$	$E[Y^1|S=3, D=0]=14$	$E[Y^1|S=3, D=1]=14$
	Observed outcomes	
$S=1$	$E[Y|S=1, D=0]=2$	
$S=2$	$E[Y|S=2, D=0]=6$	$E[Y|S=2, D=1]=8$
$S=3$	$E[Y|S=3, D=0]=10$	$E[Y|S=3, D=1]=14$

---

[20] Again, recall that we assume no measurement error in general in this book. In the presence of measurement error, some individuals might be misclassified and therefore might show up in the data with $s_i = 1$ and $d_i = 1$.

consistently as 3.35 by considering only the values for those with S equal to 2 and 3. But there is no way to consistently estimate the ATC, and hence no way to consistently estimate the unconditional ATE.

Consider now the estimated values that would be obtained with data arising from this joint distribution for a regression model specified equivalently as in Equations (6.12), (6.20), and (6.21):

$$\hat{Y} = 2.00 + 3.13(D) + 3.36(S2) + 8.64(S3). \tag{6.23}$$

In this case, the OLS estimator is still equivalent to Equation (6.22), which in an infinite sample would then be equal to Equation (6.13). But, with reference to Equation (6.13), note that the weight for the first term,

$$\frac{\text{Var}[D|S=1]\,\text{Pr}[S=1]}{\sum_S Var[D|S=s]\,\text{Pr}[S=s]},$$

is equal to 0 because $\text{Var}[D|S=1]$ is equal to 0 in the population by construction. Accordingly, the numerator of the stratum-specific weight is 0, and it enters into the summation of the denominator of the other two stratum-specific weights as 0. As a result, the regression estimator yields a coefficient on D that is 3.13, which is biased downward as an estimate of the ATT and has no relationship with the ill-defined ATE. If interpreted as an estimate of the ATT, but with supplemental conditional-variance weighting, then the coefficient of 3.13 is interpretable. But it cannot be interpreted as a meaningful estimate of the ATE in the population once one commits to the potential outcome framework and allows for individual-level heterogeneity of the treatment effect.

The importance of this demonstration is only partly revealed in this way of presenting the results. Imagine that a researcher simply observes $\{y_i, d_i, s_i\}_{i=1}^N$ and then estimates the model in Equation (6.23) without first considering the joint distribution of S and D as presented in Table 6.7. It would be entirely unclear to such a researcher that there are no individuals in the sample (or in the population) whose values for both D and S are 1. Such a researcher might therefore be led to believe that the coefficient estimate for D is a meaningful estimate of the causal effect of D for all members of the population.

All too often, regression modeling, at least as practiced in the social sciences, makes it too easy for an analyst to overlook fundamental mismatches between treatment and control cases. And, thus, one can obtain ATE estimates with regression techniques even when no meaningful ATE exists.

6.4   Regression as an Implementation of a Perfect Stratification

For completeness, in this section we make the (perhaps obvious) point that regression can be used as a technique to execute a perfect stratification. If all cells of the implicit full cross-tabulation of the adjustment variables and the causal variable are uniquely parameterized, using a saturated coding of all variables, regression can be used to carry out a perfect stratification of the data.

Consider how the estimates presented in Matching Demonstration 1 (see page 145) could have been generated by standard regression routines using a saturated coding of the causal variable and all adjustment variables. Alternatively, one could effectively estimate the ATE for Regression Demonstrations 1 and 2 (see pages 189 and 207, respectively) by enriching the parameterization of the regression model that we showed earlier does not generate a consistent and unbiased estimate of the ATE.

For the data common to these demonstrations, an analyst could specify S as two dummy variables, D as one dummy variable, and include all two-way interactions between S and D. In so doing, the analyst has enacted the same perfect stratification of the data by fitting a model that is saturated in both S and D to all of the cells of the first panel of Table 5.2:

$$\hat{Y} = 2 + 2(D) + 4(S2) + 8(S3) + 0(D \times S2) + 2(D \times S3). \qquad (6.24)$$

The values of each of the six cells of the panel are unique functions of the six estimated coefficients from the regression model.

With these coefficients, the analyst could then form differences within strata defined by all values of S and then use the marginal distribution of S to generate a consistent and unbiased estimate of the ATE (or use the conditional distribution of S given D to obtain consistent and unbiased estimates of the ATT and ATC). Although this last stratum-averaging step is not typically seen as part of a regression estimation strategy, it is nonetheless compatible with it (and now quite easy to implement, using, for example, the command *margins* after the command *regress* in Stata).

Nevertheless, for many applications, such a saturated model may not be possible, and in some cases this impossibility may be misinterpreted. For Regression Demonstration 3 (see page 213), just presented in the last section, if one were to fit the seemingly saturated model with the same six parameters as in Equation (6.24), the coefficient on D would be dropped by standard software routines. One might then attribute this to the size of the dataset and then instead use a more constrained parameterization, that is, either enter S as a simple linear term interacted with D or instead specify the model in Equation (6.23). These models must then be properly interpreted, and in no case could they be interpreted as yielding consistent and unbiased estimates of the ATE.

6.5 Regression as Supplemental Adjustment When Matching

Although we have separated our presentation of matching and regression estimators across two chapters for didactic purposes, we have been gradually working our way toward Chapter 7 on weighted regression where we will show how matching and regression can be used together very effectively. It is appropriate at this point to note that the matching literature has long recognized the utility of parametric regression as a supplemental adjustment technique that can be applied to traditional matching estimators in attempts to eliminate remaining within-sample imbalance on the matching variables. In this section, we offer a demonstration of how standard regression techniques can be used to supplement matching estimators.

Regression Demonstration 4

Recall Matching Demonstration 4 (see page 171) and consider now how regression can be used to supplement a matching algorithm. For Matching Demonstration 4, we presented matching estimates of the Catholic school effect on achievement for simulated data. We offered matching estimators under two basic scenarios, first using an incomplete specification of treatment assignment and then using a complete specification that includes a cognitive skills variable. Because both scenarios lack an adjustment for the self-selection dynamic, in which individuals select into the treatment partly as a function of their expected treatment effect, we only attempted to estimate the ATT.

In the columns labeled "Unadjusted," Table 6.8 redisplays the average bias for selected matching estimators from Table 5.7, under both specifications of the treatment

Table 6.8 Average Bias Comparisons for Selected Matching Estimates of the ATT from Matching Demonstration 4, With and Without Supplemental Regression Adjustment for the Assumed Determinants of Treatment Assignment

Method	Specification of treatment assignment variables: Incomplete specification		Complete specification	
	Unadjusted	Adjusted	Unadjusted	Adjusted
Nearest-neighbor match:				
1 without replacement (MI)	0.75	0.72	0.00	−0.10
1 with replacement (MI)	0.95	0.81	0.19	−0.12
1 with replacement and caliper = .05 SD (MI)	0.98	0.86	0.07	−0.11
5 with replacement (MI)	1.17	0.83	0.50	−0.05
5 with replacement and caliper = .05 SD (MI)	1.19	0.80	0.39	−0.13
Interval match:				
10 fixed blocks (MI)	1.68	0.84	1.68	−0.11
Optimal match (MI-opt)	1.28	0.80	0.54	0.07
Genetic match (MI-gen)	0.96	0.80	0.23	−0.07
Coarsened exact match (cem)	1.28	0.86	0.35	−0.08

Notes: See notes to Table 5.6 for software details.

assignment variables.[21] For the two columns labeled "Adjusted," the average bias is reported for the same matching estimators across the same 10 datasets, but now using a post-match regression adjustment for the matching variables.

Overall, Table 6.8 shows that supplemental regression adjustment reduces the average bias for nearly all of the matching estimators, which is consistent with the literature. The reductions occur for both the incomplete and complete specifications. This does not imply that for any single sample a supplemental regression adjustment will necessarily reduce bias, but on average such adjustments will reduce the bias that would remain if matching alone were utilized.

The reason for the reductions in bias should be obvious. Any post-match imbalance that remains in the matching variables is likely to have a component that exists as differences in mean values of the matching variables across the treatment and control cases. If we then use a parametric regression model to adjust for the matching variables in the matched datasets, the regression model yields a net estimate of the ATT after a linear adjustment for these lingering mean differences. The conditional-variance weighting property of least squares regression estimators remains, but the consequences of this property for estimates is greatly diminished when regression is used in this supplementary way, after the data have already been aligned in pursuit of the ATT.

We will discuss in Chapter 7 a variety of perspectives that suggest when and why matching and regression should be pursued together. Moving beyond regression as a supplementary procedure, we will show how weighted regression can be used to implement matching estimators, picking up on the weighting perspective on matching already introduced in Section 5.3.2.

6.6 Extensions and Other Perspectives

In this chapter, we have focused almost exclusively on the estimation of the effect of a binary cause on an interval-scaled outcome, and we have considered only least squares adjustments. Before carrying on to discuss least squares estimation of the effects of many-valued causes, we of course must concede what the reader is surely aware of: We have considered only a tiny portion of what falls under the general topic of regression modeling. We have not considered categorical outcome variables, time series analysis, nested data structures, variance-component models, and so on. One can gain a full perspective of the types of regression modeling used in just sociology and economics by consulting Agresti (2002), Allison (2009), Arminger et al. (1995), Berk (2004), Fox (2008), Hamilton (1994), Hayashi (2000), Hendry (1995), Long (1997), Powers and Xie (2000), Raudenbush and Bryk (2002), Ruud (2000), Stock and Watson (2007), Treiman (2009), and Wooldridge (2010).

In this section, we consider only one modest extension: least squares regression models for many-valued causes. This presentation then leads naturally to a discussion

---

[21] We selected the subset of the matching estimates from Table 5.7 based on whether the software allowed for supplemental regression adjustment.

that follows of what might be labeled the "all-cause correct specification" tradition of regression analysis. Informed by the demonstrations offered in this chapter, we discuss the attractiveness of the promise of this alternative perspective but also the implausibility of the perspective as a general guide for either causal analysis or regression practice in the social sciences.

6.6.1   Regression Estimators for Many-Valued Causes

We suspect that the vast majority of published regression estimates of causal effects in the social sciences are for causes with more than two values. Accordingly, as in Section 5.5.4 on matching estimators for many-valued causes, we must discuss the additional complexities of analogous regression estimators. We will again, however, restrict attention to an interval-scaled outcome.

First, again recall the basic setup for many-valued causes from Section 2.9, in which we have a set of J treatment states, a corresponding set of J causal exposure dummy variables, $\{Dj\}_{j=1}^{J}$, and a corresponding set of J potential outcome random variables, $\{Y^{Dj}\}_{j=1}^{J}$. The treatment received by each individual is Dj'.

How would one estimate the causal effect of such a J-valued cause with regression methods? The first answer should be clear from our presentation in the last section: Because regression can be seen as a form of matching, one can use the same basic strategies outlined for matching estimators of many-valued causes in Section 5.5.4. One could form a series of two-way comparisons between the values of the cause and then model each pairwise causal effect.

If the number of causal states is relatively large, then this general strategy is infeasible. Some smoothing across pairwise comparisons would be necessary, either by collapsing some of the J causal states or by imposing an ordering on the distribution of the causal effect across the J causal states. The most common parametric restriction would be to assume that the causal effect is linear in j for each individual i. For example, for a set of causal states (such as years of schooling) enumerated by values from 1, 2, 3, to J, the linearity assumption is the assumption that $y_i^{Dj} = y_i^{D1} + \beta_i(j-1)$, which requires that the difference $y_i^{Dj} - y_i^{Dj-1}$ for each individual i be equal to a constant β_i. In this case, the individual-level causal effect is then a slope β_i, rather than the simple difference in potential outcomes, δ_i, specified earlier in Equation (2.1). This setup is analogous to the dose-response models for matching estimators discussed in Section 5.5.4, but it explicitly leaves open the possibility that the dose-response relationship varies across individuals even though it remains linear.

Angrist and Krueger (1999) show in a very clear example how both a linearity assumption on the individual-specific, dose-response relationship and a fully flexible coding of adjustment variables results in an OLS weighting scheme for the average value of β_i in a sample that is even more complex than what we discussed earlier for simple binary causes (see Regression Demonstration 2). A form of conditional-variance weighting is present again, but now the weighting is in multiple dimensions because least squares must calculate average derivatives across the linearly ordered causal variable (see Angrist and Krueger 1999, equation 34). Because one cannot intuitively grasp how these weights balance out across all the dimensions of the implicit weighting scheme (at least we cannot do so), Angrist and Krueger help by offering a familiar example: an OLS estimate of the average causal effect of an additional year of schooling on labor market earnings, assuming linearity in years of schooling and using a fully flexible coding of adjustment variables for age, race, and residence location. They show that, for this example, OLS implicitly gives more weight to the causal effect of shifting from 13 to 14 years of schooling and from 14 to 15 years of schooling than for much more common differences, such as the shift from 11 to 12 years of schooling (primarily because the net conditional unexplained variance of schooling is greatest for the contrasts between 13 and 14 years and between 14 and 15 years). They also show that, for this example, the piecewise increases in average earnings happen to be largest for the years of schooling that OLS systematically weights downward. The result is a least squares estimate under the linearity constraint of .094, which is smaller than the weighted average estimate of .144 that one can calculate by dropping the linearity constraint and then averaging year-specific estimates over the marginal distribution of years of schooling.

For other examples, the weighting schemes may not generate sufficiently different estimates because the overall weighting is a complex function of the relationship between the unaccounted for variance of the causal variable within strata of the adjustment variables and the level of nonlinearity of the conditional expectation function. But the general point is clear and should be sobering: Linearity constraints across causal states may lead OLS models to generate nonintuitive (and sometimes misleading) averages of otherwise easily interpretable stratum-specific causal effects.

6.6.2 The Challenge of Regression Specification

In this section, we discuss the considerable appeal of what can be called the all-cause, complete-specification tradition of regression analysis. We argue that this orientation is impractical for most of the social sciences, for which theory is too weak and the disciplines too contentious to furnish perfect specifications that can be agreed on. At the same time, we argue that inductive approaches to discovering flawless regression models that represent all causes are mostly a form of self-deception, even though some software routines now exist that can prevent the worst forms of abuse.

Consider first a scenario in which one has a theoretical model that one believes is true. It suggests all of the inputs that determine the outcome of interest, as a set of observable variables, and it is in the form of a specific function that relates all inputs to the outcome. In this case, one can claim to have the correct specification for a regression of the outcome on some function of the variables suggested by the theoretical model. The only remaining challenges are then measurement, sampling, and observation.

The weakness of this approach is that critics can claim that the model is not true and hence that the entailed regression specification is wrong. Fighting off any such critics with empirical results can then be difficult, given that the regression specification used to generate the empirical results has been called into question.

In general, if members of a community of competing researchers assert their own true models and then offer up purportedly flawless regression models, the result may be a war of attrition in which no scientific progress is possible. It is therefore natural to ask: Can the *data* generate an all-cause, complete-specification regression model that all competing researchers can jointly adopt?

The first step in answering this question is to determine what an all-cause, complete specification would be, which is sometimes simply labeled a "correct specification."[22] In his 1978 book *Specification Searches: Ad Hoc Inference with Nonexperimental Data*, Edward Leamer lays out the following components of what he labels "The Axiom of Correct Specification":

> (a) The set of explanatory variables that are thought to determine (linearly) the dependent variable must be
>
> > (1) unique,
> > (2) complete,
> > (3) small in number, and
> > (4) observable.
>
> (b) Other determinants of the dependent variable must have a probability distribution with at most a few unknown parameters.
>
> (c) All unknown parameters must be constant. (Leamer 1978:4)

But Leamer then immediately undermines the axiom as it applies to observational data analysis in the social sciences:

> If this axiom were, in fact, accepted, we would find one equation estimated for every phenomenon, and we would have books that compiled these estimates published with the same scientific fanfare that accompanies estimates of the speed of light or the gravitational constant. Quite the contrary, we are literally deluged with regression equations, all offering to "explain" the same event, and instead of a book of findings we have volumes of competing estimates. (Leamer 1978:4)

One can quibble with Leamer's axiom (e.g., that component (a)(3) is not essential and so on), but the literature seems to provide abundant support for his conclusion. Few examples of flawless regression models suggested by true theoretical models can be found in the social science literature. One might hope for such success in the future, but the past 50 years of research do not give much reason for optimism.

Leamer instead argues that most regression models are produced by what he labels a data-instigated specification search, which he characterizes as a Sherlock Holmes form of inference wherein one refrains from developing a model or any firm hypotheses before first considering extensively all the facts of a case. Leamer argues that this approach to variable selection and specification is fraught with potential danger and invalidates traditional notions of inference.

Consider the example of the Catholic school effect on learning, and in particular the research of James Coleman and his colleagues. In seeking to estimate the effect of Catholic schooling on achievement, Coleman did not draw a complete specification for

---

[22]The literature has never clearly settled on a definition that has achieved consensus, but Leamer's is as good as any. Part of the confusion arises from the recognition that a descriptively motivated regression model can always be declared correct, no matter what its specification happens to be.

his regression models from a specific theoretical model of human learning. This decision was not because no such models existed, nor because Coleman had no appreciation for the need for such models. He was, in contrast, well aware of classic behaviorist models of learning (see Bush and Mosteller 1955, 1959) that specified complex alternative mechanisms for sequences of responses to learning trials. Although he appreciated these models, he recognized (see Coleman 1964:38) that they could not be deployed effectively in the complex environments of secondary schooling in the United States, the context of which he had already studied extensively (see Coleman 1961).

As a result, Coleman did not specify a learning model that justified the regression models that he and his colleagues presented (see Sørensen 1998; Sørensen and Morgan 2000).[23] Their basic specification strategy was instead to attempt to adjust for a sufficient subset of other causes of learning so that, net of these effects, it could be claimed that Catholic and public school students were sufficiently equivalent. The specific variables that Coleman and his colleagues chose to include in their models were based in part on Coleman's deep knowledge of what predicts learning in high school (and one could argue that Coleman was the most knowledgeable social scientist on the topic in the world at the time). But he and his colleagues also adopted an empirical approach, as indicated parenthetically at the end of the following account of their selection of adjustment variables:

> In order to minimize the effects of differences in initial selection masquerading as effects of differences in the sectors themselves, achievement subtests were regressed, by sector and grade, on a larger number of background variables that measure both objective and subjective differences in the home. Some of these subjective differences may not be prior to the student's achievement, but may in part be consequences of it, so that there may be an overcompensation for background differences. It was felt desirable to do this so as to compensate for possible unmeasured differences in family background; but of course the results may be to artificially depress the resulting levels of background-controlled achievement in Catholic and other private schools. (A few additional background variables were initially included; those that showed no effects beyond the ones listed in the following paragraph were eliminated from the analysis.) (Coleman et al. 1982:147)

Coleman and his colleagues then reported that the final list of variables included 10 they considered "clearly prior" to school sector – including family income, parents'

---

[23]When the 1982 data on seniors became available to supplement the 1980 data on sophomores, Coleman and his colleagues did move toward a stronger foundation for their specifications, providing an underlying model for the lagged achievement gain regression model that was an outgrowth of Coleman's early work on Markov chains and his proposals for longitudinal data analysis (Coleman 1964, 1981). In Hoffer et al. (1985:89–91), he and his colleagues showed that (subject to restrictions on individual heterogeneity) the lagged test score model is a linearized reduced-form model of two underlying rates (learning and forgetting) for the movement between two states (know and don't know) for each item on the cognitive test. Although the model is plausible, it is clearly constrained so that it can be estimated with simple regression techniques (see Coleman 1981:8–9 for an explanation of his modus operandi in such situations), and this is of course not the sort of constraint that one must adopt if one is truly interested in laying out the correct theoretical model of learning.

education, number of siblings, and number of rooms in the home – as well as 7 other variables that they considered "not clearly prior" to school sector – including more than 50 books in the home, owning a pocket calculator, and having a mother who thinks the student should go to college after high school.

As so often occurs in causal controversies of public importance, critics found this resulting list inadequate. From the perspective of their critics, Coleman and his colleagues had not provided a clear enough accounting of why some students were observed in Catholic schools, whereas others were observed in public schools and why levels of learning should be considered a linear function of background and the specific characteristics selected. After arguing that more would be known when follow-up data were collected and test score gains from sophomore to senior year could be analyzed, Alexander and Pallas (1983) argued that Coleman and his colleagues should have searched harder for additional adjustment variables:

> Failing this [estimating models with pretest and posttest data], another possibility would be to scout about for additional controls that might serve as proxies for student input differences that remain after socioeconomic adjustments. One candidate is the student's curriculum placement in high school. (Alexander and Pallas 1983:171)

Alexander and Pallas then laid out a rationale for this proxy approach, and they offered models that showed that the differences between public and private schools are smaller after conditioning on type of curriculum.

As this example shows, it is often simply unclear how one should go about selecting a sufficient set of conditioning variables to include in a regression equation when adopting the "adjustment for all other causes" approach to causal inference. Coleman and colleagues clearly included some variables that they believed that perhaps they should not have included, and they presumably tossed out some variables that they thought they should perhaps include but that proved to be insufficiently powerful predictors of test scores. Even so, Alexander and Pallas criticized Coleman and his colleagues for too little scouting.[24]

Leamer, as mentioned earlier, would characterize such scouting as a Sherlock Holmes–style, data-driven specification search. Leamer argues that this search strategy turns classical inference on its head:

> if theories are constructed after having studied the data, it is difficult to establish by how much, if at all, the data favor the data-instigated hypothesis. For example, suppose I think that a certain coefficient ought to be positive, and my reaction to the anomalous result of a negative estimate is to find another variable to include in the equation so that the estimate is positive. Have I found evidence that the coefficient is positive? (Leamer 1983:40)

---

[24]Contrary to the forecasts of Coleman and his critics, after the 1982 data were released, the specification debate did not end. It simply moved on to new concerns, primarily how to adjust for sophomore test scores (with or without a family background adjustment, with or without curriculum differences, with only a subset of sophomore test scores, and with or without adjustment for attenuation that is due to measurement error).

Taken to its extreme, the Sherlock Holmes regression approach may discover relationships between candidate independent variables and the outcome variable that are due to sampling variability and nothing else. David Freedman showed this possibility in a simple simulation exercise, in which he sought to demonstrate that "in a world with a large number of unrelated variables and no clear a priori specifications, uncritical use of standard [regression] methods will lead to models that appear to have a lot of explanatory power" (Freedman 1983:152). To show the plausibility of this conclusion, Freedman constructed an artificial dataset with 100 individuals, one outcome variable Y, and 50 other variables X_1 through X_{50}. The 100 values for each of these 51 variables were then independent random draws from the standard normal distribution. Thus, the data represent complete noise with only chance dependencies between the variables that mimic what any real-world sampling procedure would produce. The data were then subjected to regression analysis, with Y regressed on X_1 through X_{50}. For these 50 variables, 1 variable yielded a coefficient with a p value of less than .05 and another 14 had p values of less than .25. Freedman then ran a second regression of Y on the 15 variables that had p values of less than .25, and in this second pass, 14 of them again turned up with p values of less than .25. Most troubling, 6 of them now had p values of less than .05, and the model as a whole had an R^2 of .36. From pure noise and simulated sampling variability, Freedman produced a regression model that looks similar to any number of those published in social science articles. It had six coefficients that passed conventional standards of statistical significance, and it explained a bit more than one third of the variance of the outcome variable.[25]

The danger of data-driven specification searches is important to recognize, but not all procedures are similarly in danger, especially given developments since Leamer first presented his critique in the 1970s and 1980s. There is a new literature on data mining and statistical learning that has devised techniques to avoid the problems highlighted by Freedman's simulation (see Hastie et al. 2001). For a very clear overview of these methods, see Berk (2006, 2008). And, as we noted in Chapter 5, there are cases in which a data-driven specification search is both permissive and potentially quite useful. Consider again the causal graph in Figure 4.10 and suppose that one has a large number of variables that may be associated with both D and Y in one's dataset and that one presumes may be members of either S or X. Accordingly, one has the choice of conditioning on two different types of variables that lie along the back-door path from D to Y: the variables in S that predict D or the variables in X that predict Y. Engaging in a data-driven specification search for variables that predict Y will fall prey to inferential difficulties about the causal effect of D on Y for exactly the reasons just discussed. But a data-driven specification search for variables that predict D will not fall prey to the same troubles, because in this search one does not use any direct information about the outcome Y.

Even so, data-instigated specifications of regression equations remain a problem in practice, because few applied social scientists use the fair and disciplined algorithms in the statistical learning literature. The Catholic school example is surely a case in

---

[25]Raftery (1995) repeated Freedman's simulation experiment and obtained even more dramatic results.

which scouting led to the inclusion of variables that may not have been selected by a statistical learning algorithm. But, nonetheless, none of the scholars in the debate dared to reason backwards from their regression models in order to declare that they had inductively constructed a true model of learning. And, in general, it is hard to find examples of complete inductive model building in the published literature; scholars are usually driven by some theoretical predilections, and the results of mistaken induction are often fragile enough to be uncovered in the peer review process.[26] Milder forms of misspecification are surely pervasive.

6.7 Conclusions

Regression models, in their many forms, remain one of the most popular techniques for the evaluation of alternative explanations in the social sciences. In this chapter, we have restricted most of our attention to OLS regression of an interval-scaled variable on a binary causal variable. And, although we have considered how regression modeling can be used as a descriptive data reduction tool, we have focused mostly on regression as a parametric adjustment technique for estimating causal effects, while also presenting some of the connections between regression and matching as complementary forms of a more general conditioning estimation strategy. We conclude this chapter by discussing the strengths and weaknesses of regression as a method for causal inference from observational data.

The main strengths of regression analysis are clearly its computational simplicity, its myriad forms, its familiarity to a wide range of social scientists, and the ease with which one can induce computer software to generate point estimates and standard errors. These are all distinct advantages over the matching techniques that we summarized in Chapter 5.

But, as we have shown in this chapter, regression models have some serious weaknesses. Their ease of estimation tends to suppress attention to features of the data that matching techniques force researchers to consider, such as the potential heterogeneity of the causal effect and the alternative distributions of covariates across those exposed to different levels of the cause. Moreover, the traditional exogeneity assumption of regression (e.g., in the case of least squares regression that the independent variables must be uncorrelated with the regression error term) often befuddles applied researchers who can otherwise easily grasp the stratification and conditioning perspective that undergirds matching. As a result, regression practitioners can too easily accept their hope that the specification of plausible control variables generates an as-if randomized experiment.

Focusing more narrowly on least squares models, we have shown through several demonstrations that they generate causal effect estimates that are both nonintuitive and inappropriate when consequential heterogeneity has not been fully parameterized. In this sense, the apparent simplicity of least squares regression belies the complexity

---

[26]However, predictions about the behavior of financial markets can come close. See Krueger and Kennedy (1990) for discussion and interpretation of the apparent effect of Super Bowl victories on the stock market indices in the United States.

of how the data are reduced to a minimum mean-squared-error linear prediction. For more complex regression models, the ways in which such heterogeneity is implicitly averaged are currently unknown. But no one seems to suspect that the complications of unparameterized heterogeneity are less consequential for fancier maximum-likelihood-based regression models in the general linear modeling tradition.

Chapter 7

Weighted Regression Estimators of Causal Effects

With an Extended Example of a Weighted Regression Alternative to Matching

In the last chapter, we argued that traditional regression estimators of casual effects have substantial weaknesses, especially when individual-level causal effects are heterogeneous in ways that are not explicitly parameterized. In this chapter, we will introduce weighted regression estimators that solve these problems by appropriately averaging individual-level heterogeneity across the treatment and control groups using estimated propensity scores. In part because of this capacity, weighted regression estimators are now at the frontier of causal effect estimation, alongside the latest matching estimators that are also designed to properly handle such heterogeneity.

In the long run, we expect that weighted regression estimators will prove to be a common choice among alternative conditioning procedures that are used to estimate causal effects. In fact, we expect that weighted regression estimators will be used more frequently than the matching estimators presented in Chapter 5 when there is good overlap in the distributions of adjustment variables across the treatment and control groups. We have four primary reasons for this prediction, each of which we will explain in this chapter. First, weighted regression estimators allow the analyst to adopt the spirit of matching, and the clear thinking that it promotes, within a mode of data analysis that utilizes widely available software and that is familiar to most social scientists. Second, in contrast to their traditional unweighted analogs, weighted regression estimators can be used to generate direct estimates of the average treatment effect, average treatment effect for the treated, and average treatment effect for the controls (ATE, ATT, and ATC) in the presence of individual-level heterogeneity. Third, in contrast to matching estimators, the calculation of standard errors for average causal effect estimates is straightforward. Fourth, weighted regression estimators can be utilized when analyzing complex survey data of many types.

Nonetheless, caution is still warranted, and additional work is needed on a variety of crucial issues before our prediction can be evaluated. In addition to determining whether some advantages are real (e.g., the range of conditions under which the straightforward standard errors are also correct), some special challenges remain for determining the ultimate value of weighted regression estimators. Most importantly, as we will explain in detail below, the estimation of the weights that are passed to the regression routine is at least as difficult as the estimation of the propensity scores for matching routines. And, at present, no consensus exists on how sensitive conclusions are to variation in estimated weights, especially when some of the individual-specific weights appear unduly large.

In the remainder of this chapter, we will first introduce weighted regression estimators for the ATE, ATT, and ATC. We will then present "doubly robust" estimators that use the conditioning variables twice – first to estimate the relevant weights and then to adjust for remaining imbalance that is not eliminated by the weights. After a discussion of practical issues, including the concern over estimated weights that appear extreme, we will conclude the chapter with an extended demonstration that shows the value of weighted regression for direct estimation of the ATT and ATC.

Before offering this material, we should clarify that we will be writing narrowly about the usage of model-based weights to estimate average treatment effects. Weighted regression as a data analysis procedure is a much more general topic, and it has been used frequently for two other distinct purposes. First, regression equations can be weighted by individual-specific estimates of the inverse variance of the regression error term, which results in a weighted regression form of a feasible generalized least squares (FGLS) estimator.[1] The goal of the weighted estimator in this case is to promote efficiency of estimation, relative to unweighted ordinary least squares (OLS), in the presence of heteroscedasticity. Second, sampling weights, when not solely a function of the observed variables in the regression model, are used to weight regression equations in order to generate coefficients that can be interpreted as consistent estimates of population-level parameters. Although we will not consider FGLS estimators here, we will consider how sampling weights can be used along with the propensity-score-based weights that we will present in this chapter. Nonetheless, the justification for using sampling weights is distinct from the goals of generating consistent and unbiased estimates of the ATE, ATT, and ATC by using weights to adjust for confounding.

7.1 Weighted Regression Estimators of the ATE

In this section, we show how weighted regression estimators can be used to generate consistent and unbiased estimates of the ATE, as long as conditioning variables that satisfy the back-door criterion have been observed and properly utilized. These

---

[1] For details, see any regression textbook, such as Draper and Smith (1998) or Greene (2000).

methods have diverse and overlapping origins – inverse probability weighting in survey statistics (see Kish 1965, 1987; Thompson 2002), missing data imputation and survey nonresponse adjustment via weighted complete-case analysis (see Little 1982; Little and Rubin 2002), weighting procedures in multiple regression analysis for data from stratified samples (see DuMouchel and Duncan 1983), propensity-score models and general methods for modeling the probability of treatment assignment (see Rosenbaum and Rubin 1983b; Rubin 2006; Rubin and Thomas 2000), direct adjustment estimators (see Rosenbaum 1987, 2002), econometric evaluation estimators (Imbens 2004; Imbens and Wooldridge 2009), and inverse probability of treatment weighting in epidemiology (see Robins and Hernn 2009; Robins and Ritov 1997; van der Laan and Robins 2003).

Recall first the hypothetical example in Matching Demonstration 3 (see page 153), where matching was considered a method to weight the data in order to balance predictors of treatment assignment and thereby calculate contrasts that can be given causal interpretations. In this section and the next, we show that the three propensity-score-weighting estimators presented there as Equations (5.12) through (5.14) can be specified as three weighted regression estimators.

To estimate the ATE with weighted regression, one first must estimate the predicted probability of treatment, $\hat{p}_i$, for units in the sample, which is again the estimated propensity score. All of the methods discussed so far can be used to obtain these estimated values. Once the values for $\hat{p}_i$ are obtained, weights for the ATE are constructed as

$$\text{For } d_i = 1: \quad w_{i,\text{ ATE}} = \frac{1}{\hat{p}_i},$$

$$\text{For } d_i = 0: \quad w_{i,\text{ ATE}} = \frac{1}{1 - \hat{p}_i}. \tag{7.1}$$

These weights are equivalent in structure to survey weights that must be used to weight complex samples so that they are representative of their respective target populations. Here, the weights, $w_{i,\text{ ATE}}$, can be used to weight those in the treatment group and those in the control group so that both of these groups have the same weighted distribution on (observed) determinants of treatment assignment as the full population (where, in this case, the full population of interest is composed of both the population-level treatment group and the population-level control group).

To then estimate the ATE, a weighted bivariate regression model is estimated, where y_i are the values for the outcome, d_i are the values for the sole predictor variable, and $w_{i,\text{ ATE}}$ are the weights. No specialized software is required, and the weights are treated exactly as if they are survey weights (even though they are not survey weights, but instead weights based on estimated propensity scores that are relevant only for the estimation of the ATE).

For those accustomed to matrix representations of regression estimates, note first that the naive estimator in Equation (2.9) can be written as an OLS estimator, $(\mathbf{Q}'\mathbf{Q})^{-1}\mathbf{Q}'\mathbf{y}$, where (1) $\mathbf{Q}$ is an $n \times 2$ matrix that contains a vector of 1s in its first column and a vector of the values of d_i for each individual in its second column and (2) $\mathbf{y}$ is an $n \times 1$ column vector containing values of y_i for each individual. To estimate the ATE, a weighted regression estimator is utilized,

$$\hat{\delta}_{\text{OLS, weighted}} \equiv (\mathbf{Q'PQ})^{-1}\mathbf{Q'Py}, \tag{7.2}$$

where $\mathbf{P}$ is an $n \times n$ diagonal matrix with the corresponding value of $w_{i,\text{ ATE}}$ on the diagonal for each individual and zero elsewhere. Consider the following demonstration, which builds directly on the prior presentation of matching as weighting in Section 5.3.

Weighted Regression Demonstration 1

For Matching Demonstration 3 (see page 153), consistent and unbiased estimates of the ATE were generated from averages of estimates of the ATT and the ATC. These estimates of the ATT and ATC were differences in weighted averages of the values of y_i, constructed using weights defined by $\hat{p}_i$. In the demonstration we offer here, we will show how the same estimates of the ATE can be generated directly by weighted regression. We will dispense with the Monte Carlo setup from Matching Demonstration 3, which offers little additional insight, and we will instead shift to the simpler scenario we have used elsewhere in demonstrations, where we assume an infinite sample. We will, however, vary the setup for the potential outcomes in order to show how weighted regression can easily handle nonlinearities in the relationships that generate the propensity scores and the potential outcomes.

As shown for Matching Demonstration 3 in Equations (5.8) and (5.9), the potential outcomes were specified as functions of individual values for A and B:

$$y_i^1 = 102 + 6a_i + 4b_i + v_i^1, \tag{7.3}$$
$$y_i^0 = 100 + 3a_i + 2b_i + v_i^0, \tag{7.4}$$

where A and B are distributed as independent uniform random variables with a minimum of .1 and a maximum of 1, and where v_i^1 and v_i^0 are independent random draws from a normal distribution with expectation 0 and a standard deviation of 5. For each individual, y_i is then equal to $y_i^1(d_i) + (1 - d_i)y_i^0$, where the value of d_i is determined by a Bernoulli distribution with the probability of 1 rather than 0 as the nonlinear function in A and B that is presented in Figure 5.1 on page 154.

The first panel of Table 7.1 reproduces the true ATE for this example from the prior Table 5.5, which is 4.53. The second panel of Table 7.1 introduces a second variant on the basic setup for Matching Demonstration 3. For this variant, Equations (7.3) and (7.4) are replaced with

$$y_i^1 = 102 + 3a_i + 2b_i + 6(a_i \times b_i) + v_i^1, \tag{7.5}$$
$$y_i^0 = 100 + 2a_i + 1b_i - 2(a_i \times b_i) + v_i^0, \tag{7.6}$$

but everything else remains the same. These alternative potential outcome definitions result in a slightly more dramatic pattern for the average treatment effects (in particular, for the ATT and ATC, as will be discussed in the next demonstration where we will estimate these directly). For now, this second variant generates a slightly different value for the true ATE, which is 5.05 rather than 4.53.

For each variant of this hypothetical example, we first offer three coefficients on D from unweighted OLS regression models using three different specifications: (1) Y

Table 7.1 Weighted Regression Estimates of the ATE, Using and Extending the Data Setup for Matching Demonstration 3

	ATE
Variant I: Y^1 and Y^0 linear in A and B	
True treatment effect	4.53
OLS regression estimates:	
Y regressed on D	5.39
Y regressed on D and linear A and B	4.75
Y regressed on D and quadratic A and B	4.74
ATE-weighted regression of Y on D	4.53
Variant II: Y^1 and Y^0 nonlinear in A and B	
True treatment effect	5.05
OLS regression estimates:	
Y regressed on D	5.88
Y regressed on D and linear A and B	5.47
Y regressed on D and quadratic A and B	5.44
ATE-weighted regression of Y on D	5.05

regressed on D, (2) Y regressed on D, A, and B, and (3) Y regressed on D, A, A^2, B, and B^2. None of these unweighted OLS estimates is particularly close to its respective true ATE for either variant, and the quadratic specifications of A and B help only very little. Unweighted regression has invoked the form of implicit weighting explained in Section 6.3, and as a result even in an infinite sample the estimated parameter would not equal to the true ATE, only a variant of it that invokes supplementary weighting by the conditional variances of A and B.

For the last row of each panel, we then implemented the weighted regression models specified earlier in Equation (7.2). For these estimates, we take the estimated propensity scores from the row labeled "Perfectly specified propensity score estimates" in Table 5.5, form $w_{i,\text{ATE}}$, as in Equation (7.1), and estimate a weighted bivariate regression model, where we implicitly assemble **P** by declaring the values for $w_{i,\text{ATE}}$ as survey weights in the weighted regression routine. Both resulting estimates land exactly on the target true ATE, as will always be the case in a sufficiently large dataset if the propensity scores are estimated flawlessly. No supplemental conditional-variance weighting is introduced into the estimator because the information on A and B enters the model only through the values of $\hat{p}_i$ that then structure the weights.

Several caveats should be mentioned before we proceed. First, as with nearly all other estimators we have considered so far, these results hold only in expectation over repeated samples or in the probability limit as a single sample increases to infinity. Second, for the results to emerge as cleanly as in this demonstration, the propensity scores must be estimated flawlessly. Misspecification of the propensity-score-estimating equation will push the point estimate off of the target ATE parameter. Although the literature has not yet systematically explored how sensitive point estimates are to different types and degrees of misspecification, it is clear that a misspecified propensity-score-estimating equation will not generate weights that will balance the underlying determinants of treatment assignment for the same arguments discussed already in Chapter 5. Third, even if the weights are correct because the propensity scores are flawlessly estimated, the point estimates will be imprecise across hypothetical repeated sampling for finite datasets if large weights result after estimated propensity scores are passed to Equation (7.1).[2] Large values for $w_{i, \text{ATE}}$ will be present if the estimated propensity score, $\hat{p}_i$, is either very close to 0 or very close to 1, for treatment or control cases, respectively. Nonetheless, Imbens and Wooldridge (2009:35) offer the somewhat comforting position that "these concerns are less serious than those regarding [unweighted] regression estimators because at least the...[weighted regression] estimates will accurately reflect uncertainty." In Section 7.4 on practical issues that confront weighted regression analysis, we will discuss a literature that considers the consequences of large weights and that advocates truncation of the distribution of the weights in some situations. Next, however, we will present equivalent estimators for the ATT and ATC.

7.2 Weighted Regression Estimators of the ATT and the ATC

To estimate the ATT and ATC, a similar procedure can be adopted after constructing appropriate weights, which are again functions of the estimated propensity scores, $\hat{p}_i$. In particular, two sets of weights $w_{i, \text{ATT}}$ and $w_{i, \text{ATC}}$ are formed as

$$\text{For } d_i = 1: \quad w_{i, \text{ATT}} = 1, \tag{7.7}$$

$$\text{For } d_i = 0: \quad w_{i, \text{ATT}} = \frac{\hat{p}_i}{1 - \hat{p}_i},$$

and

$$\text{For } d_i = 1: \quad w_{i, \text{ATC}} = \frac{1 - \hat{p}_i}{\hat{p}_i}, \tag{7.8}$$

$$\text{For } d_i = 0: \quad w_{i, \text{ATC}} = 1.$$

Again, these weights are analogs to survey weights, but the target population is no longer the total population comprising the population-level treatment and control groups together.[3] Instead, when using the weight $w_{i, \text{ATT}}$, the population-level treatment group becomes the target population. Accordingly, the weight leaves the sampled

---

[2]In this case, the estimates can still be regarded as consistent for the ATE. But, for any single finite sample, the point estimate of the ATE may not be close to the true ATE because of consequential sampling variability in the tails of the distribution of the propensity score.

[3]However, they are not the same analogs. The weight $w_{i, \text{ATE}}$ is analogous to a survey weight, and in particular the survey weights known as Horvitz-Thompson (HT) weights. HT weights are equal to $1/\pi_i$, where π_i is the probability that individual i from the population will be included in the sample

treatment group unaltered (because $w_{i,\text{ATT}} = 1$ for those in the treatment group), but it attempts to turn the control group into a representative sample of the population-level treatment group (because $w_{i,\text{ATT}} = \frac{\hat{p}_i}{1-\hat{p}_i}$ for those in the control group). More specifically, there are two distinct pieces of the weight that is applied to the control group. The denominator of $w_{i,\text{ATT}}$, is the same as for $w_{i,\text{ATE}}$, $1 - \hat{p}_i$, and it therefore weights the distribution of the control cases toward the full population, giving relatively more weight to cases with larger propensity scores. The numerator of $w_{i,\text{ATT}}$, $\hat{p}_i$, then accentuates the relative weighting induced by the denominator so that the control cases are properly weighted toward the target treatment group, not the full population.[4] The weight $w_{i,\text{ATC}}$ works in the opposite direction.

As with the ATE, to estimate point values for the ATT and ATC, one estimates a weighted bivariate regression model of Y on D, specifying the corresponding weight for the parameter of interest. The same weighted estimator in Equation (7.2) is utilized, but now the $n \times n$ diagonal matrix $\mathbf{P}$ is specified with either $w_{i,\text{ATT}}$ or $w_{i,\text{ATC}}$ on its diagonal, as we show now for a continuation of the same weighted regression demonstration.

Weighted Regression Demonstration 1 (Continued)

This demonstration is a continuation of Weighted Regression Demonstration 1 (see page 229) because the same setup is utilized, and the models differ only in the weights that are specified. However, the two variants of the data setup show their differences more clearly in this continuation, because they were designed to yield different patterns of true values for the ATT and ATC.

As shown in the first panel of Table 7.2, the true ATT and ATC are 4.89 and 4.40, respectively, for Variant I. But, as shown in the second panel, the true ATT and ATC are 5.77 and 4.79 for Variant II. For this second variant, the opposite-signed parameters specified for the cross-product interactions of A and B in Equations (7.5)

(also known as the sampling probability for individual i). For the ATE weights in Equation (7.1), the values for $\hat{p}_i$ and $1 - \hat{p}_i$ are each treated as if they are the HT inclusion probabilities, and two HT weights are used to standardize the treatment and control groups to the full population (as if the treatment and control groups are samples from the population based on different designs that must then be weighted in different ways to be representative of the same original target population). In contrast, the ATT and ATC weights, $w_{i,\text{ATT}}$ and $w_{i,\text{ATC}}$ in Equations (7.7) and (7.8), are either equal to 1 or are the *odds* of values that correspond to HT sampling probabilities. The latter align the treatment or control group with the group that receives uniform weights of 1 for each target parameter. As such, the analog for the odds embedded within the weights $w_{i,\text{ATT}}$ and $w_{i,\text{ATC}}$ are ratio adjustments to base weights (either base weights equal to 1 for simple random samples or equal to HT survey weights or their generalizations for more complex designs). Accordingly, it is quite natural to multiply the weights $w_{i,\text{ATT}}$ and $w_{i,\text{ATC}}$ by any survey weights that adjust for an underlying survey design, as we will do later in this chapter when analyzing data from the 2002 and 2004 waves of the Education Longitudinal Study. For the general reasoning behind the multiplication of weights, see Levy, Lemeshow, Biemer, and Christ (2008) and Valliant, Dever, and Kreuter (2013, chapter 13).

[4] In other words, for a control case with a low propensity score, the numerator further decreases the weight because the case is unlikely to reflect the characteristics of the treatment group. For example, for $\hat{p}_i = .2$, $w_{i,\text{ATE}} = \frac{1}{1-\hat{p}_i} = \frac{1}{1-.2} = 1.25$ while $w_{i,\text{ATT}} = \frac{\hat{p}_i}{1-\hat{p}_i} = \frac{.2}{1-.2} = .25$. For a control case with a high propensity score, the numerator still decreases the weight, but it does so relatively less because the case is more likely to reflect the characteristics of the treatment group. For example, for $\hat{p}_i = .8$, $w_{i,\text{ATE}} = \frac{1}{1-\hat{p}_i} = \frac{1}{1-.8} = 5$ while $w_{i,\text{ATT}} = \frac{\hat{p}_i}{1-\hat{p}_i} = \frac{.8}{1-.8} = 4$.

Table 7.2 Weighted Regression Estimates of the ATT and ATC, Using and Extending the Data Setup for Matching Demonstration 3

	ATT	ATC
Variant I: Y^1 and Y^0 linear in A and B		
True treatment effects	4.89	4.40
OLS regression estimates:		
Y regressed on D	5.39	5.39
Y regressed on D and linear A and B	4.75	4.75
Y regressed on D and quadratic A and B	4.74	4.74
ATT or ATC weighted regression of Y on D	4.89	4.40
Variant II: Y^1 and Y^0 nonlinear in A and B		
True treatment effects	5.77	4.79
OLS regression estimates:		
Y regressed on D	5.88	5.88
Y regressed on D and linear A and B	5.47	5.47
Y regressed on D and quadratic A and B	5.44	5.44
ATT or ATC weighted regression of Y on D	5.77	4.79

and (7.6) ensure that those with high levels of A and B together have much larger individual-level treatment effects than others. For Variant I, the differential sizes of the treatment effects are separable into simple linear pieces that can be independently attributed to A and B, as shown in Equations (7.3) and (7.4).

For each panel of Table 7.2, the same three unweighted OLS regression estimates already reported for Table 7.1 are reported again. Now, they are placed in both columns because standard unweighted OLS regression estimators do not differentiate between the ATE, ATT, or the ATC. The estimator is the same for each target parameter because it is often regarded as an estimator of an implicit constant structural effect of D on Y, which would render the ATE, ATT, and ATC equal if the constant-coefficient assumption were true.

For the last row of each panel, we then implemented the weighted regression models specified earlier in Equation (7.2). For these estimates, we again took the estimated propensity scores from the row labeled "Perfectly specified propensity score estimates" in Table 5.5, but now we formed the weights $w_{i, \text{ATT}}$ or $w_{i, \text{ATC}}$ in Equations (7.7) and (7.8). Unlike the unweighted regression estimates, the ATT-weighted and ATC-weighted regression estimates land exactly on their respective target parameters for each variant. In addition, they effectively undo the complex pattern of nonlinearity for Variant II, where the propensity scores and the potential outcomes are both nonlinear in A and B.

As for the weighted regression estimates of the ATE, these results only hold over repeated samples from the same population or in the probability limit as a single sample approaches infinity. And the same caveats introduced in Section 7.1 still obtain. If the equation that estimates the propensity scores is misspecified, then the estimates of the ATT and ATC will be inconsistent and biased because the weights will not fully balance the underlying determinants of treatment assignment. In addition, disproportionately small or large weights may still emerge even if the propensity scores are estimated flawlessly, and in these cases the estimates may be imprecise (i.e., still consistent but not necessarily close to the true ATT or ATC in the single finite sample under analysis).

7.3 Doubly Robust Weighted Regression Estimators

Many perspectives exist on how matching and regression can be combined. In Section 6.5, we demonstrated how parametric regression can be used to provide supplemental adjustment for matching estimators, following several decades of practice in the applied matching literature. The rationale for this type of adjustment is that the chosen matching routine has almost certainly not eliminated all of the imbalance in the observed treatment assignment variables, and a post-match parametric regression that adjusts for the same variables while also estimating the treatment effect may eliminate at least some of the lingering imbalance.

An alternative but related perspective has also been discussed at several points, and it has more recent origins. Rather than consider regression as a supplement to an imperfect matching routine, one can consider matching as a remedy to artifactual regression results that have been produced by incautious data mining. Ho et al. (2007) suggest that the general procedure one should carry out in any multivariate analysis that aspires to generate causal inferences is to first balance one's data as much as possible with a matching routine and then estimate a regression model on the matched data. From this perspective, matching is a preprocessor, which can be used to prepare the data for subsequent analysis with something such as a regression model.

Both of these perspectives are motivated by the goal of giving the analyst two chances to "get it right," first with a matching routine and second with a parametric regression model. In this section, we present the most influential version of this encompassing strategy. For the estimators presented in Sections 7.1 and 7.2, the observed determinants of treatment assignment were used only to estimate the propensity scores. In this section, we present a related set of estimators that use these same variables a second time, as part of the regression model that generates the point estimates. Recall Equation (7.2),

$$\hat{\delta}_{\text{OLS, weighted}} \equiv (\mathbf{Q'PQ})^{-1}\mathbf{Q'Py},$$

where we defined the $\mathbf{Q}$ matrix as containing a column of 1s and a column of individual-level values d_i for the treatment variable, along with a diagonal matrix $\mathbf{P}$ that represents a set of weights specific to the target parameter. For this estimator, as used up until this section, treatment assignment variables are presumed to have been used only in a first stage to estimate the values of $\hat{p}_i$, which then structure the appropriate weights in $\mathbf{P}$. The estimators that we present and demonstrate in this section use the same variables a second time as supplementary adjustment variables. The estimator in Equation (7.2) is still used, but these variables are included in additional columns in $\mathbf{Q}$.

The idea is to offer what James Robins and his colleagues refer to as a "doubly robust" or "doubly protected" estimator (see Bang and Robins 2005; Robins and Rotnitzky 2001). Robins and Rotnitzky reflect on the fallibility of both standard regression methods, as presented in Chapter 6, as well as propensity-score-based weighting estimators, as presented in Sections 7.1 and 7.2:

> There has been considerable debate as to which approach to confounder control is to be preferred, as the first is biased if the outcome regression model is misspecified while the second approach is biased if the treatment regression, i.e., propensity, model is misspecified. This controversy could be resolved if an estimator were available that was guaranteed to be consistent ... whenever at least one of the two models was correct. ... We refer to such combined methods as doubly-robust or doubly-protected as they can protect against misspecification of either the outcome or treatment model, although not against simultaneous misspecification of both. (Robins and Rotnitzky 2001:922)

The basic motivation is to give the analyst two chances to "get it right," in hopes that misspecifications of the propensity-score-estimating equation and the final regression equation will neutralize each other. And, although Robins is credited with developing the recent asymptotic justification for a variety of specific procedures (see Robins and Hernn 2009; Robins and Ritov 1997; Robins, Rotnitzky, and Zhao 1994; Scharfstein, Rotnitzky, and Robins 1999; and van der Laan and Robins 2003), the idea of using matching and regression together is quite general and has a long legacy in applied work, as we noted earlier in Section 6.5 (see Cochran and Rubin 1973; Gelman and King 1990; Heckman, Ichimura, and Todd 1998; Hirano and Imbens 2001; and Rubin and Thomas 1996, 2000).

Because we used flawlessly estimated propensity scores for an infinite sample in Weighted Regression Demonstration 1, there would be no value in demonstrating the doubly robust estimation strategy for the same setup. We do not need a second chance to "get it right." Instead, we next offer a demonstration with doubly robust estimators, where we use simulated data and vary our capacity to "get it right" when estimating the propensity scores.

Weighted Regression Demonstration 2

For this demonstration, we revisit Matching Demonstration 4 and Regression Demonstration 4 (see pages 171 and 216, respectively). We will use the same 10 simulated datasets for the Catholic school effect on learning, modeled on the National Education Longitudinal Study (NELS) data analyzed in Morgan (2001). Recall that for this hypothetical substantive example, we have concluded that only the ATT can be estimated with any degree of confidence because we have not observed variables that would allow us to directly model self-selection on the individual-level effect of Catholic schooling.

We already showed in Regression Demonstration 4 that matching performs well when the complete specification of treatment assignment is used, along with a supplementary regression adjustment. The reason for this result is precisely the double protection argument outlined above. The varying level of imbalance that remains for each matching estimator can be seen as slight misspecification of the matching model, which is produced either because the propensity score model has not been respecified to remove as much imbalance as possible or because the matching algorithm has features that render it suboptimal for the particular application. The supplemental regression adjustments, as shown in Table 6.8, reduce the average bias in the matching estimates because they further adjust for remaining imbalance in the means of the matching variables. As such, the matching and supplementary adjustment are working together to minimize bias in the estimates of the ATT, using the double protection reasoning introduced above.

In the demonstration we offer here, we will present doubly robust estimates from ATT-weighted regression models for the same 10 simulated datasets. As shown in Table 7.3, we again use estimated propensity scores from two scenarios, where for the incomplete specification we omit the crucial cognitive skills variable as well as additional higher-order and interaction terms. We then estimate the ATT, using these two sets of propensity scores in two ways, first as simple weighted ATT estimators as for Weighted Regression Demonstration 1 but also now as doubly robust estimators as defined in this section.

For the first row of Table 7.3, we report the minimum, maximum, and average bias of ATT-weighted regression estimates under both specifications of the models that generate the values of $\hat{p}_i$ across all 10 simulated datasets. These bias values are directly comparable to those of the 19 matching estimators reported earlier in Table 5.7 for Matching Demonstration 4. As shown here, the bias is substantial for the weighted regression estimates that use the incomplete specification to estimate the propensity scores. This result is the same as for the matching estimators reported in Table 5.7, and the bias is produced mostly because of the exclusion of the variable for cognitive skills. In contrast, the average bias for the weighted regression estimates that use the complete specification to estimate the propensity score is small. In fact, the average bias of the ATT-weighted regression is in the middle of the distribution of average bias of all 19 matching estimators reported in the final column of Table 5.7.

For the second row of Table 7.3, the equivalent bias values are reported for the doubly robust estimators, where the assumed treatment assignment/selection variables are used a second time in the outcome regression equations. Perhaps surprisingly, the double protection of supplementary regression adjustment does not narrow the average bias further (and, in fact, may even increase it for these 10 samples). This result

Table 7.3 Bias for Weighted Regression Estimates of the ATT, Catholic Schooling on Achievement Across 10 Simulated Datasets Utilized for Matching Demonstration 4 and Regression Demonstration 4

	Specification of treatment assignment variables:			
	Incomplete specification		Complete specification	
Method	Min, Max	Average	Min, Max	Average
Weighted regression estimate of the ATT	−0.08, 2.17	0.82	−0.87, 1.11	−0.09
Doubly robust weighted regression estimate of the ATT	−0.05, 2.19	0.83	−0.87, 1.08	−0.10

departs from what we showed in Table 6.8 for the same 10 datasets when supplemental regression was used to eliminate some of the remaining bias. There, as opposed to here, one can see clear evidence of double protection.

The reason that the bias is relatively unchanged, for this demonstration, and hence little evidence of double protection seems to emerge, is that the ATT and ATC weights are already very well estimated in the sense that they leave very little imbalance in the data on the observed values that generate them. In particular, the logit estimation of the propensity scores matches the assumed logistic distribution for the true propensity scores. As such, no unusual weights emerge because, on average, the parametric structure of the estimated logit model does not force an unwarranted distribution onto the values of $\hat{p}_i$.

Is this result typical for weighted regression estimators? If the weights are estimated very well in the sense that they effectively represent all of the relevant information in the determinants of treatment assignment on which they are based (as can usually be discerned by an examination of balance), then supplementary adjustment by the treatment assignment/selection variables will have little or no effect on the estimates. None of the virtuous effects of double protection would be evident because the supplemental adjustment is redundant. In many cases, especially for small datasets, these conditions will not be present, and the supplementary adjustment will offer clear double protection.

As this demonstration and the last show, propensity scores can be used effectively within a weighted regression framework to estimate average causal effects. The caveats noted in Section 7.1 continue to apply, and in the next section we will discuss these. In the section that follows the next, we will offer an extended and realistic example with genuine data, which demonstrates how effective weighted regression estimates of the ATT and ATC can be, but also how carefully they must be interpreted, just like all other estimators considered in Chapters 5 through 7.

7.4 Remaining Practical Issues in Weighted Regression Analysis

In this section, we bridge the simulated results for this chapter with a related but real application. We first discuss the need to consider the consequences of having estimated weights that may appear to be unreasonably large. We then discuss how and why weighted regression estimates have a distinct advantage over matching methods in their ability to handle survey data, which is rarely generated by a simple random sample. Finally, we consider alternatives for estimating standard errors.

We will not discuss here some issues that are common to both matching and weighted regression estimators. Our prior discussions of estimation when treatment assignment is nonignorable (Section 5.5.1), estimation on the region of common support (Section 5.5.2), and estimation for many-valued treatments (Section 5.5.4) apply to weighted regression estimators as well in only slightly modified ways that should be obvious.

7.4.1 The Concern Over Extreme Weights

As we noted when introducing weighted regression estimators of the ATE, it is possible to generate large weights if the estimated propensity score, $\hat{p}$, is close to 0 or 1. Similarly, for the weights used to estimate the ATT and ATC, values of $\hat{p}$ close to 0 or 1 will generate either small or large weights. The literature on propensity-score-based weighting includes a number of decidedly skeptical pieces (e.g., Freedman and Berk 2008), some of which focus on the consequences of such weights (e.g., Lee, Lessler, and Stuart 2011). In some pieces, which we will not cite, manifestly poor weights have been used to demonstrate the obvious point that estimated propensity scores will not necessarily balance the data or warrant a causal inference. The more revelatory work demonstrates why one should proceed with caution when the estimation of the propensity scores generates disproportionately large weights, which may then give undue influence to only a very few members of a given sample. Yet, the literature has not determined precisely when small and large weights should be regarded as extreme or, more deeply, when extreme weights should be regarded as problematic.

To approach the key issues, first consider three scenarios in which extreme weights might emerge:

1. The weights are based on an empirical model of treatment assignment that uses observed variables as predictors that *do not* collectively sustain the relevant ignorability assumption.

2. The weights are based on an empirical model of treatment assignment that uses observed variables as predictors that sustain the relevant ignorability assumption, but the model is nonetheless misspecified so that the estimated distribution of the propensity score departs systematically from the distribution of the true propensity score.

3. The weights are based on an empirical model of treatment assignment that uses observed variables as predictors that sustain the relevant ignorability assumption,

and the model is thought to be well specified, after an inspection of covariate overlap and indices of balance. As a result, extreme weights must reflect either (a) a strong treatment assignment/selection regime for which some true propensity scores are very close to either 0 or 1 and/or (b) very unusual cases sampled at random from the population.

Scenario 1 is a weak position from which to estimate a causal effect, and it does not represent a unique threat to weighted regression estimators. Analogous to the discussion on matching in Chapter 5, for weighted regression estimators in this situation the analyst will necessarily pass on to the regression routine some poorly constructed weights and thereby generate estimates that are inconsistent and biased for the true treatment effects of interest. Although scenario 1 might generate extreme weights (especially if some of the misspecification issues we will discuss for the next scenario emerge), it has comparatively low probability of doing so. If back-door paths remain unblocked, it is more likely than not that the range of the estimated propensity score will be too narrow.

Scenarios 2 and 3, however, raise particular issues for weighted regression estimators. For scenario 2, the model may, for example, "overfit" the data in the specific sense that the model has been too heavily parameterized so that cases appear artificially overdetermined. Imagine the extreme (but preposterous) scenario where a researcher includes one or more individual-identifying indicator variables (e.g., a dummy variable for "individual number 566 in the dataset" based on an assumption that this person is, for reasons deeply felt by the investigator, perfectly predisposed to be in the treatment group). Any such variable will lead to a perfect prediction for the relevant case, and most software packages will happily deliver (if the relevant switch is turned on) a predicted probability of treatment assignment that is set arbitrarily close to either 0 or 1. Milder forms of overfitting can also generate estimated propensity scores that are too dispersed (see Hill et al. 2011).[5] The solution here, however, is straightforward. One should not overfit, and the protection against overfitting is to (1) ground a conditioning strategy in defendable assumptions about back-door paths and then (2) conduct balance checking while specifying the equation for propensity-score estimation.[6]

For scenario 3, everything has been done correctly. Instead, it is simply an empirical fact, as best can be discerned, that the true propensity scores are either very large or very small for some members of the population (even though they are not structural

---

[5]Matching estimators of some types may escape from some of the consequences of over-dispersed estimated propensity scores. One-to-one nearest-neighbor matching without replacement, for example, pairs up treatment and control respondents and then calculates mean differences, ignoring the scale of the propensity score. All cases are then given the same weight in the analysis. If calipers are used, however, then cases with propensity scores very near to 0 or 1 may be cast aside as unmatchable. The resulting matching estimator is then implicitly replacing unduly large weights with weights of zero, which could be deemed worthwhile under an appropriate consideration of expected mean-squared error.

[6]Another remedy, following the double protection strategy, would be to estimate weights based on an incomplete propensity-score model, where the variable that generates the extreme weights is removed from the propensity-score-estimating equation. This variable would then be used in the weighted regression model that estimates the effect. A downside of this strategy is that conditional-variance-based weighting with respect to this variable could then push the estimated effect off of its target parameter. Accordingly, this remedy should be avoided if one has reason to believe that the individual-level effect of interest varies in the levels of the variable in question.

constants equal to 0 or 1, as for Matching Demonstration 2; see page 148). What are the consequences of such very large weights, and what can be done to mitigate any negative consequences?

We noted above that the ATE, ATT, and ATC weights can be interpreted as analogs to survey weights. In the survey methodology literature, it is commonplace to evaluate the distribution of constructed survey weights and to consider procedures that may mitigate the consequences of using large weights. There are two common positions.

First, one can regard the weights as correct, even though extreme, and simply carry on (mindful that an appropriate variance estimator must, however, be utilized, as we discuss below). Levy et al. (2008) describe the distribution of weights constructed for the National Survey of Child and Adolescent Well-Being (NSCAW). These weights were constructed to adjust for the stratified sampling design and also patterns of nonresponse. They show that the constructed weights for the NSCAW had a minimum of 2 and a maximum of 8,175 and write:

> The distribution is highly skewed to the right, which is not atypical for unequal probability sample designs. This is because the units in the population that have higher probabilities of selection, and consequently smaller base weights, will dominate the sample. Some units with very small selection probabilities will still be selected in large samples... and the weights for these units can be quite large. (Levy et al. 2008:511)

In this situation, analysis can proceed even though most sample members have weights less than 50, but a few have weights greater than 4,000, with one at 8,175.

Second, one can trim the weights back so that one does not have to assume that the extreme weights are correct, as might be appropriate if one had a lack of confidence in the model that generates them. Valliant et al. (2013) consider the procedure that is used to trim the weights for the National Assessment of Educational Progress (NAEP). Here, estimated weights that are greater than 3.5 times the median weight are trimmed back to 3.5 times the median weight (and the aggregate trimmed-away weight is then redistributed equally across cases beneath the upper bound for trimming). This procedure was adopted by the data contractors with the expectation that the trimmed weights would likely impart some bias to estimated parameters of interest but would also lower the expected variance for each of them. Valliant et al. (2013:388) note that "these methods are *ad hoc* and largely theoretical," suggesting that in general mean-squared-error calculations are less influential than "agency preference or historical precedence."[7]

The same two positions are applicable to weights used to estimate the ATE, ATT, and ATC. First, an analyst who has confidence, perhaps after sufficient inspection of overlap and a specification produced by careful balance checking (as in the extended example we will offer below in Section 7.5), may choose to assert that the weights are

---

[7]A third possibility would be to smooth the weights nonparametrically across strata defined by the initial set of estimated propensity scores, pulling in the extreme values in the tails and reducing the dependence of all weight values on the particular specification of the propensity score estimating equation. See Hong (2010, 2012) for a marginal mean weighting estimator that is based on such smoothed weights.

indeed correct and worth using, regardless of their distribution. Or, second, an analyst can trim back the weights by truncating their distribution by progressively recoding them to interior percentiles of the initial estimated distribution (i.e., perhaps first back to the 1st and 99th percentiles, then the 5th and 95th percentiles, and so on).

Finally, a third option is also available, which is to focus the parameter of interest on a subset of the population. Often referred to colloquially as "moving the goalposts," one changes the parameter of interest to something for which the seemingly extreme weights are irrelevant. Rather than estimate the ATE, ATT, or ATC, one instead estimates an average effect over a specific range of the estimated propensity score. Although it would likely be hard to defend in a principled fashion, one could argue that the average treatment effect for those whose propensity scores are $\geq .1$ and $\leq .9$ is a well-defined parameter of interest. A more common strategy would be to restrict the analysis to the common support or region of overlap in the propensity score across the treatment and control groups (see Section 5.5.2 for a description of this strategy when offering matching estimates). In this case, one could, for example, still offer an estimate of the ATT that may still apply to most members of the treatment group and yet does not draw the cases with extreme weights into the analysis. The prudent approach may be to take all three of these strategies and develop conclusions that do not depend on adopting only one or another.

7.4.2 Survey Data

Relatively few of the surveys that are routinely analyzed by social scientists have simple sample designs. As explained in survey methodology texts such as Thompson (2002), most survey samples have multistage, stratified designs that entail unequal probabilities of inclusion across members of the population. Accordingly, most of the datasets from these surveys that are delivered to the desktops of observational analysts come with survey weights that have been constructed to account for their design features. Most also include weights that adjust for patterns of nonresponse, typically constructed in a series of nested steps (see Valliant et al. 2013).

Weighted regression estimators of the ATE, ATT, and ATC can incorporate these weights without difficulty. In contrast, there is no consensus position on how matching algorithms should be deployed for complex survey data. Most matching routines were designed for the analysis of simple random samples or nonsampled collections of units that can be treated as equally representative pieces of information. For the weights used for regression estimation of the ATE, ATT, and ATC, all one needs to do is (1) weight the propensity-score-estimating equation by the appropriate survey weight suggested by the data distributor and then (2) multiply the constructed weights for the ATE, ATT, and/or ATC by the same survey weight. In so doing, the analyst then passes to the regression routine a model-based weight for the relevant parameter that is modified by the probability of inclusion in the analysis sample that is being utilized. For example, the ATE weights can be formed as

$$\text{For } d_i = 1: \quad w_{i, \text{ ATE, complex}} = \text{survey weight} \times \frac{1}{\hat{p}_i},$$

$$\text{For } d_i = 0: \quad w_{i, \text{ ATE, complex}} = \text{survey weight} \times \frac{1}{1 - \hat{p}_i}. \tag{7.9}$$

The survey weights will alter the distribution of the weights that are passed to the regression routine, but they will not necessarily inflate the variance of the weight. Cases with relatively low or high estimated propensity scores may have probabilities of inclusion that place them near the mean of the constructed survey weights.[8]

7.4.3 Expected Sampling Variance and Estimated Standard Errors

For all weighted regression estimators of the type considered here, no convincing rationale exists for assuming homoscedasticity when calculating standard errors. We recommend that researchers routinely report heteroscedasticity-consistent standard errors (e.g., the *robust* option in Stata, which will be called automatically if the weights are specified as *pweights*).[9] These estimated standard errors are more likely to reflect the true expected sampling variability of weighted regression estimates, although the degree to which robust variance estimation is fully effective in finite samples (especially small ones) has not been clearly established in the literature.

In our own simulation work, we have found that heteroscedasticity-consistent standard errors have performed quite well for larger samples and the estimation of parameters that are identified. Consider the following results. Following up on Weighted Regression Demonstration 2 (see page 235), we performed the following Monte Carlo simulation. Rather than analyze the 10 simulated datasets already produced for the prior simulations (see the explanation of their generation in the introduction to Matching Demonstration 4 on page 171), we generated 10,000 new datasets, each of which included 10,000 simulated individuals. On average across all of these datasets, 1,049 cases were assigned to the treatment, and the average value for the true ATT was 6.93.

For each dataset, we then estimated the weighted regression and doubly robust weighted regression estimates of the ATT, equivalent to those labeled as the "complete specification" for Table 7.3. Across all 10,000 datasets, the average bias for each estimator was <.01 (and we assume that the average bias of each would approach 0 in the limit if the number of simulated datasets was increased from 10,000 to infinity). The standard deviations of the distributions of these estimates were .47 and .46, respectively, across all 10,000 datasets. These two values of .47 and .46 are Monte Carlo benchmarks for the true standard errors of the corresponding weighted regression estimates of the ATT.

To evaluate alternative methods for estimating standard errors for the point estimates of the ATT, we also calculated two estimated standard errors, a classical standard error that assumes homoscedasticity and a heteroscedasticity-consistent standard error (the latter using the "sandwich" package for R; see Lumley and Zeileis 2013).[10]

---

[8]The rationale for ratio multipliers to base survey weights is very general, and an analyst can further multiply ATE, ATT, and ATC weights by model-based weights of any type (such as attrition-adjustment weights in a longitudinal model, as for the demonstration in the next section based on Morgan and Todd 2008).

[9]These estimated standard errors are also sometimes referred to as Huber-White standard errors, robust standard errors, sandwich standard errors, and other variants; see Angrist and Pischke (2009, chapter 8) as well as Long and Ervin (2000).

[10]The sandwich routine for R calls a version of the estimator that is known as "HC3" and that corrects for finite sample bias in the estimated variance of the residuals (see Angrist and Pischke

The classical standard errors were too small by a wide margin, at .26 and .24 on average, respectively. However, the heteroscedasticity-consistent standard errors were close to the simulated true standard errors, at .50 and .46 on average.

As with matching estimators, more work is needed to determine when heteroscedasticity-consistent standard errors are valid. However, we are confident that they are more reasonable than the classical standard errors that are produced as defaults by most software packages. For all types of regression modeling, Angrist and Pischke (2009:307) recommend a conservative rule of thumb: For the parameter of interest, calculate both classical and heteroscedasticity-consistent standard errors and then use whichever one is larger.

7.5 An Extended Example

In this section, we offer a full analysis of the estimation of the ATT and the ATC from a weighted regression perspective, using real data where we do not know either the true treatment effects or the form of whatever underlying equation generates the true propensity scores. The substance of the demonstration is again the Catholic school effect on learning in high school.

Weighted Regression Demonstration 3

For this demonstration, we draw on Morgan and Todd (2008) and present the analysis in the steps in which it was undertaken. In particular, we will first model the effect of Catholic schooling on the math test scores of high school students in the tenth and twelfth grade in 2002 and 2004 using standard OLS regression models typical of the original research on the Catholic school effect. We will then offer weighted regression estimates of the ATT and ATC, under provisional acceptance of the identifying Assumptions 1-S and 2-S in Equations (5.1) and (5.2). However, we will conclude the demonstration with an extended set of interpretations that critically reevaluate these assumptions, wherein we will argue that, at best, only the ATT is identified. Nonetheless, we aim to convince the reader that provisional estimation of the ATC helps to explain why neither the ATC nor the ATE is identified and promotes a substantive consideration of how best to interpret the estimate of the ATT.

Data and Measures. The data for this demonstration were drawn from the 2002 base-year and 2004 follow-up waves of the Education Longitudinal Study (ELS), collected by the National Center for Education Statistics (NCES) of the United States Department of Education. The ELS is a nationally representative sample of students in public and private high schools, based on a two-stage sampling design that first

---

2009, chapter 8). Although available in Stata as well, the default that Stata calls with the commonly used vce(robust) option (or the default that is automatically called in the background when *pweights* are specified) is the original "HC0" that is biased for the variance of the residuals in finite samples. Long and Ervin (2000) offer a Monte Carlo simulation that suggests that for linear regression and samples of less than 500, the HC3 formula outperforms the HC0 formula. For their large sample size, the differences were trivial in the simulation, as we expect they would be for our simulation too.

draws a random sample of public and private high schools and then draws random within-school samples of tenth graders (typically aged 16). For the first follow-up in 2004, respondents were tracked to alternative destinations, and most respondents were twelfth graders (typically aged 18).

From among all base-year ELS participants, we restricted the analysis to respondents who were enrolled in either a Catholic school or a public school during the 2001–2002 academic school year. Table 7.4 presents means and standard deviations for the variables we will analyze on the 1,918 students who were enrolled in Catholic schools and the 12,025 students who were enrolled in public schools (for a total of 13,943 students).

Several practical features of our subsequent analysis, which have their sources in the complex nature of the ELS survey data, require detailed explanation. Because these details are not essential for understanding the main contours of the demonstration, we will provide this supplementary information in footnotes (and they are more completely specified in the appendix to Morgan and Todd 2008). Readers who are contemplating using these estimators for a project with a similar data structure – where a complex sampling design necessitates the use of a post-stratification weight that corrects for unit-level nonresponse and where some models are estimated with an adjustment for panel attrition – should consult this supplementary material for specific practical advice. In short, the weights we utilize for the ATT and ATC estimators are multiplied by additional weights that account for design features of the data. In addition, we use heteroscedasticity-consistent estimated standard errors, with an adjustment for clustering in schools, for both the ATT and ATC estimates reported below.

Step 1: Estimate a Bivariate Regression Model. To initiate the analysis with naive estimators, we first estimate three bivariate regression equations with ordinary least squares,

$$Y = \hat{\alpha} + \hat{\delta}_{\text{OLS, bivariate}} D + \varepsilon, \tag{7.10}$$

where Y is one of three interval-scaled outcome variables and D is an indicator variable equal to 1 for those who attend Catholic high schools (and equal to 0 for those who attend public high schools). The estimated coefficient $\hat{\delta}_{\text{OLS, bivariate}}$ is the estimated effect of D on Y. If regarded as a structural constant, this coefficient is an estimate of the ATE, ATT, and ATC.

The bivariate regression estimates for the Catholic school effect on achievement are presented in the first row of each panel of Table 7.5. The three panels offer analogous results for three related outcome variables: tenth grade math test scores, twelfth grade math test scores, and math gains between the tenth and twelfth grades. The estimates of $\hat{\delta}_{\text{OLS, bivariate}}$ in Equation (7.10) are 7.31, 8.45, and 2.00 for these three different outcome variables. Each of these estimates suggests that Catholic school students have higher levels of achievement on standardized tests and on growth in achievement between the tenth and twelfth grades, matching results from research since the 1980s (see Section 1.3.2, page 22).

Table 7.4 Means and Standard Deviations of the Primary Variables Used in the Demonstration

	Public		Catholic	
Variable	Mean	SD	Mean	SD
Math Test Scores				
IRT estimated number right (10th grade)	41.68	13.97	48.99	12.02
IRT estimated number right (12th grade)	47.64	15.05	56.08	12.80
Math gain (12th grade −10th grade)	4.66	6.49	6.66	6.06
Female	.50		.48	
Race (White is the reference category)				
Black	.15		.06	
Hispanic	.17		.11	
Asian	.04		.04	
Native American	.01		<.01	
Multiracial	.04		.04	
Urbanicity (Suburban is the reference category)				
Urban	.28		.58	
Rural	.21		.01	
Region (Midwest is the reference category)				
Northeast	.18		.31	
South	.34		.23	
West	.23		.17	
Family Background				
Mother's education (in years)	13.46	2.32	14.77	2.22
Father's education (in years)	13.59	2.59	15.25	2.57
SEI score of mother's occupation	44.98	12.87	50.55	12.85
SEI score of father's occupation	44.15	11.70	49.81	11.71
Family income (natural log)	10.60	1.09	11.23	.90
Family income (natural log) squared	113.61	19.83	126.99	17.04
Family income (natural log) cubed	1225.64	295.46	1441.33	267.98
Two-parent family	.75		.84	
Past History (as reported by parent)				
Learning disability	.13		.07	
Ever held back	.13		.05	
Repeated 4th grade	.01		<.01	
Years parents lived in current neighborhood	10.56	8.00	12.90	8.21

Source: Education Longitudinal Study, 2002 and 2004.

Step 2: Estimate a Multiple Regression Model by Introducing Adjustment Variables. We next estimate three multiple regression equations with ordinary least squares,

$$Y = \hat{\alpha} + \hat{\delta}_{\text{OLS, multiple}} D + X\hat{\beta} + \varepsilon, \tag{7.11}$$

Table 7.5 Catholic School Coefficients from Baseline Regression
Models Predicting Tenth Grade Math Test Scores, Twelfth Grade Math
Test Scores, and Math Test Gains

Predictor Variables	Outcome Variable: 10th Grade Math Test Score
Model 1: Dummy for Catholic school	7.31
	(.66)
Model 2: Model 1 + family background, demographics, and past history	1.48
	(.52)
	Outcome Variable: 12th Grade Math Test Score
Model 1: Dummy for Catholic school	8.45
	(.74)
Model 2: Model 1 + family background, demographics, and past history	2.13
	(.60)
	Outcome Variable: Math Gain (12th−10th)
Model 1: Dummy for Catholic school	2.00
	(.23)
Model 2: Model 1 + family background, demographics, and past history	1.27
	(.26)

Note: Heteroscedasticity-consistent standard errors in parentheses.

where X represents observed variables thought to determine D and Y (because they lie on back-door paths from D to Y, as in Figure 5.2), $\hat{\delta}_{\text{OLS, multiple}}$ is the estimated causal effect of D on Y adjusted for X, and $\hat{\beta}$ is a conformable vector of estimated coefficients that correspond to the variables in X.

Descriptive statistics for the 23 variables specified as X in Equation (7.11) are presented above in Table 7.4. The variables in X represent the most common family background, demographic, and educational history variables utilized in school effects research. The multiple regression estimates for the Catholic school effect on achievement are presented in the second row of each panel of Table 7.5. The coefficients for $\hat{\delta}_{\text{OLS, multiple}}$ are 1.48, 2.13, and 1.27 in the three panels, each of which is considerably smaller than the corresponding values of $\hat{\delta}_{\text{OLS, bivariate}}$ from the estimation of Equation (7.10). Nonetheless, the values of $\hat{\delta}_{\text{OLS, multiple}}$ suggest that Catholic school students outperform public school students even after adjustments for the variables in X, again matching results of past research since the 1980s.

Step 3: Estimate Propensity Scores and Form Weights to Estimate the ATT and ATC. In this step we iteratively estimate propensity scores in pursuit of maximum achievable balance on the variables in X and then form weights based on the final set of estimates. We first estimated a model of treatment selection/assignment using the variables in X as predictors, and using the same simple linear specification as for the multiple regression models estimated for Table 7.5.

In particular, we estimated a logit model where the variables specified as X are the same 23 variables specified as X for the multiple regression models. The estimated logit model fit the data reasonably well and delivered a chi-squared test statistic of 404 with 23 degrees of freedom. Predicted values for the estimated propensity scores, $\hat{p}_i$, were then calculated. The estimated propensity scores had a mean of .0440 and a standard deviation of .0688. The distribution was heavily skewed with a minimum of .0000182 but a maximum of .857.

We then took these estimated propensity scores and formed two sets of weights, $w_{i,\,\text{ATT}}$ and $w_{i,\,\text{ATC}}$, using the expressions in Equations (7.7) and (7.8). We next checked how effective this first set of estimated weights was in balancing the data. Again, perfect balance requires that all moments of the joint distributions of the variables in X be exactly the same in the treatment and control groups.

In this case, the raw data are substantially imbalanced, as was shown in the means and standard deviations that were reported above in Table 7.4. In general, public school students are less advantaged and are more heterogeneous with respect to the characteristics measured as X. For example, the mean of mother's education in years is 13.46 for those in public schools but 14.77 for those in Catholic schools. The mean of the log of family income is 10.60 for those in public schools but 11.23 for those in Catholic schools. Moreover, the dispersion of the log of family income is substantially different as well; its standard deviation is 1.09 for those in public schools but only .90 for those in Catholic schools.

To assess the degree of balance achieved by the weights formed in this first step, a balance criteria must be chosen. The first metric we use is the average of standardized mean differences across treatment and control groups (see Rubin 1973a, 1973b). The standardized difference of the mean for each variable in X is calculated as

$$\frac{|E_N[x_i|d_i=1] - E_N[x_i|d_i=0]|}{\sqrt{\frac{1}{2}\text{Var}_N[x_i|d_i=1] + \frac{1}{2}\text{Var}_N[x_i|d_i=0]}}, \tag{7.12}$$

which we introduced earlier as Equation (5.20). Equation (7.12) yields a scaled absolute difference in the mean of a variable in X across the treatment and control groups. These values can be combined across all variables in X in order to construct an average standardized difference of means. The average standardized difference of means can be calculated under different weighting schemes in order to compare the relative performance of alternative weights in achieving balance with respect to the target parameter (e.g., the balance of means on average for estimation of the ATT).

Because balance is not just a property of the means of variables but also of higher moments of the distributions, we used a second metric of balance for variables that are not two-valued indicator/dummy variables. For this metric, we change Equation (7.12) slightly, substituting standard deviations in the treatment and control groups for $E_N[x_i|d_i=1]$ and $E_N[x_i|d_i=0]$. The modified version of Equation (7.12) then yields a scaled absolute difference in the standard deviation of a variable in X across the treatment and control groups. Because these values are standardized, they can also be combined across alternative variables in X in order to construct an estimate of the average standardized difference in standard deviations.[11]

To set a baseline against which to measure improvements in balance, we first calculated a baseline level of imbalance for the raw data by estimating the average standardized difference of means and standard deviations for the variables in X. The means of the variables (as well as the corresponding standard deviations for variables that take on more than two values) were already reported in Table 7.4, separately for those in Catholic and public schools. To assess how much imbalance the weights eliminate, we then calculated the balance after using the two separate weights $w_{i,\text{ATT}}$ and $w_{i,\text{ATC}}$. The results showed that the weights succeeded in producing substantial balance, reducing the average standardized difference of means from .350 to .00634 when using $w_{i,\text{ATT}}$ and to .111 when using $w_{i,\text{ATC}}$. The average standardized difference of standard deviations also fell substantially from .0715 to .0391 when using $w_{i,\text{ATT}}$ and to .0287 when using $w_{i,\text{ATC}}$. The increase in balance that results from employing $w_{i,\text{ATT}}$ and $w_{i,\text{ATC}}$ is substantial, but some minor imbalance remains.

The remaining imbalance suggests that respecifying the model of treatment assignment may be worthwhile. The initial specification of the model of treatment assignment was borrowed from the simple default linear specification of the adjustment variables in X for the multiple regression models reported in Table 7.5, which is based on the typical specifications offered in the early literature. The goal is now to enrich the parameterization of the treatment assignment model in order to construct weights that further improve the balance on the variables in X when the weights are deployed. Accordingly, interactions between the variables in X not already included in the regression specification, as well as transformations of the original variables, should be considered.

Although various data mining procedures can be wedded to balancing metrics in pursuit of a best possible model (see Section 5.4.2), we chose the traditional strategy of controlled trial-and-error respecification of the propensity-score-estimating equation. We used a forward selection procedure where interactions that have some justification in theory and past research were added progressively until improvements in balance ceased to arise. We ignored the size and statistical significance of logit coefficients and only inspected improvements in balance using our chosen criteria.

As we explained just above, the original logit model fit the data reasonably well and also provided good balance by the standards of the matching literature. However, a better fit was available that also yielded weights that provided even better balance. We added 75 interaction terms to the initial logit model that predicted Catholic school attendance. The estimated propensity scores, $\hat{p}_i$, from this new model have a mean of .0440, a standard deviation of .0756, a minimum of 0, and a maximum of .892.[12]

---

[11] In principle, and as discussed in detail in Section 5.4.2, one could move on to higher moments of the distributions, assessing skewness next or beginning to consider interactions and other features of the joint distribution of X. We stop at the second moment of each variable in X. Note, however, that we consider the mean and standard deviation of log family income, its square, and its cube. Thus, for this variable, we attempt to match far more than just its expectation and variance.

[12] The model was so predictive that a number of cases were completely determined. In particular, $2,406$ public school students were given predictive values arbitrarily close to 0 by the software (although no Catholic schools students were deemed completely determined and given values arbitrarily close to 1). These public school students were mostly low-SES rural students.

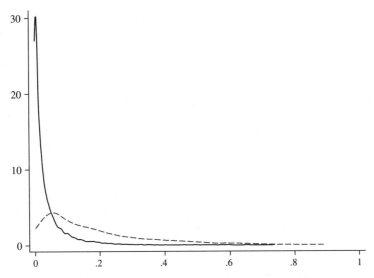

Figure 7.1 Kernel density estimates of the estimated propensity score, calculated separately for public school students (black solid line) and Catholic school students (gray dashed line).

Figure 7.1 presents kernel density estimates of these estimated propensity scores, separately for those in Catholic schools and public schools. There is substantial overlap in the estimated propensity scores, but there are no public school students with $\hat{p}_i$ greater than .738 and no Catholic school students with $\hat{p}_i$ less than .00115. If we focused in very closely on the cases within the tails of these densities, we would be able to see that there are 6 Catholic school students with $.738 < \hat{p}_i \leq .892$ who have no counterparts among public school students as well as 2,739 public school students with $0 \leq \hat{p}_i < .00115$ who have no counterparts among Catholic school students.

By the common support standards that prevail in observational data analysis, these data would be regarded as characterized by sufficient overlap for analysis to be worthwhile, since 1,912 of the 1,918 treatment cases have $\hat{p}_i$ within the range of $\hat{p}_i$ for the control cases. However, there is enough of a lack of overlap that some caution is in order, especially when making inferences about how public school students would fare if they were instead enrolled in Catholic schools. We will discuss these concerns in more detail later when offering estimates restricted to the region of overlap (i.e., the common support) where $.00115 \leq \hat{p}_i \leq .738$.

As just mentioned, the revised weights yielded slightly better balance when applied to the data. In particular, the average standardized difference of means fell further to .00437 when using $w_{i,\,\text{ATT}}$ and to .0899 when using $w_{i,\,\text{ATC}}$. Likewise, the average standardized difference of standard deviations also fell to .0166 when using $w_{i,\,\text{ATT}}$ and to .0229 when using $w_{i,\,\text{ATC}}$.

To give a better sense of how well these weights have succeeded in producing balance on a variable-by-variable basis (and for two different weighting schemes), we present the weighted means (and standard deviations where appropriate) of each of the variables in X for Catholic and public school students in Tables 7.6 and 7.7.

Table 7.6 Means and Standard Deviations of Primary Variables, Weighted by the ATT Weight from the Final Estimation of the Treatment Assignment Model

Variable	Public Mean	Public SD	Catholic Mean	Catholic SD
Female	.48		.48	
Race (White is the reference category)				
Black	.06		.06	
Hispanic	.11		.11	
Asian	.04		.04	
Native American	<.01		<.01	
Multiracial	.04		.04	
Urbanicity (Suburban is the reference category)				
Urban	.58		.58	
Rural	.01		.01	
Region (Midwest is the reference category)				
Northeast	.31		.31	
South	.23		.23	
West	.16		.17	
Family Background				
Mother's education (in years)	14.77	2.22	14.77	2.22
Father's education (in years)	15.25	2.57	15.25	2.57
SEI score of mother's occupation	50.56	12.77	50.55	12.85
SEI score of father's occupation	49.67	11.68	49.81	11.71
Family income (natural log)	11.24	.84	11.23	.90
Family income (natural log) squared	127.07	16.50	126.99	17.04
Family income (natural log) cubed	1441.92	261.94	1441.33	267.98
Two-parent family	.83		.84	
Past History (as reported by parent)				
Learning disability	.07		.07	
Ever held back	.05		.05	
Repeated 4th grade	<.01		<.01	
Years parents lived in current neighborhood	12.93	8.96	12.90	8.21

The differences between the columns in each of these two tables can be directly compared to the raw differences reported above in Table 7.4. For example, the unbalanced raw difference in mother's education between Catholic and public school students is 1.31 years (i.e., $|14.77 - 13.46|$ from Table 7.4), whereas the difference is reduced to .00 years (after rounding) when using $w_{i,\text{ATT}}$ (i.e., $|14.77 - 14.77|$ from Table 7.6) and to .12 years when using $w_{i,\text{ATC}}$ (i.e., $|13.58 - 13.46|$ from Table 7.7).

By our readings of applications of these sorts of models, the balance achieved by these weights is impressive. Some imbalance remains, and more so for the weights, $w_{i,\text{ATC}}$, that are designed to enable estimation of the ATC. The variables that proved

Table 7.7 Means and Standard Deviations of Primary Variables, Weighted by the ATC Weight from the Final Estimation of the Treatment Assignment Model

Variable	Public		Catholic	
	Mean	SD	Mean	SD
Female	.50		.53	
Race (White is the reference category)				
Black	.15		.20	
Hispanic	.17		.13	
Asian	.04		.06	
Native American	.01		.01	
Multiracial	.04		.05	
Urbanicity (Suburban is the reference category)				
Urban	.28		.33	
Rural	.21		.05	
Region (Midwest is the reference category)				
Northeast	.18		.28	
South	.34		.29	
West	.23		.16	
Family Background				
Mother's education (in years)	13.46	2.32	13.58	2.35
Father's education (in years)	13.59	2.59	13.80	2.68
SEI score of mother's occupation	44.98	12.87	45.88	13.10
SEI score of father's occupation	44.15	11.70	44.05	11.71
Family income (natural log)	10.60	1.09	10.62	1.03
Family income (natural log) squared	113.61	19.83	113.87	19.36
Family income (natural log) cubed	1225.64	295.46	1228.94	292.93
Two-parent family	.75		.71	
Past History (as reported by parent)				
Learning disability	.13		.14	
Ever held back	.13		.12	
Repeated 4th grade	.01		<.01	
Years parents lived in current neighborhood	10.56	8.00	10.72	6.89

most difficult to balance were some of the categorical variables, especially urbanicity and region because of the more limited geographic distribution of Catholic schools.

Are there general standards for how good the balance must be for analysis to proceed? Of course, perfect balance for the full distribution of X is the standard for which an analyst should strive. Even with abundant data, as in this application, the standard cannot be met. Rubin (2006) discusses the general predicament analysts will encounter, and he concludes with this advice:

> Of course, at some point, this sort of [perfect balance] assessment must terminate, because no matter how large the samples, the investigator will

almost certainly not be able to achieve this balance for all covariates and their interactions simultaneously, and higher order terms in prognostically minor covariates are clearly less important than prognostically important ones, and so scientific judgment must enter the process, just as it does when designing a randomized experiment. (Rubin 2006:462)

As discussed already, remaining imbalance can be addressed by supplemental parametric adjustment within a weighted regression framework, and we will offer such estimates later. And typically the weakness of an estimate arises not from lingering imbalance on observed variables, but rather the imbalance on unobserved variables that makes the maintenance of ignorability assumptions unreasonable. We will discuss this issue in detail later in this demonstration, when we interpret our estimates of the ATT and ATC.

Step 4: Estimate Weighted Bivariate Regression Models Using the Weights for the ATT and ATC. ATT-weighted and ATC-weighted variants of $\hat{\delta}_{\text{OLS, bivariate}}$ are reported in the three panels of Table 7.8. Notice that for all three outcome variables, the point estimate of the ATC is considerably larger than the point estimate for the ATT. Past research on the Catholic school effect has tended to show, by fitting interaction terms, that the estimated Catholic school effect is largest for low-SES and non-white students (see Bryk, Lee, and Holland 1993; Hoffer, Greeley, and Coleman 1985; Neal 1997). The results presented in Table 7.8 are consistent with this pattern. In particular, a comparison of Tables 7.6 and 7.7 shows that Catholic school students who have the profile of the treatment group have higher levels of socioeconomic status and are less likely to be black or Hispanic.

Having shown how to estimate the weights and produce point estimates for the ATT and ATC with the ELS data, we will not yet give a full interpretation to these estimates or to the common pattern of difference between the estimates of the ATT and ATC for all three outcome measures. We will save that discussion until after we have presented a series of models that assess whether this pattern of difference is robust to alternative analysis decisions.

The first issue we consider is whether allowing the models to use data off the common support of the treatment and control cases may have contributed to the difference between the estimates of the ATT and the ATC. For the estimation of the weighted regression models just reported, we used all sample members, recognizing, however, that with respect to $\hat{p}_i$, there are 6 Catholic school students who have no counterparts among public school students and 2,739 public school students who have no counterparts among Catholic school students.

Table 7.9 presents all of the models from Table 7.8 again, but this time the estimation sample is restricted to the region of overlap on the estimated propensity scores, $.00115 \leq \hat{p}_i \leq .738$. For the tenth grade math test score models, the sample size is reduced from 13,943 to 11,198. For the twelfth grade math test score and math gains models, the sample size is reduced from 10,502 to 8,469. A comparison of the results in Table 7.9 to those reported earlier in Table 7.8 shows that the point estimates of the respective average causal effects change slightly, with the ATC estimates coming in

Table 7.8 Catholic School Coefficients from ATT-Weighted and ATC-Weighted Regression Models Predicting Tenth Grade Math Test Scores, Twelfth Grade Math Test Scores, and Math Test Gains

Predictor Variables	Outcome Variable: 10th Grade Math Test Score	
	ATT Weight	ATC Weight
Model 1: Dummy for Catholic school	1.08	2.44
	(.77)	(1.17)

	Outcome Variable: 12th Grade Math Test Score	
	ATT Weight	ATC Weight
Model 1: Dummy for Catholic school	1.42	3.29
	(.89)	(1.36)

	Outcome Variable: Math Gain (12th−10th)	
	ATT Weight	ATC Weight
Model 1: Dummy for Catholic school	1.02	1.98
	(.31)	(.37)

a bit smaller. But, in general, the same pattern holds with ATT estimates remaining about the same and considerably smaller than the ATC estimates.

Step 5: Estimate Doubly Robust Weighted Regression Estimates. To demonstrate doubly robust weighted regression estimators, Table 7.10 presents weighted regression estimators analogous to those in Table 7.8 but with further regression adjustment for the 23 covariates used in Model 2 for Table 7.5. Beyond the weighted regression models presented in Table 7.8, these models are designed to adjust for any remaining imbalance in X due to misspecification of the model that was used to estimate the weights. Again, across all three outcome variables, the point estimate of the ATC is considerably larger than the point estimate for the ATT.

Given the consistency of the pattern shown in these last three tables, the next step is to ask the simple statistical question: Are the point estimates of the ATT and ATC sufficiently different that it makes sense to consider whether they support the claim that students who have the typical profile of public school students are more likely to benefit from Catholic schooling than students who have the typical profile of Catholic school students?

To answer this question, we must consider some thorny issues in statistical inference. Consider first the weighted multiple regression model for the Catholic school effect on the tenth grade math test. The estimated coefficient is 1.08 with a standard error of .56 when the $w_{i,\ \text{ATT}}$ weight is utilized. In contrast, when the $w_{i,\ \text{ATC}}$ weight

Table 7.9 Catholic School Coefficients from Weighted Regression
Models Restricted to the Region of Overlap in the Estimated
Propensity Scores

Predictor Variables	Outcome Variable: 10th Grade Math Test Score	
	ATT Weight	ATC Weight
Model 1: Dummy for Catholic school	1.02	2.18
	(.77)	(1.17)
	Outcome Variable: 12th Grade Math Test Score	
	ATT Weight	ATC Weight
Model 1: Dummy for Catholic school	1.38	2.85
	(.89)	(1.36)
	Outcome Variable: Math Gain (12th−10th)	
	ATT Weight	ATC Weight
Model 1: Dummy for Catholic school	1.03	1.82
	(.31)	(.37)

is utilized, the coefficient increases to 2.42 with a standard error of .59. Are differences such as these large enough to be considered meaningful?

A comparison of the 95 percent confidence intervals for each estimate may suggest not. Consider the regression model for the Catholic school effect on the tenth grade math test. Here, the 95-percent confidence interval is $(-.01, 2.17)$ for the estimate using $w_{i, \text{ATT}}$ and $(1.26, 3.58)$ for the estimate using $w_{i, \text{ATC}}$. Clearly, these intervals overlap. Scientific judgment, however, suggests that this overlap should not lead researchers to conclude that there are no substantive differences of importance, as we now explain.

First, the difference between the two point estimates is substantively large at 1.34; this difference suggests that the average effect of Catholic schooling is 124% larger for those who typically attend public schools than for those who typically attend Catholic schools (i.e., $\frac{2.42-1.08}{1.08} = 1.24$). The 95 percent confidence interval for the difference, based on a standard error of .81, is $(-.26, 2.92)$. This confidence interval is dominated by positive probability mass and suggests that values for the difference ≥ 2.66 are just as likely as values ≤ 0.[13]

---

[13]Moreover, the estimated standard error on which this confidence interval is based does not take into account that the two estimates were generated from the same sample. The confidence interval $(-.26, 2.92)$ is, in fact, a bit too wide (but not by much given the size of the available sample). In other applications, a same-sample correction may be more consequential.

Table 7.10 Catholic School Coefficients from Doubly Robust Weighted Regression Models

Predictor Variables	Outcome Variable: 10th Grade Math Test Score	
	ATT Weight	ATC Weight
Model 2: Model 1 + family background, demographics, and past history	1.08 (.56)	2.42 (.59)
	Outcome Variable: 12th Grade Math Test Score	
	ATT Weight	ATC Weight
Model 2: Model 1 + family background, demographics, and past history	1.60 (.66)	4.01 (.71)
	Outcome Variable: Math Gain (12th−10th)	
	ATT Weight	ATC Weight
Model 2: Model 1 + family background, demographics, and past history	1.05 (.29)	2.06 (.45)

Second, as noted just above, the difference of 1.34 is consistent with much past work on this substantive question. Evolving interpretive standards in statistical inference demand that such prior information be considered. If a full Bayesian posterior were generated, the lower end of the frequentist confidence interval, −.26, would be judged too negative as guidance for further research.

For these two reasons, we judge the difference between the doubly robust weighted regression estimates of the ATT and the ATC to be large enough to be meaningful. Before attaching meaning to this difference, we need to consider assumptions about the extent to which the observed determinants of treatment assignment can sustain alternative types of ignorability. Before offering such interpretations, we have one final type of analysis to perform, which will assess the extent to which the apparent differences between the point estimates of the ATT and the ATC are robust even to further conditioning.

Step 6: Further Assess the Robustness of the Difference Between the Estimates of the ATT and the ATC. In Step 4, we showed that the vast majority of the difference between the ATT-weighted and ATC-weighted regression estimates held when the data were restricted to the region of overlap on the estimated propensity score. The same result could also be shown for the doubly robust estimates of the same parameters. For now, we consider a different type of robustness check: whether the difference between the point estimates continues to hold after additional supplemental parametric adjustment for variables that other researchers might instead regard as essential determinants of treatment assignment/selection.

Some of the earliest literature on the Catholic school effect considered slightly different variables for regression adjustment than the variables that we included in X for this demonstration. As we discussed in Section 6.6.2 (beginning on page 219), regression adjustments were often performed with variables such as students' educational expectations and parental involvement. The original researchers, James Coleman and his colleagues, recognized that these variables were not clearly "prior to" Catholic school attendance and thus were likely influenced by the posited causal effect itself. Yet they wanted to show that, even adjusting for these variables, the apparent effects of Catholic schooling persisted in their models.

In our case, we are not primarily interested in determining how much the estimated causal effects are reduced when additional adjustment variables are entered into the various regression models (although were they to change in unexpected ways, such as vanishing entirely, one of a number of reasonable interpretations would need to be advanced). Rather, we are most interested in determining whether the inclusion of additional adjustment variables in the doubly robust weighted regressions models would cause us to revisit our preliminary decision that the difference between the estimates of the ATT and ATC is robust and deserving of interpretation.

With this goal in mind, we consider additional variables for educational expectations and parental involvement in school. As shown in Morgan and Todd (2008), students attending Catholic schools are expected to obtain more years of postsecondary schooling (nearly a year in students' own expectations and almost as much in parents' expectations). In addition, more than half of all parents of Catholic school students volunteer at their schools, which is twice as high as the rate for the parents of public school students.

Table 7.11 presents weighted regression models that utilize variables for expectations and parental involvement as supplemental adjustment variables (i.e., as variables only in the outcome regression equations, not as variables also in the propensity score estimating equation). The results in Table 7.11 show that these variables reduce the estimates of the ATT and ATC by a substantial amount. However, the coefficients remain positive and in the same pattern. Moreover, the relative difference between the estimates of the ATT and the ATC is larger for Model 3 in Table 7.11 than for Model 2 in Table 7.8.

In sum, the main pattern of results is supported by the analyses reported in Tables 7.9 through 7.11. Alternative decisions about overlap issues and supplemental regression adjustment did not alter the relative sizes of the weighted regression estimates of the ATT and ATC. As a consequence, we conclude that there is compelling evidence that these estimates differ. The deeper question is whether this pattern can be interpreted as evidence that the true ATT and ATC differ, which is the issue we consider next.[14]

---

[14] Another strategy that could be adopted to extend the analysis we report here is to attempt to model the underlying heterogeneity in a more fine-grained way. We see the comparison of estimates of the ATT and ATC as only a first step, and other methods may be a useful next step. If one believed

Table 7.11 Catholic School Coefficients from Weighted
Regression Models, Including Additional Covariates

Predictor Variables	Outcome Variable: 10th Grade Math Test Score	
	ATT Weight	ATC Weight
Model 3: Model 2 + expectations and parental involvement	.22 (.54)	1.26 (.62)

	Outcome Variable: 12th Grade Math Test Score	
	ATT Weight	ATC Weight
Model 3: Model 2 + expectations and parental involvement	.82 (.63)	2.77 (.74)

	Outcome Variable: Math Gain (12th−10th)	
	ATT Weight	ATC Weight
Model 3: Model 2 + expectations and parental involvement	.98 (.29)	1.94 (.44)

Interpretation of Results. We have generated estimates of the ATT and ATC using a treatment assignment model that includes variables we have labeled X. Recall that in the counterfactual tradition, treatment assignment patterns are represented by the propensity score, $\Pr[D=1|S]$, where S denotes *all* variables that systematically determine treatment assignment. Complete observation of S allows a researcher to assert that treatment assignment is ignorable and then consistently estimate the ATT and ATC, and, as a result, the ATE or any other average treatment effect that can be formed by weighting across the marginal distribution of the observed variables in S (see Section 4.3).

Are the ATT and ATC both identified? The crucial question for identification is whether the variables we have declared as X for our estimators in this demonstration are sufficiently complete to be regarded as equivalent to S, which we have previously labeled the perfect stratification set. To approach this question, it is useful to first reconsider the same substantive example modeled for our simulated data in Matching

---

that the estimated propensity score in a particular application has a clear interpretation (and hence is more than a tool for balancing the determinants of treatment assignment), then one could analyze directly the association between conditional average treatment effects and the estimated propensity score. Xie, Brand, and Jann (2012) and Brand and Thomas (2013) offer guidance for this sort of analysis, including a Stata add-on program. Alternatively, one could move toward direct examination of the separate response surfaces of the treatment and control groups, using a general nonparametric regression framework. Hill (2011) offers a compelling demonstration of the value of this alternative approach. Finally, as we will explain in Chapters 9 and 12, one could attempt to model some of the local average treatment effects and marginal treatment effects that constitute the ATT and ATC, although only if appropriate instrumental variables are available.

Demonstration 4, where identification of the ATT but not the ATC is clear because it was guaranteed by construction.

In the subsection on identification for that demonstration (beginning on page 173; see especially Figure 5.2), we explained why self-selection on the individual-level causal effect renders the ATC and ATE unidentified. In particular, with reference to the equations that defined the simulated data, we noted that we did not create observed variables that were sufficient to sustain an assumption of full ignorability, as defined by Equation (4.4). However, we showed that we could nonetheless consistently estimate the ATT because we created observed variables that allowed us to assert what is labeled Assumption 2-S in Equation (5.2),

$$E[Y^0|D=1,S] = E[Y^0|D=0,S].$$

Assumption 2-S is implied by a stronger assumption of partial ignorability with respect to S, such that

$$Y^0 \perp\!\!\!\perp D \,|\, S, \qquad\qquad (7.13)$$

which we introduced in Section 4.3. The basic idea is that, on average within strata defined by S, the values of y_i among those in the control group can be used to effectively estimate the counterfactual values of y_i^0 for those in the treatment group. We know by the setup of Matching Demonstration 4 that this assumption holds for the simulated data, in particular because all of the variables on the right-hand side of Equation (5.21) were declared observed variables. For the remainder of that demonstration, we then explained why the ATC is not identified, which is the same explanation for why the ATE is not identified.

Now return to the current demonstration. Because we did not simulate these data, we cannot assert that the ATT is identified with full confidence. We think it is reasonable to assert that Assumption 2-S is valid, treating X as equivalent to S for this assumption. However, it is possible that, for the cases observed in the ELS, those students most likely to benefit from public schools are those students who are least likely to attend Catholic schools, in which case Assumption 2-S does not obtain for the conditioning set X because

$$E[Y^0|D=1,X] < E[Y^0|D=0,X]. \qquad\qquad (7.14)$$

This would have been the case for Matching Demonstration 4 if we had included a term such as $-.5(\delta_i'')$ on the right-hand side of of Equation (5.21). We do not know of any research that suggests that the inequality in Equation (7.14) is true for the Catholic school effect, as would be the case if we had evidence that those who are most likely to do well in public schools are those with an aversion to educational institutions with religious foundations. On balance, it seems reasonable to assume that the ATT is identified in this demonstration.

The ATC is another matter entirely. Based on past research that maintains that self-selection is present, and the fact that we do not have any variables in the ELS that could plausibly measure students' (and their parents') own expectations for the benefits they might obtain from attending a Catholic school, we cannot accept what is labeled Assumption 1-S in Equation (5.1),

$$E[Y^1|D=1,S] = E[Y^1|D=0,S],$$

or the stronger assumption of partial ignorability,

$$Y^1 \perp\!\!\!\perp D \,|\, S, \tag{7.15}$$

that implies it. In other words, we do not believe that for the ATC the variables in X are equivalent to the more encompassing set of variables in S that defines this assumption. Operationally, we do not believe that for the ELS data we can use the values of y_i among those in the treatment group to effectively estimate the counterfactual values of y_i^1 for those in the control group, on average within strata defined by X. Sufficient evidence exists to suggest that some Catholic school students attend Catholic schools because they expect to gain from doing so. As such, we do not see any basis for regarding the ATC estimates as consistent or unbiased for the true ATC. Instead, these estimates are very likely too large because

$$E[Y^1 | D = 1, X] > E[Y^1 | D = 0, X]. \tag{7.16}$$

Although this reasoning is sufficient to convince us that the ATC is not identified, it also bears mentioning that the common support models show that 23 percent of public school students (i.e., $2,739$ out of $12,025$) have estimated propensity scores lower than the lowest estimated propensity score of any Catholic school student. Our comparison of the results in Tables 7.8 and 7.9 suggested that this lack of overlap is relatively inconsequential for the empirical results in this demonstration, and yet this provides only modest reassurance that the data can inform us at all about the likely benefit that students such as these 23 percent of public school students would obtain from attending a Catholic school.

Based on this reasoning, the ATC estimate is larger than the ATT estimate because the former is inconsistent and upwardly biased while the latter is either consistent and unbiased, or still inconsistent but far less upwardly biased. Nonetheless, we cannot know for sure that this reasoning is correct because it is contingent on the validity of Equations (7.14) and (7.16), and these are assumptions that cannot be tested.

Why might the true ATC still be larger than the true ATT? Because the reasoning just offered could be wrong, we will conclude this section by considering alternative interpretations that would hold under the most plausible alternative patterning of results where (1) the true ATC is larger than the true ATT and (2) the estimates we have reported are sufficiently unbiased to convey this basic ordering of the underlying effects.

Based on past research, there are two possible explanations for why one might expect that the true ATC would be larger than the true ATT for the effect of Catholic schooling:

1. The *common school* explanation: Catholic schools distribute opportunities for learning, such as advanced course-taking, more equitably than do public schools. This explanation was stressed by Coleman and his colleagues in their initial research and was then more comprehensively developed by Bryk, Lee, and Holland (1993). It suggests that variables such as parental education and non-minority status have positive relationships with D but negative relationships with the variation in δ among Catholic schools students. The explanation is based on the implicit claim that

$$E[Y^1 - Y^0 | D = 1, CSE = 1] > E[Y^1 - Y^0 | D = 1, CSE = 0], \tag{7.17}$$

where CSE is a "common school effect" dummy variable that indicates which particular students receive a "boost" in their individual-level treatment effect because of the comparatively egalitarian nature of Catholic schooling. Low-SES, black, and Hispanic students are more likely to receive the CSE boost. Accordingly, when the average of individual-level treatment effects is weighted to the distribution of S that characterizes the control group, the resulting average treatment effect is larger because this synthetic group of Catholic school students has more members with $CSE = 1$. (The same is also true when the weighting is performed only with respect to the observed variables in X because the relationship between S and CSE is generated by the relationship between X and CSE, which does not depend in any way on the determinants of self-selection in S that are not among the observed variables in X.)

2. The *better alternatives* explanation: Catholic schooling is particularly beneficial to those students who have poor public schooling alternatives, in particular those students from families who are not able to afford to live in school districts with the best public schools. This explanation was first fully developed by Neal (1997), and it suggests that variables such as family income and wealth have positive relationships with D but negative relationships with the variation in δ among Catholic school students. The explanation is based on an implicit claim, of the same basic structure as Equation (7.17), that

$$E[Y^1 - Y^0 | D = 1, WAP = 1] > E[Y^1 - Y^0 | D = 1, WAP = 0], \qquad (7.18)$$

where WAP is a "worse alternative public school" indicator variable equal to 1 for those Catholic school students who have particularly poor public schooling alternatives, and which is assumed for this explanation to be a function in S. However, here the claim is implicitly in two separable pieces, which makes it qualitatively different than the common school explanation. This explanation is composed of two assumptions,

$$E[Y^1 | D = 1, WAP = 1] = E[Y^1 | D = 1, WAP = 0]$$

and

$$E[Y^0 | D = 1, WAP = 1] < E[Y^0 | D = 1, WAP = 0],$$

which together imply Equation (7.18).[15] In this case, when the average of individual-level treatment effects is weighted to the distribution of S that characterizes the control group, the resulting average treatment effect is larger because this synthetic group of Catholic school students has more members with $WAP = 1$.

If either the common school explanation or the better alternatives explanation is true, then the true ATC would be larger than the true ATT, and this might also be true for the estimated ATC and ATT if the bias in the estimates is small enough not to

---

[15] We assume for simplicity of notation that the linearity of the expectations at the population level applies also at the individual level. This assumption will hold for the third explanation introduced below as well.

obscure the true difference between the target parameters (as usual, assuming that sampling error is zero).

We favor an alternative third explanation, which we think is more compatible with the underlying claim that the estimated ATC is larger than the estimated ATT in part because self-selection is very likely present and cannot be adjusted away because X is only a subset of S:

3. The *binding constraint* explanation: Differential responsiveness exists to accurate perceptions of students' likely benefits from Catholic schooling. For low-income families for whom tuition at a Catholic school represents a genuine financial sacrifice, the students who enroll in Catholic schools are much more likely to be students who are likely to benefit from enrolling. In contrast, among high-income families for whom tuition is not a substantial financial sacrifice, even students who are not likely to benefit from attending Catholic schooling instead of public schooling may enroll in Catholic schools. This explanation is discussed in Morgan (2001), and it is based on the assumption that it takes a larger individual-level value of δ to induce low-income students to enroll in Catholic schooling. As a result, variables that capture resource availability have positive relationships with D but negative relationships with the variation in δ among those who enter Catholic schooling. The implicit claim here is that

$$E[Y^1 - Y^0 | D = 1, BC = 1] > E[Y^1 - Y^0 | D = 1, BC = 0], \qquad (7.19)$$

where BC is an indicator variable equal to 1 for those students from families for whom Catholic school tuition is a genuine financial sacrifice, and which is therefore a function in S. Here, as with the better alternatives explanation, the claim is implicitly in two separable pieces,

$$E[Y^1 | D = 1, BC = 1] > E[Y^1 | D = 1, BC = 0]$$

and

$$E[Y^0 | D = 1, BC = 1] = E[Y^0 | D = 1, BC = 0],$$

which together imply Equation (7.19). In this case, when the average of individual-level treatment effects is weighted to the distribution of S that characterizes the control group, the resulting average treatment effect is larger because this synthetic group of Catholic school students has more members with $BC = 1$.

Having laid out these alternative explanations, there is no reason to assume that only one of them holds. One reasonable blended explanation would combine the better alternatives and binding constraint explanations by asserting both

$$E[Y^1 | D = 1, BC = 1] > E[Y^1 | D = 1, BC = 0]$$

and

$$E[Y^0 | D = 1, WAP = 1] < E[Y^0 | D = 1, WAP = 0],$$

and then noting that there is likely to be a strong positive association between BC and WAP in the population, conditional on S (and its typical observed subset X).

These assertions could then be combined with the claim that the difference between the estimated ATC and the estimated ATT is further increased because these assertions also lend indirect support to the claim that the estimated ATC is inconsistent and upwardly biased for the true ATC because

$$E[Y^1|D=1,X] > E[Y^1|D=0,X].$$

Other blended explanations are also possible, and we cannot rule out the possibility that Coleman and his colleagues had it correct all along: Catholic schools and public schools do have differences in their instructional practices, which at least for Coleman had deeper sources in alternative ideological beliefs about the capacities of children. These instructional differences may be particularly beneficial to students who are systematically disadvantaged in public schools.

Overall, was it worth offering provisional estimates of the ATC? Of course, we knew from the outset which assumptions likely held, and it would have been reasonable to estimate and report only the ATT. Nonetheless, we think it is clear that what are most likely inconsistent and biased estimates of the ATC still give us a valuable perspective on why the ATC is not identified, and hence enhance our explanation of why, at best, only the ATT is identified by conditioning on the observed data. In addition, the reasoning that was motivated by the need to interpret the estimated ATC may also explain why the common school explanation of Coleman and his colleagues also may reflect underlying heterogeneity that is produced by individual-level selection on the treatment effect.

7.6   Conclusions

In this chapter, we have presented weighted regression estimators of the ATE, ATT, and ATC, and we have argued that they have enough comparative advantages that they are likely to continue to grow in prominence in the coming years. We have shown how weighted regression estimators allow the analyst to adopt a matching orientation and thereby take advantage of the clarity that a matching perspective provides. We have also shown that weighted regression estimators require no specialized software and can be utilized when analyzing complex survey data of many types.

These advantages notwithstanding, it should also be clear that weighted regression estimators are no panacea, especially if practitioners fall back too casually into standard regression thinking. Instead, analysts must carefully consider the estimation of the propensity scores that generate the weights, checking balance and then examining the consequences of large weights. Only thereafter should one calculate point estimates of the average treatment effects of interest.

The ultimate value of weighted regression estimators in comparison to other types of conditioning estimators has not yet been fully determined. As with the matching estimators presented Chapter 5, this is an area of active methodological scholarship, and it is not easy to predict the developments of the coming years. Our current prediction is that weighted regression will be at the center of an emergent consensus, but our prediction is accompanied by substantial uncertainty.

With this chapter, we have concluded our presentation of techniques that estimate causal effects by conditioning on variables that block back-door paths. In the next section of the book, we present techniques that can be used effectively when simple conditioning will not identify causal effects because crucial variables that lie along back-door paths have not been observed. We will consider general strategies first, elaborating on the causal graphs we have used so far in this book. Thereafter, we will present instrumental variable estimators, front-door conditioning estimators grounded in assumptions about causal mechanisms, and approaches that utilize over-time observations of the outcome variable.

Part IV: Estimating Causal Effects When Back-Door Conditioning Is Ineffective

Chapter 8

Self-Selection, Heterogeneity, and Causal Graphs

In this chapter, we will lay the groundwork for our presentation of three strategies to estimate causal effects when simple conditioning on observed variables that lie along back-door paths will not suffice. These strategies will be taken up in Chapters 9, 10, and 11, where we will explain instrumental variable estimators, front-door identification with causal mechanisms, and conditioning estimators that use data on pretreatment values of the outcome variable. Under very specific assumptions, these three strategies will identify average causal effects of interest, even though selection is on the unobservables and treatment assignment is nonignorable.

In this chapter, we will first review the related concepts of nonignorable treatment assignment and selection on the unobservables, using the directed graphs presented in prior chapters. To deepen the understanding of these concepts, we will then demonstrate why the usage of additional posttreatment data on the outcome of interest is unlikely to aid in the point identification of the treatment effects of most central concern. One indirect goal of this demonstration is to convince the reader that oft-heard claims such as "I would be able to establish that this association is causal if I had longitudinal data" are nearly always untrue if the longed-for longitudinal data are additional measurements taken only after treatment exposure. Instead, longitudinal data are most useful, as we will later explain in detail in Chapter 11, when pretreatment measures are available for those who are subsequently exposed to the treatment.

The extended example we will use to make this point involves simple confounding (stable unobserved common causes of the treatment variable and the outcome variable) and more subtle confounding (direct self-selection into the treatment based on accurate perceptions of the individual-level treatment effect). Accordingly, this example will serve as a bridge to the second half of this chapter, where we will expand upon our introduction to causal graph methodology in order to explain how heterogeneity of effects can be accommodated. In Chapter 2, we introduced the potential outcome model while taking the position that individual-level heterogeneity of treatment effects is pervasive. In particular, we have implicitly assumed, and often explicitly stated, that, in general, individual-level treatment effects do not all equal the average treatment effect (ATE) (i.e., $\delta_i \neq E[\delta]$ for all i). We have offered many examples where individual-level treatment effects are patterned in consequential nonrandom ways, and we have explained the consequences of such heterogeneity for conditioning estimators in Chapters 5, 6, and 7.

From this chapter onward, we will consider the interconnections between self-selection and individual-level heterogeneity of treatment effects more directly, as these patterns combine to generate selection on the unobservables. We ended our presentation of conditioning by considering these complications, especially for the interpretation of the results for Weighted Regression Demonstration 3 (see pages 257–262). In this chapter, we approach these complications more directly by first showing how to use causal graphs with latent class variables to jointly represent patterns of self-selection and heterogeneity.

8.1   Nonignorability and Selection on the Unobservables Revisited

As demonstrated in Sections 4.3.1 and 4.3.2, the concept of ignorable treatment assignment is closely related to the concept of selection on the observables. In many cases, they can both be represented by the same directed graph. Recall Figure 4.8(a), in which there are two types of paths between D and Y: the causal effect of D on Y represented by $D \rightarrow Y$ and an unspecified set of back-door paths represented collectively by the bidirected edge in $D \leftarrow\!-\!-\!-\!\rightarrow Y$. If this graph can be elaborated, as in Figure 4.8(b), by replacing $D \leftarrow\!-\!-\!-\!\rightarrow Y$ with a fully articulated back-door path $D \leftarrow S \rightarrow Y$, then the graph becomes a full causal model. Observation of S ensures that the conditioning strategies of the prior three chapters can be used to generate consistent estimates of the causal effect of D on Y. In this scenario, selection is on the observables – the variables in S – and the remaining treatment assignment mechanism is composed only of random variation that is ignorable.

If, in contrast, as shown in Figure 4.9(b) rather than Figure 4.8(b), only a subset Z of the variables in S is observed, then selection is on the unobservables because some components of S are now embedded in U. Conditioning on Z in this graph leaves unblocked back-door paths represented by $D \leftarrow U \leftarrow\!-\!-\!-\!\rightarrow Y$ untouched. And, as a result, any observed association between D and Y within strata defined by Z cannot be separated into the genuine causal effect of D on Y and the back-door noncausal association between D and Y that is generated by the unobserved determinants of selection, U.

As explained in Chapter 4, and then as shown in demonstrations in Chapters 5 through 7, the concepts of ignorability and selection on the observables are a bit more subtle when potential outcomes are introduced. Weaker forms of conditional independence can be asserted about the joint distribution of Y^1, Y^0, D, and S than can be easily conveyed as in Figures 4.8 and 4.9, in which the observable variable Y is depicted instead of the underlying potential outcome variables Y^1 and Y^0. For example, the average treatment effect for the treated (ATT) can be estimated consistently

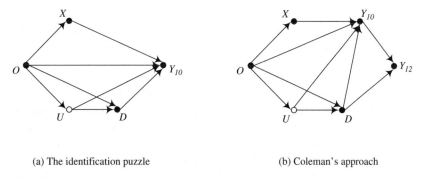

(a) The identification puzzle   (b) Coleman's approach

Figure 8.1 Coleman's strategy for the identification of the causal effect of Catholic schooling on achievement.

by asserting only that Y^0 is independent of D conditional on S, even though full ignorability does not hold; see the discussion of Equation (4.4) and Section 4.3.3.[1] These possibilities are not fully revealed by conventional directed graphs, given that for these graphs potential outcome variables are typically not represented (see the discussion of Figure 3.9).

8.2   Selection on the Unobservables and the Utility of Additional Posttreatment Measures of the Outcome

In this section, we present the challenges of modeling causal effects when selection is on unobservables, using as an example the regression equations offered by James Coleman and his colleagues for the estimation of the causal effect of Catholic schooling on high school achievement. The basic estimation challenge that Coleman and his colleagues confronted is depicted in Figure 8.1(a). Although somewhat simplified, the graph represents the basic types of causal variables that Coleman and his colleagues contemplated in their original analyses and then in debate with their critics (Coleman et al. 1982; see also our prior discussion in Section 6.6.2). For Figure 8.1(a), Y_{10} is an observed score on a standardized achievement test given in the tenth grade to second-year high school students. The variable D is again an observed dichotomous causal variable equal to 1 for those who attend Catholic schools and 0 for those who attend public schools. A primary goal of the original research was to estimate the causal effect of D on Y_{10}, conceptualized as the average causal effect of Catholic schooling on achievement.

The remaining variables in the graph represent the basic types of variables that Coleman and his colleagues decided should be specified as adjustment variables in

---

[1]Furthermore, as discussed in Section 5.2.1 in the presentation of matching techniques, one need not assume full conditional independence of Y^0 in order to offer consistent and unbiased estimates of the ATT. An equality of the two conditional expectations, $E[Y^0|D=1,S] = E[Y^0|D=0,S]$, will suffice; see discussion of Assumption 2-S in Equation (5.2).

their regression equations. The variables in X are determinants of achievement test scores that have no direct causal effects on school sector selection. The variables in O are ultimate background factors that determine all other variables in the graph (i.e., X, U, D, and Y_{10}). Coleman and his colleagues designated many variables that they believed were contained in O and X, although they were unsure of whether to consider them as members of O or X. Finally, the variables in U are the crucial variables that Coleman and his colleagues recognized were probably unobserved. These variables – intrinsic motivation to learn, subtle features of the home environment, and anticipation of the causal effect itself – were thought to determine both school sector choice and achievement. Because these variables were assumed to be unobserved, they are represented collectively in Figure 8.1(a) by the variable U with a node that is a hollow circle ∘.

In the initial research, the analysis strategy of Coleman and his colleagues was to condition on observed variables in O and X in order to attempt to remove as much of the confounding of $D \rightarrow Y_{10}$ as possible. Reinterpreted with the aid of directed graphs, to identify the causal effect they needed to block five separate back-door paths from D to Y_{10}:

1. $D \leftarrow U \rightarrow Y_{10}$,

2. $D \leftarrow U \leftarrow O \rightarrow Y_{10}$,

3. $D \leftarrow U \leftarrow O \rightarrow X \rightarrow Y_{10}$,

4. $D \leftarrow O \rightarrow Y_{10}$, and

5. $D \leftarrow O \rightarrow X \rightarrow Y_{10}$.

According to the back-door criterion, conditioning on the variables in O and X would block paths 2, 3, 4, and 5 but would leave path 1 unblocked. Coleman and his colleagues recognized that the variables in U rendered the causal effect of D on Y formally unidentified, and they could not convince their critics that their estimates from models that conditioned only on O and X were sufficiently close to the ATE (or the ATT) to offer the conclusion that they did.

Their solution to this predicament is presented in Figure 8.1(b), which became possible when the second round of survey data was released two years later (see Coleman and Hoffer 1987; Hoffer et al. 1985). Rather than focus on explaining variation in the tenth grade test score variable, Y_{10}, they reclassified it as a conditioning variable and instead designated the new twelfth grade test score variable, Y_{12}, as the outcome of interest. They then argued that Y_{10} could serve to screen off the effects of the variables in U on Y_{12}. In particular, their strategy was equivalent to asserting that conditioning on O, X, and Y_{10} blocks all five of the back-door paths between D and Y_{12} in Figure 8.1(b):

1. $D \leftarrow U \rightarrow Y_{10} \rightarrow Y_{12}$,

2. $D \leftarrow U \leftarrow O \rightarrow Y_{10} \rightarrow Y_{12}$,

3. $D \leftarrow U \leftarrow O \rightarrow X \rightarrow Y_{10} \rightarrow Y_{12}$,

4. $D \leftarrow O \rightarrow Y_{10} \rightarrow Y_{12}$, and

5. $D \leftarrow O \rightarrow X \rightarrow Y_{10} \rightarrow Y_{12}$.

Indeed, all of these back-door paths are blocked by conditioning on Y_{10}, and four of them are blocked by supplemental but unnecessary conditioning on O and X as well.[2]

However, notice that Y_{10} does not satisfy Condition 2 of the back-door criterion. It lies on a directed path that begins at the causal variable and reaches the outcome variable, $D \rightarrow Y_{10} \rightarrow Y_{12}$. Conditioning on Y_{10} adjusts away some of the total causal effect of Catholic schooling on achievement measured in the twelfth grade. As a result, by adopting this analysis strategy, Coleman and his colleagues changed the average causal effect of interest from the total effect of Catholic schooling on achievement to a net direct effect, $D \rightarrow Y_{12}$. This effect is best thought of as the average gain in achievement between the tenth and twelfth grades attributable to Catholic schooling, although the specific interpretation depends on the model that is estimated.

Holding aside the fundamental issue of whether narrowing the focus to this particular net direct effect is helpful for the research questions at hand, consider now the graph presented in Figure 8.2, which suggests two more basic criticisms of this approach. The first was seized on by their critics: the variables in U that confound the causal effect of D on Y_{12} are not effectively screened off by Y_{10}. More generally, Figure 8.2 should be seen as more plausible than Figure 8.1(b) because it is unreasonable to rule out three direct effects: $O \rightarrow Y_{12}$, $X \rightarrow Y_{12}$, and $U \rightarrow Y_{12}$ (or, as Coleman and colleagues asserted, that these effects are indirect and transmitted through $Y_{10} \rightarrow Y_{12}$). Because O and X are observed, the claims that $O \rightarrow Y_{12}$ and $X \rightarrow Y_{12}$ should be included in the graph can be accommodated in subsequent analysis. However, the inclusion of $U \rightarrow Y_{12}$ renders back-door adjustment with the observed variables ineffective. The rationale for the inclusion of $U \rightarrow Y_{12}$ in the graph is the following. If the parents of highly motivated students were more likely to pay tuition to enroll in Catholic schools, then enhanced motivation would contribute directly to learning in both the tenth grade and the twelfth grade. Similarly, it is unreasonable to assume that the motivation that is correlated with willingness to pay tuition would exert only a one-time boost early in the Catholic school careers of students, which would then structure subsequent achievement only by way of a twelfth grade knock-on effect from the initial boost in achievement in the tenth grade. Taken together, the critics argued, in effect, that Coleman and his colleagues mistakenly ignored a back-door path, $D \leftarrow U \rightarrow Y_{12}$, that remains unblocked after conditioning on O, X, and Y_{10} because they failed to acknowledge that U causes Y_{12} directly.

The second criticism is more subtle. For Figure 8.2, we have omitted the effect $Y_{10} \rightarrow Y_{12}$ that Coleman and his colleagues seem to have asserted for their models, and which we depicted in Figure 8.1(b). Whether this effect should be included in the graph depends on the richness of the variables in O, X, and U. If these variables, when combined with D, constitute full models of achievement for public and Catholic school students, then no direct relationship between Y_{10} and Y_{12} needs to be represented in

---

[2]Coleman and his colleagues sometimes wrote as if they allowed for effects such as $O \rightarrow Y_{12}$ and $X \rightarrow Y_{12}$. The back-door paths generated by these additional direct effects would all be blocked by conditioning on O, X, and Y_{10}. The key assumption is their implicit claim that U does not have an effect on Y_{12}, except through Y_{10}.

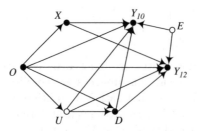

Figure 8.2 Criticism of Coleman's estimates of the effect of Catholic schooling on learning.

the graph. The critics did not take a position on this issue, but we suspect that they would not have been willing to maintain that the measured variables in O and X were rich enough to eliminate the need for including $Y_{10} \to Y_{12}$ in such a graph. Instead, they would have likely allowed for such an effect so that the causes embedded in $e_{Y_{10}}$ are allowed to contribute to Y_{12} via Y_{10}. The critics only emphasized the implausibility of interpreting models under the faulty assumption that $D \leftarrow U \to Y_{12}$ does not exist, and this point of criticism stands regardless of what position one takes on the plausibility of $Y_{10} \to Y_{12}$.

Yet, once one begins to think about the relationship between Y_{10} and Y_{12}, it may seem implausible to assume that these variables are only associated because of their common dependence on O, X, U, and D. Is it reasonable to assume that the structural error terms $e_{Y_{10}}$ and $e_{Y_{12}}$, which are viewable only under magnification, are independent of each other? For Figure 8.1(b), they were assumed to be so. For Figure 8.2, instead it is assumed that they are only independent of each other after we allow for unobserved common direct causes of both Y_{10} and Y_{12}, which are represented by E in Figure 8.2. Although the common cause relationship $Y_{10} \leftarrow E \to Y_{12}$ is quite general, for now consider E to be an unobserved and completely random characteristic of students that measures their taste for tests. Independent of all else, some students enjoy taking tests and perform well on them.[3] If such a common cause of Y_{10} and Y_{12} exists, it creates four new back-door paths from D to Y_{12} through E:

1. $D \leftarrow U \to Y_{10} \leftarrow E \to Y_{12}$,

2. $D \leftarrow O \to Y_{10} \leftarrow E \to Y_{12}$,

3. $D \leftarrow U \leftarrow O \to X \to Y_{10} \leftarrow E \to Y_{12}$,

4. $D \leftarrow O \to X \to Y_{10} \leftarrow E \to Y_{12}$.

For these back-door paths, Y_{10} is a collider. In the absence of conditioning, all four of these paths are blocked and do not create additional noncausal associations between D and Y_{12}. But, when O, X, and Y_{10} are used as adjustment variables, the first path

---

[3]Of course, if this particular variable in E exists, it is unlikely to be independent of both O and X. Were we to add edges to also specify $O \to E$ or $X \to E$ as additional causal effects in Figure 8.2, all of what is written in the main text would be the same. These additional back-door paths would be blocked by conditioning on O and X, but the same back-door path, $D \leftarrow U \to Y_{10} \leftarrow E \to Y_{12}$, that is unblocked after conditioning on Y_{10} would remain.

becomes unblocked because conditioning on Y_{10} induces an association between U and E (see Chapter 4). The second and third back-door paths remain blocked by the simultaneous conditioning on the observed variables O and X, even though conditioning on Y_{10} induces an association between O and E as well as between X and E. In short, if Figure 8.2 is the appropriate representation of the causal system, Coleman and his colleagues unblocked an already blocked back-door path, thereby creating a new net association between D and Y_{12}. This induced noncausal association is then mixed in with the noncausal association generated by $D \leftarrow U \rightarrow Y_{12}$, which critics argued had been mistakenly assumed not to exist. The following demonstration, which considers the simplest possible causal model consistent with Figure 8.2, explains these complications together, while also leading into the explicit consideration of heterogeneity that we will take up in the section that follows it.

Panel Data Demonstration 1

For this simulated example of the analysis of Catholic school effects by Coleman and his colleagues, we will use the panel data notation introduced in Section 2.8. Nonetheless, because we will consider only posttreatment data on the outcome and because our other variables are fixed in time, we can suppress some of the subscripting and other fine distinctions introduced there.[4] In order to focus on the essential identification issues, we consider only linear specifications and, except when noted otherwise, restrict all causal effects to be constants or to vary across individuals in completely random ways independent of all else in the models. These conditions are consistent with the original research by Coleman and his colleagues, even though they are inconsistent with our extensive treatment of this research question using the more recent literature (as shown in Chapters 5 through 7, where nonlinearities and heterogeneity of effects were both shown to be vital considerations when modeling the Catholic school effect on achievement).

For this demonstration, the potential outcome variables are Y_t^1 and Y_t^0, where $t = \{10, 11, 12\}$ for the three grades that occur during the observation window from the tenth grade through the twelfth grade. Because treatment selection occurs before $t = 10$, we have observed data only for posttreatment time periods. The treatment indicator variable, D, is equal to 1 if a student is enrolled in a Catholic school and 0 if enrolled in a public school. (We again ignore other types of private schools.) The observed outcome variable, Y_t, is defined as $DY_t^1 + (1-D)Y_t^0$ for $t = \{10, 11, 12\}$.

For simplicity, our other variables O, X, and U should be interpreted as indices of many underlying variables, which we will scale as normally distributed composite variables. Consistent with our discussion of Figures 8.1 and 8.2, X is a composite determinant of achievement test scores, Y_t, that has no direct effect on whether students select Catholic schooling, D. U is a composite variable of unobserved factors that determines both D and Y_t. And O is a composite variable of ultimate background

---

[4]Most importantly, for this demonstration we do not need to make a distinction between D_{it} (the treatment exposure indicator) and D_i^* (the treatment group indicator). Because we have no pretreatment observations and we ignore (as did Coleman and his colleagues) students who switch between public and Catholic schools between the tenth and twelfth grade, $D_{it} = D_i^*$ for all t. We therefore use our customary D without subscripts for either i or t.

factors that determines U, X, D, and Y_t. To give these composite variables distributions that are familiar, O is a standard normal random variable with mean of 0 and variance of 1. Having defined O as an exogenous root variable, we then set $X = O + e_X$ and $U = O + e_U$, where e_X and e_U are independent standard normal random variables with mean of 0 and variance of 1.

We consider four data setup scenarios, defined as a cross-classification of two binary conditions: (1) the presence of self-selection on accurate expectations of individual-level effects of Catholic schooling and (2) the presence of supplemental dependence between the potential outcomes that is produced by an exogenous variable E (see Figure 8.2). For all scenarios, the probability of Catholic school enrollment is specified as a logistic distribution

$$\Pr[D = 1 | O, U] = \frac{\exp(-3.8 + O + U)}{1 + \exp(-3.8 + O + U)}, \tag{8.1}$$

where O and U are as defined above. The probabilities defined by Equation (8.1) are then specified as the parameters for draws from a binomial distribution, yielding the indicator variable D for Catholic schooling defined above. As explained below, we introduce self-selection into two of the four scenarios by allowing U to structure the potential outcomes, such that those with higher values for U have higher average individual-level treatment effects. Because of the presence of U in Equation (8.1), students with higher values for U are then more likely to enter Catholic schooling and more likely to benefit from doing so.

For the scenarios without supplemental dependence on E, Y_t^0 is defined as

$$\begin{aligned}
Y_{10}^0 &= 100 + O + U + X + v_{10}^0, \\
Y_{11}^0 &= 101 + O + U + X + v_{11}^0, \\
Y_{12}^0 &= 102 + O + U + X + v_{12}^0,
\end{aligned} \tag{8.2}$$

where the v_t^0 are independent normal random variables with mean of 0. For the scenarios with supplemental dependence, Y_t^0 is instead defined as

$$\begin{aligned}
Y_{10}^0 &= 100 + O + U + X + E + v_{10}^0, \\
Y_{11}^0 &= 101 + O + U + X + E + v_{11}^0, \\
Y_{12}^0 &= 102 + O + U + X + E + v_{12}^0,
\end{aligned} \tag{8.3}$$

where E is also a normal random variable with mean of 0.[5] On average, Y_t^0 follows a linear time path as determined by the intercept values of 100, 101, and 102. However, the levels of these potential outcomes for individuals are set by the time-invariant values of O, U, X, and E as well as by time-specific shocks to their outcomes, v_t^0.

---

[5]To yield a covariance structure similar to real test score data, for Equation (8.2) we specify v_t^0 as time-period-specific standard normal random variables multiplied by 10. For Equation (8.3), we specify E as a standard normal random variable multiplied by 5 and v_t^0 as time-period-specific standard normal random variables multiplied by 8.6.

To specify a treatment effect that increases linearly in time, Y_t^1 is defined as

$$
\begin{aligned}
Y_{10}^1 &= Y_{10}^0 + \delta' + \delta'', \\
Y_{11}^1 &= Y_{11}^0 + (1 + \delta') + \delta'', \\
Y_{12}^1 &= Y_{12}^0 + (2 + \delta') + \delta'',
\end{aligned}
\tag{8.4}
$$

where δ' is a baseline individual-level causal effect, specified as a normal random variable with mean of 10 and variance of 1. The constitution of δ'' depends on whether self-selection is present. In the no-self-selection scenarios, δ'' is additional random individual-level variation, specified as a standard normal random variable with mean of 0 and variance of 1. In the self-selection scenarios, δ'' varies systematically over individuals. The values of δ'' are set as single draws from individual-specific standard normal random variables with expectation equal to each individual's realized value u_i of U.

The first panel of Table 8.1 presents the ATE, ATT, and ATC for the effect of Catholic schooling on test scores in the tenth and twelfth grades, as implied by the data setup just detailed. For the first two columns, students do not self-select into Catholic schooling based on accurate expectations of the individual-level gains from doing so. Accordingly, the ATE, ATT, and ATC are all the same, and they all increase from 10.00 to 12.00 between the tenth and twelfth grades, as determined by Equation (8.4). For the last two columns, students self-select on the causal effect, as explained above. The ATT is larger than the ATE and the ATC. But, again, the average gain is 2.00 for the ATE, ATT, and ATC.

The second panel of Table 8.1 presents estimates for the coefficient on D from a series of regression models, first for cross-sectional estimators that use outcome data from only one point in time and then, as in the research of Coleman and colleagues, for panel data models that use outcome data from two posttreatment points in time. These latter regression models, as we will explain in Chapter 11, are labeled panel data analysis of covariance models. We use generic ordinary least squares (OLS) regression estimation, as in the original research of Coleman and his colleagues, but we have given OLS estimation a best case data setup where linear specifications are reasonable and the pattern of individual-level heterogeneity is simple.

Consider first the results for the regression models in the no-self-selection scenarios. For the first column, we also do not assume that E is a common cause of both Y_{10} and Y_{12}, which does not render Y_{10} a collider on any back-door paths from D to Y_{12}. For the second column, E is a common cause. This distinction for the data setup produces no differences for the cross-sectional regression estimates of the coefficient on D because Y_{10} is not in the conditioning set for these models.

For both sets of cross-sectional estimators in the no-self-selection scenarios, the same pattern of results is present for models that have Y_{10} as the outcome variable and for models that have Y_{12} as the outcome variables. The naive estimator is too large, and adjustment for O and X eliminates only some of the confounding.[6]

---

[6] Additional adjustment for U would yield estimates that are equal to the ATE, ATT, and ATC. This last result is merely a benchmark, given that we have assumed that U is unobserved, as in the research of Coleman and his colleagues.

Table 8.1 Simulated Results for the Identification Approach Adopted by Coleman and Colleagues

Setup conditions:				
Self-selection on the causal effect	No	No	Yes	Yes
E is a common a cause of Y_t^0	No	Yes	No	Yes
True average treatment effects:				
ATE, tenth grade	10.00	10.00	10.00	10.00
ATT, tenth grade	10.00	10.00	11.85	11.85
ATC, tenth grade	10.00	10.00	9.81	9.81
ATE, twelfth grade	12.00	12.00	12.00	12.00
ATT, twelfth grade	12.00	12.00	13.85	13.85
ATC, twelfth grade	12.00	12.00	11.81	11.81
	Estimated Coefficient on D			
Cross-sectional estimators:				
Regression of Y_{10} on D	14.75	14.75	16.60	16.60
Regression of Y_{10} on O, X, and D	10.80	10.80	12.58	12.58
Regression of Y_{12} on D	16.75	16.75	18.60	18.60
Regression of Y_{12} on O, X, and D	12.80	12.80	14.58	14.58
Panel data analysis of covariance estimators:				
Regression of Y_{12} on Y_{10}, O, X, and D	12.68	9.98	14.41	11.27

For the analysis of covariance models with Y_{12} as the outcome variable, the coefficient estimates on D do not equal the ATE, ATT, or ATC for the twelfth grade, which are all 12.00. The coefficient estimate of 12.68 is too large when E is not a common cause because the unblocked back-door path through U biases the coefficient upward. The estimate is then too small when E is a common cause because the upward bias from the unblocked back-door path through U is then joined by a larger downward bias produced by conditioning on Y_{10}, which is a collider along the new back-door paths through E (see the discussion that precedes the demonstration for an accounting of these back-door paths).

As shown in the third and fourth columns, the basic pattern for the estimated models changes only modestly when self-selection is present. The last two columns are analogous to the first two, after allowing the self-selection term, δ'', to vary with U, as explained above for Equation (8.4). For these last two scenarios, none of the cross-sectional estimators yield estimates equal to any of the average causal effects, either in the tenth grade or the twelfth grade.[7] The analysis of covariance models have

---

[7]Even models that adjust for U would not suffice. For the self-selection scenarios, the individual-level effect of Catholic schooling varies with U, such that those with higher average values of U have

the same pattern as in the no-self-selection scenarios. When E is not a common cause of Y_{10} and Y_{12}, the estimate is larger than the ATE, ATT, and ATC. When E is a common cause, rendering Y_{10} a collider, the estimate is smaller than the ATE, ATT, and ATC.

Finally, notice also that the analysis of covariance estimates are nowhere near to the average gain in achievement between the tenth and twelfth grades, which is equal to 2.00, and which by construction is equal for both Catholic and public school students and does not depend on whether self-selection is present. As such, under this setup, which is consistent with Figure 8.2, the analysis of covariance models do not estimate a net direct effect of Catholic schooling on achievement in the twelfth grade that can be given an interpretation as an estimate of an average gain in achievement.

Many analysis puzzles and ensuing causal controversies in the social science literature have the same basic features as this demonstration. The plausibility of the criticism in Figure 8.2 should serve as a caution because it demonstrates how additional posttreatment observations of the outcome variable are unlikely to resolve the fundamental identification challenge created by selection on unobservables and non-ignorability of treatment assignment. We consider models of this type in detail in Chapter 11, including a more complete discussion of the sorts of processes that generate over-time dependencies, such as the one produced by E in Figure 8.2. However, we also have some more encouraging results to offer in Chapter 11. If, as was the case for this demonstration, the causal effect is evolving in time, such that the trajectory for the outcome in the absence of treatment has the same time path before and after treatment exposure, then it is possible to obtain consistent estimates of some average treatment effects of interest.

Before offering a full explanation of analysis strategies when data are available from pretreatment time periods, we will consider two other types of estimators – instrumental variable estimators in Chapter 9 and mechanism-based estimators in Chapter 10. To motivate our presentation of these methods in subsequent chapters, we conclude this chapter in the next section by enriching our presentation of directed graph methodology, explaining how graphs can be used to represent full patterns of self-selection and heterogeneity using latent class variables.

---

larger treatment effects. As a by-product, the individual-level causal effect also varies with O and X, given that these variables are positively associated with U because of the common-cause path $U \leftarrow O \rightarrow X$. The conditional-variance weighting that is implicit in OLS regression averages these heterogeneous individual-level effects, giving more weight to those whose implicit propensity scores are near to .5. Given the data setup, where only 9 percent of students end up enrolled in Catholic schools – because of the intercept of -3.8 for the logistic distribution in Equation (8.1) – the implicit weighting would move the estimated average effects toward the ATT relative to the ATC. Yet, the coefficient estimates would not equal the ATT. As explained in Section 6.3, these coefficients are best interpreted as estimates of the ATE with supplemental conditional-variance-based weighting, which, for this demonstration, would push the estimates toward the ATT because of the distribution of D and the positive association between U and D. In practice, of course, the variable U would not be available for analysis in the first place.

8.3   Causal Graphs for Complex Patterns of Self-Selection and Heterogeneity

We conclude this chapter with a presentation of how patterns of self-selection and heterogeneity can be represented with directed graphs.[8] We have two goals in this section. First, we want to demonstrate and explain the generality of the graphical approach to the representation of causal relationships, which we will utilize in subsequent chapters. Second, we want to make the following obvious (but frequently forgotten) point: The best solution to an unobserved variable problem is to develop and deploy additional new measures. Too often, the literature implies that self-selection patterns are so complex that estimation in their presence will forever remain infeasible. We see more promise in developing methods to directly confront the limitations of available data than the literature often implies, and we want to make this case explicitly before offering explanations in the next three chapters of estimation strategies that may be feasible while the limitations of available data remain unaddressed.

In this section, we use the charter school effect on learning as the focal example, which we introduced in Section 1.3.2 (see page 24). This example shares some of the same complications as the example of the Catholic school effect that we have considered extensively already, but it offers some new complications as well (and serves as a bridge to the school voucher instrumental variable example that will be considered in Chapter 9). In the material that follows, we start with a simple latent class model of heterogeneity that allows families of different types to differ in their likelihood of choosing charter schools. We then build toward a full directed graph that shows the estimation challenge clearly, at which point we will then elaborate the back-door paths in order to discuss feasible estimation strategies that would be possible if new measures were constructed that would enable direct modeling of choice behavior. We conclude with a graph-aided interpretation of extant empirical research on charter school effects.

8.3.1   A Starting Point: Separate Graphs for Separate Latent Classes

Consider the two causal graphs in Figure 8.3, and suppose that the population is partitioned into two latent classes, each of which has its own graph in panel (a) or panel (b). Suppose that P is a family's parental background, D is charter school attendance, and Y is a score on a standardized test given to all fifth graders in a large metropolitan school district. For these two graphs, the subscripts refer to latent classes, which are also indicated by a latent class membership variable G that takes on values of 1 and 2.[9]

Although surely a gross oversimplification, suppose nonetheless that the population is composed of fifth graders who have been raised in two types of families. Families with $G = 1$ choose schools predominantly for lifestyle reasons, such as proximity to the home and tastes for particular school cultures, assuming that all schools are similar

---

[8]This section draws on material previously published in Morgan and Winship (2012).

[9]The lower case values x, d, and y for the two causal graphs are meant to connote that these are realized values of X, D, and Y that may differ in their distributions across the two latent classes.

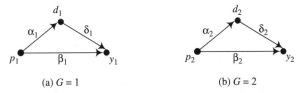

(a) $G = 1$ (b) $G = 2$

Figure 8.3 Separate causal graphs for two groups of individuals ($G = 1$ and $G = 2$) where the effects of parental background (P) and charter schools (D) on test scores (Y) may differ for the two groups.

in instructional impact because achievement is largely a function of individual effort. Families with $G = 2$ choose elementary schools for their children by selecting the school, subject to constraints, that they feel will maximize the achievement of their children, assuming that schools differ in quality and that their children may learn more in some schools than in others. Accordingly, these families are attentive to the national press coverage of educational policy, in which many policymakers have argued for increasing the number of charter schools in the country because some research has claimed that charter schools are more effective than traditional public schools. As a consequence, the second group of families is more likely to send their children to charter schools, such that the mean of D is higher for those families with $G = 2$ than $G = 1$.

Finally, suppose that parents with higher levels of education are more likely to value distinctive forms of education, and as a result are more likely to send their children to charter schools (independent of whether or not highly educated parents are more likely to be found in the latent class for whom $G = 2$, which we will discuss later). They are also more likely to be able to support children in completing homework and otherwise making the most of the educational opportunities that are offered to their children. Accordingly, suppose that in both groups the causal effects $P \to D$ and $P \to Y$ are positive and substantial (i.e., that α_1, β_1, α_2, and β_2 in Figure 8.3 are positive and substantial).

The question for investigation is whether the effect of D on Y is positive for both groups and, if so, whether it is the same size for both groups. If one is willing to assume, as some of the literature suggests, that the second group of families is correct in the sense that school quality does matter for student learning, and further that charter schools are higher quality, then we should expect that both δ_1 and δ_2 are more likely positive than not. And, if one believes that parents with $G = 2$ have some sense that this is correct, then not only will more of them send their children to charter schools, they will also sort their children more effectively into charter and non-charter schools. In other words, they will also be more likely to continue to enroll their children in regular public schools if they feel that their children will not benefit from the distinctive characteristics of available charter schools (e.g., if the charter schools that have openings have instructional themes that their children find distasteful). Because both of these self-selection effects are reinforcing, is it likely that $\delta_2 > \delta_1$.[10]

---

[10]These target parameters, δ_2 and δ_1, are defined implicitly as the average effect of charter schooling for all students from families with $G = 2$ and $G = 1$, respectively.

If this plausible scenario is true in reality, what would happen if a researcher ignored the latent classes (either by mistake or, more realistically, because the membership variable G is unobserved) and simply assumed that a single graph prevailed? In this case, a researcher might estimate the effect of D on Y for each value of P and then average these effects over the distribution of P, yielding a population-level estimate δ. At best, this estimate would be uninformative about the underlying pattern of heterogeneity that suggests that $\delta_2 > \delta_1$. At worst, this estimate would be completely wrong as an estimate of the average causal effect of D on Y, which we have referred to as the ATE in prior chapters. For example, if P predicts latent class membership G, and G predicts the size of the effect of D on Y, then P-stratum-specific effects mix together individual-level causal effects that vary with the conditional distribution of G within the strata of P. Combining P-stratum-specific effects by calculating an average effect across only the distribution of P does not properly weight the G-stratum-specific effects that are embedded in differential patterns within the strata of P.

In order to consider these possibilities, we need to have a model of selection into D that is informed by a model of the traits of families that would cause them to be found in underlying latent classes. It is most natural to pursue such a model in a single causal graph that explicitly represents the latent classes by including the variable G as a node within it.

8.3.2   A Single Graph That Represents All Latent Classes

Consider Figure 8.4, which contains a standard causal triangle where D has an effect on Y but where both D and Y share a common cause P. To this triangle, the latent class membership variable G is introduced as an explicit cause of D. The variable G is given a hollow node, $\circ$, to indicate that it is unobserved.[11]

The arrow from G to D is present because there are alternative groups of families, coded by the alternative values of the unobserved variable G, that approach differently the decision of whether to send their children to charter schools. As a result, G predicts charter school attendance, D. Although we will continue to write as if G only takes on two values that identify two latent classes, this restriction is no longer necessary. G may take on as many values as there are alternative groups of families who differ systematically in how they approach and enact the decision of whether to send their children to a charter school. (We considered only two values for G in Figure 8.3 to limit the number of G-specific graphs that needed to be drawn.)

The corresponding structural equations for the graph in Figure 8.4 are then

$$P = f_P(e_P), \tag{8.5}$$

$$G = f_G(e_G), \tag{8.6}$$

$$D = f_D(P, G, e_D), \tag{8.7}$$

$$Y = f_Y(P, D, e_Y). \tag{8.8}$$

---

[11]Here, we follow Elwert and Winship (2010) and introduce a latent class variable G to represent effect heterogeneity.

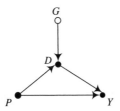

Figure 8.4 A graph where groups are represented by an unobserved latent class variable (G) in a single graph.

The latent class membership variable G only enters these structural equations in two places, on its own in Equation (8.6) and then as an input to $f_D(.)$ in Equation (8.7). Recall also that the unspecified structure of the functions such as $f_Y(.)$ permits the sort of interactions discussed in the last subsection, where, for example, the effect of D depends on the level of P. (See Section 3.3.2 for a full discussion.)

To accept Figure 8.4 as a full representation of the causal relationships between G and the other variables P, D, and Y, we must be able to assume that G shares no common causes with P, D, or Y that have been mistakenly suppressed. The necessary assumptions are that students who have parents with higher levels of education are no more likely to know of the educational policy dialogue that claims that charter schools have advantages and also are no more likely to think that school quality has any effects on achievement. We must also be willing to assume that, within values of P and D, G has no causal effect on Y, which with our example is tantamount to assuming that those who attempt to maximize the learning of their children by selecting optimal schools (1) do not manage to do so well enough so that the obtained effect is any larger on average, conditional on other factors, for their own children than for those who do not attempt to select on the causal effect of schooling and (2) do not do anything else that helps their children to benefit from the learning opportunities provided to them in school or in the home that is not already captured by the direct effect $P \to Y$. This would require that the impulse to select into charter schools based on beliefs about the size of the charter school effect for one's own child is a completely ignorable process, since it does not result in any actual selection on the variation in the causal effect nor generate any reinforcing behavior that might complement the charter school effect. None of the literature on charter school effects supports such a dismissal of the power of self-selection.

Accordingly, for Figures 8.5(a) and 8.5(b), we add an arrow from G to Y to the graph presented earlier in Figure 8.4. For Figure 8.5(a), which includes only this one additional arrow, the structural equations are

$$P = f_P(e_P), \tag{8.9}$$
$$G = f_G(e_G), \tag{8.10}$$
$$D = f_D(P,G,e_D), \tag{8.11}$$
$$Y = f_Y(P,D,G,e_Y). \tag{8.12}$$

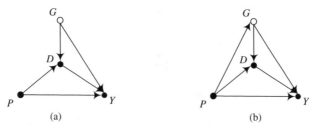

Figure 8.5 Two graphs where selection into charter schools (D) is determined by group (G) and where selection renders the effect of D on Y unidentified as long as G remains unobserved.

For Figure 8.5(b), we drop the assumption that G is independent of P. This elaborated graph now includes an arrow from P to G. As a result, $f_G(e_G)$ is no longer the appropriate function for G. Equation (8.10) must be replaced by

$$G = f_G(P, e_G) \qquad (8.13)$$

so that family background is an explicit cause of latent class membership. It is likely that parents with high socioeconomic status are more likely to select on the possible causal effect of charter schooling, which is how the latent classes were discussed for Figure 8.3. Still, how these latent classes emerge is not sufficiently transparent in Figure 8.5. A more explicit causal model that gives structure to the causal pathway from P to G may help to clarify the self-selection dynamic, as we show next.[12]

8.3.3   Self-Selection into the Latent Classes

Suppose that the latent class membership variable G is determined at least in part by a variable that measures a family's subjective expectation of their child's likely benefit from attending a charter school instead of a regular public school. Although we could enter this variable into a graph with a single letter, such as S or E, for Figure 8.6 we use a full mnemonic representation as a variable labeled $\text{Exp}(D \to Y)$, which represents "the family's subjective expectation of the specific effect of D on Y for their particular child." For Figure 8.6(a), which is a direct analog to Figure 8.5(a), this subjective

---

[12]In some work, latent class membership variables such as G are not regarded as causal variables. We regard G as a nominal causal variable because we can conceive of an intervention that could move a family from one value of G to another. The goal of introducing G into this graph is to provide a representation of consequential individual-level heterogeneity, and accordingly we allow G to have nominal causal effects on the outcome of interest through $G \to Y$. We will explain later why a fully elaborated graph that locates the sources of all such heterogeneity in structural variables that lie on directed paths that reach G would thereby render G as a redundant carrier of these fully specified determinants of heterogeneity. In this case, G could then be removed from the graph, as there would be no need for a nominal causal variable to transmit the heterogeneity produced by its genuine determinants.

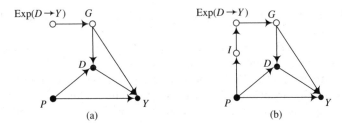

Figure 8.6 Two graphs where selection on the unobservables is given an explicit representation as self-selection on subjective expectations of variation in the causal effect of D on Y. For panel (b), these expectations are determined by information (I) that is differentially available to families with particular parental backgrounds (P).

expectation is the sole determinant of G. The structural equations are then

$$P = f_P(e_P), \tag{8.14}$$
$$\text{Exp}(D \to Y) = f_{\text{Exp}}(e_{\text{Exp}}), \tag{8.15}$$
$$G = f_G[\text{Exp}(D \to Y), e_G], \tag{8.16}$$
$$D = f_D(P, G, e_D), \tag{8.17}$$
$$Y = f_Y(P, D, G, e_Y). \tag{8.18}$$

Note that $\text{Exp}(D \to Y)$ is determined solely by e_{Exp} in Equation (8.15). Thus, the graph in Figure 8.6(a) would be an accurate representation of the system of causal relationships if subjective expectations were either completely random or instead based solely on characteristics of families that are independent of the family background variables in P.

Given what we have written about the likelihood that families with different patterns of P will end up in different latent classes represented by G, it seems clear that Figure 8.6(a) is not the proper representation for the research scenario we have already specified. Accordingly, in Figure 8.6(b), e_{Exp} is joined by an unspecified additional input I into the subjective expectation of the child-specific causal effect of D on Y, which is then presumed to be caused, in part, by family background. As a result, there is now a path from P to G through I and $\text{Exp}(D \to Y)$. The structural equations are now augmented as

$$P = f_P(e_P), \tag{8.19}$$
$$I = f_I(P, e_I), \tag{8.20}$$
$$\text{Exp}(D \to Y) = f_{\text{Exp}}(I, e_{\text{Exp}}), \tag{8.21}$$
$$G = f_G[\text{Exp}(D \to Y), e_G], \tag{8.22}$$
$$D = f_D(P, G, e_D), \tag{8.23}$$
$$Y = f_Y(P, D, G, e_Y). \tag{8.24}$$

In sociology, the causal effect of P on $\text{Exp}(D \to Y)$ via I follows from the position that privileged positions in social structure are occupied by advantaged families. From these positions, individuals acquire information I that allows them to recognize benefits that are available to them.[13]

The directed path $P \to I \to \text{Exp}(D \to Y) \to G$ carries one important systematic source of self-selection through to the latent class variable G. However, this path is not the only directed path that is present, as could be seen under magnification. The differences in information about the charter school effect that are independent of parental background have their effect on G through $e_I \to I \to \text{Exp}(D \to Y) \to G$. Likewise, differences in expectation formation processes that are not based on information and are independent of parental background have their effects on G through $e_{\text{Exp}} \to \text{Exp}(D \to Y) \to G$.

By building the full graph progressively from Figures 8.4 through 8.6(b), we have explicitly elaborated what is often presumed in models that incorporate self-selection. Background variables in P are related to the cause D by way of a set of latent classes in G that encode subjective evaluations of the individual-specific causal effect of D on Y. These expectations are functions, in part, of characteristics in P by way of the information I that is differentially available to families that differ on P. Yet, even though we now have an elaborate representation of self-selection, we still have not brought what some would label "contextual effects" into the model, such as neighborhood of residence. We consider this complication next, which is clearly important to consider when modeling the effects of charter schools on learning.

8.3.4   Self-Selection into the Treatment and a Complementary Context

How hard is the task of allowing for contextual effects? Consider Figure 8.7, which incorporates a contextual variable N into the causal graph in Figure 8.6(b). N represents all causes of student achievement that can be conceptualized as either features of a student's residential neighborhood or features of a charter school's surrounding neighborhood. The component variables in N might be access to resources not measured by P or specific local cultures that may or may not promote student achievement.[14]

---

[13]In addition, it may be that there are also additional common causes of P and I, which could be represented by a bidirected edge between P and I in the graph. This would be reasonable if informational advantages that structure expectations for optimal school choices are determined by deeper structural factors that also confer socioeconomic advantages on parents before they arrive at the decision point of whether or not to send their children to charter schools. It is also possible that parental background determines expectation formation processes to some degree, such that different families process information differently, as would be case if the directed path $P \to \text{Exp}(D \to Y) \to G$ were added to the graph. We do not include these additional causal pathways in the graph because they would add additional back-door paths to the subsequent discussion without changing the core identification results.

[14]If the latter are only diffuse cultural understandings that only weakly shape local norms about the appropriateness of enacting the role of achievement-oriented student, then such variables may be difficult to observe. In this case, N might then be coded as a series of neighborhood dummy identifier variables. Analysis of these effects would then only be possible if there were sufficient numbers of students to analyze from within each neighborhood studied. Without such variation, the potential effects of N could not be separated from individual characteristics of students and their families. And, if modeled in this way, only the total effects of N would be identified, since the dummy variables for N would not contain any information on the underlying explanatory factors that structure the neighborhood effects that they identify.

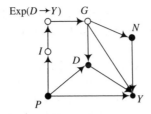

Figure 8.7 A graph where self-selection on the causal effect of charter schooling also triggers self-selection into consequential and interactive neighborhood contexts (N).

With the addition of N, the function for Y is now $f_Y(P,D,G,N,e_Y)$. Recall, again, that N is not restricted by any functional form assumption for $f_Y(.)$. As a result, the causal effect of N can modify or interact with the effects of G, D, and P on Y.[15]

Figure 8.7 also allows for even more powerful effects of self-selection. Suppose that self-selection into the latent classes in G is associated with self-selection into N as well. We see two separate and countervailing tendencies. Parents attuned to the potential benefits of charter schooling are also more likely to choose neighborhood contexts that best allow them to encourage their children to study hard in school. At the same time, after obtaining an attendance offer from a charter school, a family may also decide to move to an alternative neighborhood in the catchment area of a suboptimal regular public school, since attendance at such a school may no longer be a consideration in the family's residential decision. If either effect is present, then the function for N is equal to $f_N(G,e_N)$, and we then have seven structural equations as

$$P = f_P(e_P), \tag{8.25}$$

$$I = f_I(P,e_I), \tag{8.26}$$

$$\text{Exp}(D \rightarrow Y) = f_{\text{Exp}}(I,e_{\text{Exp}}), \tag{8.27}$$

$$G = f_G[\text{Exp}(D \rightarrow Y), e_G], \tag{8.28}$$

$$D = f_D(P,G,e_D), \tag{8.29}$$

$$N = f_N(G,e_N), \tag{8.30}$$

$$Y = f_Y(P,D,G,N,e_Y). \tag{8.31}$$

The nonparametric nature of these structural equations allows for fully interactive effects. Most importantly, the function for Y, $f_Y(P,D,G,N,e_Y)$, allows for the effects of D and N to vary within each distinct combination of values between them, as would be the case if the charter school effect varied based on the neighborhood within which students lived. We should also note that we could enrich the causal model further by drawing from the literature that posits deeper causal narratives for the joint determination of P and N, as well as other causal pathways that link P to N. We see the graph in Figure 8.7 as sufficiently rich to motivate the discussion we have offered so far as well as for the analysis of the extant empirical research to which we turn next.

---

[15]See VanderWeele (2009b) for an incisive analysis of the difference between an interaction and an effect modification. Our interest, conceptually at least, is in instances of genuine causal interaction, although much of what we write would hold under simpler structures of only effect modification.

8.3.5   A Graph-Aided Interpretation of Extant Empirical Models of Charter School Effects

There are two basic goals of writing down a causal graph: (1) to represent the set of causal relationships implied by the available state of knowledge and (2) to assess the feasibility of alternative estimation strategies. Figure 8.7 represents a causal graph that is a reasonable representation of the causal structure that generates the charter school effect. This is a matter of judgment, and one might contend, for example, that the claim that self-selection on the charter school effect generates movement between neighborhoods is overly complex.

Suppose that one has access to observational data, such as the National Assessment of Education Progress (NAEP) data analyzed by Lubienski and Lubienski (2003), that provide information on standardized test scores, school type, and some family background characteristics. For the sake of our exposition, suppose further that these NAEP data had even better measures of family background and neighborhood characteristics, so that we could conclude that high-quality data are available for all of the variables in Figure 8.7 with solid nodes: Y, D, N, and P. Yet, no data are available for the variables with hollow nodes: I, $\text{Exp}(D \to Y)$, and G. The primary goal of analysis is to estimate the average causal effect of D on Y, as this effect interacts with the complementary causal effect of N on Y. Can one adjust for confounding in order to estimate these effects?

The first question to consider is the following: Is there a set of observed variables in Figure 8.7 that satisfies the back-door criterion with respect to the causal effect of D on Y and the causal effect of N on Y? The relevant back-door paths are,

for D,

1. $D \leftarrow P \to Y$,
2. $D \leftarrow P \to I \to \text{Exp}(D \to Y) \to G \to Y$,
3. $D \leftarrow P \to I \to \text{Exp}(D \to Y) \to G \to N \to Y$,
4. $D \leftarrow G \to N \to Y$,
5. $D \leftarrow G \to Y$;

and

for N,

6. $N \leftarrow G \to Y$,
7. $N \leftarrow G \to D \to Y$,
8. $N \leftarrow G \leftarrow \text{Exp}(D \to Y) \leftarrow I \leftarrow P \to Y$,
9. $N \leftarrow G \leftarrow \text{Exp}(D \to Y) \leftarrow I \leftarrow P \to D \to Y$,
10. $N \leftarrow G \leftarrow \text{Exp}(D \to Y) \leftarrow I \leftarrow P \to D \leftarrow G \to Y$.

How many of these paths can be blocked by conditioning on the observed data? For models that estimate the effect of D on Y, paths 1 through 4 can be blocked by conditioning on P and N. However, path 5 remains unblocked. Likewise, paths 7 through 10 can be blocked by conditioning on P and D, but path 6 remains unblocked. For the two unblocked paths, the same problematic effect is present: $G \rightarrow Y$. This effect transmits the effects of exogenous causal determinants of I and $\text{Exp}(D \rightarrow Y)$, which are e_I and e_{Exp}, through the directed path $I \rightarrow \text{Exp}(D \rightarrow Y) \rightarrow G \rightarrow Y$. Thus, if there is selection on the causal effect itself, independent of parental background, then it enters the model through G and confounds the conditional associations between D and Y and between N and Y. This confounding cannot be eliminated by conditioning on the observed data, and therefore back-door conditioning for the effects of D and N on Y is infeasible.

This conclusion is hardly revelatory for readers who already know the literature on self-selection bias and/or the empirical literature on charter school effects. Nonetheless, we would argue that there is substantial didactic and communicative value in seeing this result expressed with a causal graph to which the back-door criterion is then applied. To effectively use back-door conditioning to identify all average causal effects of primary interest – the ATC, ATT, and ATE – we would need to specify a more elaborate set of causes of G, which would generate G through and/or alongside the structural equations that determine I and $\text{Exp}(D \rightarrow Y)$. It would be sufficient to have a model for G, $f_G(.)$, where all inputs other than e_G are observed and where these inputs are sufficiently rich such that e_G can be regarded as a constant that applies to everyone. Equivalently, we would need a set of observed measures as variables that point to G in the graph that would allow us to declare that there are no more latent classes precisely because G is a deterministic function of the observed variables in the graph. In this case, all sources of the noncausal association between D and Y and between N and Y would be indexed by the observed variables that determine G, and we could safely remove G from the graph and allow those variables to point directly to D, N, and Y.

Without access to such observed variables, which prevent us from explaining away the existence of the latent classes indexed by G, how can analysis proceed? There are two main choices. First, the analyst can concede that self-selection on the causal effect is present (because G cannot be eliminated from the graph), which may even generate neighborhood-based selection as a by-product. In these circumstances, the presence of the causal relationship $G \rightarrow Y$ renders estimates of D and N on Y unidentified, either in interactive fashion or when averaging one over the other. In this case, analysis must then be scaled back, and we would recommend that set identification results be pursued, as we will explain later in Chapter 12. The new goal would be to estimate an interval within which the average causal effect of interest must fall.

In the actual empirical literature on the charter school effect, this humble option has not been pursued by any of the main combatants in the debates. Instead, this research is a good example of what Manski (2013b) has labeled "policy analysis with incredible certitude." The desire to provide point estimates of causal effects has been too strong, even though it would seem clear to many outside readers that the debate over charter school effects persists simply because the point estimate of the ATE is unidentified. Consider the two dominant positions in the extant literature.

One group of researchers has used the lottery-induced structure of charter school enrollments in order to estimate effects. In cities where the number of charter schools is insufficient to meet demand, most schools are required by their chartering authorities to perform random lotteries and offer admission first to those who win their lotteries. This admissions mechanism delivers two comparable groups: lottery winners and lottery losers. If the lottery winners attend charter schools and the lottery losers do not, then observation of achievement trajectories following the lottery enables estimation of average treatment effects for those who participate in the lottery.[16]

The researchers who use this research design typically define the charter school effect of interest as solely the ATT: the effect of charter schooling among those who would self-select into charter schooling by applying for the lottery. This is entirely appropriate for the interpretation of lottery-based results. The problem is that these same scholars too frequently forget the bounded nature of their conclusions. For example, in their study of charter schools in New York City, Hoxby, Murarka, and Kang (2009:vii) introduce their results in their "Executive Summary" with three bullet points:

- "Lottery-based analysis of charter schools' effects on achievement is, by far, the most reliable method of evaluation. It is the only method that reliably eliminates 'selection biases' which occur if students who apply to charter schools are more disadvantaged, more motivated, or different in any other way than students who do not apply."

- "On average, a student who attended a charter school for all of grades kindergarten through eight would close about 86 percent of the 'Scarsdale-Harlem achievement gap' in math and 66 percent of the achievement gap in English. A student who attended fewer grades would improve by a commensurately smaller amount."

- "On average, a lotteried-out student who stayed in the traditional public schools for all of grades kindergarten through eight would stay on grade level but would not close the 'Scarsdale-Harlem achievement gap' by much. However, the lotteried-out students' performance does improve and is better than the norm in the U.S. where, as a rule, disadvantaged students fall further behind as they age."

Nowhere in their "Executive Summary" is it stated that these comparisons across lotteried-in and lotteried-out students are only informative about those who self-select into the relevant charter school lottery. Selection bias is not eliminated, as claimed in the first bullet point; the lottery-generated data simply provide a reasonable comparison among those who self-select into the charter school lottery.

More generally, it is not conceded that the results are uninformative about two first-order policy questions:

1. What is the expected charter school effect for a randomly chosen student from New York City, assuming the supply of charter schools is held constant?

2. What is the expected charter school effect for the subset of students in public schools who would be induced to apply to charter schools if a policy intervention were implemented to expand the supply of charter schools?

---

[16]For simplicity, in this discussion we ignore non-comparability generated by random variation, initial compliance with the lottery, and selective attrition over the observation window following the lottery. In most cases, the researchers who have modeled these effects have carefully adjusted their data to neutralize these threats to inference.

To answer these questions, the authors would need a full model of the charter school effect for students who never applied to charter schools, and lottery-based studies use no data on these students. The ATT is the appropriate parameter to estimate only when addressing a second-order policy question that is not the focus of the debate: "Should charter schools in New York City continue to receive support because those students who have chosen to attend them have learned more than they would have if they had attended regular public schools?" The answer to this question appears to be yes, and yet Hoxby et al. (2009) chose not to limit their conclusions only to this supportable position, at least when putting forward their primary conclusions.

If this were not troubling enough, the alternative is worse. One can simply assume away the unblocked paths that include $G \to Y$, which is tantamount to assuming that self-selection does not exist. The study of the Center for Research on Education Outcomes (2009) is closer to this position. Its authors offer a complex set of conclusions based on national results where charter school students are matched to students from traditional public schools based on observed characteristics. Their overall conclusion is that charter schools are generally ineffective for a majority of students, even though charter schools may be effective for a minority of students who are not well served by regular public schools:

> In our nationally pooled sample, two subgroups fare better in charters than in the traditional system: students in poverty and ELL [English Language Learner] students. ... These findings are particularly heartening for the charter advocates who target the most challenging educational populations or strive to improve education options in the most difficult communities. Charter schools that are organized around a mission to teach the most economically disadvantaged students in particular seem to have developed expertise in serving these communities. ... The flip-side of this insight should not be ignored either. Students not in poverty and students who are not English language learners on average do notably worse than the same students who remain in the traditional public school system. Additional work is needed to determine the reasons underlying this phenomenon. Perhaps these students are "off-mission" in the schools they attend. (Center for Research on Education Outcomes 2009:7)

These conclusions are offered based on models that match students on observable characteristics, leaving unobserved selection on the causal effect unaccounted for and almost entirely ignored in the write-up of the results. The report is written as if variation in the average treatment effect for different types of students is the central interest, and that matching justifies estimation of all such conditional average effects.

It bears noting that the pattern presented in this study by CREO is consistent with the one that we favored for the extended example that concluded Chapter 7, which we see more generally as the signature of an underlying pattern of self-selection. In this case, students from families who are living in poverty but who make their way into charter schools are fleeing poor alternatives in their own neighborhood schools and, furthermore, have extra amounts of motivation to succeed in school. At the same time, it is likely that students from more advantaged families are more likely to be attending charter schools solely for lifestyle reasons. In fact, they may be trading off academic opportunities in high-quality schools that they have found distasteful for medium-quality charter schools with peer cultures or instructional programs that are more appealing.

8.4   Conclusions

In this chapter, we have made the transition from "easy" to "hard" instances of causal effect estimation. No longer does simple conditioning on the observed determinants of the cause or all other direct causes of the outcome allow for identification and consistent estimation of the causal effect of interest. Instead, we have considered examples in which important variables that might have been used to mount an effective conditioning strategy are unobserved. Thus, selection of the treatment of interest is on unobserved characteristics of individuals that have unknown but suspected relationships to the outcome. We have also expanded our usage of directed graphs to show how patterns of self-selection and heterogeneity can be given explicit representations, and we have made the implicit case that omitted variable problems are best addressed, in the long run, by developing and deploying new measures.

We turn in the next three chapters to a presentation of additional methods to identify causal effects with observed variables when one cannot move beyond the available data: instrumental variable estimators in Chapter 9, causal mechanisms in Chapter 10, and panel data models that utilize pretreatment measures of the outcome variable in Chapter 11. We will explain these strategies in detail, discussing their strengths and weaknesses along the way.

Chapter 9

Instrumental Variable Estimators of Causal Effects

If a perfect stratification cannot be enacted with the available data, and thus neither matching nor regression nor any other type of basic conditioning technique can be used to effectively estimate a causal effect of D on Y, one possible solution is to find an exogenous source of variation that determines Y only by way of the causal variable D. The causal effect is then estimated by measuring how Y varies with the portion of the total variation in D that is attributable to the exogenous variation. The variable that indexes this variation in D is an instrumental variable (IV).

In this chapter, we orient the reader by first presenting IV estimation with a binary instrument and a simple demonstration. We then return to the origins of IV techniques, and we contrast this estimation strategy with the perspective on regression that was presented in Chapter 6. We then develop the same ideas using the potential outcome model, showing how the counterfactual perspective has led to a new literature on how to interpret IV estimates. This new literature suggests that IV techniques are more effective for estimating narrowly defined local average causal effects than for estimating more general average causal effects, such as the average treatment effect (ATE) and the average treatment effect for the treated (ATT). We also consider causal graphs that can represent this recent literature and conclude with a discussion of marginal treatment effects identified by local IVs.

9.1 Causal Effect Estimation with a Binary IV

We begin our presentation with the simplest scenario in which an instrumental variable can be used to effectively estimate a causal effect. Recall the causal regression setup in Equation (6.3):

$$Y = \alpha + \delta D + \varepsilon, \tag{9.1}$$

where Y is the outcome variable, D is a binary causal exposure variable, α is an intercept, δ is the causal effect of D on Y, and ε is a summary random variable

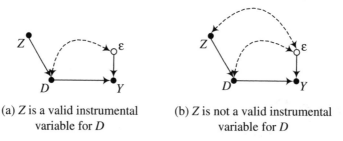

(a) Z is a valid instrumental   (b) Z is not a valid instrumental
    variable for D variable for D

Figure 9.1 Two graphs in which Z is a potential instrumental variable.

that represents all other causes of Y. As noted above, when Equation (9.1) is used to represent the causal effect of D on Y, the parameter δ is usually considered an invariant, structural causal effect that applies to all members of the population of interest. We will maintain this traditional assumption in this section.[1]

Suppose that the probability that D is equal to 1 rather than 0 is a function of a binary variable Z that takes on values of 0 and 1. Figure 9.1 presents two possible ways in which the variable Z could be related to both D and Y. Note first that, for both graphs, the back-door paths represented by $D \leftarrow\!\text{-}\!\text{-}\!\text{-}\!\rightarrow \varepsilon \rightarrow Y$ prevent a least squares regression of Y on D from generating a consistent or unbiased estimate of the effect of D on Y. More generally, no conditioning estimator would effectively estimate the causal effect of D on Y for these graphs because no observed variables satisfy the back-door criterion.

Now consider how Z, which we have labeled a "potential instrumental variable," is related to the outcome variable Y according to the alternative structures of these two graphs. For both Figures 9.1(a) and 9.1(b), Z has an association with Y because of the directed path $Z \rightarrow D \rightarrow Y$. In addition, the paths collectively represented by $Z \rightarrow D \leftarrow\!\text{-}\!\text{-}\!\text{-}\!\rightarrow \varepsilon \rightarrow Y$ do not contribute to the association between Z and Y because D is a collider variable along all of them. However, for Figure 9.1(b), the additional paths represented by $Z \leftarrow\!\text{-}\!\text{-}\!\text{-}\!\rightarrow \varepsilon \rightarrow Y$ contribute to the association between Z and Y because of the common causes that determine both Z and ε.[2] This last difference is the reason that Z can be used to estimate the causal effect of D on Y for Figure 9.1(a) but not for Figure 9.1(b), as we will now explain.

If we continue to maintain the assumption that the effect of D on Y is a constant structural effect δ, then it is not necessary to relate all of the variation in D to all of the variation in Y in order to obtain a consistent estimate of the causal effect. Under this assumption, if we can find a way of isolating the covariation in D and Y that is causal, then we can ignore the other covariation in D and Y that is noncausal because it is generated by the common causes of D and ε. For Figure 9.1(a), the variable Z represents an isolated source of variation across which the analyst can examine the covariation in D and Y that is causal. In this sense, Z is an "instrument" for

---

[1] In Pearl's framework, we are assuming that $f_Y(D, e_Y)$ is defined as the linear function $\alpha + \delta D + \varepsilon$ and where α and β are scalar constants.

[2] We have drawn the two bidirected edges separately for Figure 9.1(b) for simplicity. The same reasoning holds when some of the common causes of Z and ε are also causes of D (and so on).

examining an isolated slice of the covariation in D and Y. For Figure 9.1(b), Z does not provide an isolated source of variation. The analyst can still examine the portion of the covariation in D and Y that can be calculated across levels of Z, but some of this covariation must be noncausal because D and Y are dependent on common causes embedded in the bidirected edge in the back-door paths collectively represented by $D \leftarrow Z \leftarrow\text{--}\text{--} \rightarrow \varepsilon \rightarrow Y$.

To see this result without reference to graphs, take the population-level expectation of Equation (9.1), $E[Y] = E[\alpha + \delta D + \varepsilon] = \alpha + \delta E[D] + E[\varepsilon]$, and rewrite it as a difference equation in Z:

$$E[Y|Z=1] - E[Y|Z=0] \tag{9.2}$$
$$= \delta(E[D|Z=1] - E[D|Z=0]) + (E[\varepsilon|Z=1] - E[\varepsilon|Z=0]).$$

Equation (9.2) is now focused narrowly on the variation in Y, D, and ε that exists across levels of Z.[3] Now, take Equation (9.2) and divide both sides by $E[D|Z=1] - E[D|Z=0]$, yielding

$$\frac{E[Y|Z=1] - E[Y|Z=0]}{E[D|Z=1] - E[D|Z=0]} \tag{9.3}$$
$$= \frac{\delta(E[D|Z=1] - E[D|Z=0]) + (E[\varepsilon|Z=1] - E[\varepsilon|Z=0])}{E[D|Z=1] - E[D|Z=0]}.$$

If the data are generated by the set of causal relationships depicted in Figure 9.1(a), then Z has no linear association with ε, and $E[\varepsilon|Z=1] - E[\varepsilon|Z=0]$ in Equation (9.3) is equal to 0. Consequently, the right-hand side of Equation (9.3) simplifies to δ:

$$\frac{E[Y|Z=1] - E[Y|Z=0]}{E[D|Z=1] - E[D|Z=0]} = \delta. \tag{9.4}$$

Under these conditions, the ratio of the population-level linear association between Y and Z and between D and Z is equal to the causal effect of D on Y. This result suggests that, if Z is in fact associated with D but not associated with ε (or with Y, except through D), then the following sample-based estimator will equal δ in an infinite sample:

$$\hat{\delta}_{\text{IV,WALD}} \equiv \frac{E_N[y_i|z_i=1] - E_N[y_i|z_i=0]}{E_N[d_i|z_i=1] - E_N[d_i|z_i=0]}. \tag{9.5}$$

As suggested by its subscript, this is the IV estimator, which is known as the Wald estimator when the instrument is binary. Although the Wald estimator is consistent for δ in this scenario, the assumption that δ is an invariant structural effect is crucial for this result.[4] From a potential outcomes perspective, in which we generally assume that

---

[3] Equation (9.2) is generated in the following way. First, write $E[Y] = \alpha + \delta E[D] + E[\varepsilon]$ conditional on the two values of Z, yielding $E[Y|Z=1] = \alpha + \delta E[D|Z=1] + E[\varepsilon|Z=1]$ and $E[Y|Z=0] = \alpha + \delta E[D|Z=0] + E[\varepsilon|Z=0]$. Note that, because α and δ are considered constant structural effects for this traditional motivation of IV estimation, they do not vary with Z. Now, subtract $E[Y|Z=0] = \alpha + \delta E[D|Z=0] + E[\varepsilon|Z=0]$ from $E[Y|Z=1] = \alpha + \delta E[D|Z=1] + E[\varepsilon|Z=1]$. The parameter α is eliminated by the subtraction, and δ can be factored out of its two terms, resulting in Equation (9.2).

[4] As for the origin of the Wald estimator, it is customarily traced to Wald (1940) by authors such as Angrist and Krueger (1999). As we discuss later, the Wald estimator is not generally unbiased in a finite sample and instead is only consistent.

causal effects vary meaningfully across individuals, this assumption is very limiting and quite likely unreasonable. Explaining when and how this assumption can be relaxed is one of the main goals of this chapter.

For completeness, return to consideration of Figure 9.1(b), in which Z has a nonzero association with ε. In this case, $E[\varepsilon|Z=1] - E[\varepsilon|Z=0]$ in Equations (9.2) and (9.3) cannot be equal to 0, and thus Equation (9.3) does not reduce further to Equation (9.4). Rather, it reduces only to

$$\frac{E[Y|Z=1] - E[Y|Z=0]}{E[D|Z=1] - E[D|Z=0]} = \delta + \frac{E[\varepsilon|Z=1] - E[\varepsilon|Z=0]}{E[D|Z=1] - E[D|Z=0]} . \tag{9.6}$$

In this case, the ratio of the population-level linear association between Y and Z and between D and Z does not equal the causal effect of D on Y but rather the causal effect of D on Y plus the last term on the right-hand side of Equation (9.6). The Wald estimator in Equation (9.5) is not consistent for δ in this case. Instead, it converges to the right-hand side of Equation (9.6), which is equal to δ plus a bias term that is a function of the net association between Z and ε.

More generally, an IV estimator is a ratio that is a joint projection of Y and D onto a third dimension Z. In this sense, an IV estimator isolates a specific portion of the covariation in D and Y. For that selected covariation to be the basis of a valid causal inference for the effect of D on Y, it cannot be attributable to any extraneous common causes that determine both Z and Y. And, to justify the causal effect estimate generated by a subset of the covariation in D and Y as a consistent estimate of the population-level causal effect of D on Y, it is typically assumed in this tradition that δ is a constant for all members of the population.[5] Consider the following hypothetical demonstration, which is based on the school voucher example introduced in Section 1.3.2 (see page 23).

IV Demonstration 1

Suppose that a state education department wishes to determine whether private high schools outperform public high schools in a given metropolitan area, as measured by the achievement of ninth graders on a standardized test. For context, suppose that a school voucher program is operating in the city and that the state is considering whether to introduce the program in other areas in order to shift students out of public schools and into private schools.

To answer this policy question, the state department of education uses a census of the population in the metropolitan area to select a random sample of 10,000 ninth graders. They then give a standardized test to each sampled ninth grader at the end of the year, and they collect data as $\{y_i, d_i\}_{i=1}^{10,000}$, where Y is the score on the standardized

---

[5]There is one trivial way around this assumption. If the naive estimator of D on Y is consistent for the average causal effect of D on Y, then D is its own IV. The subset of the covariation in D and Y that is projected onto Z is then the full set of covariation in D and Y because Z is equal to D. In this case, no extrapolation is needed and the constant treatment effect assumption can be avoided. As we will discuss later, there are slightly less trivial ways to avoid the assumption as well. One alternative is to assert that δ is a constant among the treated and then stipulate instead that the IV identifies only the ATT. Although plausible, there are better ways to handle heterogeneity, as we will discuss as this chapter unfolds.

Table 9.1 The Distribution of Voucher Winners by School Sector for IV Demonstration 1

		Public school $d_i=0$	Private school $d_i=1$
Voucher loser	$z_i=0$	8000	1000
Voucher winner	$z_i=1$	800	200

test and D is equal to 1 for students who attend private high schools and equal to 0 for students who attend public high schools.

After the data are collected, suppose that the values of y_i are regressed on the values of d_i and that a predicted regression surface is obtained:

$$\hat{Y} = 50.0 + 9.67(D). \tag{9.7}$$

The state officials recognize that private school students typically have more highly educated parents and therefore are more likely to have higher test scores no matter what curriculum and school culture they have been exposed to. Accordingly, they surmise that 9.67 is likely a poor causal effect estimate, or at least not one that they would want to defend in public as equal to the ATE.

The state officials therefore decide to merge the data with administrative records on the school voucher program in the area. For this program, all eighth graders in the city (both in public and private schools) are entered into a random lottery for $3,000 school vouchers that are redeemable at a private high school. By mandate, 10 percent of all eligible students win the voucher lottery.

After merging the data, the state officials cross-tabulate eighth grade lottery winners (where $z_i = 1$ for those who won the lottery and $z_i = 0$ for those who did not) by school sector attendance in the ninth grade, d_i. As shown in Table 9.1, 1,000 of the 10,000 sampled students were voucher lottery winners.[6] Of these 1,000 students, 200 were observed in private schools in the ninth grade. In comparison, of the 9,000 sampled students who were not voucher lottery winners, only 1,000 were observed in private schools.

The researchers assume that the dummy variable Z for winning the voucher lottery is a valid IV for D because they believe that (1) the randomization of the lottery renders Z independent of ε in the population-level causal regression equation $Y = \alpha + \delta D + \varepsilon$ and that (2) Z has a causal effect on Y only through D. They therefore estimate $\hat{\delta}_{\text{IV,WALD}}$ in Equation (9.5) as

$$\frac{E_N[y_i|z_i=1] - E_N[y_i|z_i=0]}{E_N[d_i|z_i=1] - E_N[d_i|z_i=0]} = \frac{51.600 - 51.111}{.200 - .111} = 5.5, \tag{9.8}$$

and conclude that the true causal effect of private schooling on ninth grade achievement is 5.5 rather than 9.67. Operationally, the Wald estimator takes the average

---

[6]For simplicity, we have assumed that the sample percentage of lottery winners is the same as the population percentage. Of course, some random variation would be expected in any sample.

difference in test scores among those students who have won a voucher and those who have not won a voucher and divides that difference by a corresponding difference in the proportion of high school students who attend private schools among those who have won a voucher and the proportion of high school students who attend private schools among those who have not won a voucher. The numerator of Equation (9.8) is equal to $51.600 - 51.111$ by an assumed construction of the outcome Y, which we will present later when we repeat this demonstration in more detail in Section 9.3.1 (as IV Demonstration 2, beginning on page 309). The denominator, however, can be calculated directly from Table 9.1. In particular, $E_N[d_i = 1|z_i = 1] = 200/1000 = .200$, whereas $E_N[d_i = 1|z_i = 0] = 1000/9000 = .111$.

For this demonstration, Z is a valid instrument by the traditional assumptions maintained in this section of the chapter; it is randomly assigned to students and has no association with Y except for the one produced by the directed path $Z \rightarrow D \rightarrow Y$. The resulting estimator yields a point estimate that can be given a traditional causal interpretation based on the position that the casual effect of interest is a constant structural effect. However, as we will show later in this chapter when we reintroduce this demonstration as IV Demonstration 2, the particular causal effect that this IV identifies is quite a bit different when individual-level causal effect heterogeneity is not assumed away. The IV does not identify the ATE or the ATT. Instead, it identifies only the average causal effect for the subset of all students who would attend a private school if given a voucher but who would not attend a private school in the absence of a voucher. This means, for example, that the IV estimate is uninformative about the average causal effect among those who would enroll in private high schools in the absence of a voucher. This group of students represents the vast majority of private school students (in this example $1,000/1,200$ or 83 percent). The potential outcome literature has provided the insight that allows such a precise causal interpretation and clarifies what a valid IV estimate does not inform. Before presenting that newer material, we return to a more complete accounting of the traditional IV literature.

9.2   Traditional IV Estimators

As detailed by Goldberger (1972), Bowden and Turkington (1984), Heckman (2000), and Bollen (2012), IV estimators were developed first in the 1920s by biologists and economists analyzing equilibrium price determination in market exchange (see E. Working 1927; H. Working 1925; and Wright 1921, 1925). After subsequent development in the 1930s and 1940s (e.g., Wright 1934; Schultz 1938; Reiersøl 1941; and Haavelmo 1943), IV estimators were brought into widespread use in economics by researchers associated with the Cowles commission (see Hood and Koopmans 1953; Koopmans and Reiersøl 1950).[7] The structural equation tradition in sociology shares

---

[7]The canonical example for this early development was the estimation of price determination in markets. For a market in equilibrium, only one price and one quantity of goods sold is observable at any point in time. To make prospective predictions about the potential effects of exogenous supply-and-demand shocks on prices and quantities for a new market equilibrium, the shape of latent supply-and-demand curves must be determined. To estimate points on such curves, separate variables are

similar origins to that of the IV literature (see Duncan 1975). The most familiar deployment of IV estimation in the extant sociological research is as the order condition for identification of a system of structural equations (see Bollen 1989, 1995, 1996a, 1996b, 2001, 2012; Fox 1984).

9.2.1 The Linear Structural Equation IV Estimator

Consider the same basic ideas presented earlier for the Wald estimator in Equation (9.5). Again, recall the causal regression setup in Equation (9.1):

$$Y = \alpha + \delta D + \varepsilon, \tag{9.9}$$

and again assume that we are in the traditional setup where δ is an invariant, structural causal effect that applies to all members of the population of interest. The ordinary least squares (OLS) estimator of the regression coefficient on D is

$$\hat{\delta}_{\text{OLS, bivariate}} \equiv \frac{\text{Cov}_N(y_i, d_i)}{\text{Var}_N(d_i)}, \tag{9.10}$$

where $\text{Cov}_N(.)$ and $\text{Var}_N(.)$ denote unbiased, sample-based estimates from a sample of size N of the population-level covariance and variance.

Now, again suppose that the back-door association between D and ε renders the least squares estimator biased and inconsistent for δ in Equation (9.9). If least squares cannot be used to effectively estimate δ, an alternative IV estimator can be attempted, with an IV Z, as in

$$\hat{\delta}_{\text{IV}} \equiv \frac{\text{Cov}_N(y_i, z_i)}{\text{Cov}_N(d_i, z_i)}, \tag{9.11}$$

where Z can now take on more than two values. If the instrument Z is linearly associated with D but unassociated with ε, then the IV estimator in Equation (9.11) is consistent for δ in Equation (9.9).[8]

One way to see why IV estimators yield consistent estimates is to again consider the population-level relationships between Y, D, and Z, as in Equations (9.1)–(9.4). Manipulating Equation (9.1) as before, the covariance between the outcome Y and the instrument Z can be written as

$$\text{Cov}(Y, Z) = \delta \text{Cov}(D, Z) + \text{Cov}(\varepsilon, Z), \tag{9.12}$$

again assuming that δ is a constant structural effect. Dividing by $\text{Cov}(D, Z)$ then yields

$$\frac{\text{Cov}(Y, Z)}{\text{Cov}(D, Z)} = \frac{\delta \text{Cov}(D, Z) + \text{Cov}(\varepsilon, Z)}{\text{Cov}(D, Z)}, \tag{9.13}$$

---

needed that uniquely index separate supply-and-demand shocks. Usually based on data from a set of exchangeable markets or alternative well-chosen equilibria for the same market from past time periods, these IVs are then used to identify different points of intersection between a shifted linear supply/demand curve and a fixed linear demand/supply curve.

[8]Notice that substituting d_i for z_i in Equation (9.11) results in the least squares regression estimator in Equation (9.10). Thus, the least squares regression estimator implicitly treats D as an instrument for itself.

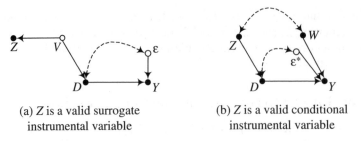

(a) Z is a valid surrogate
instrumental variable

(b) Z is a valid conditional
instrumental variable

Figure 9.2 Two graphs in which Z is a valid IV.

which is directly analogous to Equation (9.3). When $\mathrm{Cov}(\varepsilon, Z)$ is equal to 0 in the population, then the right-hand side of Equation (9.13) simplifies to δ. This suggests that

$$\frac{\mathrm{Cov}_N(y_i, z_i)}{\mathrm{Cov}_N(d_i, z_i)} \xrightarrow{p} \delta \qquad (9.14)$$

if $\mathrm{Cov}(\varepsilon, Z) = 0$ in the population and if $\mathrm{Cov}(D, Z) \neq 0$. This would be the case for the causal diagram in Figure 9.1(a). But here the claim is more general and holds for cases in which Z is many-valued (and, in fact, for cases in which D is many-valued as well, assuming that the linear, constant-coefficient specification in Equation (9.9) is appropriate).

Our prior graphical presentation was also more restricted than it needed to be. As we discussed above, if Z is related to D and Y as in Figure 9.1(a), then IV estimation with Z is feasible.[9] Figure 9.2(a) presents another graph in which Z can be used as a valid instrumental variable for the effect of D on Y. In this case, the instrument V is unobserved. What we have instead is a surrogate instrumental variable, Z, which is not a direct cause of D but which has an association with D because both Z and D mutually depend on the unobserved instrument V. A complication, which we will discuss below, is that the association between Z and D may be very weak in this case, creating a particular set of statistical problems.[10]

Figure 9.2(b) represents perhaps the most common setup in the traditional applied literature. In this case, Z is not a "clean" instrument that has no association with Y other than the association that is generated by a directed path that begins at the instrument and ends at the outcome. Nonetheless, all of the additional associations between Z and Y are generated by back-door paths from the instrument Z to the outcome Y that can be blocked by simultaneous conditioning on observed variables, such as W in Figure 9.2(b). In this case, Z is sometimes referred to as a conditional

---

[9]The only difference in this section is that we are now allowing Z to take on more than two values. Because of this relaxation, in order for the graph to be compatible with the traditional setup, we need to make the further assumption that Z has a linear causal effect on D, which neither varies over individuals nor in any piecewise fashion across the levels of Z.

[10]As elsewhere up until this point in the chapter, in order to stay within the traditional setup we must now assume that linearity holds throughout, so that the effect of V on Z generates a linear relationship between Z and D.

instrumental variable, and estimation must proceed with a conditional version of Equation (9.11), usually the two-stage least squares estimator that we will discuss below. Consider the following examples of IV estimation from the applied literature.

9.2.2 Examples of IV Estimation

Among social scientists, demographic, and health researchers, economists are the most prone to the adoption of IV estimation.[11] Consider the causal effect of education on labor market earnings, as introduced earlier in Section 1.3.1 and as used as a focal example in prior chapters. As reviewed in pieces such as Card (1999) and Angrist and Krueger (2001), the causal effect of years of schooling on subsequent earnings has been estimated with a variety of IVs, including proximity to college, regional and temporal variation in school construction, tuition at local colleges, temporal variation in the minimum school-leaving age, and quarter of birth. The argument in these applications is that the IVs predict educational attainment but have no direct effects on earnings. For the quarter-of-birth instrument, which is one of the most widely discussed IVs in the literature, Angrist and Krueger (1991:981–82) reason:

> If the fraction of students who desire to leave school before they reach the legal dropout age is constant across birthdays, a student's birthday should be expected to influence his or her ultimate educational attainment. This relationship would be expected because, in the absence of rolling admissions to school, students born in different months of the year start school at different ages. This fact, in conjunction with compulsory schooling laws, which require students to attend school until they reach a specified birthday, produces a correlation between date of birth and years of schooling.... Students who are born early in the calendar year are typically older when they enter school than children born late in the year.... Hence, if a fixed fraction of students is constrained by the compulsory attendance law, those born in the beginning of the year will have less schooling, on average, than those born near the end of the year.

The results of Angrist and Krueger (1991) were sufficiently persuasive that many additional researchers have since been convinced to use related IVs. Braakmann (2011), for example, uses an IV based on month of birth to estimate the effect of education on health-related behaviors. Other researchers have used IVs that index over-time variation in compulsory schooling laws, such as Oreopoulos (2006) for the effect of education on earnings, Machin, Marie, and Vuji (2011) for the effect of education on crime, and Kemptner, Jürges, and Reinhold (2011) for the effect of education on health.[12]

---

[11]Some of the examples we introduce in this section adopt the traditional setup presented above, wherein the causal effect of interest is a structural constant. Many of the more recent examples adopt the alternative setup that we will introduce later in the chapter, where it is assumed as a first principle that treatment effects vary at the individual level. For now, we offer these examples only to demonstrate the typical identifying assumptions presented in Figures 9.1(a) and 9.2.

[12]And, in turn, each of these effects has been modeled with alternative IVs. For example, in the case of health and health behaviors, Lindahl (2005) used lottery winnings as an IV for the effect of income

For another example, consider the literature on the effects of military service on subsequent labor market outcomes, a topic that both economists and sociologists have studied for several decades. Some have argued that military service can serve as an effective quasi-job-training program for young men likely to otherwise experience low earnings because of weak attachment to more traditional forms of education (Browning, Lopreato, and Poston 1973) or as a more general productivity signal that generates a veteran premium (De Tray 1982). In one of the earliest studies, which focused primarily on veterans who served around the time of the Korean War, Cutright (1974) concluded that

> two years in the military environment, coupled with the extensive benefits awarded to veterans, do *not* result in a clear cut, large net positive effect of service. ... Therefore, one may question the likely utility of social programs that offer minority men and whites likely to have low earnings a career contingency in a bridging environment similar to that provided by military service. (Cutright 1974:326)

Following the war in Vietnam, attention then focused on whether any military service premium had declined (see Rosen and Taubman 1982; Schwartz 1986; see also Berger and Hirsch 1983). Angrist (1990) used the randomization created by the Vietnam-era draft lottery to estimate the veteran effect. Here, the draft lottery turns date of birth into an IV, in much the same way that compulsory school entry and school-leaving laws turn date of birth into an IV for years of education. Angrist determined that veteran status had a negative effect on earnings, which he attributed to a loss of labor market experience.[13]

The quarter-of-birth and draft-lottery IVs are widely discussed because they are determined by processes that are very unlikely to be structured by any of the unobserved determinants of the outcome of interest. Many IVs in the economics literature do not have this feature. As discussed in Section 1.3.2, much of the early debate on the effectiveness of private schooling relative to its alternatives was carried out with observational survey data on high school students from national samples of public and Catholic high schools. In attempts to resolve the concerns over selection bias, Evans and Schwab (1995), Hoxby (1996), and Neal (1997) introduced plausible IVs for

---

on health and mortality, while Cawley, Moran, and Simon (2010) used an over-time discontinuity in Social Security benefits as an IV for the effect of income on the obesity of the elderly.

[13]We do not mean to imply that Angrist's work on this issue ended in 1990. Not only was his work very important for understanding what IV estimators accomplish in the presence of causal effect heterogeneity (see next section), but he also continued with subsequent substantive work on the military service effect. Angrist and Krueger (1994) estimated the World War II veteran effect, and, in an attempt to reconcile past estimates, concluded: "Empirical results using the 1960, 1970, and 1980 censuses support a conclusion that World War II veterans earn no more than comparable nonveterans, and may well earn less. These findings suggest that the answer to the question 'Why do World War II veterans earn more than nonveterans?' appears to be that World War II veterans would have earned more than nonveterans even had they not served in the military. Military service, in fact, may have reduced World War II veterans' earnings from what they otherwise would have been" (Angrist and Krueger 1994:92). Then, Angrist (1998) assessed the effects of voluntary service in the military in the 1980s, and he found mixed evidence. In Angrist and Chen (2011), he revisited the Vietnam-era veteran effect, analyzing 2000 census data, arguing that the effect on education is positive because of the GI bill but is nonetheless close to zero for earnings.

Catholic school attendance. Hoxby and Neal argued that the share of the local population that is Catholic is a valid IV for Catholic school attendance, maintaining that exogenous differences in population composition influence the likelihood of attending a Catholic school (by lowering the costs of opening such schools, which then lowers the tuition that schools need to charge and to which parents respond when making school sector selection decisions). Variation in the number of Catholics in each county was attributed to lagged effects from past immigration patterns and was therefore assumed to have no direct effect on learning.[14]

These examples of IV estimation in economics are but the tip of an iceberg of applications. Some economists have argued that the attention given to IV estimation in the applied literature in the past two decades has been excessive, and a thriving debate is now under way on the future direction of observational research in economics, as revealed by a comparison of Angrist and Pischke (2010) and Imbens (2010) to Deaton (2010), Heckman and Urzua (2010), and Leamer (2010). We will discuss this debate at several points in this chapter and the next.

Moving beyond research in economics, examples of IV estimation follow more varied patterns. For political science, Dunning (2012), Sekhon and Titiunik (2012), and Sovey and Green (2011) review the growing list of applications that draw on "natural experiments," some of which are instrumental variable designs. In sociology, Bollen (2012) shows that IV estimation is uncommon but still utilized in models with a structural equations foundation.[15] Finally, epidemiologists and health researchers have not utilized instrumental variables with substantial frequency, to some extent heeding the warning of Hernán and Robins (2006b:364) about the "risk [of] transforming the methodologic dream of avoiding unmeasured confounding into a nightmare of conflicting biased estimates."

9.2.3 The IV Identifying Assumption Cannot Be Tested

The basic identification assumption underlying all of the studies just summarized – that Z has no net association with Y except for the one generated by the directed path $Z \to D \to Y$ – is a strong and untestable assumption. Some researchers believe mistakenly that this assumption is empirically testable. In particular, they believe that the assumption that Z has no direct effect on Y implies that there is no association

---

[14]Evans and Schwab (1995) used a student's religious identification as an IV. This IV has proven to be rather unconvincing to many scholars (see Altonji, Elder, and Taber 2005a, 2005b) because the assumed exclusion restriction appears unreasonable. Neal (1997) also used this IV but only selectively in his analysis.

[15]As noted, IV estimation is an inherent part of structural equation modeling because point estimates of coefficients are not infrequently generated by IV-based order identification in this tradition of analysis. For example, Messner, Baumer, and Rosenfeld (2004) estimate a two-equation model, with communities as the unit of analysis, where (1) the effect of community-level social trust on the homicide rate is identified by an instrument for community-level subjective alienation and (2) the effect of community-level social activism on the homicide rate is identified by instruments for the community-level average amount of television watched and extreme political attitudes. This type of application relies quite heavily on the theoretical rationale for instrument selection, more so than for the "natural experiment" instruments prized in economics where the prima facie case for the independence of the instrument Z from causes of Y other than D is stronger.

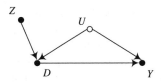

Figure 9.3 A graph with an unblocked back-door path and a valid IV.

between Z and Y conditional on D. Thus, if an association between Z and Y is detected after conditioning on D, then the assumption must be incorrect.

Causal graphs show clearly why individuals might believe that the identifying assumption can be tested in this way and also why it cannot. Consider the simplest possible causal graph, presented in Figure 9.3, in which (1) an unblocked back-door path between D and Y exists because of the unobserved common cause U, but (2) Z is a valid IV that identifies the causal effect of D on Y.[16] Here, the instrument Z is valid because it causes D, has no causal effect on Y other than through D, and is unconditionally unassociated with U.

Why is the suggested test faulty? Again, the rationale for the test is that conditioning on D will block the indirect relationship between Z and Y through D. Accordingly, if the only association between Z and Y is indirect through D, then it is thought that there should be no association between Z and Y after conditioning on D. If such a net association is detected, then it may seem reasonable to conclude that the IV identifying assumption must be false.

Although this rationale feels convincing, it is incorrect. It is certainly true that if the IV assumption is invalid, then Z and Y will be associated after conditioning on D. But the converse is not true. In fact, Z and Y will always be associated after conditioning on D when the IV assumption is valid. The explanation follows from the fact that D in Figure 9.3 is a collider that is mutually caused by both Z and U. As discussed extensively in this book, conditioning on a collider variable creates dependence between the variables that cause it. Accordingly, conditioning on D in this graph creates dependence between Z and U, even though the IV identifying assumption is valid. And, as a result, Z and Y will always be associated within at least one stratum of D even if the IV is valid. The faulty test yields an association between Z and Y when conditioning on D regardless of whether the IV identifying assumption is valid.[17]

---

[16] For our purposes here, this graph is isomorphic with Figure 9.1(a). We have replaced the bidirected edge from D to ε by representing the association between D and ε as a single back-door path between D and Y generated by an unobserved common cause U. We have then eliminated the remainder of the distribution of ε from the graph because it is unconditionally unassociated with Z and D (and is, in Pearl's framework, now the usual error term e_Y that comes into view only under magnification).

[17] For completeness, consider what the faulty test reveals when the IV assumption is invalid. Suppose that Figure 9.3 is augmented by an unobserved cause E and then two edges $E \rightarrow Z$ and $E \rightarrow U$. In this case, Z and Y would be associated within levels of D for two reasons: (1) conditioning on the collider D generates a net association between Z and U (and hence Z and Y) and (2) the common cause E of Z and U generates an unconditional association between Z and Y.

9.2.4   Recognized Pitfalls of Traditional IV Estimation

The traditional IV literature suggests that, as long as there is an instrument that predicts the causal variable of interest but is linearly unrelated to the outcome variable except by way of the causal variable, then an IV estimator will effectively estimate the causal effect. Even within this traditional setup, however, there are some recognized pitfalls of an IV estimation strategy. First, the assumption that an IV does not have a net direct effect on the outcome variable is often hard to defend. Second, even when an IV does not have a net direct effect on the outcome variable, IV estimators are biased in finite samples. Moreover, this bias can be substantial when an instrument only weakly predicts the causal variable. We discuss each of these weaknesses here.

Even "natural experiments" that generate compelling IVs are not immune from criticism. Consider date of birth as an instrument for veteran status in estimating the effect of military service in Vietnam on subsequent earnings (Angrist 1990). The case for excluding date of birth from the earnings equation that is the primary interest of the study is the randomization of the draft lottery, in which draft numbers were assigned at random to different dates of birth. Even though a type of randomization generates the IV, this does not necessarily imply that date of birth has no net direct effect on earnings. After the randomization occurs and the lottery outcomes are known, employers may behave differently with respect to individuals with different lottery numbers, investing more heavily in individuals who are less likely to be drafted. As a result, lottery number may be a direct, though probably weak, determinant of future earnings (see Heckman 1997; Moffitt 1996). IVs that are not generated by randomization are even more susceptible to causal narratives that challenge the assumption that the purported IV does not have a net direct effect on the outcome of interest.

Even in the absence of this complication, there are well-recognized statistical pitfalls. By using only a portion of the covariation in the causal variable and the outcome variable, IV estimators use only a portion of the information in the data. This represents a direct loss in statistical power, and as a result IV estimators tend to exhibit substantially more expected sampling variance than other estimators. By the criterion of mean-squared error, a consistent and asymptotically unbiased IV estimator can be outperformed by a biased and inconsistent regression estimator.

The problem can be especially acute in some cases. It has been shown that instruments that only weakly predict the causal variable of interest should be avoided entirely, even if they generate point estimates with acceptably small estimated standard errors (see Bound, Jaeger, and Baker 1995). In brief, the argument here is fourfold: (1) in finite samples, IV point estimates can always be computed because sample covariances are never exactly equal to zero; (2) as a result, an IV point estimate can be computed even for an instrument that is invalid because it does not predict the endogenous variable in the population (i.e., even if $\mathrm{Cov}(D, Z) = 0$ in the population, rendering Equation (9.13) undefined because its denominator is equal to 0); (3) at the same time, the formulas for calculating the standard errors of IV estimates fail in such situations, giving artificially small standard errors (when in fact the true standard error for the undefined parameter is infinity); and (4) the bias imparted by a small violation of the assumption that the IV affects the outcome variable only by way of the causal variable can explode if the instrument is weak.[18] To see this last result, consider Equation (9.6), which depicts the expected bias in the Wald estimator for a binary IV as the term

$$\frac{E[\varepsilon|Z=1] - E[\varepsilon|Z=0]}{E[D|Z=1] - E[D|Z=0]}. \tag{9.15}$$

When the identifying assumption is violated, the numerator of Equation (9.15) is nonzero because Z is associated with Y through ε. The bias is then an inverse function of the strength of the instrument; the weaker the instrument, the smaller the denominator and the larger the bias. If the denominator is close to zero, even a tiny violation of the identifying assumption can generate a large amount of bias. And, unfortunately, this relationship is independent of the sample size. Thus, even though a weak instrument may suggest a reasonable (or perhaps even intriguing) point estimate, and one with an acceptably small estimated standard error, the IV estimate may contain no genuine information whatsoever about the true causal effect of interest (see Hahn and Hausman 2003; Small and Rosenbaum 2008; Staiger and Stock 1997; Wooldridge 2010).[19]

Beyond these widely recognized pitfalls of standard IV estimation in economics, a third criticism is emerging of current practice, especially in economics. As explained by Angrist and Krueger (2001) and Angrist and Pischke (2009, 2010), the ways in which IVs are used has changed in the past 40 years. Because it has been hard to achieve consensus that particular no-net-direct-effect assumptions are credible, IVs that arise from naturally occurring variation have become more popular. Genuine "gifts of nature," such as variation in the weather and natural boundaries, have become the most prized sources of IVs (see Dunning 2012 and Rosenzweig and Wolpin 2000 for lists of such instruments).

Not all economists see this shift in IV estimation techniques toward naturally occurring IVs as necessarily a step in the right direction. Rosenzweig and Wolpin (2000) offer one of the most cogent critiques (but see also Deaton 2010; Heckman 2000, 2005; Heckman and Urzua 2010). Rosenzweig and Wolpin make three main points. First, the variation on which naturally occurring IVs capitalize is often poorly explained and/or does not reflect the variation that maintained theories posit should be important. As a result, IV estimates from natural experiments have a black-box character that lessens their appeal as informative estimates for the development of theory or policy guidance. Second, the random variation created by a naturally occurring experiment does not necessarily ensure that an IV has no net direct effect on the outcome. Other causes of the outcome can respond to the natural event that generates the IV as well. Third, naturally occurring IVs typically do not estimate structural parameters

---

[18]Complications (1), (2), (3), and (4) are all closely related. Situation (4) can be considered a less extreme version of the three-part predicament depicted in (1)–(3).

[19]There are no clear guidelines on how large an association between an IV and a treatment variable must be before analysis can proceed safely. Most everyone agrees that an IV is too weak if it does not yield a test statistic that rejects a null hypotheses of no association between Z and D. However, a main point of this literature is that the converse is not true. If a dataset is large enough, the small association between Z and D generated by a weak IV can still yield a test statistic that rejects a null hypotheses of no association. Even so, the problems in the main text are not vitiated, especially the explosion of the bias generated by a small violation of the identifying assumption.

of fundamental interest, which can and should be defined in advance of estimation based on criteria other than whether a naturally occurring IV is available. Rosenzweig and Wolpin contend that natural experiments tend to lead analysts to ignore these issues because natural experiments appear falsely infallible. Reflecting on the same set of issues, Deaton (2010:432) writes, "The general lesson is once again the ultimate futility of trying to avoid thinking about how and why things work."

We will explain these points further in this chapter, after introducing IV estimation in the presence of causal effect heterogeneity. (We will also then revisit this critique in the next chapter on mechanisms.) The new IV literature, to be discussed next, addresses complications of the constant coefficient assumption implied by the stipulated constant value of δ in Equations (9.1) and (9.9). The issues are similar to those presented for regression estimators in Chapter 6, in that heterogeneity invalidates traditional causal inference from IV estimates. But IV estimators do identify specific narrow slices of average causal effects that may be of distinct interest, and as a result they represent a narrowly targeted estimation strategy with considerable appeal.

9.3 Instrumental Variable Estimators in the Presence of Individual-Level Heterogeneity

Following the adoption of a counterfactual perspective, a group of econometricians and statisticians has clarified what IVs identify when individual-level causal effects are heterogeneous. In this section, we will emphasize the work that has developed the connections between traditional IV estimators and potential-outcome-defined treatment effects (Angrist and Imbens 1995; Angrist, Imbens, and Rubin 1996; Imbens and Angrist 1994; Imbens and Rubin 1997). The key innovation here is the definition of a new treatment effect parameter: the local average treatment effect (LATE). We will also discuss other important work that has further clarified these issues, and some of this literature is more general than the LATE literature that we introduce first (see Heckman 1996, 1997, 2000, 2010; Heckman, Tobias, and Vytlacil 2003; Heckman, Urzua, and Vytlacil 2006; Heckman and Vytlacil 1999, 2000, 2005; Manski 2003; Vytlacil 2002).[20]

9.3.1 IV Estimation as LATE Estimation

Consider the following motivation of the Wald estimator in Equation (9.5), and recall the definition of Y presented in Equation (4.5):

$$Y = Y^0 + (Y^1 - Y^0)D \tag{9.16}$$
$$= Y^0 + \delta D$$
$$= \mu^0 + \delta D + \upsilon^0,$$

---

[20]Given the rapidity of these developments in the IV literature, some disagreement on the origins of these ideas pervades the literature. Heckman and Robb (1985, 1986) did provide extensive analysis of what IV estimators identify in the presence of heterogeneity. Subsequent work by others, as we discuss in this section, has clarified and extended these ideas, even while Heckman and his colleagues continued to refine their ideas.

where $\mu^0 \equiv E[Y^0]$ and $\upsilon^0 \equiv Y^0 - E[Y^0]$. Note that δ is now defined as $Y^1 - Y^0$, unlike its structural representation in Equations (9.1) and (9.9) where δ was implicitly assumed to be constant across all individuals.

To understand when an IV estimator can be interpreted as an average causal effect estimator, Imbens and Angrist (1994) developed a framework to classify individuals into those who respond positively to an instrument, those who remain unaffected by an instrument, and those who rebel against an instrument. Their innovation was to define potential treatment assignment variables, $D^{Z=z}$, for each state z of the instrument Z. When D and Z are binary variables, there are four possible groups of individuals in the population.[21] These can be summarized by a four-category latent variable C for compliance status:

$$\text{Compliers } (C=c): D^{Z=0} = 0 \text{ and } D^{Z=1} = 1,$$
$$\text{Defiers } (C=d): D^{Z=0} = 1 \text{ and } D^{Z=1} = 0,$$
$$\text{Always takers } (C=a): D^{Z=0} = 1 \text{ and } D^{Z=1} = 1,$$
$$\text{Never takers } (C=n): D^{Z=0} = 0 \text{ and } D^{Z=1} = 0.$$

Consider the private schooling example presented earlier in IV Demonstration 1 (see page 294). Students who would enroll in private schools only if offered the voucher are *compliers* $(C=c)$. Students who would enroll in private schools only if not offered the voucher are *defiers* $(C=d)$. Students who would always enroll in private schools, regardless of whether they are offered the voucher, are *always takers* $(C=a)$. And, finally, students who would never enroll in private schools are *never takers* $(C=n)$.

Analogous to the definition of the observed outcome, Y, the observed treatment indicator variable D can then be defined as

$$D = D^{Z=0} + (D^{Z=1} - D^{Z=0})Z \qquad (9.17)$$
$$= D^{Z=0} + \kappa Z,$$

where $\kappa \equiv D^{Z=1} - D^{Z=0}$.[22] The parameter κ in Equation (9.17) is the individual-level causal effect of the instrument on D, and it varies across individuals if $D^{Z=1} - D^{Z=0}$ varies across individuals (i.e., if observed treatment status varies as the instrument is switched from "off" to "on" for each individual). If the instrument represents encouragement to take the treatment, such as the randomly assigned school voucher in IV Demonstration 1, then κ can be interpreted as the individual-level compliance inducement effect of the instrument. Accordingly, $\kappa = 1$ for compliers and $\kappa = -1$ for defiers. For always takers and never takers, $\kappa = 0$ because none of these individuals respond to the instrument.

---

[21] These four groups are considered principal strata in the framework of Frangakis and Rubin (2002).

[22] Note also that D has now been counterfactually defined with reference to D^Z. Accordingly, the definition of Y in Equation (9.16) is conditional on the definition of D in Equation (9.17). Furthermore, although there is little benefit in doing so, it bears noting that this definition of D could be structured analogously to the definition of Y in Equation (9.16) so that the last line would become $D = \zeta + \kappa Z + \iota$, where $\zeta \equiv E[D^{Z=0}]$ and $\iota \equiv D^{Z=0} - E[D^{Z=0}]$. The parameter ζ would then be the expected probability of being in the treatment if all individuals were assigned to the state "instrument switched off," and ι would be the individual-level departure from this expected value, taking on values $1 - E[D^{Z=0}]$ and $-E[D^{Z=0}]$ to balance the right-hand side of Equation (9.17) so that D is equal to either 1 or 0.

Given these definitions of potential outcome variables and potential treatment variables, a valid instrument Z for the causal effect of D on Y must satisfy three assumptions in order to identify a LATE:

$$\text{Independence assumption: } (Y^1, Y^0, D^{Z=1}, D^{Z=0}) \perp\!\!\!\perp Z, \qquad (9.18)$$

$$\text{Nonzero effect of instrument assumption: } \kappa \neq 0 \text{ for all } i, \qquad (9.19)$$

$$\text{Monotonicity assumption: either } \kappa \geq 0 \text{ for all } i \text{ or } \kappa \leq 0 \text{ for all } i. \qquad (9.20)$$

The independence assumption in Equation (9.18) is analogous to the assumption that $\text{Cov}(Z, \varepsilon) = 0$ in the traditional IV literature; see the earlier discussion of Equation (9.13).[23] It stipulates that the instrument must be independent of the potential outcomes and potential treatments. Knowing the value of the instrument for individual i must not yield any information about the potential outcome of individual i under either treatment state. Moreover, knowing the realized value of the instrument for individual i must not yield any information about the probability of being in the treatment under alternative hypothetical values of the instrument. This latter point may well appear confusing, but it is exactly analogous to the independence assumption of potential outcomes from observed treatment status, discussed earlier for Equation (2.6). A valid instrument predicts observed treatment status (D), but it does not predict potential treatment status ($D^{Z=z}$).

The assumptions in Equations (9.19) and (9.20) are assumptions about individual responses to shifts in the instrument. The assumption of a nonzero effect of Z on D is a stipulation that the instrument must predict treatment assignment for at least some individuals. There must be at least some compliers or some defiers in the population of interest. The monotonicity assumption then further specifies that the effect of Z on D must be either weakly positive or weakly negative for all individuals i. Thus, there may be either defiers or compliers in the population but not both.[24]

If these three assumptions obtain, then an instrument Z identifies the LATE: the average treatment effect for the subset of the population whose treatment selection is induced by the instrument.[25] If $\kappa \geq 0$ for all i, then the Wald estimator from Equation

---

[23] The stable unit treatment value assumption (SUTVA) must continue to hold, and now it must apply to potential treatments as well. In addition, as we noted earlier, our presentation here follows Imbens and Angrist (1994), Angrist and Imbens (1995), and Angrist et al. (1996). As we note later, IVs can be defined in slightly different (in some cases more general) ways. But for now, we restrict attention to the LATE literature, in which assumptions such as complete independence of the instrument are utilized.

[24] Manski defines this assumption as a monotone treatment selection (MTS) assumption in order to distinguish it from his monotone treatment response (MTR) assumption (see Manski 1997; Manski and Pepper 2000). Vytlacil (2002) establishes its connections to the index structure model laid out in Heckman and Vytlacil (1999), as we discuss in Section 9.3.4. Heckman (2000, 2010), Heckman and Vytlacil (2005, 2007), Heckman et al. (2006), and Heckman and Urzua (2010) provide a broad accounting of the relationships between alternative IV estimators.

[25] The LATE is often referred to as the "complier average causal effect" to signify that the descriptor local is really defined by the restricted applicability of the LATE estimate to the average causal effect of compliers. The following example does not use the compliance language explicitly, except insofar as those who respond to the voucher are labeled compliers. The LATE literature that we cite provides the explicit connections between LATE and the large literature on noncompliance in randomized experiments (see in particular Imbens and Rubin 1997 and citations therein).

(9.5) converges to a particular LATE:

$$\hat{\delta}_{\text{IV},\text{WALD}} \xrightarrow{p} E[\delta|C=c], \qquad (9.21)$$

which is equal to $E[Y^1 - Y^0|D^{Z=1}=1, D^{Z=0}=0]$ and is therefore the average treatment effect among compliers. In contrast, if $\kappa \leq 0$ for all i, then the Wald estimator from Equation (9.5) converges to the opposite LATE:

$$\hat{\delta}_{\text{IV},\text{WALD}} \xrightarrow{p} E[\delta|C=d], \qquad (9.22)$$

which is equal to $E[Y^1 - Y^0|D^{Z=1}=0, D^{Z=0}=1]$ and is therefore the average treatment effect among defiers. In either case, the treatment effects of always takers and never takers are not informed in any way by the IV estimate.

In the next section, we will explain the claims in Equations (9.21) and (9.22) in considerable detail through a more elaborate version of IV Demonstration 1. But the basic intuition is straightforward. A valid IV is nothing more than an exogenous dimension across which the treatment and outcome variables are analyzed jointly. For a binary instrument, this dimension is a simple contrast, which is the ratio presented earlier; see Equations (9.5) and (9.8):

$$\frac{E_N[y_i|z_i=1] - E_N[y_i|z_i=0]}{E_N[d_i|z_i=1] - E_N[d_i|z_i=0]}.$$

The numerator is the naive estimate of the effect of Z on Y, and the denominator is the naive estimate of the effect of Z on D.

To give a causal interpretation to this ratio of differences across the third dimension indexed by Z, a model of individual treatment response must be specified and then used to interpret the causal effect estimate. The model of individual response adopted for a LATE analysis with a binary IV is the fourfold typology of compliance, captured by the latent variable C defined earlier for always takers, never takers, compliers, and defiers. Within this model, the always takers and never takers do not respond to the instrument (i.e., they did not choose to take part in the "experiment" created by the IV, and thus their treatment assignment is not determined by Z). This means that they are distributed in the same proportion within alternative values of the instrument Z. And, as a result, differences in the average value of Y, when examined across Z, are not a function of the outcomes of always takers and never takers.[26]

In contrast, defiers and compliers contribute all of the variation that generates the IV estimate because only their behavior is responsive to the instrument. For this reason, any differences in the average value of Y, when examined across Z, must result from treatment effects for those who move into and out of the causal states represented by D. If compliers are present but defiers are not, then the estimate offered by the ratio is interpretable as the average treatment effect for compliers. If defiers are present but compliers are not, then the estimate offered by the ratio is interpretable as the average treatment effect for defiers. If both compliers and defiers are present, then the estimate generated by the ratio does not have a well-defined causal interpretation. In the following demonstration, we will consider the most common case in the

---

[26]In a sense, the outcomes of always takers and never takers represent a type of background noise that is ignored by the IV estimator. More precisely, always takers and never takers have a distribution of outcomes, but the distribution of these outcomes is balanced across the values of the instrument.

LATE literature for which the monotonicity condition holds in the direction such that compliers exist in the population but defiers do not.

IV Demonstration 2

Recall the setup for IV Demonstration 1 (see page 294). In an attempt to determine whether private high schools outperform public high schools, a state education department assembles a dataset on a random sample of $10,000$ ninth graders, $\{y_i, d_i, z_i\}_{i=1}^{10,000}$, where Y is a standardized test, and D is equal to 1 for students who attend private high schools and 0 for students who attend public high schools. Likewise, Z is equal to 1 for those who win a lottery for a \$3,000 school voucher and 0 for those who do not.

As noted for IV Demonstration 1, a bivariate regression of the values of y_i on d_i yielded a treatment effect estimate of 9.67; see discussion of Equation (9.7). An alternative IV estimate with Z as an instrument for D yielded an estimate of 5.5; see Equation (9.8). The hypothetical state officials relied on the IV estimate rather than on the regression estimate because they recognized that private school students have more advantaged social backgrounds. And they assumed that the randomization of the voucher lottery, in combination with the relationship between D and Z shown in Table 9.1, established the voucher lottery outcome as a valid IV.

We stated just after our discussion of IV Demonstration 1 that the estimate of 5.5 is properly interpreted as a particular LATE: the average causal effect for the subset of all students who would attend a private school if given a voucher but would not attend a private school in the absence of a voucher. To explain the reasoning behind this conclusion, we will first explain how the observed values for D and Y are generated as a consequence of variation in Z. Then we will introduce potential outcomes and use the treatment response model in order to explain why the IV estimate is interpretable as the average treatment effect for compliers.

We reported the frequency distribution of D and Z in Table 9.1. The same information is presented again in Table 9.2, but now also as probability values (where, for example, the term $\Pr_N[.,.]$ in the upper left cell is equal to $\Pr_N[d_i = 0, z_i = 0]$ by plugging the row and column headings into the joint probability statement). We also now report the expectations of the outcome variable Y, conditional on D and Z.

First, consider how the least squares and IV estimates are calculated. The coefficient of 9.67 on D in Equation (9.7) is equal to the naive estimator, $E_N[y_i|d_i = 1] - E_N[y_i|d_i = 0]$. One can calculate these two conditional expectations from the elements of Table 9.2 by forming weighted averages within columns:

$$E_N[y_i|d_i = 1] = \frac{.1}{.1 + .02}60 + \frac{.02}{.1 + .02}58 = 59.667,$$

$$E_N[y_i|d_i = 0] = \frac{.8}{.8 + .08}50 + \frac{.08}{.8 + .08}50 = 50.0.$$

As shown earlier in Equation (9.8), the IV estimate of 5.5 is the ratio of two specific contrasts:

$$\frac{E_N[y_i|z_i = 1] - E_N[y_i|z_i = 0]}{E_N[d_i|z_i = 1] - E_N[d_i|z_i = 0]} = \frac{51.6 - 51.111}{.2 - .111} = 5.5. \tag{9.23}$$

Table 9.2 The Joint Probability Distribution and
Conditional Expectations of the Test Score for Voucher
Winner by School Sector for IV Demonstrations 1 and 2

		Public school $d_i = 0$	Private school $d_i = 1$
Voucher loser	$z_i = 0$	$N = 8000$	$N = 1000$
		$\Pr_N [.,.] = .8$	$\Pr_N [.,.] = .1$
		$E_N[y_i|.,.] = 50$	$E_N[y_i|.,.] = 60$
Voucher winner	$z_i = 1$	$N = 800$	$N = 200$
		$\Pr_N [.,.] = .08$	$\Pr_N [.,.] = .02$
		$E_N[y_i|.,.] = 50$	$E_N[y_i|.,.] = 58$

Both of these contrasts are calculated within the rows of Table 9.2 rather than the columns. The contrast in the numerator is the naive estimate of the effect of Z on Y. It is calculated as the difference between

$$E_N[y_i|z_i = 1] = \frac{.08}{.08 + .02}50 + \frac{.02}{.08 + .02}58 = 51.6 \quad \text{and}$$

$$E_N[y_i|z_i = 0] = \frac{.8}{.8 + .1}50 + \frac{.1}{.8 + .1}60 = 51.111.$$

The contrast in the denominator is the naive estimate of the effect of Z on D. It is calculated as the difference between

$$E_N[d_i = 1|z_i = 1] = \frac{.02}{.08 + .02} = .2 \quad \text{and}$$

$$E_N[d_i = 1|z_i = 0] = \frac{.1}{.8 + .1} = .111.$$

Thus, calculating the IV estimate in this case is quite simple and does not require any consideration of the underlying potential outcomes or potential treatments.

But, to interpret the IV estimate when causal effect heterogeneity is present, the potential outcome and potential treatment framework are needed. Consider the three identifying assumptions in Equations (9.18) through (9.20). Because the voucher lottery is completely random, the voucher instrument Z is independent of the potential outcome and potential treatment variables. Also, as just shown, Z predicts D, thereby sustaining the nonzero effect assumption. Thus, the first two assumptions are satisfied as explained for IV Demonstration 1.

Only the monotonicity assumption in Equation (9.20) requires a new justification. Fortunately, there is no evidence in the school choice literature that students and their parents rebel against vouchers, changing their behavior to avoid private schooling only

Table 9.3 The Distribution of Never Takers, Compliers, and Always Takers for IV Demonstration 2

		Public school $d_i=0$	Private school $d_i=1$
Voucher loser	$z_i=0$	7200 Never takers 800 Compliers	1000 Always takers
Voucher winner	$z_i=1$	800 Never takers	111 Always takers 89 Compliers

when offered a voucher.[27] Accordingly, it seems reasonable to assume that there are no defiers in this hypothetical population and hence that the monotonicity assumption obtains.

The joint implications of independence and monotonicity for the four groups of individuals in the cells of Table 9.2 should be clear. Monotonicity allows us to stipulate that defiers do not exist, while independence ensures that the same distribution of never takers, always takers, and compliers is present among those who win the voucher lottery and those who do not. As a result, the proportion of always takers can be estimated consistently from the first row of Table 9.2 and the proportion of never takers can be estimated consistently from the second row of Table 9.2 as

$$\frac{\Pr_N[d_i=1, z_i=0]}{\Pr_N[z_i=0]} \xrightarrow{p} \Pr[C=a], \tag{9.24}$$

$$\frac{\Pr_N[d_i=0, z_i=1]}{\Pr_N[z_i=1]} \xrightarrow{p} \Pr[C=n]. \tag{9.25}$$

For this example, the proportion of always takers is $1,000/9,000 = .111$, and the proportion of never takers is $800/1,000 = .8$. Because no defiers exist in the population with regard to this instrument, these two estimated proportions can be subtracted from 1 in order to obtain the proportion of compliers:

$$1 - \frac{\Pr_N[d_i=1, z_i=0]}{\Pr_N[z_i=0]} - \frac{\Pr_N[d_i=0, z_i=1]}{\Pr_N[z_i=1]} \xrightarrow{p} \Pr[C=c]. \tag{9.26}$$

For this example, the proportion of compliers in the population is $1 - .111 - .8 = .089$. Applying this distribution of always takers, never takers, and compliers (and a bit of rounding) to the frequencies from Table 9.2 yields the joint frequency distribution presented in Table 9.3.

For Table 9.3, notice the symmetry across rows that is generated by the independence of the instrument: $1,000/7,200 \approx 111/800$ (subject to rounding) and $800/9,000 \approx 89/1,000$ (again, subject to rounding). Of course, there is one major difference between

---

[27]Some parents might reason that private schooling is no longer as attractive if it is to be flooded with an army of voucher-funded children. Thus, defiers might emerge if the vouchers were widely distributed, and the monotonicity condition would then fail. In this case, however, SUTVA would also fail, necessitating deeper analysis in any case.

the two rows: The compliers are in private schools among voucher winners but in public schools among voucher losers.

Before continuing, two important points should be noted. First, it is important to recognize that the calculations that give rise to the distribution of never takers, always takers, and compliers in Table 9.3 are not determined solely by the data. In a deeper sense, they are entailed by maintenance of the monotonicity assumption. In the absence of that assumption, an unspecified number of defiers would be in the table as well, making the calculation of these proportions impossible.

Second, not all students in the dataset can be individually identified as always takers, never takers, or compliers. Consider the private school students for this example. Of these 1,200 students, 1,000 students are known to be always takers, as they have observed values of $d_i = 1$ and $z_i = 0$. The remaining 200 private school students are observationally equivalent, with observed values of $d_i = 1$ and $z_i = 1$. We know, based on the maintenance of the monotonicity and independence assumptions, that these 200 students include 111 always takers and 89 compliers. But it is impossible to determine which of these 200 students are among the 111 always takers and which are among the 89 compliers. The same pattern prevails for public school students. Here, we can definitively identify 800 students as never takers, but the 8,000 public school students who are voucher losers cannot be definitively partitioned into the specific 7,200 never takers and the 800 compliers.

Now, consider why the IV estimator yields 5.5 for this example, and then why 5.5 is interpretable as the average effect of private schooling for those who are induced to enroll in private schools because they have won vouchers. As noted already, because Z is independent of Y^1 and Y^0, the same proportion of always takers and never takers is present among both voucher winners and voucher losers. The difference in the expectation of Y across the two rows of Table 9.2 must arise from (1) the existence of compliers only in public schools in the first row and only in private schools in the second row and (2) the existence of a nonzero average treatment effect for compliers.

To see this claim more formally, recall Equation (9.21), in which this particular LATE is defined. By the linearity of expectations and the definition of an individual-level causal effect as a linear difference between y_i^1 and y_i^0, the average causal effect among compliers, $E[\delta|C=c]$, is equal to the expectation $E[Y^1|C=c]$ minus the expectation $E[Y^0|C=c]$. To obtain a consistent estimate of the average causal effect for compliers, it is sufficient to obtain consistent estimates of $E[Y^1|C=c]$ and $E[Y^0|C=c]$ separately and then to subtract the latter from the former.

Fortunately, this strategy is feasible because the contribution of these two conditional expectations to the observed data can be written out in two equations and then solved. In particular, $E[Y^1|C=c]$ and $E[Y^0|C=c]$ contribute to the expectations of the observed outcome Y, conditional on D and Z, in the following two equations:

$$E[Y|D=1, Z=1] = \frac{\Pr[C=c]}{\Pr[C=c]+\Pr[C=a]} E[Y^1|C=c]$$

$$+ \frac{\Pr[C=a]}{\Pr[C=c]+\Pr[C=a]} E[Y^1|C=a],$$

(9.27)

$$E[Y|D=0,Z=0] = \frac{\Pr[C=c]}{\Pr[C=c]+\Pr[C=n]} E[Y^0|C=c]$$

$$+ \frac{\Pr[C=n]}{\Pr[C=c]+\Pr[C=n]} E[Y^0|C=n]. \tag{9.28}$$

These two equations are population-level decompositions of the conditional expectations for the observed data that correspond to the two cells of the diagonal of Table 9.2. These are the only two cells in which compliers are present, and thus the only two cells in which the observed data are affected by the outcomes of compliers.

How can we plug values into Equations (9.27) and (9.28) in order to solve for $E[Y^1|C=c]$ and $E[Y^0|C=c]$ and thereby obtain all of the ingredients of a consistent estimate of $E[\delta|C=c]$? We have already shown from applying the convergence assertions in Equations (9.24)–(9.26) that the terms $\Pr[C=c]$, $\Pr[C=a]$, and $\Pr[C=n]$ can be consistently estimated. And, in fact, these are given earlier for the example data as .089, .8, and .111. Thus, to solve these equations for $E[Y^1|C=c]$ and $E[Y^0|C=c]$, the only remaining pieces that need to be estimated are $E[Y^1|C=a]$ and $E[Y^0|C=n]$, which are the average outcome under the treatment for the always takers and the average outcome under the control for the never takers. Fortunately, the independence and monotonicity assumptions guarantee that voucher losers in private schools represent a random sample of always takers. Thus, $E[Y^1|C=a]$ is estimated consistently by $E_N[y_i|d_i=1,z_i=0]$, which is 60 for this example (see the upper right-hand cell in Table 9.2). Similarly, because voucher winners in public schools represent a random sample of never takers, $E[Y^0|C=n]$ is estimated consistently by $E_N[y_i|d_i=0,z_i=1]$, which is equal to 50 for this example (see the lower left-hand cell in Table 9.2). Plugging all of these values into Equations (9.27) and (9.28) then yields

$$58 = \frac{.089}{.089+.111} E[Y^1|C=c] + \frac{.111}{.089+.111} 60, \tag{9.29}$$

$$50 = \frac{.089}{.089+.8} E[Y^0|C=c] + \frac{.8}{.089+.8} 50. \tag{9.30}$$

Solving Equation (9.29) for $E[Y^1|C=c]$ results in 55.5, whereas solving Equation (9.30) for $E[Y^0|C=c]$ results in 50. The difference between these values is 5.5, which is the average causal effect for the subset of all students who would attend a private school if given a voucher but would not attend a private school in the absence of a voucher.[28] The value of 5.5 yields no information whatsoever about the effect of private schooling for the always takers and the never takers.[29]

---

[28]For completeness, consider how the naive estimate and the LATE would differ if all remained the same except the stipulated value of 50 for $E_N[d_i=0,z_i=0]$ in Table 9.2. If $E_N[d_i=0,z_i=0]$ were instead 50.25, then the naive estimate would be 9.44 and the LATE estimate would be 3.00. And, if $E_N[d_i=0,z_i=0]$ were instead 50.5, then the naive estimate would be 9.21 and the LATE estimate would be .5. Thus, for the example in the main text, compliers on average do no worse in public schools than never takers. But, for these two variants of the example, compliers on average do slightly better in public schools than never takers. As a result, the calculations in Equation (9.29) remain the same, but the values of Equation (9.30) change such that $E[Y^0|C=c]$ is equal to 52.5 and 55.0, respectively. The LATE estimate is therefore smaller in both cases because the performance of compliers in public schools is higher (whereas the performance of compliers in private schools remains the same).

[29]We can estimate $E[Y^1|C=a]$ and $E[Y^0|C=n]$ consistently with $E_N[y_i|d_i=1,z_i=0]$ and $E_N[y_i|d_i=0,z_i=1]$. But we have no way to effectively estimate their counterfactual analogs: the

Of course, the Wald estimate is also 5.5, as shown in Equations (9.8) and (9.23). And thus, in one sense, the Wald estimator can be thought of as a quick alternative method for calculating all of the steps just presented to solve exactly for $E[\delta|C = c]$. Even so, this correspondence does not explain when and how the Wald estimator can be interpreted as the average causal effect for compliers. For this example, the correspondence arises precisely because we have assumed that there are no defiers in the population, based on the substance of the application and the treatment response model that we adopted. As a result, the Wald estimate can be interpreted as a consistent estimate of the average effect of private schooling for those who comply with the instrument because it is equal to that value under the assumed model of treatment response we are willing to adopt.[30]

LATE estimators have been criticized because the identified effect is defined by the instrument under consideration. As a result, different instruments define different average treatment effects for the same group of treated individuals. And, when this is possible, the meanings of the labels for the latent compliance variable C depend on the instrument, such that some individuals can be never takers for one instrument and compliers for another. Deaton writes:

> Without explicit prior consideration of the effect of the instrument choice on the parameter being estimated, such a procedure is effectively the opposite of standard statistical practice in which a parameter of interest is defined first, followed by an estimator that delivers that parameter. Instead, we have a procedure in which the choice of the instrument ... is implicitly allowed to determine the parameter of interest. This goes beyond the old story of looking for an object where the light is strong enough to see; rather, we have at least some control over the light but choose to let if fall where it may and then proclaim that whatever it illuminates is what we were looking for all along. (Deaton 2010:429)

Although from one perspective the instrument-dependent nature of the LATE is a weakness, from another perspective it is the most attractive feature of the LATE. For IV Demonstration 2, the IV estimate does not provide any information about the average effect for individuals who would attend private schooling anyway (i.e., the always takers) or for those who would still not attend the private schools if given a voucher (i.e., the never takers). Instead, the IV estimate is an estimate of a narrowly defined average effect only among those induced to take the treatment by the voucher policy intervention. But, for IV Demonstration 2, this is precisely what should be of interest to the state officials. If the policy question is "What is the effect of vouchers

---

mean outcome in public schools for always takers and the mean outcome in privates schools for never takers.

[30]In other words, the Wald estimate of 5.5 is also a quick method for calculating an entirely different causal effect under a different set of assumptions. If monotonicity cannot be defended, then the IV estimate can be given a traditional structural interpretation, under the assumption that the causal effect is constant for all individuals. In this sense, because the Wald estimate has more than one possible causal interpretation, merely understanding how it is calculated does not furnish an explanation for how it can be interpreted.

on school performance?" then they presumably care most about the average effect for compliers.

The limited power of the LATE interpretation of an IV estimate is thus, in some contexts, beneficial because of its targeted clarity. Moreover, when supplemented by a range of additional IV estimates (i.e., different voucher sizes and so on), complementary LATE estimates may collectively represent an extremely useful set of parameters that describe variation in the causal effect of interest for different groups of individuals exposed to the cause for alternative (but related) reasons. Before summarizing the marginal treatment effect literature that more completely specifies the interrelationships among all types of average causal effect estimators, we first lay out the implications of the LATE perspective for traditional IV estimation and consider the relevant graphs for representing IVs that identify LATEs.

9.3.2 Implications of the LATE Perspective for Traditional IV Estimation

The LATE literature specifies a set of assumptions under which it is permissible to give IV estimates an average causal effect interpretation using the potential outcome model. In this sense, the new framework is mostly a set of guidelines for how to interpret IV estimates. As such, the LATE perspective has direct implications for traditional IV estimation, as introduced earlier in this chapter.

Monotonicity and Assumptions of Homogeneous Response

An important implication of the LATE framework is that many conventional IV estimates lack a justifiable average causal effect interpretation if the IV does not satisfy a monotonicity condition. In the presence of causal effect heterogeneity and in the absence of monotonicity of response to the instrument, a conventional IV estimator yields a parameter estimate that has no clear interpretation, as it is likely an unidentifiable mixture of the treatment effects of compliers and defiers.

For IV Demonstration 2, we showed that the estimate of 5.5 is applicable to students whose families would change their child's enrollment choice from a public school to a private school for a \$3,000 voucher. Can an assumption be introduced that allows the estimate of 5.5 to be be interpreted as informative about other students who do (or who would) attend private schools?

Two variants of the same homogeneity assumption allow for such extrapolated inference: the assumption that the causal effect is a structural effect that is (1) constant across all members of the population or (2) constant across all members of the population who typically take the treatment.[31] In its stronger form (1), the assumption simply asserts that the causal effect estimate is equally valid for all members of the population, regardless of whether or not the group of students whose enrollment status would change in response to the voucher is representative of the population of students as a whole. In its weaker form (2), the assumption pushes the assumed constancy of the effect only half as far, stipulating that the IV estimate is valid as an estimate

---

[31] These assumptions are known as constant-coefficient assumptions, homogeneous response assumptions, or shifted outcome assumptions (see Angrist and Pischke 2009; Manski 1995, 2003).

of the ATT only. For IV Demonstration 2, the weaker variant of the homogeneity assumption is equivalent to asserting that the IV estimate provides information only about the achievement gains obtained by private school students. Although weaker, the homogeneity assumption (2) is still quite strong, in that all individuals in private schools are considered homogeneous with respect to the size of the treatment effect. In examples such as IV Demonstration 2, it is clear that there are two distinct groups within the treated: always takers and compliers. And there is little reason to expect that both groups respond in exactly the same way to private schooling. Thus, for examples such as this one, Manski (1995:44) argues that this homogeneity assumption "strains credibility" because there is almost certainly patterned heterogeneity in the effect among treated individuals.

One traditional way to bolster a homogeneity assumption is to condition on variables in a vector X that can account for all such heterogeneity and then assert a conditional homogeneity of response assumption. To do so, the Wald estimator must be abandoned in favor of a two-stage least squares (2SLS) estimator. As shown in any econometrics textbook (e.g., Greene 2000; Wooldridge 2010), the endogenous regressors D and X are embedded in an encompassing $\mathbf{X}$ matrix, which is $n \times k$, where n is the number of respondents and k is the number of variables in X plus 2 (one for the constant and one for the treatment variable D). Then, a matrix $\mathbf{Z}$ is constructed that is equivalent to X, except that the column in $\mathbf{X}$ that includes the treatment variable D is replaced with its instrument Z. The 2SLS estimator is then

$$\hat{\delta}_{\text{IV,2SLS}} \equiv (\mathbf{Z}'\mathbf{X})^{-1}\mathbf{Z}'\mathbf{y}, \qquad (9.31)$$

where $\mathbf{y}$ is an $n \times 1$ vector containing the outcomes y_i. The strategy is to attempt to condition out all of the systematic variability in the observed response and then to simultaneously use the instrument Z to identify a pure net structural effect that can be regarded as an invariant constant.

Is this strategy a feasible solution? Probably not. If good measures of all of the necessary variables in X are available, a simple OLS estimator probably would have been feasible in the first place. Rarely would all possible necessary variables be available, except for a single variable that has a net additive constant effect on the outcome that can then be estimated consistently by an available IV.

Other Challenges for Interpretation

IV estimates are hard, and sometimes impossible, to interpret as LATE estimates when the instrument measures something other than an incentive to which individuals can consciously respond by complying or defying. Instruments based on exogenous field variation (as championed in Angrist and Krueger 2001 but criticized in Rosenzweig and Wolpin 2000 and Deaton 2010) can be particularly hard to interpret, because the shifts in costs and benefits that the natural variation is supposed to induce generally remain unspecified, thereby weakening a main link in the narrative that explains why some individuals take the treatment in response to the instrument.

Moreover, if two or more IVs are available, then the traditional econometric literature suggests that they should both be used to "overidentify" the model and obtain a more precise treatment effect estimate by the 2SLS estimator in Equation (9.31).

Overidentified models, in which more than one instrument is used to identify the same treatment effect, generate a mixture-of-LATEs challenge.

Consider the Catholic school example discussed earlier. Suppose that two IVs are used: the share of the county that identifies as Catholic and a student's religious identification (as in Neal 1997). Even if these potential IVs have no net direct effects on test scores (and, further, that the weak instrument problem discussed earlier is not applicable), can a theoretically meaningful LATE interpretation be given to the effect that the two instruments in this example jointly identify? For the religious identification instrument, the implied LATE is the average effect of Catholic schooling among those who are likely to attend Catholic schools only because they are Catholic. When overidentified with the IV represented by the share of the local population that is Catholic, this first LATE is mixed in with a second LATE: the average effect of Catholic schooling among those who attend Catholic schools only because of the small difference in tuition that a high proportion of Catholics in the local population tends to generate. As a result, the overidentified causal effect that is estimated by the 2SLS estimator would be an average across two very different groups of hypothetical individuals, both of which likely deserve separate attention.[32]

It is sometimes possible to deconstruct 2SLS estimates into component LATEs when multiple IVs are used, but explaining how to do so is beyond the scope of this book. We advise readers who are in this situation to first consult Angrist and Pischke (2009, section 4.5) and the work cited therein. Our position is that there is comparatively little value in estimating a single treatment effect parameter using a 2SLS model in these cases. Usually, if more than one LATE is identified, then these can and should be estimated separately.

In sum, if causal effect heterogeneity is present, then a constant-coefficient interpretation of an IV estimate is implausible. However, if the instrument satisfies a monotonicity condition and can be conceptualized as a proximate inducement to take the treatment, then IV estimates can be given LATE interpretations. These fine-grained interpretations can be very illuminating about particular groups of individuals, even though they may provide no information whatsoever about other groups of individuals in the population (including other individuals who typically choose to take the treatment). Thus, the limited nature of IV estimators when interpreted as LATE estimators shows both the potential strengths and weaknesses of IV estimation in general.

9.3.3 Graphs for IVs That Identify LATEs

In this section, we will explain how to represent instrumental variables that identify LATEs as observed variables in directed graphs. The primary complication is that compliance with the instrumental variable is a latent class variable, across which it

---

[32]There is also the possibility that, contra the justification of Hoxby (1996), the individuals who are induced to attend Catholic schooling because a greater share of the population is Catholic are not at all the same as those who are supposedly at the margin of a tuition-based cost calculation. There may be another mechanism that generates any observed association between Z and D, such as the possibility that the Catholic schools are simply better in these communities and hence are more attractive in general. This is one basic criticism that Rosenzweig and Wolpin (2000) level against all such naturally occurring IVs: There is often no evidence that the IV is inducing a group of individuals to take the treatment according to an assumed cost-benefit set of choices.

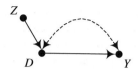

Figure 9.4 Instrumental variable identification of the causal effect of charter schools (D) on test scores (Y), where Z is the instrument.

must be assumed that heterogeneity of effects is present. Recall the charter schools example, as introduced in Section 1.3.2 and then as analyzed at length in Section 8.3. In that prior discussion, we showed how causal graphs can be used to represent complex patterns of self-selection and heterogeneity using a latent class variable. In particular, reconsider Figure 8.6(b), which we used to explain why back-door conditioning for the effect of charter schooling D on educational achievement Y was infeasible because of back-door paths through G.[33]

For simplicity, suppose now that the parental background confounder P is also unobserved. As a result, the analyst is left with no way to even begin to enact a back-door conditioning strategy for the effect of charter schools. Suppose, instead, that an instrumental variable Z is observed, as in Figure 9.4.

Are there any plausible instrumental variables for charter school attendance? The geographic distance between the student's residence and the charter school site is similar to the sorts of potential IVs used in the traditional economics literature. The rationale would be that the location of the treatment site is arbitrary but has an effect on enrollment propensity because of the implicit costs of traveling to the site, which are typically borne by the parents, not by the school district. For the charter school example, it is unclear whether such an instrument would have any chance of satisfying the relevant assumptions, and it would depend crucially on the extent to which charter schools are located in arbitrary places. Our reading of the literature is that charter schools tend to be located nearer to students most likely to benefit from charter schooling, both because many have missions to serve disadvantaged students who are thought to benefit most from having a charter school opportunity and also because parents may then move to neighborhoods that are closer to the charter schools that they select for their children. It is possible that these threats to the identifying assumption could be mitigated by conditioning on other determinants of the location of charter schools within the district and also obtaining family residence data before students entered charter schools.

For the sake of methodological clarity in our presentation, we will use as our example a more convincing but unlikely instrumental variable, in the sense that it has never

---

[33]For a real example with a structure that is not too dissimilar from ours, Jin and Rubin (2009) eschew graphs and adopt an alternative principal stratification approach to represent latent classes for types of compliance as well as average effects within these classes. By omission, they demonstrate that graphs are not needed in order to offer a sensible representation of underlying heterogeneity and to focus on compliance types of particular interest. The utility of the principal stratification approach for compliance latent classes, first laid out in Frangakis and Rubin (2002), is unrelated to a more recent debate on its utility for interpreting direct and indirect causal effects (see Joffe 2011; Pearl 2011; VanderWeele 2008, 2011a).

yet become available and is unlikely to become available in the future. Suppose that in New York City conditional cash transfers are offered to families that send their children to charter schools. Suppose that this program is modeled on New York City's recent Opportunity NYC program, which was justified by the position that families should be given incentives to make decisions that promote their children's futures.

Suppose that for the new hypothetical program \$3,000 in cash is offered each year to families for each child that they enroll in a charter school. Since charter schools do not charge tuition, families can spend the \$3,000 per child however they see fit. Suppose further that, because of a budget constraint, cash transfers cannot be extended to all eligible families. For fairness, it is decided that families should be drawn at random from among all families resident in New York City with school-age children. Accordingly, a fixed number of letters is sent out notifying a set of winning families.

It is later determined that 10 percent of students in charter schools received cash transfers. A dataset is then compiled with performance data on all students in the school district, and the cash transfer offer is coded as a variable Z, which is equal to 1 for those who were offered a cash transfer and 0 for those who were not. A quick analysis of the data shows that some families who received offers of cash transfers turned them down and chose to send their children to regular public schools. Moreover, it is then assumed that at least some of the charter school students who received cash transfers would have attended charter schools anyway, and they were simply lucky to have also received a cash transfer.

By the standards typical of IV applications, Z would be considered a valid instrument. It is randomly assigned in the population, and it is reasonable to assume that it has a direct causal effect on D because it is an effective incentive for charter school attendance. (We have also assumed in the setup that the data show that Z predicts D.) Again, the crucial assumption is that the entire association between Z and Y is attributable solely to the directed path, $Z \to D \to Y$. As we will discuss below in this section, this assumption is debatable in this case because the subsidy is cash and, without further restrictions, could be used by families of charter school students to purchase other goods that have effects on Y. Any such alternative uses of the cash transfer would open up additional causal pathways from Z to Y that are not intercepted by D. For now, however, we will provisionally accept this identification assumption.

What parameter does Z identify? Suppose that a monotonicity assumption is valid whereby the cash transfers do not create a disincentive for anyone to enter charter schools (i.e., defiers with respect to Z do not exist in the population). This assumption allows us to abandon the constant coefficient assumption and instead assert that Z identifies the following LATE: the average effect of charter schooling among those who enter charter schools in response to the offer of a conditional cash transfer.

Figure 9.5 shows one way to represent estimators of this type. For these two graphs, the population can be partitioned into two mutually exclusive groups, compliers and noncompliers (as explained in Section 9.3, and assuming defiers do not exist). Figure 9.5(a) is the graph for compliers. No back-door paths connect D to Y in this graph because compliers, by definition, decide to enter charter schools solely based on

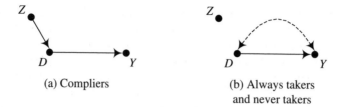

(a) Compliers   (b) Always takers
                                              and never takers

Figure 9.5 Instrumental variable identification of the causal effect of charter schools (D) on test scores (Y), where separate graphs are drawn for compliers and noncompliers.

whether they are offered the conditional cash transfer.[34] Analogous to the distribution calculated for Table 9.3, compliers are present in both regular public schools and charter schools.

Figure 9.5(b) is the graph for noncompliers. Always takers enter charter schools regardless of whether they receive the offer of a cash transfer, and never takers do not enter charter schools regardless of whether they receive the offer of a cash transfer. As a result, Z does not cause D for either always takers or never takers. The analyst can therefore place Z within the graph, but the causal effect $Z \to D$ must be omitted. Instead, the analyst includes $D \leftarrow\text{-}\text{-}\text{-}\text{-}\to Y$ to represent the unobserved joint causes of D and Y, as it is these factors that suggest why identification via an instrumental variable is needed.

Given that a central theme of this book is the power of graphs to represent causal relationships, we will conclude by addressing a final question: Can the clarity of Figure 9.5 be represented in a single causal graph, akin to the move from Figure 8.3 to Figure 8.4 in Section 8.3? Yes, but readers may not agree that the clarity is preserved.

Figure 9.6 is a combined representation of Figure 9.5, which now applies to the full population. The representation of compliance-based heterogeneity is accomplished by augmenting the graph with a latent class variable, C, which signifies whether an individual is a complier or not.[35] In particular, C takes on three values, one for compliers, one for always takers, and one for never takers (and we assume that defiers do not exist in the population). Most importantly, C interacts with Z in determining D, and then C interacts with D in determining Y.[36]

Now, to make the connection to the fully elaborated Figure 8.7 (see page 285), consider Figure 9.7, which includes all of the relevant back-door paths between D

---

[34] We thank Peter Steiner for pointing out to us that, contrary to figure A2 in Morgan and Winship (2012), Figure 9.5(a) should not include $D \leftarrow\text{-}\text{-}\text{-}\text{-}\to Y$. None of the common causes of D and Y among noncompliers have analogous effects for compliers because D is determined solely by Z for compliers.

[35] An alternative and more compact graph could be used for Figure 9.6. Because C is unobserved, one could simply declare that it is a member of the set of variables that generate the bidirected edge in Figure 9.6 (or as a member of the set of variables in V that will be introduced below for Figure 9.7). We give C its own pair of explicit causal effects on D and Y for clarity, even though it makes the graph more complex than it needs to be.

[36] It is possible that one could assume that C does not determine Y. This would be the case, for example, if one had reason to believe that the LATE is equal to the ATE, which would seem to be very unlikely in social science applications.

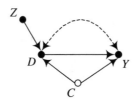

Figure 9.6 A combined graph for Figures 9.5(a)–(b), where Z is the instrument and compliance is represented as an unobserved latent class variable (C).

and Y represented as $D \leftarrow\!\!-\!\!-\!\!-\!\!\rightarrow Y$ for Figure 9.6. Still, for simplicity we have replaced the path $I \rightarrow \text{Exp}(D \rightarrow Y) \rightarrow G$ from Figure 8.7 with the single variable V.[37] The conditional cash transfer is then represented as an instrumental variable Z, which has a sole causal effect in the graph on D because we have assumed that the cash transfer offer does not have other effects on Y. Finally, we then add the additional back-door path from D to Y through their new common cause C.

With the addition of $D \leftarrow C \rightarrow Y$ to the graph, two sources of confusion may arise for some readers. First, it remains true that we cannot use back-door conditioning to estimate the effect of D on Y because of the unblockable back-door path through $D \leftarrow V \rightarrow Y$. However, it is important to remember that the similarly structured back-door path $D \leftarrow C \rightarrow Y$ does not present any problems for an IV estimator because it is not a back-door path from Z to Y, nor part of a directed path that carries the effect of Z to Y. It only represents an additional unblocked back-door path from D to Y. Second, nothing in the causal graph itself explains why a resulting IV estimator delivers an average causal effect that applies only to compliers. To understand this point, it may be more helpful to draw two separate causal graphs, as in Figure 9.5. The single causal graph does not reveal that there is an implicit interaction between Z and C as causes of D. In particular, the instrument Z does not cause D for noncompliers, and C does not cause D for those who do not receive the offer of a cash transfer. Only the co-occurrence of Z and C switches some members of the population from $D = 0$ to $D = 1$.

Now, to conclude the discussion of the estimation of the charter school effect, consider two final points. It is likely that Figure 9.7 improperly omits the likely causal effect $P \rightarrow C$. The parental background variable P implicitly includes within it a variable for family income. Students from families with high incomes should be less likely to switch from regular public schools to charter schools because of an offer of a modest conditional cash transfer to their parents. Adding such a path, however, would not harm the feasibility of the IV estimator, since it does not generate an unblockable path from Z to Y. In fact, the effect $P \rightarrow C$ helps to explain who compliers likely are, because it suggests that they are more likely to be lower income families. In this sense, recognizing the likely presence of this effect helps to interpret the LATE that the IV identifies.[38] In addition, this type of effect reveals a distinct advantage of a

---

[37]This simplification is permissible under the assumption that an implicit error term e_V contains all of the information in the error terms e_I, e_{Exp}, and e_G in the causal graph in Figure 8.7.

[38]For situations such as these, Angrist and Fernandez-Val (2013) propose alternative methods for using multiple IVs and covariate information to put forward estimates of broader parameters based on identified LATEs (see also Aronow and Carnegie 2013).

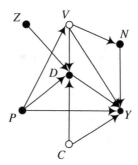

Figure 9.7 Identification of the LATE using an instrument (Z) for the charter school graph presented earlier in Figure 8.7. The unobserved variable V is a composite for the causal chain that generates self-selection in Figure 8.7 through information access and selection on the subjective evaluation of the individual-level causal effect.

single population-level graph that represents compliance as a node within it. Separate graphs for compliance classes do not permit a representation of the effect of P on compliance.

But, of course, not all additional causal effects will help to clarify the IV estimator. Suppose, for example, that Z generates an effect on N because the cash transfer is used to pay higher rent in another neighborhood for some families. As a result, a direct effect from Z to N is opened up. Conditioning on the observed variable N will block the new directed path $Z \rightarrow N \rightarrow Y$. But, because N is a collider on another path, $Z \rightarrow N \leftarrow V \rightarrow Y$, conditioning on N opens up this pathway by inducing a relationship between Z and V. Thus, conditioning away self-selection into neighborhoods then allows self-selection on the causal effect of charter schooling to confound the IV estimate of the LATE.

9.3.4   Local IVs and Marginal Treatment Effects

In this final section, we discuss one additional perspective on the identifying power of IVs in the presence of individual-level heterogeneity, which shows how a perfect instrument can help to identify a full pattern of causal effect heterogeneity. Heckman and Vytlacil (1999, 2000, 2005, 2007), building upon Heckman (1997), have shown that LATEs and many other average treatment effects can be seen as weighted averages of more fundamental marginal treatment effects.[39] Although the generality of their perspective is captivating, and the methodological content of the perspective unifies many strands of the literature in causal effect estimation, we summarize it only briefly here because the demands on data are quite substantial. The all-powerful IV that is needed to estimate a full schedule of marginal treatment effects will rarely be available to researchers.

The marginal treatment effect (MTE) perspective can be easily grasped with only a slight modification to the setup of IV Demonstration 2. Instead of 10 percent of students receiving a voucher that is exactly equal to $3,000$, suppose instead that

---

[39]See also Heckman et al. (2006) and Heckman and Urzua (2010).

these 10 percent receive a voucher that is a random draw from a uniform distribution with a minimum of $1 and a maximum equal to the tuition charged by the most expensive private school in the area.

For Heckman and his coauthors, the size of each student's voucher is a valid instrument Z, maintaining the same assumptions as we did for IV Demonstration 2 (i.e., Z is randomly assigned, Z has a nonzero effect on D, and the effect of Z on D is monotonic). The monotonicity assumption is a little more complex than before, but it stipulates that the true probability of taking the treatment is higher for all individuals with values of Z equal to z'' rather than z' if $z'' > z'$. This fits cleanly into the notation introduced earlier in this chapter by simply allowing Z to be many-valued.

Heckman and his coauthors then define two related concepts: a local instrumental variable (LIV) and an MTE. An LIV is the limiting case of a component binary IV drawn from Z in which z'' approaches z' for any two values of Z such that $z'' > z'$. Each LIV then defines a marginal treatment effect, which is the limiting form of a LATE, in which the IV is an LIV.

Consider the more elaborate version of IV Demonstration 2 just introduced here. One could form LIVs from Z by stratifying the data by the values of Z and then considering adjacent strata. Given a large enough sample for a large enough voucher program, LIVs could be constructed for each dollar increase in the voucher. Each LIV could then be used to estimate a LATE, and these LIV-identified LATEs could then be considered MTEs.

Heckman and Vytlacil (2005) show that most average causal effect estimates can be represented as weighted averages of MTEs, identified by LIVs. But the weighting schemes differ based on the parameter of interest, some of which, as was the case in our regression chapter, may have no inherent interest. Heckman and Vytlacil therefore propose a more general strategy. They argue that researchers should define the policy-relevant treatment effect (PRTE) based on an assessment of how a contemplated policy would affect treatment selection. Then, MTEs should be estimated with LIVs and weighted appropriately to obtain the PRTE that is of primary interest.

There is much to recommend in this approach, and in fact it should not be considered an approach relevant only to policy research. The approach is quite easily extended to targeted theory-relevant causal effects, for which one wishes to weight marginal causal effects according to a foundational theoretical model. But, in spite of this appeal, the entire approach may well come to represent a gold standard for what ought to be done rather than what actually can be done in practice. If it is generally recognized that IVs satisfying the LATE assumptions are hard to find, then those that satisfy LIV assumptions for all MTEs of interest must be harder still.[40]

---

[40]Partly for this reason, Carneiro, Heckman, and Vytlacil (2010) elaborate the PRTE perspective and focus attention on a limiting form of the PRTE that they label marginal policy-relevant treatment effects (MPRTEs). They argue that it is easier to estimate MPRTEs and that these may be all that are needed to evaluate proposed policy changes.

9.4   Conclusions

The impressive development of the IV literature in econometrics and statistics in the past two decades suggests a variety of recommendations for practice that differ from those found in the older IV literature:

1. Weak instruments yield estimates that are especially susceptible to finite sample bias. Consequently, natural experiments should be avoided if the implied IVs only very weakly predict the causal variable of interest. No matter how seductive their claims to satisfy identification assumptions may be, resulting point estimates and standard errors may be very misleading.

2. If individual-level causal effect heterogeneity is present, a potential IV should be used only if it satisfies a monotonicity condition. IV estimates should then be interpreted as LATE estimates defined by the instrument.

3. If individual-level causal effect heterogeneity is present, IVs should probably not be combined in a 2SLS model (except in the rare cases in which measures of all of the variables that account for the causal effect heterogeneity are available). Instead, IV estimates should be offered for those IVs that satisfy monotonicity conditions. These alternative estimates should then be interpreted as LATE estimates and reconciled with each other based on a narrative about why the causal effect varies for different types of individuals who are exposed to the cause (and/or different levels of the cause) for different reasons.

4. When possible, IVs should be used to examine general patterns of causal effect heterogeneity. Using IVs to estimate only the ATE or ATT is too narrow of a purpose because there is likely a good deal of variation in the treatment effect that is amenable to analysis with complementary IVs. The possibilities for this type of analysis are most clearly developed in the literature on the identification of MTEs using LIVs.

Properly handled, there is much to recommend in the IV estimation strategy for causal analysis. But, of course, IVs may not be available. We now turn to other techniques that may allow for the point identification and estimation of a causal effect when a complete model of causal exposure cannot be formulated because selection is determined by relevant unobserved variables.

Chapter 10

Mechanisms and Causal Explanation

Social scientists have recognized for decades that the best explanations for how causes bring about their effects must specify in empirically verifiable ways the causal pathways between causes and their outcomes. This valuation of depth of causal explanation applies to the counterfactual tradition as well. Accordingly, it is widely recognized that a consistent estimate of a counterfactually defined causal effect of D on Y may not qualify as a sufficiently deep causal account of how D effects Y, based on the standards that prevail in a particular field of study.

In this chapter, we first discuss the dangers of insufficiently deep explanations of causal effects, reconsidering the weak explanatory power of some of the natural experiments discussed already in Chapter 9. We then consider the older literature on intervening variables in the social sciences as a way to introduce the mechanism-based estimation strategy proposed by Pearl (2009). In some respects, Pearl's approach is completely new, and it shows in a novel and sophisticated way how intervening variables can be used to identify causal effects, even when unblocked back-door paths between a causal variable and an outcome variable are present. In other respects, however, Pearl's approach is refreshingly familiar, as it helps to clarify the appropriate usage of intervening variables when attempting to deepen the explanation of a causal claim.

Independent of Pearl's important work, a diverse group of social scientists has appealed recently for the importance of mechanisms to all explanation in social science research. Although some of these appeals are not inconsistent with the basic counterfactual approach to observational data analysis, some of the more extended appeals argue against the utility of the counterfactual model. After presenting these positions, we argue instead that there is no incompatibility between the elaboration of generative causal mechanisms and the counterfactual approach to observational data analysis. Finally, we draw on Machamer, Darden, and Craver (2000) and introduce their concepts of *mechanism sketch*, *mechanism schema*, and the process of *bottoming out* in mechanistic explanation. This terminology helps to frame our final discussion of how mechanisms can be used to sustain and deepen causal explanation, which draws together Pearl's front-door identification strategy with standards for sufficient causal depth.

10.1   The Dangers of Insufficiently Deep Explanations

Before considering how mechanisms can be used to identify causal effects, we first discuss the importance of explanatory depth in causal analysis. To do so, we return to the critical work of Rosenzweig and Wolpin (2000) on the limited appeal of many natural experiments. As discussed in Chapter 9, the natural experiment literature in economics uses naturally occurring forms of randomness as instrumental variables (IVs) in order to identify and then estimate causal effects of long-standing interest. Initially, this literature was heralded as the arrival of a new age of econometric analysis, in which it appeared possible to consistently estimate some of the causal effects of greatest interest to economists, such as the effect of human capital investments in education on earnings. Looking back on these bold claims, Rosenzweig and Wolpin (2000:829–30) conclude, "The impression left by this literature is that if one accepts that the instruments are perfectly random and plausibly affect the variable whose effect is of interest, then the instrumental-variables estimates are conclusive." They then argue that these estimates are far from conclusive and, in fact, are far more shallow than typically recognized.

To somewhat overstate the case, the initial overconfidence of the natural experiment movement was based on the mistaken belief that the randomness of a natural experiment allows one to offer valid causal inference in the absence of any explicit theory. One might say that some econometricians had been seduced by a position implicit in some writing in statistics: We do not need explicit theories in order to perform data analysis. Rosenzweig and Wolpin (2000) counter this position by arguing that the theory that underlies any model specification is critical to the interpretation of an estimate of a causal effect, and almost all examples of estimation by natural experiments have model specifications that make implicit theoretical claims (see also Deaton 2010 and our prior discussion in Section 9.3).

Here, we provide an informal presentation of two of the examples analyzed by Rosenzweig and Wolpin – the effect of education on earnings and the effect of military service on earnings – each of which was already introduced and discussed briefly in Chapter 9. We will use directed graphs here in order to demonstrate the issues involved.

Angrist and Krueger (1991, 1992) address the ability-bias concern for the estimation of the causal effect of schooling on subsequent labor market earnings. They assert that the quarter in which one is born is random but nonetheless predicts one's level of education because of compulsory school entry and dropout laws (see the quotation in Section 9.2 that gives the rationale). Angrist and Krueger's estimates of the increase in log earnings for each year of education fall between .072 and .102, values that are consistent with those found by others using different methods (see Card 1999).

As discussed already in Chapter 9, the first limitation of their results is that their IV estimates apply to only a narrow segment of the population: those individuals whose

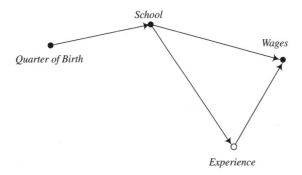

Figure 10.1 A directed graph for compliers with quarter of birth as an IV for years of schooling.

schooling would have changed if their birthdate was in a different quarter. Again, this is a local average treatment effect (LATE) estimate that applies to those individuals whose years of schooling are responsive to school entry and dropout laws. No one maintains that this narrow subpopulation is simply a random sample of all individuals in the population of interest, and most would argue that Angrist and Krueger estimated the returns to schooling only for disadvantaged youth prone to dropping out of high school for other reasons. There are, however, complications of how to interpret their estimates even for this subpopulation.

Consider the directed graph in Figure 10.1, which is based loosely on the critique offered by Rosenzweig and Wolpin (2000) and is a summary of the debate that unfolded in the years following the publication of Angrist and Krueger (1991). The graph is not meant to represent a full causal account of how education determines wages, and it is restricted only to the causal effect among compliers. As discussed in Chapter 9, for this example, compliers are those members of the population whose education is (or would be) responsive to a switch in their quarter of birth. Thus, this graph takes it for granted that this particular LATE is the only parameter that is informed by this analysis.[1]

For Figure 10.1, schooling has both direct and indirect effects on wages. Most important, as is maintained in the human capital literature, schooling is thought to have a negative indirect effect on wages through work experience; going to school longer reduces the amount of work experience one acquires by any particular age. Accordingly, there is a causal pathway from schooling to wages via work experience. The quarter-of-birth IV does not provide a separate estimate of this distinct pathway because a change in schooling in response to one's birthdate also changes work experience. Instead, the quarter-of-birth IV estimates only the total effect of schooling on wages, not its direct effect.[2] The IV yields a LATE estimate that likely mixes together two

---

[1] This position is equivalent to assuming that a more encompassing graph exists that is applicable to the entire population and that Figure 10.1 applies only to compliers. See our prior discussion of Figure 9.5(a).

[2] Angrist and Krueger could deal with this problem by conditioning on a measure of work experience, but in their data no such variable is available. The only alternative with their data, which

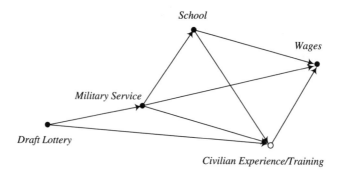

Figure 10.2 A directed graph for compliers with the Vietnam draft lottery as an IV for military service.

distinct and countervailing causal pathways: a positive direct effect of schooling on wages and a negative indirect effect via work experience. Given the long-standing interest of economists in the interaction between investments in formal schooling and the provision of on-the-job training, this total effect estimate can be regarded as an insufficiently deep causal account of the effect of education on earnings (even if one is convinced, as we are, that learning about compliers in this case is still illuminating).

For a second example, consider the effect of military service on lifetime earnings, as analyzed by Angrist (1990) and introduced in Section 9.2. The question of interest here is whether military service provides important training that increases later earnings in civilian life or rather whether military service is simply a period of lost civilian work experience. Military service, however, is not random. On the one hand, individuals must pass a mental ability test and a health examination in order to enlist. On the other hand, individuals with attractive schooling opportunities and civilian labor market opportunities are less likely to enlist. To deal with the problem of nonrandom selection into the military, Angrist (1990) considers individuals who were potentially eligible to be drafted into the military by a lottery during the later stages of the Vietnam War. To ensure that the draft was fair, the government decided to draft individuals based on a lottery that selected birthdates (see our earlier discussion in Section 9.2).

Consider now the directed graph in Figure 10.2, again based on the summary of critiques of the study compiled by Rosenzweig and Wolpin (2000). Here, there are three issues to consider. First, as we noted for the last example, the draft lottery identifies a LATE, and thus it is not applicable to the causal effect for always takers (those who voluntarily enlisted) and never takers (those who were draft dodgers, those who failed the mental ability test, and those who failed the physical examination). Accordingly, as for Figure 10.1, the graph in Figure 10.2 also applies to compliers only.

Second, note that there is a potential path from the draft lottery to civilian experience/training. If this path exists, then the draft lottery is not a valid IV for military

---

is common in the literature, is to attempt to untangle the effects by conditioning on the number of years since the respondent completed his or her highest level of education. However, this form of conditioning does not completely explain away the association between schooling and work experience, as is widely recognized (see Card 1999).

service. As we noted in Chapter 9, Heckman (1997) argued that employers would be likely to invest less in individuals with unfavorable lottery numbers. For example, it is plausible that employers may have given less on-the-job training to those most likely to be drafted and/or may have assigned such individuals to short-run tasks that did not require the accumulation of skill to master. If this effect exists, then the draft lottery would be an invalid instrument for military service.

Third, and most important for our consideration here, there are four separate causal pathways between military service and wages. In addition to the direct causal effect of military service on wages, there are two directed paths solely mediated by civilian experience and schooling, respectively. Here, it is generally thought that military service reduces civilian labor force experience and schooling, both of which then reduce wages. But there is then a countervailing effect of military service that snakes through the graph via a fourth directed path: Military service reduces schooling, which then increases work experience, and which then increases wages. Because all four of these pathways are activated by the shock induced by the draft lottery, Angrist's only resort is to assert that his estimates are for the total effect of military service on wages. But, given the inherent interest in untangling how the military service effect interacts with both schooling and the accumulation of civilian work experience, a total effect estimate is insufficiently deep to end all future research, even though this natural experiment was informative and a very important contribution to the literature.

Rosenzweig and Wolpin (2000) consider many other examples, all of which emphasize the same basic points as these two examples. IV analyses typically provide total causal effect estimates, often in substantive areas in which scholars have an inherent interest in the separable causal pathways that generate the outcome in response to the cause. Understanding the separable causal pathways that make up these total effects requires an explicit specification of additional intervening and mediating variables, which together compose the full causal mechanism of interest.

Now consider these two issues more generally, moving away from IV estimates toward more general estimates of average causal effects. Consider an optimistic scenario for which one obtains what all agree is a consistent estimate of the average treatment effect (ATE) for the effect of D on Y (as warranted, for example, by a consensus that one has conditioned on all variables that block all back-door paths from D to Y). Even in this scenario, in which the causal claim is valid by the standards of the counterfactual model, there are two related ways in which such an estimate can be regarded as insufficiently deep.

First, the ATE may not be a parameter of any fundamental interest. This point has been made most forcefully by Heckman (2000, 2005, 2010). If treatment effects are heterogeneous, which often is the case, as we have noted in prior chapters, then what is of fundamental interest are conditional average treatment effects of some form (perhaps even just the average treatment effect for the treated (ATT) and the average treatment effect for the controls (ATC)). The overall ATE is simply a weighted average of any such underlying effects. Heckman and many others have noted that the ATE may be of limited use for predicting the outcomes of policy interventions, either for new populations or in different contexts, because the ATE is tied to the current configuration of the population and prevailing pattern of causal exposure. Often, social scientists desire explanations for outcomes that can be modified in straightforward ways when populations shift or contexts change in ways that are separable from conditional average causal effects (especially if there is reason to believe that the conditional average casual effects are more fundamental in the sense that they can be expected to remain invariant under any such shifts and changes).

The second issue, which is our primary focus in the remainder of this chapter, is that a consistent estimate of the ATE, or any other average causal effect, does not necessarily entail any particular mechanism that explains how D brings about Y. If a theory suggests why and how D brings about Y, then merely providing evidence of the amount that Y can be expected to change in response to an intervention on D does not then provide any support for the underlying theory. If an assessment of the support for the underlying theory is desired, which is often a primary goal of social science research, then a more narrowly focused analysis of the putative causal pathways that relate D to Y must be undertaken.

In the next section, we present Pearl's approach to this type of analysis. We will make the case that his approach is consistent with decades of prior research in the social sciences and, furthermore, that it provides a particularly clear guide for future research in the same mode of analysis. To set the stage for Pearl's perspective, we first bring this long history of practice to the foreground, presenting the classic literature on the importance of intervening variables in causal explanation.

10.2   The Front-Door Criterion and Identification of Causal Effects by Mechanisms

For decades, social scientists have considered the explication of mechanisms through the introduction of intervening and mediating variables to be essential to sound explanatory practice in causal analysis. Duncan and his colleagues wrote in their 1972 book, *Socioeconomic Background and Achievement*:

> much of the scientific quest is concerned with the search for intervening
> variables that will serve to interpret or explain gross associations presumed
> to reflect a causal relationship. (Duncan, Featherman, and Duncan 1972:12)

The words "interpret" and "explain" are implicit references to the language of social research that is usually associated with Paul Lazarsfeld. More than two decades earlier, Kendall and Lazarsfeld (1950) distinguished between alternative types of elaboration that can be carried out when investigating an association between a causal variable X and an outcome variable Y.[3] Each type of elaboration involves the introduction of a

---

[3] The citations in the text are for a retrospective discussion of Samuel Stouffer's research in *The American Soldier* (which we also discussed in Chapter 1). Patricia Kendall was the lead author for the 1950 piece we cite, and yet she is almost never credited in the derivative literature as a contributor to this language. Because it is often written that Lazarsfeld presented and discussed this basic typology in many places, it is possible that it is fair to credit him disproportionately for these ideas. Our reading of the literature, however, does not yield a clear interpretation. It does suggest to us that Kendall has received less credit than she deserved. Indeed, in the version of these ideas published in Hyman's widely read text *Survey Design and Analysis*, Hyman (1955:275) notes only in a footnote, "The major

test factor (or test variable) T, after which the association between the causal variable X and the outcome variable Y is calculated for each value of T.

Kendall and Lazarsfeld considered two types of M-elaboration, which arise when the partial association between X and Y within strata defined by T is smaller than the original total association between X and Y.[4] Assuming that both X and T precede Y in time, the two types of M-elaboration are determined by whether T also precedes X in time:

1. If T follows X in time, then an M-type elaboration can be represented by a chain of mediation, $X \rightarrow T \rightarrow Y$ (Kendall and Lazarsfeld 1950:157). They refer to this type of elaboration as an *interpretation* of the association between X and Y, using a test factor that is an "intervening" variable.

2. If T precedes X in time, then an M-type elaboration can be represented by a fork of mutual dependence, $X \leftarrow T \rightarrow Y$ (Kendall and Lazarsfeld 1950:157). They refer to this type of elaboration as an *explanation* of the association between X and Y, using a test factor that is an "antecedent" variable.

Although Kendall and Lazarsfeld did not argue that causality can be established by what they call "interpretation," and although they use the word "explanation" in a rather limited sense (and with a different usage than Duncan and colleagues, as just quoted), Kendall and Lazarsfeld were clearly interested in mechanistic accounts that reveal how causes bring about their effects. For example, they wrote:

> When we interpret a result we try to determine the process through which the assumed cause is related to what we take to be its effect. How did the result come about? What are the "links" between the two variables? Answers to these questions are provided in the interpretation of the result. (Kendall and Lazarsfeld 1950:148)

A long tradition of empirical social science exists that follows these basic ideas, in which a central goal of investigation is the search for variables – intervening, mediating, or processual variables – that can account for an association that is thought to be causal (see MacKinnon 2008; Winship and Harding 2008). Even so, as we noted in Chapter 6, applications of the all-cause regression specification approach gradually obscured this goal, leading from the 1970s through the 1990s to more frequent attempts to offer single-equation models in which all effects of all observed putative causes of an outcome are estimated simultaneously.

Fortunately, Pearl (2009) has developed an "inside-out" perspective on the identifying power of mechanisms that can help the social sciences recapture the inclination to pursue mechanistic explanation of causal effects in empirical research. Pearl's approach

---

sections of this chapter have been written by Patricia Kendall and represent an enlargement of an earlier analytic schema."

[4] They also lay out a third form of elaboration, referred to as P-type elaboration (or specification). Here, the goal is to focus "on the relative size of the partial relationship [between X and Y within strata of the test factor T] in order to specify the circumstances under which the original relation is more or less pronounced" (Kendall and Lazarsfeld 1950:157). This type of elaboration is appropriate when the test factor is related to either X or Y but not both. We ignore this form of elaboration here, as we focus on variables that are related to both the causal and outcome variables.

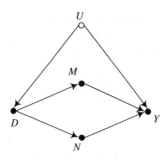

Figure 10.3 A directed graph in which M and N represent an exhaustive and isolated identifying mechanism for the causal effect of D on Y.

goes well beyond Lazarsfeldian elaboration. Instead, he shows how one can consistently estimate the effect of a causal variable on an outcome variable by estimating the effect as it propagates through an exhaustive and isolated mechanism. He labels this strategy the "front-door" identification of a causal effect (see Pearl 2009:81–85).

Consider Figure 10.3, where an unblockable back-door path $D \leftarrow U \rightarrow Y$ exists from D and Y because the variable U is unobserved. As a result, identification using Pearl's back-door criterion, as presented in Chapter 4, is infeasible. Nonetheless, the variables M and N intercept the full causal effect of D on Y, and for Pearl these two variables represent an identifying mechanism for the effect of D on Y. (Again, recall that there is no assumption of linearity in a directed graph of this type, and thus this model is consistent with any nonlinear structural equations for the generation of D, M, N, and Y.)

For this graph, observation of M and N identifies the causal effect of D on Y by a two-part consideration of the back-door criterion:

1. For $D \rightarrow M$ and $D \rightarrow N$, two back-door paths exist: $D \leftarrow U \rightarrow Y \leftarrow M$ and $D \leftarrow U \rightarrow Y \leftarrow N$, respectively.[5] Because Y is a collider variable along both of these back-door paths, each path is blocked by Y in the absence of any conditioning. Accordingly, these back-door paths do not contribute to the observed associations between D and either M or N. And, as a result, the causal effects $D \rightarrow M$ and $D \rightarrow N$ can be estimated consistently by their simple unconditional associations.

2. For $M \rightarrow Y$ and $N \rightarrow Y$, two back-door paths exist for each causal effect. For $M \rightarrow Y$, they are $M \leftarrow D \leftarrow U \rightarrow Y$ and $M \leftarrow D \rightarrow N \rightarrow Y$. For $N \rightarrow Y$, they are $N \leftarrow D \leftarrow U \rightarrow Y$ and $N \leftarrow D \rightarrow M \rightarrow Y$. All four of these back-door paths generate noncausal associations between M and Y and also between N and Y

---

[5] Both $D \leftarrow U \rightarrow Y \leftarrow M$ and $D \leftarrow U \rightarrow Y \leftarrow N$ satisfy the definition of a back-door path, even though they do not end with $\rightarrow M$ and $\rightarrow N$, respectively. They are both paths between pairs of causally ordered variables (D and M, on the one hand, and D and N, on the other), and they both begin with an arrow pointing to the first variable (i.e., $D \leftarrow$). As such, they are back-door paths for $D \rightarrow M$ and $D \rightarrow N$, respectively, even though they do not resemble most of the other examples of back-door paths presented in this book. We have focused our earlier examples on back-door paths that must be blocked by conditioning. However, see our discussion of Figure 4.12 in the appendix to Chapter 4, where we discuss a similar back-door path.

because none of these paths includes collider variables that block them in the absence of any conditioning. Fortunately, all four of these back-door paths can be blocked by conditioning on D. And, as a result, the casual effects $M \to Y$ and $N \to Y$ can be estimated consistently after conditioning on the observed variable D.

In short, for this graph the unblocked back-door path $D \leftarrow U \to Y$ does not prevent consistent estimation of the effect of D on Y as long as both M and N are observed. Because one can obtain consistent estimates of the casual effects of D on both M and N and, in turn, of both M and N on Y, one can calculate the full causal effect of D on Y by combining these causal effect estimates. Pearl (2009:83) gives the general estimation formula in the underlying probability distributions. For the ATE, the simplest case that is also consistent with the structure of Figure 10.3 is one where (1) D is a binary variable, (2) M, N, and Y are interval-scaled variables, (3) all structural equations are linear and additive, e.g., $Y = f_Y(M, N, e_Y) = a_Y + b_M M + b_N N + e_Y$, and (4) all individual-level causal effects do not vary in the levels of the variables in the mechanism. In this case, the ATE can be estimated consistently by a sum of products,

$$\left(\hat{b}_{D \to M} \times \hat{b}_{M \to Y}\right) + \left(\hat{b}_{D \to N} \times \hat{b}_{N \to Y}\right),$$

where $\hat{b}_{D \to M}$ is the estimated coefficient for M in a bivariate regression of Y on M, $\hat{b}_{D \to N}$ is the estimated coefficient for N in a bivariate regression of Y on N, and $\hat{b}_{M \to Y}$ and $\hat{b}_{N \to Y}$ are the estimated coefficients for M and N for a multiple regression of Y on M, N, and D.[6]

As shown by this example, the front-door identification strategy can be very simple, as it involves nothing more than a straightforward two-step consideration of the back-door criterion introduced in Chapter 4 (see our presentation beginning on page 109). Pearl formalizes this identification strategy by offering a complement to his back-door criterion, which he labels the front-door criterion (Pearl 2009:82):

Front-Door Criterion

If one or more unblocked back-door paths connect a causal variable to an outcome variable, the causal effect is identified by conditioning on a set of observed variables, $\{M\}$, that make up an identifying mechanism if

Condition 1 (exhaustiveness). The variables in the set $\{M\}$ intercept all directed paths from the causal variable to the outcome variable;

and

---

[6]In this case, Pearl's front-door adjustment formula reduces to this sum of products for the causal contrast he would label $E[Y|do(D = 1)] - E[Y|do(D = 0)]$. When conditions (1) through (3) do not obtain, the computation is more complex because one needs to propagate the effects through the probability distributions of the mechanistic variables, conditional on the causal variable, and then calculate average effects for the chosen contrast (see Pearl 2009). We omit these details because we are focusing in this chapter on the essential identification issues.

Condition 2 (isolation). No unblocked back-door paths connect
the causal variable to the variables in the set $\{M\}$, and all back-
door paths from the variables in the set $\{M\}$ to the outcome
variable can be blocked by conditioning on the causal variable.[7]

Notice that the front-door criterion does not give direct guidance on how deep an identifying mechanism must be in order to qualify as a sufficiently deep causal explanation.
It only specifies the features that a mechanism must have in order to be used to identify an average effect of a causal variable on an outcome variable. Any such identifying
mechanism can be considered shallow by the standards of a particular research group
or field of interest. Before addressing this issue, we clarify in the remainder of this
section what the requirements of exhaustiveness and isolation entail.

The Assumptions That the Mechanism Is Exhaustive and Isolated

The crucial assumptions of this approach, as we have noted already, are that the identifying mechanism is exhaustive and isolated. To see the importance of isolation first,
consider the alternative graph presented in Figure 10.4. Here, the variable U is again
unobserved, and thus the back-door path $D \leftarrow U \rightarrow Y$ cannot be blocked by conditioning on any observed variables in the graph. In this case, the variable M intercepts the
full causal effect of D on Y, and therefore the identifying mechanism is exhaustive,
satisfying Condition 1 of Pearl's front-door criterion. However, the identifying mechanism M is not isolated because U has a direct causal effect on M. Corresponding
to both components of Condition 2 of the front-door criterion, two problems arise.
The causal effect $U \rightarrow M$ generates a new back-door path from M to Y, which is
$M \leftarrow U \rightarrow Y$, as well as a new back-door path from D and M, which is $D \leftarrow U \rightarrow M$.
Neither of the effects that compose the casual pathway $D \rightarrow M \rightarrow Y$ can be estimated
consistently because no observed variables are available to block these new back-door
paths by conditioning. With reference to the structure of Pearl's front-door criterion,
the problems are that (1) an unblocked back-door path connects the causal variable
to M ($D \leftarrow U \rightarrow M$) and (2) a back-door path from M to Y cannot be blocked by
conditioning on D ($M \leftarrow U \rightarrow Y$).

This example also shows what is crucial in the criterion of isolation. The mechanistic variables must be isolated from otherwise unblocked back-door paths so that
the back-door criterion can be applied in order to recover the full causal effect from
the data. For Figure 10.4, if U were an observed variable, then the back-door paths
$D \leftarrow U \rightarrow M$ and $M \leftarrow U \rightarrow Y$ could be blocked by conditioning on U. In this case, we
could then use the front-door criterion in a conditional variant, estimating the effect
of D on Y by estimating the effects $D \rightarrow M$ and then $M \rightarrow Y$ while conditioning
on U.

More generally, isolation comes in strong and weak forms, both of which are sufficient. For the strong form, none of the variables that lie on back-door paths from D

---

[7]In our accounting of Pearl's front-door criterion, we label his (i) as our Condition 1, which we
then characterize as an exhaustiveness condition. We also combine his (ii) and (iii) into our Condition
2, which we characterize as isolation.

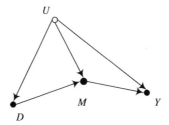

Figure 10.4 A directed graph in which M is not an isolated mechanism for the causal effect of D on Y.

to Y can have causal effects on the mechanistic variables, except indirectly through directed paths that pass through D. For Figure 10.3, the only directed paths from U that terminate at M and N pass through D. In this case, conditioning on D alone is sufficient. For the weak form, some of the variables that lie along back-door paths from D to Y can have causal effects on the mechanistic variables through directed paths that do not pass through D, but it must still be possible to block any relevant back-door paths by conditioning on D *and* other observed variables in the graph. The weak form of isolation makes it clear that one must be concerned only about the dependence of the variables in the mechanism on components of back-door paths between D and Y that cannot be blocked by conditioning on observed variables.

An implication of the necessity of assuming isolation is that a very good understanding of the back-door paths between the causal variable and the outcome variable is needed. If the isolation assumption cannot be maintained, then the mechanistic variables are similarly affected by the same set of dependencies that invalidate basic back-door conditioning as a strategy to identify the causal effect.

Now return to Condition 1 of the front-door criterion, which requires that the identifying mechanism is exhaustive. For Figure 10.5(a), suppose, again, that there is an unblocked back-door path and also two distinct causal pathways from D to Y, which are represented by the two variables M and N. Finally, unlike for Figure 10.3, suppose that N is unobserved.

In this case, the causal effects $D \rightarrow N$ and $N \rightarrow Y$ cannot be estimated, and thus the full causal effect of D on Y cannot be estimated by front-door conditioning. If one were to assert a mistaken assumption that M is an identifying causal mechanism (for example, by substituting a back-door path $D \leftarrow N \rightarrow Y$ for the genuine causal pathway $D \rightarrow N \rightarrow Y$), then one can obtain a causal effect estimate. But the causal effect of D on Y will then be underestimated (assuming all causal effects are linear, positive, etc.) because the part of the causal effect generated by the pathway $D \rightarrow N \rightarrow Y$ is attributed to a mistakenly asserted back-door path $D \leftarrow N \rightarrow Y$.

Relaxing the Assumption That the Mechanism Is Exhaustive

For the example depicted in Figure 10.5(a), a mechanism-based empirical analysis may still be useful because an isolated piece of the causal effect can still be estimated. Even though N is unobserved, the causal effect of D on M is identified by their observed

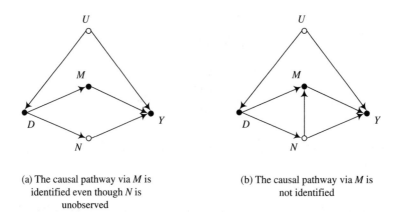

(a) The causal pathway via M is
identified even though N is
unobserved

(b) The causal pathway via M is
not identified

Figure 10.5 Directed graphs in which one pathway in an exhaustive and isolated mechanism is unobserved.

association because the back-door path from D to M, which is $D \leftarrow U \rightarrow Y \leftarrow M$, is blocked by the collider Y in the absence of any conditioning. And, as before, the causal effect of M on Y is identified because both back-door paths from M to Y, which are $M \leftarrow D \leftarrow U \rightarrow Y$ and $M \leftarrow D \rightarrow N \rightarrow Y$, can be blocked by conditioning on D. One can thereby obtain a consistent estimate of the causal effect of M on Y by conditioning on D, which guarantees that the part of the causal effect that travels through the pathway $D \rightarrow M \rightarrow Y$ can be estimated consistently. This is an important result, especially for practice, because identifying and consistently estimating a distinct causal pathway can be very useful, even if one cannot in the end offer a causal effect estimate of the full effect of the causal variable on the outcome variable.[8]

Knight and Winship (2013:290–91) consider an example with the structure of Figure 10.5(a) and where the population of interest is women in the paid labor force. For their example, D is whether or not women have children, M is the number of hours per week worked in a paid job, N is the amount of bias in compensation practices against women with children, and Y is wages paid to women. Even if one cannot estimate the effect of bias against women with children on wages received, being able to estimate the wage differences that are attributable to the reduction of work hours that often results from having children may be important to establish the case for providing better childcare to support working women.

---

[8]Note, further, that one can then assess the portion of the total variation in D and Y that can be attributed to the pathway through M. This calculation can be useful, but it does not identify the amount of the causal effect that is explained by M. The problem is that the association between D and Y that is not attributable to M cannot be apportioned across the two unblocked paths $D \leftarrow U \rightarrow Y$ and $D \rightarrow N \rightarrow Y$. In addition, if the underlying structural equations are nonlinear, then additional complications arise because of the possibility of inherent interactions between the causal variable and mechanistic identifying variables in the production of the outcome variables. For full explanations of these complications, see Pearl (2012a, 2012b), VanderWeele (2009b, 2012, in press), and Wang and Sobel (2013). We also discuss these complications in the appendix to chapter 11, where we introduce a framework for defining and estimating net direct effects.

But, even though this partial estimation result is very useful, there is an important implicit assumption that is hidden by this example: Variables on the unobserved causal pathways cannot have direct or indirect causal effects on any of the variables that lie along the observed causal pathways. To see the complications that such dependence could produce, consider the graph in Figure 10.5(b). For this graph, the mechanism is isolated, but the unobserved variable N has a causal effect on M. In this situation, the effect of M on Y is not identified. Three back-door paths between M and Y are still blocked by D: $M \leftarrow D \leftarrow U \rightarrow Y$, $M \leftarrow D \rightarrow N \rightarrow Y$, and $M \leftarrow N \leftarrow D \leftarrow U \rightarrow Y$. However, a fourth back-door path, $M \leftarrow N \rightarrow Y$, cannot be blocked by conditioning on D or any other observed variable. This path remains unblocked because its only intermediate variable N is unobserved and is not a collider. Not only is it impossible to estimate the effect of D on the unobserved variable N, but the variable N may transmit to both M and Y its own exogenous variation that is completely unrelated to D and U. Without recognition of this outside source of dependence, one could mistakenly infer that the causal pathway $D \rightarrow M \rightarrow Y$ is much more powerful than it is (assuming all causal effects are linear, positive, etc.). Knight and Winship (2013:293) elaborate their example, summarized above, to show this result as well. They allow unobserved bias N to affect work hours M. In this case, the analyst can no longer advocate for increased childcare support by building support for the causal pathway $D \rightarrow M \rightarrow Y$ because the reductions in wages may all result from the direct effect of bias on wages and the indirect effect of bias on wages through work hours (which could result either from differential work assignments from biased supervisors or the protective withdrawal on the part of working women who confront a hostile workplace).

As these simple graphs show, if one wants to use a mechanism-based strategy to identify the full effect of a causal variable on an outcome variable, one must put forward an identifying mechanism that is exhaustive and isolated. The exhaustiveness requirement can be relaxed if one is satisfied with identifying only part of the causal effect of interest. But, to secure partial estimation of this form, one must be willing to assume that the observed portion of the mechanism is not affected by any of the variables in the unobserved portion of the mechanism. And, to assert this assumption, one typically needs to have a very specific theoretical model of the complete identifying mechanism, even though parts of it remain unobserved.

In sum, Pearl's front-door criterion represents a powerful and original set of ideas that clarifies the extent to which a mechanism, composed of one or more intervening variables, can be used to identify and estimate a causal effect. And, by approaching the causal inference predicament from the inside out, the approach helps to shape the question of whether a causal claim qualifies as a sufficiently deep explanation. Rather than estimating a causal effect by back-door conditioning or with a naturally occurring experiment and then wondering whether a mechanism can be elaborated to show how the effect comes about, Pearl instead shows that, if we can agree on the variables that constitute an exhaustive and isolated mechanism, then we can estimate the causal effect from its component causal pathways.

The approach cannot be taken out of context, though, lest one claim too much explanatory power for a nonexhaustive and/or nonisolated mechanism. And, at the same time, it must be acknowledged that even a front-door-identified causal effect may not be explained deeply enough by the exhaustive and isolated mechanism that identifies it, for any such mechanism may still be too shallow for a particular substantive area. To consider these issues further, we now turn to recent social scientific writing on mechanisms, wherein the connections among theory construction, generative mechanisms, and modes of explanation have been carefully examined.

10.3   The Appeal for Generative Mechanisms

To some extent inspired by the work of Jon Elster (e.g., Elster 1989), but more generally by their reading of past successes and failures in explanatory inquiry in the social sciences since the 1940s, Peter Hedström and Richard Swedberg convened a group of leading scholars to discuss a reorientation of explanatory practices in the social sciences. A collection of papers was then published in 1998 as *Social Mechanisms: An Analytical Approach to Social Theory*. At the same time, John Goldthorpe developed his case for grounding all causal modeling in the social sciences on the elaboration of generative mechanisms. His proposal was published in 2001 as "Causation, Statistics, and Sociology," which then received comments from a variety of scholars, including the statisticians David Cox and Nanny Wermuth (whose response we will discuss below).[9] Subsequently, Peter Hedström laid out in considerable detail his full program for a mechanism-based social science in his 2005 book, *Dissecting the Social: On the Principles of Analytical Sociology*. Along with Peter Bearman, Hedström then compiled *The Oxford Handbook on Analytical Sociology*, which is a collective effort to demonstrate the power of this approach (Hedström and Bearman 2009). To orient the reader to this push for mechanism-based social science, we will first present the slightly different proposals of Goldthorpe (2001) and then Hedström (2005).[10]

---

[9]Goldthorpe's argument was republished as a chapter in both the first and second editions of his programmatic collection, *On Sociology* (e.g., Goldthorpe 2007, vol 1., ch. 9). Our citations are to the original 2001 journal article.

[10]We will not trace the full history of mechanistic theorizing in the social sciences, as the pieces just cited cover much of the same territory (see also Hedström and Ylikoski 2010; Knight and Winship 2013). However, it does seem to us that the novelty of recent mechanism-based theorizing has been oversold at least a bit. At the same time that Kendall and Lazarsfeld were directing researchers to build mechanistic interpretations of causal effects in empirical research, while holding up Stouffer as a successful practitioner, mechanism-based modes of conjecturing were widespread among theorists. Hedström and Udehn (2009) lionize Robert Merton's proposal for the pursuit of middle-range theory, which they argue is implicitly mechanism-based. Like many others, they contrast Merton's work with the "grand" system-level theoretical program of Talcott Parsons. But even Parsons wrote frequently of mechanisms, including those for socialization, social control, and the equilibration of the social system (see Parsons 1951, chapters 6 and 7). As an example, Parsons (1951:206) wrote, "A mechanism of social control, then, is a motivational process in one or more individual actors which tends to counteract a tendency to deviance from the fulfillment of role-expectations, in himself or in one or more alters. It is a re-equilibrating mechanism." Explaining his usage of the term "mechanism," Parsons (1951:22) wrote, "Motivational dynamics in sociological theory, then, must take the form in the first instance of the formulation of mechanisms which 'account for' the functioning of social systems, for the maintenance or breakdown of given structural patterns, for a typical process of transition from one structural pattern to another. Such a mechanism is always an empirical generalization about the operation of motivational 'forces' under the conditions stated." It is clear from this work that Parsons saw mechanistic theorizing as a way to develop the sort of conjectures that Craver (2007) would regard as "how-possible" or "how-plausible" models and that Hedström and Bearman (2009:7) label "semigeneral mechanisms." It is also clear that Parsons' mechanisms came to be regarded as insufficiently deep by subsequent sociologists. Indeed, a great deal of the empirical work that has been

Goldthorpe develops his proposal by first attempting to drive a wedge between the counterfactual model, which he labels "causation as consequential manipulation," and his alternative mechanistic model of causality. To develop his argument, he advances the claim that the counterfactual model is irretrievably tied to actual experimental manipulations of causal variables. Goldthorpe (2001:6) writes, "the crux of the matter is of course the insistence of Rubin, Holland, and others that causes must be manipulable, and their unwillingness to allow causal significance to be accorded to variables that are not manipulable, at least in principle."[11] (See our discussion of this issue in Chapter 13, in which we present the positions of Woodward (2003) and others that this argument is incorrect.) Goldthorpe also argues, citing the work of statisticians Cox and Wermuth (1996; also Cox 1992), that the counterfactual model too easily settles for causal claims of insufficient depth that do not account for how causes bring about their effects (a point that is clearly supported by our presentation of the natural experiment literature discussed earlier in this chapter).

With this criticism of the counterfactual model in mind, Goldthorpe writes:

> The approach to causal analysis that is here proposed ... is presented in the form of a three phase sequence: (i) establishing the phenomena that form the *explananda*; (ii) hypothesizing generative processes at the level of social action; and (iii) testing the hypotheses. (Goldthorpe 2001:10)

Goldthorpe then proceeds to lay out the basic contours of the first two steps, arguing that the most useful mechanisms in the social sciences are those that are based on rational choice theory (because these are sufficiently microscopic and focus on the actions and beliefs of individuals). He then explicates the third step of his approach, which is his requirement that one must test hypotheses suggested by the generative mechanism that one has proposed.[12] And here, he takes the position that generative processes of the sort that he prefers, which are "specified at a 'deeper,' or 'more microscopic' level, than that of the data that constitute the *explananda*" (15), can almost never be directly evaluated. Instead, Goldthorpe advises that analysts spend their effort determining whether the hypothesized generative mechanisms can be indirectly supported through repeated validation of entailed hypotheses. These entailed hypotheses may be at the level of the "microscopic" mechanism, as would be the case for conjectured patterns among correlates of the unobserved true preferences and micro-interactions that,

---

published since the 1950s can be interpreted as attempts to deepen the mechanistic claims that were overly abundant in the work of Parsons, Merton, and other scholars of their era.

[11]It is unclear from this passage what Goldthorpe means by "in principle." However, on the next page of the article, he offers an example that reveals that his conception of a manipulation can be very narrow. Following an example of Holland (1986), he introduces mechanistic variables and argues that in so doing he has produced a causal narrative that has nothing to do with hypothetical manipulations/interventions and hence cannot be represented in a counterfactual framework. The argument seems to be that the mechanistic variables are nonmanipulable because they are typically voluntary decisions of an individual.

[12]By our reading, Goldthorpe ends up back at the counterfactual model at this point. His defense against such a reading is this: "while it might seem that, at this stage, attention does after all come to focus on the effects of – given – causes rather than on the causes of effects, this is within the context not of randomized experimental design but of (what should be) a theoretically informed account of a generative process that is subject to ongoing evaluation." (Goldthorpe 2001:13)

according to the hypothesized generative mechanism, determine the action that generates the outcome. More realistically, these entailed hypotheses would be at the higher level where data are available on the outcome of interest, as would be the case when analyzing conjectured patterns of the outcome itself at the population level (based on further assumptions about how the generative mechanism produces a distribution of outcomes across subgroups that can be identified by their measured characteristics).[13]

Consider now the more extended appeal for mechanism-based social science offered by Peter Hedström (2005). Unlike Goldthorpe, Hedström does not confront the counterfactual causality literature directly. He instead develops an argument that suggests the following orienting principle: "The core idea behind the mechanism approach is that we explain not by evoking universal laws, or by identifying statistically relevant factors, but by specifying mechanisms that show how phenomena are brought about" (Hedström 2005:24). Hedström arrives at this position from two different directions. First, he signs on to the dominant position in the philosophy of science that the best explanation of how a phenomenon is brought about must necessarily be a causal account of some form that does not rely on unrealistic presuppositions of the existence of invariant general (or covering) laws (see Hedström 2005:15–20). But he also justifies the principle by criticizing regression-based causal models in social science (see Hedström 2005:20–23):

> Such statistical analysis is often described as a form of "causal analysis." If a factor appears to be systematically related to the expected value or the conditional probability of the outcome, then the factor is often referred to as a (probabilistic) "cause" of the outcome. Although it makes little sense to quibble over words, I would like to reserve the word *cause* for a less casual notion of causality. (Hedström 2005:23)

Like Goldthorpe, Hedström also lays out a script for the construction of mechanisms, which is similar in its appeal to the principles of methodological individualism (although he steps back from Goldthorpe's stronger preference for forms of rational choice theory). Most importantly, Hedström does not advocate the testing of mechanisms in the way that Goldthorpe does. Instead, he defends the explanatory utility of mechanisms in a much more general way:

> Mechanisms should be seen as theoretical propositions about causal tendencies, not as statements about actualities. An explanation may be perfectly

---

[13]Blossfeld (2009) aims to build on Goldthorpe's orienting program and argues for the centrality of event history processes in generating relationships that can be interpreted as causal. Blossfeld (2009:101) writes, "We can only hope to make sensible causal statements about how a given or (hypothesized) change in variable Y_t^A (e.g., pregnancy/birth) in the past affects the probability of a change in variable $Y_{t'}^B$ (e.g., marriage) in the future. Correspondingly, the basic causal relation becomes: $\Delta Y_t^A \to \Delta Pr(Y_{t'}^B)$, $(t < t')$. In other words, a change in the time-dependent covariate Y_t^A will change the probability that the dependent variable $Y_{t'}^B$ will change in the future $(t < t')$." If Blossfeld's "given or (hypothesized)" distinction can be interpreted as consistent with potential outcomes tied to the states A and not-A, then this is simply a restatement of the potential outcome definition of a causal effect. If not, and this definition is meant to dispense with counterfactual dependence, then the only "causal relations" that are well defined for Blossfeld are those between values of observed variables that change in time during the observation window. If it is the latter, then this is just the sort of robust dependence relationship rejected by Goldthorpe (and Blossfeld), although now expressed as a robust-dependence-in-time perspective.

correct if understood as a proposition about a causal tendency, and yet it may be inadequate for predicting actual outcomes if other processes are also at work.... Since it is the rule rather than the exception that concretely observed phenomena are influenced by several different processes, testing a theory by examining the accuracy of its predictions is likely to conflate the truth or relevance of the postulated mechanism with the importance of other processes, and this may lead us to mistakenly reject perfectly appropriate causal accounts. (Hedström 2005:108)

What, then, is the explanatory status of mechanisms that are "theoretical propositions about causal tendencies"? Here, Hedström seems to argue three basic points: (1) The best explanations are causal accounts; (2) causal accounts must include mechanisms; (3) mechanisms may or may not explain actualities. The direct implication of this position is that the best explanations in social science do not necessarily have to explain actualities, but they must explain causal tendencies.[14]

In our view, Goldthorpe and Hedström have developed a wonderful appeal for the need to construct sufficiently deep theories of social processes. Their work is filled with much insight on a variety of promising ways to formulate plausible theoretical mechanisms. And they have helped to further convince many social scientists that specifying the social processes that account for how causes bring about their effects is a central goal of analysis. It is therefore unsurprising that this new wave of scholarship has inspired a scholarly movement of sorts, which we and many others regard as both promising and very healthy for theory construction in the social sciences.[15]

But, to be seen as blueprints for explanatory causal analysis, both of these particular perspectives need to be augmented. And we would argue that they can be usefully extended by embracing the counterfactual model of causality more directly.[16] We explain this position in the remainder of this section, first by offering a slightly

---

[14]It is unclear from the foregoing passage what the problematic "other processes" are meant to be. Clearly, they must come in two different forms: (1) other mechanisms that are completely independent of the postulated mechanism of primary interest and (2) other mechanisms that interact with the postulated mechanism of primary interest. If one is estimating the effects of a cause, it is only the latter that are problematic for the evaluation of mechanisms, and hence for the evaluation of causal accounts based on mechanisms. See our earlier discussion of partial identification by the front-door criterion.

[15]But we are somewhat more traditional in favoring mechanisms that are formal models, as we are less convinced of the utility of many simulation-based methods of theory construction; see Hedström (2005:76–87, 131–36, 149) on the appeal of such techniques. We agree with Humphreys' (2004:132) caution: "Agent-based models are a powerful addition to the armory of social scientists, but as with any black-box computational procedures, the illusion of understanding is all too easy to generate." Scholars such as Manzo (2011) disagree sharply with this position, and we welcome efforts that will convince us to change our position.

[16]In more recent work, Hedström and Udehn (2009:42) write that the mechanism-based approach to social science "should not necessarily be seen as an alternative to the counterfactual approach, but rather as adding further requirements. As emphasized by Woodward (2002) and Morgan and Winship (2007) there is nothing in the counterfactual approach as such which guarantees sufficient causal depth, because perceptions of sufficient depth are discipline-specific while the counterfactual approach is not." We assume that the reference here is to the argument that we offered in this chapter in the first edition of this book.

different reading of recent developments in the philosophy of science and then by raising the issue of how social scientists can retain the capacity to adjudicate between competing mechanistic accounts of causal effects.

Philosophy of Science and Explanation by Mechanisms

As we noted at the beginning of this chapter, the appeal to mechanisms as elaborations of causal claims has been entrenched in the sociological literature for many decades, and we suspect that this emphasis is true of all of the social sciences. Where perhaps the focus on generative mechanisms is new is in the proposed elevation of their status to the object of primary investigation. This elevation is not unrelated to shifting currents in the philosophy of science.

Some proponents of a mechanism-based social science have drawn inspiration from the demise of covering law models of explanation in the philosophy of science (e.g., Gorski 2004; Hedström 2005). The covering law model, as most famously explicated by Carl Hempel, maintains that all valid explanations can be formulated as logico-deductive entailment from invariant general laws.[17] Since 1970 at least, the covering law model has received near-continuous challenges to its basic premises (see Godfrey-Smith 2003; Salmon 1989; and Woodward 2003). The presupposition that general and exceptionless laws exist and that they can be used to warrant all valid causal claims cannot be sustained in most scientific disciplines, and certainly not in the social sciences.

In response, a variety of alternative models of scientific explanation have arisen, with realist models attracting the largest number of adherents (see Psillos 1999). In general, realist models grant ontological status to unobserved quantities (sometimes conferring provisional truth status on them). Most realist models reject causal nihilism and affirm that valid scientific explanation must necessarily be grounded in causal accounts of some form. Especially when invoked for the social sciences, realist models admit the possibility of inherent heterogeneity of causal relationships – in time, space, and within populations. But the depth of the appeal to unobservables varies across types of realist models, as does the level of provisional truth status conferred upon unobservables.

What we wish to emphasize in this section is how uneasily the focus on generative mechanisms sits within this new terrain, especially when seen as a call for the construction of mechanisms that explain only causal tendencies rather than observable actualities. As we have just noted, the appeal for mechanistic explanation in the social sciences is attractive partly because it is claimed that it does not rest on the existence of general laws. But some philosophers who endorse the mechanisms position disagree with this claim, arguing that laws are still essential, perhaps even constituting the defining characteristic of what a mechanism is. For example, after having written on the role of mechanisms in scientific explanation for several decades, Mario

---

[17]We do not attempt to give a full summary of the fall of covering law models, as many accessible texts exist in philosophy (see citations in the main text) and others that are written for social scientists alone (e.g., Gorski 2004; Hedström 2005). Nor do we give a full explication of the variety of positions that have replaced the covering law model, as these too are well summarized elsewhere (see citations in the main text for realist models in particular, as well as the variety of naturalist positions that contend with them).

Bunge (2004:207) concluded, "No law, no possible mechanism; and no mechanism, no explanation."[18]

Hedström (2005) considers this complication carefully, and he emphasizes that his mechanism-based approach to explanation is less reliant on invariant laws than the covering law models that he rejects. This position is certainly correct, but one must still consider how to grapple with what seem to be three fairly common declarative statements on the relationships between laws and mechanisms: (1) invariant covering laws do not exist, (2) mechanisms depend to some degree on the existence of laws that are weaker than general, invariant laws, (3) mechanisms are nested within each other. Accepting all three statements seems to delimit laws to the smallest possible configuration within each level of a set of nested mechanisms.

If a mechanism is designed to explain how an actuality comes about, then all seems to be fine with this perspective.[19] But if one instead maintains, as does Hedström (2005), that mechanisms are the key to the explanation of causal tendencies only – such that the validity of a mechanism cannot be undermined by its inability to explain anything in particular – then this line of thought leads all too easily to the critical realist perspective on mechanisms. We suspect that most empirical social scientists would find critical realism uninspiring. Critical realism's pioneer, Roy Bhaskar, writes with regard to mechanisms:

> The world consists of mechanisms not events. Such mechanisms combine to generate the flux of phenomena that constitute the actual states and happenings of the world. They may be said to be real, though it is rarely that they are actually manifest and rarer still that they are empirically identified by men. They are the intransitive objects of scientific theory. They are quite independent of men – as thinkers, causal agents and perceivers. They are not unknowable though knowledge of them depends upon a rare blending of intellectual, practico-technical and perceptual skills. They are not artificial constructs. But neither are they Platonic forms. For they can become manifest to men in experience. Thus we are not imprisoned in caves, either of our own or of nature's making. We are not doomed to ignorance. But neither are we spontaneously free. This is the arduous task of science: the production of the knowledge of those enduring and continually active mechanisms of nature that produce the phenomena of the world. (Bhaskar 1998[1997]:34–35)

If the social sciences sign on to the idea that mechanisms are general and transcendentally valid explanations that may not explain any actualities or particularities, we will be led inevitably to a fundamental premise of critical realism: The mechanisms that constitute causal explanations are irreducible, even if they are nested in each

---

[18]Given Bunge's position, one then necessarily wonders, How does one discover these crucial explanatory mechanisms? Bunge's (2004:200) guidance is this: "There is no method, let alone logic, for conjecturing mechanisms. True, Peirce wrote about the 'method of abduction,' but 'abduction' is synonymous with 'conjecturing,' and this – as Peirce himself warned – is an art, not a technique. One reason is that, typically, mechanisms are unobservable, and therefore their description is bound to contain concepts that do not occur in empirical data."

[19]As best we can tell, Woodward (2003, section 4.6) takes this position, noting the need for only a rather limited "backing" relationship between causal claims and laws.

other. This position is summarized by Andrew Collier (2005:335): "Critical realism defends the idea that reality is many-layered, and each level has its own kind of laws, irreducible to those of any other layer."

For the social sciences, one might argue that such an irreducibility presumption could be unifying in offering protection against incursions from biology and physics.[20] But, if accepted, it would undermine the work of social scientists who have at least some interest in developing causal claims that unite levels of analysis. If irreducibility were accepted, how then could methodological individualists such as Goldthorpe criticize those who seek to develop macrolevel causal claims with only minimally sufficient reliance on proximate actors and institutions (e.g., Alexander 2003)?[21] Gorski (2004), for example, lays out a constructive realist model of explanation, built up from the causal process perspective of Wesley Salmon's early work, that is completely at odds with the perspective of Goldthorpe (2007).[22] But Goldthorpe (2007, chapter 2) questions the explanatory utility of all secondhand historical analysis, in essence rejecting the capacity of historical analysis, as practiced in sociology, to sustain causal claims of any form. If Goldthorpe were to sign on to irreducibility, which we doubt he would, he could not thereby criticize macrosocial claims of causal relationships.

Given that these extreme positions on mechanisms in the philosophy of science are likely to be unhelpful to practicing social scientists, and given that Hedström and others in the generative mechanisms movement seem to agree (see Hedström 2005:70–74), which type of philosophy of science gives the appropriate backing for causal analysis? If anything, it is the philosophical writing on the counterfactual model that provides a solid and pragmatic foundation, as best represented for the causal modeling tradition by Woodward (2003). The key to understanding why this is the case is to consider alternative ways to adjudicate between the rival mechanisms proposed by alternative investigators, which we turn to next.

Adjudication Between Rival Mechanisms

Imagine that social scientist A and social scientist B have proposed distinct mechanisms for how X brings about Y. How do they determine whose mechanism is supported? According to Goldthorpe (2001, 2007), each scholar is expected to derive entailed hypotheses, typically indirect, and test them with data. One might hope that scholars A and B will be able to agree on a clear critical test that could tip the scales in favor of one mechanism or the other. Unfortunately, the mechanisms of the two

---

[20] Of course, nontranscendental protection is available as well, as we explain below when discussing the notion of "bottoming out" in mechanistic explanation. Even Strevens (2008:472), while defending reducibility to fundamental physics, can concede that many lower-level details may be suppressed when they play no role in producing the outcome of interest.

[21] Methodological individualism is the basic position of Goldthorpe (2001, 2007) and Hedström (2005), as influenced heavily by the scholarship of Raymond Boudon (see Boudon 1998 and citations therein).

[22] Gorski (2004) endorses the causal process model of Wesley Salmon, as developed in Salmon's work from the 1970s and early 1980s (see Salmon 1984). Given the ways in which Salmon's work developed later in his career, turning completely toward causal mechanical ideas based on the notion of conserved quantities, his updated ideas seem completely irrelevant to what we take to be Gorski's main position: "Social science *is* 'nothing but history.' The real error was ever to think it could be anything more" (Gorski 2004:30; emphasis in original).

scholars may be so different that no such critical test can be derived and agreed on (as would often be the case if scholar A is a sociologist and scholar B is an economist, for example). If no such agreement can be found, the two scholars may end up expending effort seeking to affirm their own entailed indirect hypotheses, producing results that are then regarded as irrelevant by each other.

Consider the reaction that Goldthorpe's proposal elicited from the statisticians David Cox and Nanny Wermuth, whose prior work Goldthorpe had used to develop his proposal:

> Goldthorpe (2001) has argued for this ... view of causality as the appropriate one for sociology with explanation via rational choice theory as an important route for interpretation. To be satisfactory there needs to be evidence, typically arising from studies of different kinds, that such generating processes are not merely hypothesized. Causality is not to be established by merely calling a statistical model causal. (Cox and Wermuth 2001:69)

By our interpretation, Cox and Wermuth take the position that generative mechanisms must be directly evaluated, not evaluated only by indirect entailed hypotheses. Anything short of this analysis strategy could result in a flourishing of mechanisms in the social sciences, without an attendant sense of which ones have sufficient empirical support to command attention.

The alternative to any such mechanism anarchy could be even worse: mechanism warlordism. The mechanisms of the most industrious scholars – those who can dream up the largest number of hypotheses to affirm, who can recruit the largest number of students to do the same, and who can attract the largest amount of funding to collect the data on their hypotheses – could receive the most affirmation. The only defense for out-of-favor mechanisms might then be to appeal to the hidden structures of alternative mechanisms, which one would be tempted to claim cannot be evaluated because of a lack of data.

In sum, the generative mechanisms movement in the social sciences is an admirable call for theory construction.[23] The appeal for finely articulated "semigeneral" mechanisms to fill the toolkit of social theory (see Hedström and Bearman 2009:6–7) is consistent with the appeal to forms of abduction that generate what Craver (2007) has labeled "how-possibly" and "how-plausibly" models:

> *How-possibly models* are often heuristically useful in constructing and exploring the space of possible mechanisms, but they are not adequate explanations. *How-actually models*, in contrast, describe real components, activities, and organizational features of the mechanism that in fact produces the phenomenon. Between these extremes is a range of *how-plausibly models* that are more or less consistent with the known constraints on the components, their activities, and their organization. (Craver 2007:111–12).

We see the generative mechanisms movement as a call for the construction of how-possible and how-plausible models. Such activity is worthwhile in its own right, and it

---

[23]In fact, the first author found it convincing and inspiring when writing Morgan (2005).

may help to inform empirical research. We see causal analysis, in contrast, as a directed effort to evaluate the support for specific claims implied by how-actually models.

Finally, as long as the generative mechanisms movement does not verge into critical-realist transcendentalism in the future, it will remain a very useful call for the pursuit of sufficiently deep causal accounts. An elegant mechanism, if how-plausible, may be a crucial step forward in convincing a field that an accepted causal claim is supported by an underlying mechanistic account that should be seen as too shallow. Thus, even though theory construction is not causal analysis, theory construction may help to establish our goals for explanatory depth. We take the position that genuine causal depth must be secured by empirical analysis, and that such analysis is often best grounded on the counterfactual model. In the next section, we work our way back to Pearl's front-door criterion, after introducing a language of mechanism sketches and mechanism schemas that is helpful for classifying alternative representations of mechanisms.

10.4   The Pursuit of Explanation with Mechanisms That Bottom Out

Amid the resurgence of writing on mechanisms, we find one statement more helpful than many others, the 2000 article "Thinking About Mechanisms" written by Machamer, Darden, and Craver and published in *Philosophy of Science*. In their article, Machamer, Darden, and Craver develop two particularly helpful lines of thought: (1) the distinctions between a fully articulated mechanism, a mechanism sketch, and a mechanism schema, and (2) the process of "bottoming out" in mechanistic model building and subsequent explanation. To develop these concepts, Machamer et al. (2000:12) first note that "in a complete description of [a] mechanism, there are no gaps that leave specific steps unintelligible; the process as a whole is rendered intelligible in terms of entities and activities that are acceptable to a field at a time." But they then explain that explanatory inquiry using mechanisms is not an all-or-nothing affair, in which every step is always specified. Variability in the representation of mechanisms is possible because

> mechanisms occur in nested hierarchies. ... The levels in these hierarchies should be thought of as part-whole hierarchies with the additional restriction that lower level entities, properties, and activities are components in mechanisms that produce higher level phenomena...." (Machamer et al. 2000:13)

In spite of such nesting, there is a natural *bottoming out* of mechanism-based explanations:

> Nested hierarchical descriptions of mechanisms typically *bottom out* in lowest level mechanisms. These are the components that are accepted as relatively fundamental or taken to be unproblematic for the purposes of a given scientist, research group, or field. Bottoming out is relative: Different types of entities and activities are where a given field stops when constructing

mechanisms. The explanation comes to an end, and description of lower-level mechanisms would be irrelevant to their interests. (Machamer et al. 2000:13)

Then, by thinking through the complexity of the nesting of mechanisms, and how scholars represent mechanisms to each other, they develop two related concepts. A *mechanism schema* is a representation of a mechanism in which some known details (or known nested levels) are suppressed for the sake of simplicity. Or, as Machamer et al. (2000:15) state, "a *mechanism schema* is a truncated abstract description of a mechanism that can be filled with descriptions of known component parts and activities." In contrast, a *mechanism sketch* is quite different:

> A sketch is an abstraction for which bottom out entities and activities cannot (yet) be supplied or which contains gaps in its stages. The productive continuity from one stage to the next has missing pieces, black boxes, which we do not yet know how to fill in. A sketch thus serves to indicate what further work needs to be done in order to have a mechanism schema. Sometimes a sketch has to be abandoned in the light of new findings. In other cases it may become a schema, serving as an abstraction that can be instantiated as needed. (Machamer et al. 2000:18)

Within this framework, one can conceive of causal analysis in the social sciences as the pursuit of explanations that bottom out.[24] Although there will inevitably be differences of opinion on how deep an explanation must be to bottom out, it would seem uncontroversial to state that a valid explanation that invokes a mechanism must bottom out at least as far as the observables in the data at hand.[25]

This position, although we think it uncontroversial, is still not specific enough for our tastes. In the remainder of this section, we discuss two entangled issues: (1) when deep accounts cannot be provided because nominal causal states are too coarsely defined and (2) how the pursuit of deep causal accounts should proceed when the specification of causal states will allow it.

Coarse Causal States Impede the Pursuit of Causal Depth. When we introduced in Section 2.1 the causal states that define potential outcomes, we took the position that causal states should be finely articulated and motivate comparisons that are local and reasonable. We also noted, in the subsection beginning on page 39,

---

[24]See also the discussions of modularity in Woodward (2003, chapter 7) and Knight and Winship (2013).

[25]A critical realist could escape from this position in a variety of ways: asserting irreducibility and transcendentalism and then, more specifically, by arguing in the end that the data that one is forced to consider are but a poor reflection of the phenomena that the mechanism truly does explain. For all of these reasons, the lack of observed explanatory power for observed events would then be argued to be untroubling. This position, however, then becomes a variant of an appeal to a hidden but presupposed valid underlying structure, which Woodward convincingly argues cannot be an acceptable explanatory strategy for any field that hopes to resolve its explanatory controversies because "the appeal to hidden structure makes it too easy to protect one's favored theory of explanation from genuine counterexamples" (Woodward 2003:175). Moreover, if the particularities in the data are merely a poor reflection of the phenomenon that the mechanism is supposed to explain, then presumably whatever generates the mismatch can be encoded in the mechanism that explains both the genuine phenomenon of interest and the process that generates the misleading data.

that we see value in regarding each state of each treatment as a nominal state with constitutive features (e.g., entities, activities, and relations) that are jointly capable of producing the outcome of interest. We then provided examples where the causal states are composed of constitutive features. For example, the types of schools that define the causal states for many of the examples in this book are composed of teachers, classrooms, curricula, administrators, normative environments, affiliated institutions, and networks of peers and parents. We maintained that causal effects can still be defined with reference to these nominal states because we can conceive of differences in outcomes that would result from alternative exposure in toto to the constitutive features of each nominal state.

We will not repeat the pragmatism-based argument of Section 2.1 here, but the discussion there ends on a crucial question relevant here. After presenting, as an example, the literature that has focused on the particular features that may give Catholic schools an advantage in the production of learning, we asked what should be made of any such lower-level causal claims. More generally, we asked: When should nominal causal states be decomposed into component causal states in pursuit of more finely articulated causal explanations? If Catholic schools and public schools are to be regarded as alternative nominal causal states relative to each other, and yet one has some basis for claiming that Catholic schools produce a learning advantage because of a greater endowment of one particular constitutive feature – dense parental networks, for example – then should we dispense with the nominal causal states and redefine the research question solely to enable an examination of this particular feature?

If interest exists in understanding what really would happen if we exposed individuals to alternative nominal causal states, then this interest alone is sufficient justification for not doing so. Many policy evaluation studies take this position. Investigators may not be able to discern which particular features of an intervention program are effective, but they still want to know whether the program, altogether, has an effect on average, and, furthermore, whether that effect varies across types of individuals who could be exposed to it. More generally, when nominal causal states are composed of many constitutive features, only some of which contribute to the production of the outcome, the relevant scholarly community may recognize that only shallow effects are feasible to estimate. If data are not available to analyze all of the constitutive features – either to assess their autonomous causal capacities or the processes that imbue them with joint and interactive causal capacities – then the causal claims based on the nominal states alone may be regarded as insufficiently deep and yet still be as deep as can be sustained by an empirical analysis.

Overall, when these challenges arise in practice, we see inherent tensions, and hence can offer no general guidance, on whether one should (1) preserve the nominal causal states as originally defined and settle for a shallow account of how the cause produces its effect, (2) preserve the nominal causal states as originally defined but then also attempt to deepen the shallow account by specifying a full mechanism for how the cause produces its effect, or (3) decompose the original nominal causal states into component causal states and then specify separable mechanisms that can account for the partial effects attributable to particular constitutive features, now redefined as lower-level nominal states. Making this choice surely depends on the state of relevant theory about the process under analysis, whether the constitutive features can be meaningfully separated and then measured, whether data with such measures are available, and what standards prevail in the field for which the study is intended.

The Pursuit of Causal Depth by Investigation of Nested Mechanisms. As we noted at the beginning of this chapter with illustrative quotations from Duncan et al. (1972) and Kendall and Lazarsfeld (1950), for many decades social scientists have used intervening variables in empirical analysis in attempts to explain how causes bring about their effects. In this section, we will apply the language of Machamer et al. (2000) to this tradition of analysis, using causal graphs and making explicit connections to our presentation of Pearl's front-door criterion.

Recall that for Machamer et al. (2000) the difference between a mechanism sketch and a mechanism schema is the source of the abstraction in each. For a schema, all of the parts of the mechanism are known, and the abstraction from them is enacted for presentation purposes. For a sketch, some of the parts are unknown, and thus the abstraction is a representation of the analyst's epistemic limitations.

Before considering directed graphs of the sort utilized in this book, we should note how easy it is to find arrow-based figurative representations that can be interpreted as mechanism abstractions. Here are five examples from widely regarded work in sociology where →'s are utilized liberally:

1. In *Foundations of Social Theory*, Coleman (1990:637, see figure 23.4) offers a representation of a feedback process for social policy research where five actors and activities – "Government," "Intermediaries," "Policy recipients," "Research on social policy," and "Political representation" – are connected by eight →'s and five --→'s.

2. In *Class Counts*, Wright (1997:262, see figure 10.1) offers a model for the effects of class location on class identity where "Direct class location" and "Mediated class location" are connected to "Class identity" through seven →'s that are connected to "Patterns of daily work interaction," "Material class interests," "Production-centered class experiences," and "Consumption-centered class experiences."

3. In "A Plea for Mechanisms," Elster (1998:63, see figure 3.1) offers mechanisms for the interaction of democracy and religion where "Democracy," "Religion," "Irreligion," "Desires," "Opportunities," and "Action" are connected by six →'s.

4. In *Interaction Ritual Chains*, Collins (2004:48, see figure 2.1) offers a representation where Durkheimian "Collective effervescence" converts four "Ritual ingredients" (group assembly, barrier to outsiders, mutual focus of attention, and shared mood) into four "Ritual outcomes" (group solidarity, emotional energy in the individual, symbols of social relationship, and standards of morality), via a relation] → [.

5. In *A Theory of Fields*, Fligstein and McAdam (2012:20, see figure 1.1) offer a representation of "three linked mechanisms" for episodes of contention where two types of actors "Incumbent" and "Challenger" respond to "Destabilizing changes" and "Escalation of perceived uncertainty" by enacting "Attribution of threat/opportunity," which leads to "Social appropriation," and then to "Innovative collective action." These mechanisms of action are connected by 12 →'s,

Many more examples exist.[26] Our point in offering these five examples is to highlight
the revealed presentation value of mechanism abstractions that use arrows.

Now, consider the sort of finely structured causal graphs we have considered extensively in this book. Suppose that one uses some form of back-door conditioning to
identify a causal effect of D on Y. Suppose, furthermore, that one has done so from
within the counterfactual framework, settling on the ATE as the parameter of first-
order interest. One then necessarily confronts the question that we raised in the
beginning of this chapter: Does a counterfactually defined and consistent estimate
of the causal effect of D on Y by itself meet the standard of an explanation that
bottoms out?

The answer to this question is clear: It depends on what D and Y are, who is
conducting the study, and for what purposes. If one wishes to know only how a hypothetical intervention on D would shift Y, and hence has no interest in anything else
whatsoever, then bottoming out has been achieved in some minimalist way, as we
noted above. The analysis yields up a consistent estimate of the average causal effect
of D on Y that holds for the population within which both are observed. In this case,
the model $D \to Y$ is then regarded as merely a mechanism sketch, which suffices to be
treated as a sufficiently deep explanation for the purposes at hand.

However, it will often be the case that an estimate of a warranted causal effect of D
on Y, grounded in the potential outcomes framework, is properly considered to be an
insufficiently deep explanation of how D brings about Y. Such a judgment would be
appropriate when the interest of the field is in understanding both how interventions
on D and interventions on the mechanistic intervening variables that link D to Y
would shift Y. In this case, the model $D \to Y$ is a mechanism sketch that, for the
purposes at hand, cannot be regarded as a sufficiently deep explanation. The arrow in
the sketch $D \to Y$ is a black box that must be filled in through further analysis.

If a warranted claim of a counterfactually defined causal effect is properly regarded
as insufficiently deep, the recourse is not to abandon the counterfactual model but
rather to investigate the nested mechanism that intercepts the effect of D on Y. Such
analysis may initially take the form of a set of alternative conjectured mechanisms.
But, ultimately, any such analysis must return to the particular observed relationship
between D and Y in the population of interest (or more broadly in multiple well-defined
populations for which the question of interest is relevant). Thus, although theoretical
creativity and new measurement techniques may be required, opening up black boxes
is, at least in part, an empirical pursuit. At some point, one must specify the causal
states for the variables that constitute the mechanisms, and each link thereby specified
to give rise to an effect must be submitted to its own causal analysis.

Consider this process abstractly with reference to a causal graph, after which we
will introduce examples from the published literature. Suppose that one pursues further
theoretical conjecturing and subsequent empirical analysis, and one then determines
that the variables A, B, and C constitute an exhaustive and isolated mechanism that
identifies the causal effect of D on Y by Pearl's front-door criterion. At this point, one
may be tempted to declare that a sufficiently deep causal explanation of the effect of

---

[26]An early (and particularly obtuse) one is Parsons (1937:741, footnote 1), where a "web of inter-
woven strands" is used to explain how means-end chains evolve in time and structure unit acts.

D on Y has been secured. Such a claim may well be true, but it is not guaranteed. It could be that further nested mechanistic variables exist, such that, for example, three additional variables M, N, and O (each of which is a concept of considerable interest to one's peers) are then found to elaborate the casual pathway $D \to A \to Y$. In this case, the causal pathway $D \to A \to Y$ is then itself best regarded in hindsight as merely a component of a mechanism sketch. When M, N, and O are then observed, $D \to A \to Y$ is replaced in the mechanism sketch with, for example, one or more related causal pathways, such as $D \to M \to A \to N \to Y$ and $D \to A \to O \to Y$. In this example, A, B, and C may well identify the causal effect by the front-door criterion, but they do not qualify as a sufficiently deep causal account of how D brings about Y.

As we noted in the first part of this chapter, the progressive deepening of causal explanation through the modeling of intervening processes is entirely consistent with social science tradition. And yet we also claimed that Pearl's front-door criterion can help guide sharpened analysis practices in this regard. To see what we mean, consider the example of Duncan's research on status attainment processes again. As we noted in Chapter 1, Blau and Duncan (1967) deepened the causal account of how parental social status determines offsprings' social status by specifying what most scholars now regard as the most important link: levels of educational attainment.

Thereafter, Duncan and his colleagues then supported and encouraged further work on the process of educational attainment, most importantly the Wisconsin model of status attainment that we introduced in Section 1.3.1. This model is a direct extension of Blau and Duncan's research, in which the causal pathways between parental status and offspring's attainment were elaborated by the introduction of intervening variables for significant others' influence and educational aspirations. In fact, in the most important article in this tradition, Sewell et al. (1969) used mechanistic language to introduce the contribution of their study:

> we present theory and data regarding what we believe to be a logically consistent social psychological model. This provides a plausible causal argument to link stratification and mental ability inputs through a set of social psychological and behavioral mechanisms to educational and occupational attainments. One compelling feature of the model is that some of the inputs may be manipulated through experimental or other purposive interventions. This means that parts of it can be experimentally tested in future research and that practical policy agents can reasonably hope to use it in order to change educational and occupational attainments. (Sewell et al. 1969:84)

The Wisconsin model was very favorably received in sociology, as it was considered to be consistent with the basic features of the model of Blau and Duncan (1967) and yet had a claim to greater causal depth.[27]

Even so, as we also noted in Section 1.3.1, critics emerged immediately (see also Morgan 2005, chapter 2, for a more extended summary). The basic argument was that, even if significant others' influence and educational aspirations have causal effects on

---

[27]Another widely read example of the time is the earlier mechanistic elaboration in the research of Kendall and Wolf, contributed by Kendall to Hyman (1955:324–27).

educational attainment, they are both grounded in part in sources outside of parental status and mental ability (a point the authors of the Wisconsin model recognized). Thus, although significant others' influence and educational aspirations may be helpful to some extent in offering an interpretation of some of the causal process that generates intergenerational correlations of educational attainment, these intervening variables do not qualify as an exhaustive and isolated mechanism that fully accounts for the effects of parental status on offsprings' status.

In this regard, the Blau and Duncan model can be regarded as a mechanism sketch for the status attainment process, and the Wisconsin model can then be regarded as a mechanism-based attempt to deepen its implied explanation. The Wisconsin model was therefore an important step forward, but it was not conclusive and did not settle all further research.

More than four decades later, it is now clear that the Wisconsin model itself is a mechanism sketch. The research community of inequality scholars in sociology seems to have concluded that its pathways have not bottomed out, and much research continues on the processes that generate educational aspirations (as well as whether or not the relationship between aspirations and attainment is sufficiently explanatory to be useful). Moreover, some scholars (e.g., Goldthorpe 2007) have produced entirely different mechanism sketches for the relationship between parental status and educational attainment. The future of this research tradition is clearly careful empirical analysis that can adjudicate between these rival mechanism sketches, which will be decisive only when alternative mechanism sketches are pushed down to lower-level entities on which critical tests can then be performed.

10.5   Conclusions

Pearl's front-door strategy for the identification of a causal effect is a powerful and illuminating perspective on the explanatory power of mechanisms. It clearly shows that the identification of a causal effect by a mechanism requires that the mechanism be exhaustive and isolated and that its variables be observed. For such a mechanism to count as a sufficiently deep explanation, its causal pathways must be finely enough articulated that it meets whatever standard of bottoming out is maintained in the relevant field of study. If such a standard is not reached, then the causal effect is identified even though it is not accompanied by a sufficiently deep explanation. Instead, the identifying causal pathways represent a mechanism sketch that demands further analysis.

Considering this chapter and the strategies for causal effect estimation from prior chapters, we have come back full circle to our initial presentation of causal modeling options in Chapter 1. We noted there, with reference to Figure 1.3, that a causal effect that is identified by both the back-door criterion and an IV is best explained when it is also identified by an exhaustive and isolated mechanism. This is the highest standard for an explanatory causal analysis, at least until a field decides that a crucial linkage within a mechanism must then be opened up and subjected to its own analysis.

In the next chapter, we turn in a different direction to consider the extent to which over-time data on an outcome variable can be used to identify and estimate a causal effect. One often hears presentations in which scholars remark, "I cannot get at causality because I do not have longitudinal data." We will argue in the next chapter that longitudinal data, although very helpful in many cases, are not the panacea that such statements seem to imply. Moreover, we will show that some long-standing techniques that are thought to reveal causal effects are strongly dependent on assumptions that are often entirely inappropriate and sometimes completely unrecognized.

Chapter 11

Repeated Observations and the Estimation of Causal Effects

As discussed in previous chapters, the fundamental challenge of causal inference is that an individual cannot be simultaneously observed in both the treatment and control states. In some situations, however, it is possible to observe the same individual or unit of observation in the treatment and control states *at different points in time*. If the potential outcomes do not evolve in time for reasons other than the treatment, then the causal effect of a treatment can be estimated as the difference between an individual's observed outcome in the control state at time 1 and the same individual's observed outcome in the treatment state at time 2. The assumption that potential outcomes are stable in time (and thus age for individuals) is often heroic. If, however, potential outcomes evolve in a predictable way, then it may be possible to use the longitudinal structure of the data to predict the counterfactual outcomes of each individual.

We begin our discussion with the interrupted time series (ITS) design, which we introduced already with the example of the year of the fire horse in Section 2.8.1. The ITS design is the simplest case where the goal is to determine the degree to which a treatment shifts the underlying trajectory of an outcome. It is simple because the analysis is based only on data for a single individual or unit of analysis observed at multiple time points. We also consider the regression discontinuity design. Although not a case in which we have repeated observations on a single individual or unit, the structure of the regression discontinuity (RD) design is sufficiently similar to that of the ITS that it is useful to present it here as well. For the RD design, we also consider the case of fuzzy assignment, which amounts to using instrumental variable (IV) methods to correct for possible imprecision in the treatment assignment criterion.

We then transition to a full consideration of panel data: multiple observations over time on multiple individuals or units. We first examine the adequacy of traditional two-period adjustment strategies, building on our brief introduction to these models in Section 8.2, where we demonstrated how little insight can be gained from additional posttreatment data. We show in this chapter that such methods, even when used with pretreatment data, are inadequate for making causal inferences unless one is willing to make strong and usually untestable assumptions about how the outcome evolves over time across individuals.

We then consider a more comprehensive model-based approach. The key requirements of this approach are assumptions about the evolutionary dynamics of the outcome and how selection into the treatment depends on these dynamics. This type of strategy typically requires data from multiple pretreatment time periods. With data over a sufficient number of time periods, it is possible to test the appropriateness of different models.

Although for the main body of this chapter we will assume that the time period at which the treatment occurs is fixed and has no dynamic structure, in an appendix to this chapter we will consider scenarios in which the treatment can be repeated and the specific timing of each treatment instantiation is endogenous. These scenarios are considerably more complex because a treatment indicator must be modeled for every time period, recognizing that selection of the treatment in any single time period is not only a function of individual characteristics but also of previous decisions and expectations of future decisions.

11.1 Interrupted Time Series Models

To estimate the treatment effect for a study with an ITS design, a time series model is typically offered:

$$Y_t = f(t) + D_t b + e_t, \tag{11.1}$$

where Y_t is some function in time (which is represented by $f(t)$ on the right-hand side), D_t is a dummy variable indicating whether the treatment is in effect in time period t, and e_t is time-varying noise. The basic strategy of an ITS analysis is to use the observed trajectory of Y_t prior to the treatment to forecast the future trajectory of Y_t in the absence of the treatment (see the introductions in Marcantonio and Cook 1994, McDowall, McCleary, Meidinger, and Hay 1980, and Shadish et al. 2001).

Consider, as in our prior example in Section 2.8.1 on the year of the fire horse, how potential outcome notation can be used to understand the ITS design. For setup, suppose that we have discrete intervals of time t that increase from 1 to T. The outcome variable Y_t in Equation (11.1) then has observed values $\{y_1, y_2, y_3, ..., y_T\}$. The two-state causal variable, D_t, is equal to 1 if the treatment is in place during a time period t and is equal to 0 otherwise. For the ITS design, it is typically assumed that once the treatment is introduced in time period t^*, its effect persists through the end of the observation window, T.

Analogous to (but a bit simpler than) our general setup in Section 2.8.1, the ITS observed data are defined in terms of potential outcome variables as

1. Before the treatment is introduced (for $t < t^*$):[1]

$$D_t = 0$$
$$Y_t = Y_t^0$$

2. After the treatment is in place (from t^* through T):

$$D_t = 1$$
$$Y_t = Y_t^1$$
$$Y_t^0 \text{ exists but is counterfactual.}$$

The causal effect of the treatment is then

$$\delta_t = Y_t^1 - Y_t^0 \tag{11.2}$$

for time periods t^* through T. By the definition of the potential outcomes, Equation (11.2) is equal to

$$\delta_t = Y_t - Y_t^0, \tag{11.3}$$

again for time periods t^* through T.

The crucial identifying assumption for the ITS design is that the observed values of y_t before t^* can be used to specify $f(t)$ for all time periods, including time periods from t^* to T.[2] Equivalently, the primary weakness of the ITS design is that the evolution of Y_t prior to the introduction of the treatment may not be a sufficiently good predictor of how Y_t would evolve in the absence of the treatment. In other words, even though the pretreatment evolution of Y_t is by definition a perfect reflection of the evolution of Y_t^0 before t^*, it may be unreasonable to extrapolate to posttreatment time periods in order to estimate treatment effects defined by Equation (11.3).

Consider the trajectory of Y_t in the hypothetical example depicted in Figure 11.1. The solid line represents the observed data on Y_t, and the time of the introduction of the treatment is indicated on the horizontal axis, which we defined above as t^*. The true counterfactual evolution of Y_t^0 in the absence of treatment is represented by the dashed line. Clearly, this counterfactual trajectory would be poorly predicted by a straightforward linear extrapolation from the observed data before the treatment. In fact, in this case, assuming that the counterfactual trajectory followed such a linear trajectory would result in substantial overestimation of the treatment effect in all posttreatment time periods.

For estimation, no issues beyond those relevant to a standard time series analysis arise for an ITS model. The key statistical concern is that the errors e_t are likely to be correlated over time. If we use least squares regression, the parameter estimates will be consistent, but the standard errors and any hypothesis tests based on them will be incorrect. This problem can be especially acute when the number of data points in a

---

[1]Again, as in Section 2.8.1, counterfactual values for Y_t^1 exist in pretreatment time periods, but these values are not typically considered in an ITS analysis.

[2]A secondary assumption, which we do not emphasize, is the common position that the parameter b in Equation (11.1) is a structural constant that does not vary with t. This assumption can be relaxed, allowing b to vary from t^* through T as some function in t.

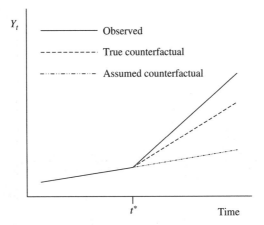

Figure 11.1 Trajectories of the observed outcome as well as the true and assumed counterfactual outcomes for a faulty ITS model.

time series is small. We will not address any of the issues involved in estimating time series models, as there are many books that cover the topic in depth (e.g., Hamilton 1994; Hendry 1995).

Instead, we will illustrate the basic thinking behind an ITS analysis with an example from Braga, Kennedy, Waring, and Piehl (2001), which is presented in Figure 11.2. Braga and his colleagues were interested in evaluating whether an innovative program, "Operation Ceasefire," initiated by the Boston Police Department in June 1996, prevented youth homicides. Figure 11.2 presents the trend in the monthly youth homicide rate in Boston between June 1991 and May 1998.

Operation Ceasefire involved meetings with gang-involved youth who were engaged in gang conflict. Gang members were offered educational, employment, and other social services if they committed to refraining from gang-related deviance. At the same time, the police made it clear that they would use every legal means available to see that those who continued to be involved in violent behavior were sent to prison (see Kennedy 1997 for a more detailed description).

The vertical line in Figure 11.2 marks the date at which Operation Ceasefire was initiated. The two horizontal lines indicate, respectively, the mean level of youth homicides before and after June 1996. As can be seen in Figure 11.2, there appears to be an abrupt and large drop in the level of youth homicide in Boston immediately after the implementation of Operation Ceasefire.

Braga and his colleagues carried out a more formal analysis using standard time series techniques (but because their dependent variable is a count variable – number of youth homicides in a month – they used a Poisson regression model). In their first model, they adjusted only for seasonal effects by using dummy variables for month and a linear term for time. Inclusion of the time trend is particularly important. Although it is not clear in Figure 11.2, there is a slight downward time trend in the homicide rate in the pretreatment time period, which it seems reasonable to assume would have continued even if Operation Ceasefire had not been implemented. For this model,

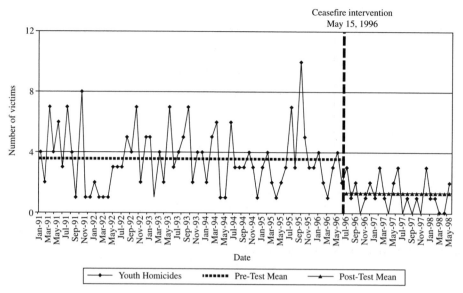

Figure 11.2 Monthly youth homicide rates in Boston, 1991–1999.
Source: Braga et al. (2001), figure 2.

Braga and his colleagues reported a large negative and statistically significant effect of Operation Ceasefire on youth homicides.

In general, a researcher would usually prefer a situation in which the underlying time trend and the treatment effect are in the opposite directions. In such a case, disentangling the treatment effect and the time trend is far easier, strengthening any warrant for a causal inference. However, for this example, the underlying time trend and the expected program impact move in the same direction. Thus, a researcher should be concerned about the ability to accurately estimate the size of the program effect in the presence of the underlying trend. Recall that Figure 11.1 illustrated a similar but more problematic situation. Not only does the time trend move in the same direction as the effect in that hypothetical example, but there is a change in the time trend in the same direction as the treatment effect, making accurate estimation of the effect impossible with an ITS model.

The research of Braga et al. (2001) represents a very high-quality example of how to use an ITS design to investigate a treatment effect. They offered four types of supplemental analysis, each of which is broadly applicable to all ITS designs, and each of which can be used to strengthen the warrant for a causal claim. First, they considered additional dependent variables that Operation Ceasefire should have affected. Specifically, they analyzed whether the program affected the number of gun assaults and reports of gun shots fired. For these dependent variables, they found an even larger program impact. Their analysis would have been strengthened further if they also had considered dependent variables that Operation Ceasefire should not have affected as much (e.g., number of robberies and incidents of domestic violence). Here, evidence of an impact would suggest that factors other than the program itself at least partially accounted for the observed drop in youth homicides.

Second, they focused their hypothesis and considered it within meaningful subgroups. In particular, they analyzed the level of gun assaults in police district B-2, the district where gang violence was the highest in the early 1990s and Operation Ceasefire was most active. As they predicted, the program effect was larger in district B-2 than in the city as a whole. If there had been districts with high levels of youth violence where Operation Ceasefire was inactive, it would have been useful to have tested for a program impact. If evidence were found that the program had an impact in these districts, it would suggest that something other than the program was responsible for the observed decline in youth homicides, gun assaults, and gun shots fired. Unfortunately, at least for the analyst, almost all youth violence in Boston occurred in three adjacent police districts, all districts in which Operation Ceasefire was active. As a result, such an analysis was not possible.

Third, they included additional adjustment variables in their time series models in order to capture the underlying time trend as well as the year-by-year variability before and after the introduction of the treatment. These time-varying covariates included unemployment rates, the size of the youth population, the robbery rate, the homicide rate for older victims, and the drug-related arrest rate. The advisability of adjusting for the latter three variables is questionable, given that these variables are also likely to have been affected by Operation Ceasefire to at least some degree. Nonetheless, conditioning on these additional variables produced little change in their estimate of the program impact on any of their dependent variables.

Finally, they compared the time trend in homicides in Boston with the time trend in 41 other cities where no targeted interventions for homicide rates were implemented. Their goal was to determine whether other cities experienced declines in rates as abrupt as the one observed in Boston. The explanation of Braga and his colleagues for the abruptness of the decline in Boston – in fact, a decline that emerged in only two months – was that word got out on the street quickly that the police meant business. For many of the other cities considered, homicide rates fell throughout the 1990s. With the exception of New York City, the declines were substantially smaller than in Boston. Braga and his colleagues then showed that abrupt declines did occur in five other cities, but the exact timing of these abrupt declines was different than in Boston. This evidence raises perhaps the greatest doubt about the assertion that Operation Ceasefire reduced youth homicide rates in Boston because there is no clear explanation for why these abrupt declines occurred elsewhere either. And, because it may be implausible that Operation Ceasefire's effect could have fully taken hold in as short as two months, the possibility exists that the decline in homicide rates and the introduction of Operation Ceasefire were coincidental.

If Braga and his colleagues had carried out their evaluation a number of years later, they could have implemented one additional strategy. In 1999, Operation Ceasefire was cut back and then terminated. If they had performed their analysis for at least a few years beyond 1999, they could have examined whether the termination of the program resulted in an increase in homicide rates. In fact, after 1999, the youth homicide rate did increase such that, by the summer of 2006, it was at nearly the same level as in the early 1990s (see Braga, Hureau, and Winship 2008). The subsequent increase provides additional evidence for the impact of Operation Ceasefire while it was in place.

This example nicely illustrates the variety of general strategies that are often available to strengthen an ITS analysis:

1. Assess the effect of the cause on multiple outcomes that should be affected by the cause.

2. Assess the effect of the cause on outcomes that should not be affected by the cause.

3. Assess the effect of the cause within subgroups across which the causal effect should vary in predictable ways.

4. Adjust for trends in other variables that may affect or be related to the underlying time series of interest.

5. Compare the focal time trend with the time trend for other units or populations to determine whether breaks in the time series are likely to occur in the absence of the cause.

6. Assess the impact of the termination of the cause in addition to its initiation.

These strategies are often available for other types of analysis, and they are also widely applicable to all forms of data analysis that attempt to infer causation from over-time relationships. Unless one has a case as dramatic as the year of the fire horse, these strategies are essential for building support for a causal inference.

11.2   Regression Discontinuity Designs

A regression discontinuity (RD) design is very similar to an ITS design, except that the treatment assignment pattern is a function of the values of a variable rather than the passage of time. An RD design is especially appropriate in situations where treatment assignment is sharply discontinuous in the values of a variable, and it has been applied to a variety of problems: the effect of student scholarships on career aspirations (Thistlewaite and Campbell 1960), the effect of unemployment benefits for former prisoners on recidivism (Berk and Rauma 1983), the effect of financial aid on attendance at a particular college (Van der Klaauw 2002), the effect of class size on student test scores (Angrist and Lavy 1999), and the willingness of parents to pay for better schools (Black 1999).

Campbell was the first to propose the RD design (see Shadish et al. 2001; Trochim 1984), but it has evolved considerably since then (see Bloom 2012; Hahn, Todd, and Van der Klaauw 2001; Imbens and Lemieux 2008). It is most easily understood with an example. Here we consider the example of Mark and Mellor (1991), which is discussed also by Shadish et al. (2001). Mark and Mellor were interested in examining the effect that an event of high personal relevance may have on hindsight bias – the claim that an event was foreseeable after it occurred. Selecting all members of a large manufacturing union, they examined the specific effect of being laid off from work on retrospective

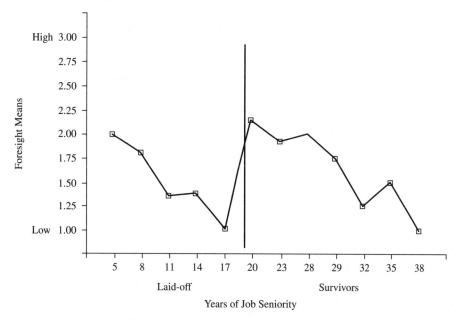

Figure 11.3 Foreseeability of a layoff as an example of an RD design.
Source: Mark and Mellor (1991), figure 1.

foresight claims (measured as agreement with statements such as "I saw it coming all the way"). The study took place just after workers with fewer than 20 years of seniority were laid off. Figure 11.3 shows the relationships between retrospective claims of foresight and both layoff status and seniority.

As shown in Figure 11.3, there is an abrupt discontinuity in the relationship between retrospective foresight claims and seniority at the point in seniority where individuals were laid off. All workers, regardless of whether they were laid off, were asked whether they had expected the layoffs that occurred among union members. Those who were not laid off (i.e., individuals with 20 or more years of seniority) were on average more likely to claim that the layoffs in the union were expected, even though they were not themselves laid off. At the same time, those who were laid off were less likely to claim that the layoffs were expected. Notice also that the association between seniority and retrospective claims of foresight is in the opposite direction of the estimated treatment effect: Individuals with more seniority were less likely to claim that the layoffs were expected, even among those who were subsequently laid off. Because the layoff effect and the underlying seniority association are in the opposite directions, Mark and Mellor had strong evidence that being laid off decreased the likelihood that an individual who was laid off would claim that the layoffs were expected. This finding strengthened their overall interpretation that the personal relevance of a negative event decreases the likelihood that an individual will claim that the event was expected. They concluded that this pattern reflects a type of self-protection, according to which individuals seek to avoid blaming themselves for not having been sufficiently prepared to mitigate the negative consequences of an event.

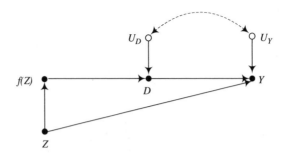

Figure 11.4 An example of a fuzzy RD design.

In general, an RD model can be estimated in the same way as an ITS model because most of the same issues apply. One key and helpful difference is that in most RD designs individuals are sampled independently. As a result, the problem of correlated errors in an ITS design is absent in an RD design.

A generalization of the RD design, known as the fuzzy RD design, has received recent attention.[3] Here, the treatment variable is a function of an assignment process in which there is error that is associated with Y. Consider the graph in Figure 11.4, where $f(Z)$ represents the intended assignment rule to D as a function in Z. However, in this case the assignment is imperfect in the specific sense that the other determinants of assignment, U_D, cannot be assumed to be independent of the unobserved determinants of Y, U_Y. Instead, U_D and U_Y are assumed to be determined by common causes represented by the dashed, bidirected edge in $U_D \leftarrow\!\text{-}\text{-}\text{-}\!\rightarrow U_Y$.

For fuzzy RD design, the assignment rule is not deterministic because D is a function of both $f(Z)$ and U_D, and, furthermore, because U_D and U_Y are determined by common causes. For a fuzzy RD analysis, the investigator conditions on Z so that $f(Z)$ can then be used as an instrument for D (because it is mean-independent of both U_D and U_Y and affects Y only through D).[4] Conditioning on Z is necessary to eliminate the back-door path $f(Z) \leftarrow Z \rightarrow Y$ in order to establish $f(Z)$ as a valid IV; see our prior discussion of conditional IVs in relation to Figure 9.2(b). A fuzzy RD analysis is possible only if $f(Z)$ is some nonlinear function of Z, so that $f(Z)$ and Z are not linearly dependent.

Angrist and Lavy (1999) use the fuzzy RD design to study the effects of class size on student test performance in Israel. In the Israeli public school system during the period of their study, an additional class was added to a grade within a school when the existing classes crossed a threshold size of 40 students in response to an increase in the overall school enrollment. This policy created a discontinuity in the distribution of class sizes, which allowed Angrist and Lavy to create a nonlinear function of enrollment that could then be used as an instrument for class size. They found that class size has

---

[3]Another type of generalization is toward the consideration of multiple assignment variables (see Wong, Steiner, and Cook 2013).

[4]The situation is directly analogous to an experiment with noncompliance in which noncompliance is thought to be nonrandom. As Imbens and Rubin (1997) show, one can deal with noncompliance by treating the intention-to-treat indicator as an instrument, Z, for the actual treatment, D.

a substantial effect on test performance for fourth and fifth graders, but not for third graders.

As shown by these examples, RD designs are simple and can be convincing. But the same weaknesses of ITS designs are present. Counterfactual values must be extrapolated from observed data below and above the value that triggers the introduction of the treatment. If the assumptions built into the chosen method of extrapolation are unreasonable, then causal effect estimates will be incorrect. Caughey and Sekhon (2011) present a critique of RD analyses where adoption of the assumptions requires substantial additional conditioning, precisely of the sort that RD designs are meant to avoid. In other cases, where the assumptions are reasonable, RD designs can be very powerful (see Berk, Barnes, Ahlman, and Kurtz 2010; Shadish, Galindo, Wong et al. 2011).

11.3 Panel Data

A severe limitation of time series data is that we have data on only a single unit over time. Because we do not observe the treated unit of analysis in the control state after the treatment is introduced, the only way to estimate the counterfactual outcome is to assume that the future can be predicted accurately from the past. This assumption is generally untestable.

Panel data, where multiple individuals or units are observed over time, may solve this problem. Assuming that each individual's time series is relatively long, separate ITS analyses could be conducted for each individual and then pooled to form an average causal effect estimate. Moreover, because individuals receive the treatment at different times or do not receive the treatment at all, it is possible to observe how Y_t^0 changes over time for some individuals after others have received the treatment. To the degree that Y_t^0 evolves similarly over time for individuals in the treatment and control groups, it may be possible to make reasonable predictions about the counterfactual values of Y_t^0 for individuals in the treatment group after they are exposed to the treatment.

For the remainder of this chapter, we will adopt the panel data notation introduced in Section 2.8.2 to differentiate between quantities that vary only over individuals (subscripted by i), quantities that vary only over time (subscripted by t), and quantities that vary over individuals and time (subscripted by it). In most cases, the subscripting for i is redundant and is utilized only for additional clarity. For example, in prior chapters, we have represented the average treatment effect (ATE) as $E[\delta]$, recognizing that the argument of the expectation, δ, can be regarded as a random variable that takes on values that vary across individuals, which we specified was possible when defining the individual-level causal effect as $\delta_i = y_i^1 - y_i^0$. Our notation in the remainder of this chapter requires that we write the same time-constant ATE as $E[\delta_i]$, because for a time-varying ATE, we would instead need to subscript for t as well, writing $E[\delta_{it}]$.

As explained in Section 2.8.2, we will distinguish between two different treatment indicator variables: D_{it} is a time-varying dummy variable that indicates whether individual i receives the treatment in time period t, and D_i^* is a time-constant dummy variable that indicates whether individual i ever receives the treatment at any point in the time span under study. D_{it} is best thought of as a treatment exposure indicator variable, and D_i^* is best thought of as a treatment group indicator variable. Observed and potential outcomes are related to each other by time-specific relations, $Y_{it} = D_{it}Y_{it}^1 + (1 - D_{it})Y_{it}^0$, where Y_{it}^1, Y_{it}^0, and D_{it} all vary over i and t. Finally, individual-level treatment effects vary in time, such that $\delta_{it} = y_{it}^1 - y_{it}^0$ for all t. As noted above, the ATE is now likewise time-specific and can be written as $E[\delta_{it}] = E[Y_{it}^1 - Y_{it}^0]$. The average treatment effect for the treated (ATT) and the average treatment effect for the controls (ATC), and any other conditional average treatment effect one might be interested in estimating, are defined analogously.

11.3.1   Traditional Adjustment Strategies

The most common situation in panel data analysis consists of nonequivalent treatment and control groups and two periods of data, where the first wave of data is from pretreatment time period $t-1$ and the second wave of data is from posttreatment time period t. Such two-period, pretreatment-posttreatment panel data are sometimes thought to be a panacea for not having a randomized experiment. Typically, it is assumed that changes over time in the control group can be used to adjust the changes observed for the treatment group, with the net change then representing a consistent and unbiased estimate of the causal effect of the treatment.

Unfortunately, the situation is far more complicated. There are an infinite number of ways to adjust for differences in gains between the treatment and control groups, and alternative methods of adjustment give estimates that sometimes differ dramatically. Specifically, as we will show in this section, by choosing a particular adjustment technique when analyzing two-period, pretreatment-posttreatment data, any estimate that a researcher may want can be obtained.

Consider the two most common methods, usually referred to as *change score* and *analysis of covariance* models. The change score model is often referred to as a panel data variant of a difference-in-difference model, especially in the economics literature; see Imbens and Wooldridge (2009). These two models are equivalent to estimating the following two equations with least squares regression:

$$\text{Change score:} \qquad Y_{it} - Y_{it-1} = a + D_i^* c + e_i, \tag{11.4}$$

$$\text{Analysis of covariance:} \qquad Y_{it} = a + Y_{it-1}b + D_i^* c + e_i. \tag{11.5}$$

These two equations provide different means of adjustment for Y_{it-1}. In the change score model, one adjusts Y_{it} by subtracting out Y_{it-1}. For the analysis of covariance model, one adjusts Y_{it} by regressing it on Y_{it-1}.[5] Recall that we introduced the analysis of covariance model already in Panel Data Demonstration 1 (see page 273), where we considered its utility when analyzing posttreatment-only data.

---

[5] Also, it bears mentioning that when we present alternative equations, such as Equations (11.4) and (11.5) in this chapter, we give generic notation – such as a, b, and c – to regression parameters, such as intercepts and treatment effect estimates. We do the same, in general, for regression residuals, and so on. We do not mean to imply that such quantities are equal across equations, but it is cumbersome to introduce distinct notation for each coefficient across equations to make sure that we never imply equality by reusing generic characters such as a, b, c, and e.

Consider now an example that shows how these two models can yield different results with pretreatment-posttreatment data. After decades of studying the environmental and genetic determinants of intelligence, considerable controversy remains over their relative effects on lifecourse outcomes. As discussed in Devlin, Fienberg, Resnick, and Roeder (1997) and other similar collections, these debates resurfaced after the publication of *The Bell Curve: Intelligence and Class Structure in American Life* by Herrnstein and Murray in 1994. Even though existing reviews of the literature emphasized the malleability of IQ (see Ceci 1991), Herrnstein and Murray concluded in their widely read book:

> Taken together, the story of attempts to raise intelligence is one of high hopes, flamboyant claims, and disappointing results. For the foreseeable future, the problems of low cognitive ability are not going to be solved by outside interventions to make children smarter. (Herrnstein and Murray 1994:389)

As discussed in Winship and Korenman (1997), the weight of evidence supports the claim that education determines measured intelligence to some degree, even though debate remains on how best to measure intelligence.

Consider now a very specific question associated with this controversy: What is the causal effect of a twelfth year of education on measured IQ? The following results, based on data from the National Longitudinal Survey of Youth, show how different the results from change score and analysis of covariance models can be (see Winship and Winship 2013 for additional details). For both models, IQ is measured before the twelfth grade for all individuals who meet various analysis-sample criteria ($N = 1,354$). IQ is then measured after high school completion, and high school completion is designated the treatment variable. A change score model yields a treatment effect estimate of 1.318 (with a standard error of .241) and an analysis of covariance model yields a treatment effect estimate of 2.323 (with a standard error of .217).[6] These estimates are quite different: The analysis of covariance model suggests that the effect of a twelfth year of schooling on IQ is 76 percent larger than that of the change score model (i.e., $[2.323 - 1.318]/1.318$). Which estimate should one use? Before we discuss how (and if) one can choose between these two types of traditional adjustment, consider a more general, but still simple, hypothetical example.

Panel Data Demonstration 2

For this demonstration, we will again consider the Catholic school effect analyzed by Coleman and his colleagues. Departing from the setup of Panel Data Demonstration 1 (see page 273), in this demonstration we will consider how alternative estimators perform assuming a world in which (1) no Catholic elementary schools or middle

---

[6]For completeness, we report additional features of these models here. Each was estimated with three other covariates: age, year of the test, and a standardized measure of socioeconomic status. The coefficients on these three variables were $-.18$, $-.76$, and $-.43$ for the change score model and $-.97$, $-.50$, and 2.05 for the analysis of covariance model. The R^2 was .06 for the change score model and .68 for the analysis of covariance model, in which the lag coefficient on IQ in the twelfth grade was .62.

schools exist, (2) all students consider entering either public or Catholic high schools after the end of the eighth grade, and (3) a pretreatment achievement test score is available for the eighth grade. Such a world does not exist, because (1) and (2) are not true (and, furthermore, Coleman and colleagues were not fortunate enough to have (3) either). We offer a demonstration assuming such a world for didactic purposes.

Basic Setup. As for Panel Data Demonstration 1, we will consider only linear specifications and, except when otherwise detailed, restrict all causal effects to be constants or to vary across individuals in completely random ways independent of all else in the models. This setup gives traditional panel data regression estimators the best chance of succeeding. Even so, recall that we showed through demonstrations in Chapters 5, 6, and 7 that least squares regression estimators invoke implicit weighting that is unlikely to effectively deal with the nonrandom individual-level heterogeneity of the Catholic school effect on achievement. We hold these additional complications aside in this demonstration in order to focus narrowly on the potential value of traditional panel data estimators of causal effects.

The potential outcome variables are Y_{it}^1 and Y_{it}^0, where now $t = \{8, 9, 10\}$ for the three grades that occur during the assumed observation window from the eighth grade through the tenth grade. Because treatment selection occurs before $t = 9$, we have observed data only for one pretreatment time period, and we will assume that Coleman and his colleagues had the tenth grade data as well (and that no data were collected in the ninth grade). The treatment group indicator variable, D_i^*, is equal to 1 if the student enrolls in a Catholic high school and 0 if the student enrolls in a public high school.

The observed outcome variable, Y_{it}, is defined with reference to the time-specific treatment exposure indicator variable, D_{it}. Because no one can be exposed to Catholic schools in the eighth grade (i.e., we have assumed that Catholic middle schools do not exist for this demonstration), $D_{i8} = 0$ for all students. As a result, $Y_{i8} = Y_{i8}^0$ for all students, and the eighth grade test score is therefore a pretreatment outcome that we observe for all students. However, for $t = \{9, 10\}$ the observable outcome is equal to the relevant potential outcome defined by our usual definition: $D_{it}Y_{it}^1 + (1 - D_{it})Y_{it}^0$. We will assume that those observed to be in Catholic or public schools in the tenth grade were in the same type of school in the ninth grade as well. Accordingly, for this demonstration, $D_i^* = D_{i10}$, and the definition of the observed outcome in the tenth grade can be written either as $Y_{i10} = D_{i10}Y_{i10}^1 + (1 - D_{i10})Y_{i10}^0$ or as $Y_{i10} = D_i^* Y_{i10}^1 + (1 - D_i^*)Y_{i10}^0$.[7]

Figure 11.5 presents a directed graph for the core features of this demonstration, which we will elaborate below when introducing alternative patterns of treatment selection. The graph has the same basic structure as Figure 8.2, but the first achievement test (now Y_8) is a pretreatment outcome, while the second achievement test (now Y_{10}) is a posttreatment outcome. We are interested in estimating the effect of D on Y_{10}, as in the research of Coleman and his colleagues. In comparison to Panel Data

---

[7] As is typical in this type of analysis, students who switch treatments between the ninth and tenth grades receive no special consideration because we do not observe their type of school enrollment in the ninth grade. See the appendix to this chapter, where we introduce the literature on dynamic treatment regimes that attempts to model all combinations of the effects of time-varying treatments.

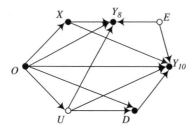

Figure 11.5 A directed graph for the effect of Catholic schooling on tenth grade achievement when a measure of eighth grade achievement is also available.

Demonstration 1, we have more reason to be optimistic. We can compare posttreatment outcomes for the Catholic school students to their pretreatment outcomes when in public schools and also to the outcomes of students enrolled in public schools in the posttreatment time period.

For simplicity, our other variables O, X, and U are specified as indices of many underlying variables, which we again scale as normally distributed composite variables. X is a composite determinant of achievement test scores in all years that has no direct effect on whether students select Catholic schooling. U is a composite variable of unobserved factors that determine both Catholic school attendance and achievement tests in all years. O is a composite variable of ultimate background factors that has effects on U and X, as well as direct effects on Catholic school attendance and test scores in all years. To give these composite variables distributions that are familiar and that align with their counterparts for Panel Data Demonstration 1, O is a standard normal random variable with mean of 0 and variance of 1. Having defined O as an exogenous root variable, we then set $X = O + e_X$ and $U = O + e_U$, where e_X and e_U are independent standard normal random variables with mean of 0 and variance of 1. Finally, E is a normal random variable with mean of 0 and variance of 1 that is a common cause of test scores in all years and is independent of all else in the graph.

Scenarios for Analysis. We will consider eight separate scenarios, defined as a cross-classification of two patterns of between-group differences in the trajectories of outcomes (parallel or divergent trajectories) and four treatment selection patterns (no-self-selection, self-selection on the individual-specific treatment effect, positive selection on the pretreatment outcome, and negative selection on the pretreatment outcome). For the four parallel-trajectory scenarios, Y_{it}^0 is defined as

$$Y_{i8}^0 = 98 + O_i + U_i + X_i + E_i + v_{i8}^0, \tag{11.6}$$
$$Y_{i9}^0 = 99 + O_i + U_i + X_i + E_i + v_{i9}^0,$$
$$Y_{i10}^0 = 100 + O_i + U_i + X_i + E_i + v_{i10}^0,$$

where the v_{it}^0 terms are values from independent normal random variables with mean of 0 and variance of 10. On average, Y_{it}^0 follows a linear time path as determined by the intercept values of 98, 99, and 100.[8] However, the levels of these potential outcomes for individuals are set by the time-invariant values of O_i, U_i, X_i, and E_i, as well as by time-specific shocks to their outcomes, v_{it}^0.

To specify a treatment effect that increases in time, Y_{it}^1 is defined as

$$Y_{i9}^1 = Y_{i9}^0 + \delta_i' + \delta_i'', \tag{11.7}$$
$$Y_{i10}^1 = Y_{i10}^0 + (1 + \delta_i') + \delta_i'',$$

where δ_i' is a baseline individual-level treatment effect, specified as values for each individual from a normal random variable with mean of 9 and variance of 1. The values of δ_i'' are a separate source of individual-level variation in the treatment effect, specified as values for each individual from a normal random variable with mean of 0 and variance of 1. In the no-self-selection scenarios, δ_i'' is additional random individual-level variation. In the scenarios that specify self-selection on the treatment effect, δ_i'' will be an input into treatment selection decisions, as explained below.[9]

For the four divergent-trajectory scenarios, we specify group-specific intercepts for Equation (11.6) so that the trajectory of $E[Y_{it}^0]$ differs across the treatment and control groups after the onset of treatment. For the treatment group ($D_i^* = 1$), the intercepts are specified as 98 for $t = 8$ and 99 for $t = 9$, but then 100.5 for $t = 10$. For the control group ($D_i^* = 0$), the intercepts are specified also as 98 for $t = 8$ and 99 for $t = 9$, but then only 99.5 for $t = 10$. Taken together, by the end of the observation window in the tenth grade, the treatment group's value for $E[Y_{i10}^0]$ is higher by 1 than for the control group (i.e., $100.5 - 99.5 = 1$). For this "fan spread" pattern, in the absence of treatment the test scores of those in the treatment group increase faster after the onset of treatment than the test scores of those in the control. In other words, students who select into Catholic schools would have had a boost in achievement even if they had remained in public schools, net of all other determinants of Y_{it}^0 in Equation (11.6).[10]

We consider four types of treatment selection. For the first, treatment selection is on fixed characteristics of individuals unrelated in any way to the outcomes before or after the treatment. Accordingly, the probability of Catholic school enrollment is specified as a logistic distribution

$$\Pr[D_i^* = 1 | O_i, U_i] = \frac{\exp(-3.8 + O_i + U_i)}{1 + \exp(-3.8 + O_i + U_i)}, \tag{11.8}$$

---

[8]To mimic real data, the intercept values such as 98, 99, and 100 in this demonstration will always be expected values of individual-specific intercepts, where the variation in the intercepts is independent of all else on the right-hand sides of the relevant equations. We suppress this fact in the main text for simplicity of exposition. To be precise, we set up individual-specific time trends in y_{it}^0 by specifying individual-specific multipliers (from uniform distributions with strictly positive probability and mean of 1), which we then apply to the common time trends that define the expectations $E[Y_{it}^0]$, sometimes differentially with respect to D_i^*.

[9]Y_{i8}^1 is counterfactual for all individuals, and we exclude it from Equation (11.7) because we have assumed for this demonstration that Catholic middle schools do not exist. Explicitly allowing for it would suggest otherwise.

[10]We take no position on why fan spread occurs, although in this demonstration it is equivalent to assuming that another variable, Q, exists that generates this pattern by structuring Y_{it}^0 in interaction with D_i^* after the onset of treatment. In many situations in education research, it is assumed that fan spread exists because learning is a cumulative process. We consider this type of thinking later in this chapter, when considering dynamic scenarios where Y_{it} is structured directly by its prior values.

where O_i and U_i are as defined above. As in prior demonstrations, the probabilities defined by Equation (11.8) are then set as the parameters for draws from a binomial distribution, yielding the indicator variable D_i^* for Catholic schooling defined above.

For the second type of treatment selection, we introduce self-selection on the individual-specific treatment effect, assuming that students and their parents are able to forecast and then choose based on accurate beliefs about how much they would benefit from attending a Catholic school. The treatment selection probability is specified as

$$\Pr[D_i^* = 1 | O_i, U_i] = \frac{\exp(-7.3 + O_i + U_i + 5\delta_i'')}{1 + \exp(-7.3 + O_i + U_i + 5\delta_i'')}, \tag{11.9}$$

where δ_i'' is as defined above. For the directed graph in Figure 11.5, this type of self-selection is equivalent to adding an additional bidirected edge, $D \leftarrow\!\!-\!-\!-\!\!\rightarrow Y_{10}$.[11]

For the final two types of treatment selection, we specify the treatment selection probability as

$$\Pr[D_i^* = 1 | O_i, U_i] = \frac{\exp(-3.8 + O_i + U_i + k(Y_{i8} - E[Y_{i8}]))}{1 + \exp(-3.8 + O_i + U_i + k(Y_{i8} - E[Y_{i8}]))}, \tag{11.10}$$

where $(Y_{i8} - E[Y_{i8}])$ is the individual deviation from the expectation of the pretreatment test. These individual-level values are scaled by k, which is set to either .05 or $-.05$. For the positive value of k, students and their parents are selecting the treatment assuming that those with higher pretreatment test scores will be the most likely to benefit from Catholic schooling, perhaps because they believe that Catholic schools have a more challenging curriculum from which only high achievers will benefit. For the negative value of k, they are assuming the opposite, perhaps because they believe that Catholic schools can compensate for lower achievement in the past. For the directed graph in Figure 11.5, this type of selection is equivalent to adding an additional direct causal effect, $Y_8 \rightarrow D$.

Results from Traditional Panel Data Estimators.

Consider first the four parallel-trajectory scenarios in the first panel of Table 11.1. For the first column, selection patterns are simple and based only on the fixed characteristics of individuals, as specified above in Equation (11.8). The true ATE is 10.00, which is equal to the ATT and ATC by construction.

For this scenario, the naive estimator yields 14.75 (on average across repeated samples), and this value is upwardly biased because O and U are positively associated with both D and Y_{10}. We could use a cross-sectional estimator to block the back-door paths through the observed characteristics of individuals, O, which are shown in Figure 11.5. Unfortunately, the unobserved variable U generates a back-door path $D \leftarrow U \rightarrow Y_{10}$ that remains unblocked after conditioning on O.

Can traditional panel data estimators solve this problem? The change score estimator yields a value of 10.00, which is equal to the true ATE, ATT, and ATC. In

---

[11]Or, as with the charter school example in Section 8.3, we could add a fully elaborated back-door path using latent classes and attempting to capture the inputs into the self-selection decision itself (see Figures 8.5 through 8.7).

Table 11.1 Change Score and Analysis of Covariance Estimates of the Catholic School Effect in the Tenth Grade

Setup conditions:				
Self-selection on the causal effect	No	Yes	No	No
Positive self-selection on the pretest	No	No	Yes	No
Negative self-selection on the pretest	No	No	No	Yes

	Parallel Trajectories			
True average treatment effects:				
ATE	10.00	10.00	10.00	10.00
ATT	10.00	11.51	10.00	10.00
ATC	10.00	9.83	10.00	10.00
Estimated coefficients for D^*:				
Naive estimator:				
Regression of Y_{10} on D^*	14.75	13.86	15.92	13.25
Change score estimator:				
Regression of $(Y_{10} - Y_8)$ on D^*	10.00	11.51	7.96	12.26
Analysis of covariance estimator:				
Regression of Y_{10} on D^*, Y_8, O, and X	10.51	11.75	10.49	10.52

	Divergent Trajectories			
True average treatment effects:				
ATE	10.00	10.00	10.00	10.00
ATT	10.00	11.51	10.00	10.00
ATC	10.00	9.83	10.00	10.00
Estimated coefficients for D^*:				
Naive estimator:				
Regression of Y_{10} on D^*	15.75	14.86	16.88	14.30
Change score estimator:				
Regression of $(Y_{10} - Y_8)$ on D^*	11.00	12.51	8.92	13.31
Analysis of covariance estimator:				
Regression of Y_{10} on D^*, Y_8, O, and X	11.52	12.75	11.50	11.53

contrast, the analysis of covariance estimator instead yields a value of 10.51, which is upwardly biased for the ATE, ATT, and ATC. The bias is smaller than for the naive estimator because conditioning on O has removed some of the back-door confounding.[12] We know that 10.00 is the correct answer and therefore can favor the estimate from the change score model. If we did not know the correct answer ahead of time and/or were uncertain about the correct directed graph for the generation of the data,

---

[12] At the same time, it is unclear from these results what the total consequences are of conditioning further on X and Y_8. The latter is a collider that, when conditioned on, induces a back-door association between D and Y_{10} by unblocking the path $D \leftarrow U \rightarrow Y_8 \leftarrow E \rightarrow Y_{10}$. Notice that conditioning on Y_8 would also unblock other back-door paths in the graph, but all of these remain blocked by simultaneous conditioning on O and X.

we would have a hard time picking between these two estimates. Before offering an explanation for this result, it is helpful to consider the next scenario for comparison.

The second column of the first panel presents the same results for the scenario where treatment selection is on fixed characteristics as well as on the causal effect itself, as specified above in Equation (11.9). As with past demonstrations, the true ATT is now greater than both the ATE and the ATC. The naive estimator is again upwardly biased for the ATE, ATT, and ATC. The analysis of covariance estimator is also upwardly biased for all three. In contrast, the change score model yields an estimate of 11.51, which is equal to the ATT, but not to the ATE or to the ATC.

Taken together, the first two columns suggest that the change score estimator yields values that will on average equal the ATT. When self-selection is absent, the ATT will equal the ATE and the ATC, and as a result a consistent and unbiased estimate of the ATT will also be a consistent and unbiased estimate of the ATE and the ATC.

We will explain this result more formally following this demonstration. The core of the explanation is that the change score model subtracts out the effects of all fixed characteristics – observed and unobserved – and can then generate a consistent and unbiased estimate of the ATT in scenarios such as these two. If one is willing to assume that no self-selection on the causal effect is present, as is the case for the scenario in the first column, then this estimate of the ATT is also consistent and unbiased for the ATE and ATC.

These results may suggest that the change score model is most commonly the best choice for these situations, and one might also be encouraged that even in the presence of self-selection it can be used to effectively estimate the ATT. The third and fourth columns were constructed to temper any such enthusiasm. For these two columns, individuals do not self-select on the treatment effect itself, and thus the ATE, ATT, and ATC are all equal. However, treatment selection is on the pretreatment outcome, as specified by Equation (11.10), where individuals choose Catholic schooling either as a positive or negative function of the eighth grade test score.

For the third and fourth columns of the first panel, the naive estimator is again upwardly biased for the ATE, ATT, and ATC. More important for our consideration, the change score model now yields values that are either too small or too large, depending on whether selection is a positive or negative function in the eighth grade test score. In fact, the estimates yielded by the analysis of covariance model are on average much closer to the ATE, ATT, and ATC. In these scenarios, the change score model continues to generate estimates based on average differences between eighth and tenth grade tests scores within the treatment group, but the average of these differences is no longer a consistent or unbiased estimate of the ATT. If the distribution of the individual-level treatment effects does not vary across the treatment and control groups, the average difference, $Y_{i10} - Y_{i8}$, within the treatment group will be too small if those with high values for Y_{i8} select into Catholic schooling and will be too large if those with low values of Y_{i8} select into Catholic schooling. The analysis of covariance estimator offers a less extreme adjustment for the eighth grade test score and is thus closer to the true ATE, ATT, and ATC. Nonetheless, the analysis of covariance estimator generates values that do not equal the target parameters, and this suggests that the adjustment may not be correct, which we will explain below is usually the case (i.e., except in the very rare case that the estimated regression coefficient exactly adjusts for a regression to the mean effect that is generated by the behavior of individuals).

The second panel of Table 11.1 presents four scenarios for diverging trajectories of Y_{it}^0, using the same four patterns of treatment selection. The true ATE, ATT, and ATC are all the same as for the first panel because the underlying fan spread in the potential outcomes only applies to the potential outcome in the control state in the tenth grade, Y_{i10}^0. The treatment effect continues to be defined by Equation (11.7), which does not vary across the scenarios for parallel and divergent trajectories. To focus on this key point, recall from our discussion above that even for this set of scenarios the trajectories of the potential outcome are the same for the treatment and control groups from $t = 8$ to $t = 9$ because

$$E[Y_{i8}^0|D_i^* = 1] - E[Y_{i8}^0|D_i^* = 0] = 98 - 98 = 0$$

and

$$E[Y_{i9}^0|D_i^* = 1] - E[Y_{i9}^0|D_i^* = 0] = 99 - 99 = 0.$$

The difference of 1 emerges between $t = 9$ and $t = 10$ because we specified that

$$E[Y_{i10}^0|D_i^* = 1] - E[Y_{i10}^0|D_i^* = 0] = 100.5 - 99.5 = 1.$$

These divergent trajectories are inconsequential for the true ATE, ATT, and ATC because these target parameters continue to be structured in the same way for both groups. More specifically, the term $(1 + \delta_i') + \delta_i''$ in Equation (11.7) does not vary by group.

Consider the first column in the second panel, which is for the treatment selection pattern where selection is on the fixed characteristics of individuals, O_i and U_i. As was the case for the parallel-trajectory scenarios, the naive estimator and analysis of covariance estimator yield values that are too large. Now, however, the change score estimator fails as well. The change score estimator yields a coefficient on D^* that is equal to 11, assuming an infinite sample (or averaged over repeated samples). In particular, with reference to Equation (11.4), it yields

$$\begin{aligned} Y_{i10} - Y_{i8} &= \hat{a} + D_i^*\hat{c} \\ &= 1.5 + D_i^*11. \end{aligned}$$

As we explain in the next section, the intercept is equal to

$$\begin{aligned} \hat{a} &= E[Y_{i10}^0|D_i^* = 0] - E[Y_{i8}^0|D_i^* = 0] \\ &= E[Y_{i10}|D_i^* = 0] - E[Y_{i8}|D_i^* = 0] \\ &= 99.5 - 98, \end{aligned}$$

which is the difference in the observed outcome in the absence of treatment for public school students. In addition, the treatment effect estimate is equal to

$$\hat{c} = \left\{ E[Y_{i10}^1 | D_i^* = 1] - E[Y_{i8}^0 | D_i^* = 1] \right\} - \left\{ E[Y_{i10}^0 | D_i^* = 0] - E[Y_{i8}^0 | D_i^* = 0] \right\}$$
$$= \left\{ E[Y_{i10} | D_i^* = 1] - E[Y_{i8} | D_i^* = 1] \right\} - \left\{ E[Y_{i10} | D_i^* = 0] - E[Y_{i8} | D_i^* = 0] \right\}$$
$$= \left\{ 110.5 - 98 \right\} - \left\{ 99.5 - 98 \right\},$$

which is the expected gain in the observed outcome for the students who enrolled in Catholic schools minus the expected gain in the observed outcome for the students who remained in public schools. The change score estimator delivers an upwardly biased estimate because it assumes implicitly (but incorrectly) that the observed average gain among public school students in test scores between the eighth and tenth grades is the same gain that those who enter Catholic schools would have experienced if they had instead remained in public schools. This assumption was correct for the parallel-trajectory scenario presented in the first panel of Table 11.1, but it is incorrect, by construction, for the scenarios we are considering now. For our divergent-trajectory scenarios, we have set the (counterfactual) gain to be equal to 2.5 for the treatment group and the factual and observed gain to be 1.5 for the control group (i.e., $100.5 - 98 = 2.5$ in contrast to $99.5 - 98 = 1.5$).

For completeness, consider the final three columns in the second panel briefly. As shown in the second column, the change score estimator remains upwardly biased for the ATT when self-selection is present, and the magnitude of the bias is exactly the same because the trajectory-induced bias is unrelated to selection on the treatment effect. As shown in the third and fourth columns, when selection is on the pretreatment outcome, the change score estimator will yield values that are either too small or too large for the ATT, ATE, and ATC for the same reasons as in the first panel for parallel trajectories. The analysis of covariance models show the same basic patterns as for the parallel-trajectory scenarios.

Altogether, this demonstration suggests that the change score estimator will offer consistent and unbiased estimates of the ATT when selection is not a function of the pretreatment outcome and when the unobserved average trajectory for the potential outcome in the absence of treatment for the treatment group is equal to the observed trajectory for the control group.[13] If individuals do not self-select on the treatment effect, then the ATT will be equal to the ATE and ATC by definition; the change score estimator is therefore consistent and unbiased for all three. If, however, selection is on the pretreatment outcome or the trajectories are not parallel, then the change score estimator is no longer consistent and unbiased for the ATT (or the ATC or ATE). Finally, although we have yet to fully explained why, in this demonstration the analysis of covariance model never appears to be consistent or unbiased for any of the target parameters, even though it appears to be less sensitive to departures from

---

[13] In this demonstration, the parallel trajectories are also linear. Linearity is not required. Parallelism exists if $E[Y_{it}^0 | D_i^* = 1] - E[Y_{it}^0 | D_i^* = 0] = k$ for all t, where k is any constant that does not vary in t. For the demonstration, we set $k = 0$. Parallelism in all values of t is sufficient, but it is not necessary. As we explain below, we only need k to be the same for the two time periods in which the pretreatment and posttreatment outcomes are observed.

the parallelism of the first four scenarios. The explanation for this outcome will follow the demonstration.

In conclusion, we should note four additional points. First, only because we constructed the data for this demonstration is it clear when the change score estimator outperforms the analysis of covariance estimator. In observational research, the true values for the ATE, ATT, and ATC are unknown and thus no benchmark for comparison is available. Second, it is of course possible to have a situation where selection is a function of both the pretreatment outcome and also accurate expectations of the individual-level treatment effect. In this case, the results of the demonstration are as implied: Neither the change score estimator nor the analysis of covariance estimator would deliver estimates that are consistent or unbiased for any of the average treatment effects. Third, with data from only two points in time, there is no way to evaluate whether selection is on the pretreatment outcome using the observed data. This point should be obvious from a consideration of Figure 11.5. If we add the effect $Y_8 \rightarrow D$ to the graph, we have no way to analyze the data to separate this casual effect from the association generated by the unobserved variable in $Y_8 \leftarrow U \rightarrow D$. Thus, any argument in favor of the change score model would have to rest entirely on an argument grounded in theory or past research. Finally, if the fan spread pattern emerges in the same basic pattern considered here, where it only emerges at the same time as the treatment, analysis will be extremely difficult. However, if the trajectories differ but can be effectively modeled as a function of the pretreatment data from more than one time period, then the model-based strategy we introduce in the final section of this chapter may be effective.

Return to the question that motivated this demonstration, where we are not in the fortunate situation of knowing the true values for the ATE, ATT, or ATC. All we have are alternative treatment effect estimates suggested by a change score model and an analysis of covariance model. How should one choose between them? There are at least three possible ways to decide:

1. Choose the method that gives us the results we want.

2. Choose based on the nature of the problem. As Allison (1990) suggests: If selection is based on fixed characteristics, use change score analysis. If selection is based on the dependent variable, use an analysis of covariance.

3. Use the data to determine which model, if either, is appropriate (as in Heckman and Hotz 1989).

We hope that the first approach to the decision is not a serious consideration. The second is a better option, in that it at least suggests that one should begin to think through the specific nature of the problem of interest. The third appears most promising, at least at face value. However, in some cases (perhaps most where these two estimators are utilized), we have data from only two points in time. This is the situation for the estimate of the causal effect of a twelfth year of schooling on IQ. It is also the case for the demonstration that we have just offered. Unfortunately, with only two time periods, the data cannot be used to test whether one of the two models is more appropriate. As we will explain when we discuss model-based approaches in the next section, we need at least two periods of pretreatment data in order to carry out an informative test. For example, we would be able to perform a test for the demonstration above only if we also had test score data from the seventh grade.

For now, consider the case in which we continue to have data from only one pretreatment time point, $t-1$, and one posttreatment time point, t. Suppose also that no self-selection on the individual-level treatment effect is present so that the ATT is equal to the ATE. Consider the implicit assumptions that are made if it is asserted that either a change score model or an analysis of covariance model is a consistent and unbiased estimator of the ATT or the ATE:

- The change score model assumes that, *in the absence of treatment*, any difference between the expectations of the outcome for those in the treatment group and those in the control group remains constant over time. With potential outcome notation, the required assumption for two-period, pretreatment-posttreatment data is that $E[Y_{it-1}^0|D_i^*=1] - E[Y_{it-1}^0|D_i^*=0] = k$ and $E[Y_{it}^0|D_i^*=1] - E[Y_{it}^0|D_i^*=0] = k$, where k is the same constant in both time periods $t-1$ and t. The constant k can be equal to 0, as in the parallel-trajectories scenarios for Panel Data Demonstration 2. In this case, there are no differences in $E[Y_{it}^0]$ between the treatment and control groups in time periods $t-1$ and t.

- The analysis of covariance model assumes that, *in the absence of treatment*, any difference between the expectations of the outcome for those in the treatment group and those in the control group shrinks by a multiplicative factor r in each subsequent time period. An implication of this assumption is that, after enough time, the analysis of covariance model assumes that there would be no difference in the expected outcomes for the treatment and control groups if the treatment is not introduced. With potential outcome notation, the required assumption for two-period, pretreatment-posttreatment data is that any difference $E[Y_{it-1}^0|D_i^*=1] - E[Y_{it-1}^0|D_i^*=0] = k$ in the pretreatment time period $t-1$ is equal to $E[Y_{it}^0|D_i^*=1] - E[Y_{it}^0|D_i^*=0] = k \times r$ in the posttreatment time period t, where k is the same constant in both time periods and where r is the amount of between-group shrinkage that is assumed to occur in each and every time period. Any remaining difference between the two groups approaches 0 in the limit so that by time period $t = \infty$, $E[Y_{it=\infty}^0|D_i^* = 1] = E[Y_{it=\infty}^0|D_i^* = 0]$. In addition, the analysis of covariance model assumes that r is equal, in an infinite sample, to the least squares coefficient on Y_{it-1} in a regression equation $Y_{it} = a + Y_{it-1}b + D_i^*c + e_i$ (or $Y_{it} = a + Y_{it-1}b + D_i^*c + X_iq + e_i$ if additional adjustment variables in X_i are also specified).

The key difference between these two models is therefore their implicit assumptions about the evolution of the difference between $E[Y_{it}^0]$ for the treatment group and for the control group.

Consider a general equation that can be used to represent the value of the ATE for the posttreatment time period t (again assuming that the ATE is equal to the ATT because no self-selection is present):

$$E[\delta_{it}] = \left(E[Y_{it}^1|D_i^*=1] - E[Y_{it-1}^0|D_i^*=1]\right)$$
$$- \alpha\left(E[Y_{it}^0|D_i^*=0] - E[Y_{it-1}^0|D_i^*=0]\right) \tag{11.11}$$

for some unknown value α.[14] The term in the first set of parentheses is equal to the average difference in the observed outcome between time periods $t-1$ and t for the treatment group. Given the definition of the observed outcome, this difference is equal to $E[Y_{it}|D_i^*=1] - E[Y_{it-1}|D_i^*=1]$, which is the observed gain in the treatment group. The second term is an adjustment factor. It has two pieces: an unspecified value, α, and a term in parentheses that is the difference between time periods $t-1$ and t in the potential outcome in the absence of the treatment for the control group. The latter is equal to $E[Y_{it}|D_i^*=0] - E[Y_{it-1}|D_i^*=0]$, which is the observed gain in the control group. Equation (11.11) can therefore be rewritten as

$$E[\delta_{it}] = (E[Y_{it}|D_i^*=1] - E[Y_{it-1}|D_i^*=1]) \tag{11.12}$$
$$- \alpha(E[Y_{it}|D_i^*=0] - E[Y_{it-1}|D_i^*=0]),$$

and its right-hand side can then be written even more simply with words as

$$(\text{treatment group gain in } Y) - \alpha\,(\text{control group gain in } Y) \tag{11.13}$$

The change score model and the analysis of covariance model can be seen as alternative methods that make very different and very rigid assumptions about the value of α in Equations (11.11)–(11.13). The change score model implicitly assumes that $\alpha = 1$. In contrast, the analysis of covariance model implicitly assumes that $\alpha = r$, where r is the intraclass correlation between Y_{it} and Y_{it-1} (i.e., r is the correlation coefficient for Y_{it} and Y_{it-1}). In other contexts, this correlation coefficient r is known as the reliability of Y. If other covariates are included in the model, then the analysis of covariance model assumes that the coefficient on Y_{it-1} is a conditional variant of the intraclass correlation (i.e., the intraclass correlation of residualized variants of Y_{it} and Y_{it-1}, from which their common linear dependence on the covariates has been purged).

Researchers often believe that, because the coefficient on Y_{it-1} is estimated from the data, an analysis of covariance model is superior to a change score model. This position is incorrect. To be sure, an analysis of covariance model does estimate a coefficient on Y_{it-1}, and this coefficient can be interpreted as an intraclass correlation coefficient. This fact is irrelevant to the more fundamental issue of whether r, or a conditional variant of it, is the correct adjustment factor for generating consistent estimates of average treatment effects. In other words, if the goal is to estimate the ATE or ATT, researchers who favor the analysis of covariance model because it allows the data to

---

[14]In this section, we will consider values for α that would be appropriate for estimating the ATE because this is the typical scenario in which researchers use change score models and analysis of covariance models. We could offer an analogous explanation for the ATT in the presence of self-selection, and here the values for α would be different if self-selection were present. The overall argument would have the same structure but would begin with an analogous expression for $E[\delta_{it}|D_i^*=1]$ instead of Equation (11.11). We avoid having to do this by stating above that no self-selection is present for this explanation, so that the ATE is equal to the ATT.

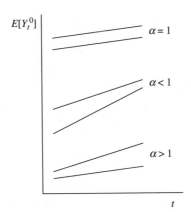

Figure 11.6 Examples of possible trajectories for $E[Y_{it}^0]$ for the treatment group (the upper line of each pair) and the control group (the lower line of each pair) where the correct adjustment favor, α, varies.

determine the coefficient on Y_{it-1} are assuming implicitly that α in Equations (11.11)–(11.13) should be equal to r, or a conditional variant of it that is determined by the relationship between Y_{it-1} and any covariates that are specified.

Consider the graph presented in Figure 11.6, which presents three scenarios for changes over time in the expected value of the outcome in the absence of the treatment.[15] For each pair of lines, the upper line is $E[Y_{it}^0]$ for the treatment group, and the lower line is $E[Y_{it}^0]$ for the control group. For the first pair of lines, the correct adjustment factor, α, is equal to 1. In this situation, a change score model is appropriate. For the second pair of lines, the correct adjustment factor is less than 1. In this situation, the change score model is inappropriate, but it is conceivable that the analysis of covariance model can deliver an estimated coefficient on Y_{it-1} that is equal to the correct adjustment factor. It is not guaranteed to do so because, even in an infinite sample, the coefficient on Y_{it-1} is simply the partial slope for the best linear predictor of Y_{it}. Finally, this graph shows a third possibility where the correct adjustment factor is greater than 1. In this case, neither the change score model nor the analysis of covariance model is appropriate. Here, in the absence of treatment, $E[Y_{it}^0]$ for the treatment group and for the control group diverge over time. There are many possible examples of this situation, but the most famous is represented by situations described in the Gospel of Matthew, where it is written, "Unto every one that hath shall be given, and he shall have abundance: but from him that hath not shall be taken away even that which he hath" (quoted in Merton 1968, who is credited with introducing this version of the idea into the social sciences).

The key point is that different methods of estimation make different implicit assumptions about how the difference in the expectations between the treatment and control groups would change with time in the absence of the treatment (which, in the

---

[15]See Judd and Kenny (1981, figure 6.4) for a similar figure and explanation that does not use potential outcomes.

counterfactual tradition, are different assumptions about treatment and control group differences in the evolution of $E[Y_{it}^0]$). Researchers typically use either change score models or analysis of covariance models without taking note of these assumptions. Nevertheless, these assumptions can be very consequential, as we now show in a more general way than for Panel Data Demonstration 2.

Our claim that any assumption about α is potentially reasonable can be justified by consideration of the following model for the generation of the potential outcome variables:

$$Y_{it}^0 = \lambda_i + T\tau_i, \tag{11.14}$$
$$Y_{it}^1 = Y_{it}^0 + \delta_i, \tag{11.15}$$

where λ_i is an individual-varying intercept, the variable T identifies the time period, and τ_i is an individual-varying coefficient on time T. For this model, the following equality holds:

$$E[Y_{it}^0|D_i^* = 1] - E[Y_{it}^0|D_i^* = 0] = (E[\lambda_i|D_i^* = 1] - E[\lambda_i|D_i^* = 0]) \tag{11.16}$$
$$+ T\left(E[\tau_i|D_i^* = 1] - E[\tau_i|D_i^* = 0]\right),$$

where the T on the right-hand side is set equal to the value of t in the subscript on the left-hand side. Without loss of generality, assume for the moment that $(E[\lambda_i|D_i^* = 1] - E[\lambda_i|D_i^* = 0]) > 0$. In this case, note that whether the initial difference in $E[Y_{it-1}^0]$ between those in the treatment group and those in the control group remains the same, grows, or shrinks, respectively, is a function of whether $(E[\tau_i|D_i^* = 1] - E[\tau_i|D_i^* = 0])$ is equal to 0, is greater than 0, or is less than 0.

If we assume that $(E[\lambda_i|D_i^* = 1] - E[\lambda_i|D_i^* = 0]) = 0$, then the appropriate adjustment factor, α, equals

$$\alpha = 1 + (E[\tau_i|D_i^* = 1] - E[\tau_i|D_i^* = 0]). \tag{11.17}$$

If $(E[\tau_i|D_i^* = 1] - E[\tau_i|D_i^* = 0]) = 0$ (i.e., $E[Y_{it}^0]$ changes on average over time at the same rate for individuals in the treatment and control group), then $\alpha = 1$ in Equation (11.17), and the assumptions of the change score model are appropriate. If $(E[\tau_i|D_i^* = 1] - E[\tau_i|D_i^* = 0]) = (r - 1)$, which is necessarily nonpositive (because $0 < r < 1$), then $\alpha = r$ in Equation (11.17), and the assumptions of the analysis of covariance model are appropriate instead.

Of course, there is no reason that $(E[\tau_i|D_i^* = 1] - E[\tau_i|D_i^* = 0])$ should necessarily be equal to either 0 or $r - 1$. Thus, it is possible that neither the change score model nor the analysis of covariance model provides the correct adjustment. To bring this point home, consider Table 11.2, in which the stipulated true causal effect is 1. The table reports different estimates of the average treatment effect for different combinations of correct and assumed adjustment factors. Equivalently, because $\alpha = 1 + (E[\tau_i|D_i^* = 1] - E[\tau_i|D_i^* = 0])$, the table reports estimates for different actual and assumed values of the difference in slopes for the treatment and control groups.

Note first that all of the diagonal elements of Table 11.2 are equal to 1. If the assumed adjustment factor equals the correct adjustment factor, we get the correct estimate of the causal effect, 1. Below the diagonal are cases where we have overadjusted (that is, in which the assumed adjustment factor is greater than the correct

Table 11.2 Estimated Average Treatment Effects for Different Combinations of Correct and Assumed Adjustment Factors, Where the True Effect Is Equal to 1

		Assumed α				
		2	1.5	1	.5	0
	2	1	1.5	2	2.5	3
	1.5	.5	1	1.5	2	2.5
Correct α	1	0	.5	1	1.5	2
	.5	−.5	0	.5	1	1.5
	0	−1	−.5	0	.5	1

one). As a result, we get estimates of the causal effect that are too low, ranging from .5 to −1, including an estimate of no effect at all. Above the diagonal, we have cases where we have underadjusted (that is, the assumed adjustment factor is smaller than the correct one). As a result, our estimates of the causal effect are too high, ranging from 1.5 to 3.

For this example, the true average treatment effect is 1 by construction. Across Table 11.2, we have estimates ranging from −1 to 3. We could easily expand the range of these estimates by considering a broader range of correct and assumed adjustment factors. And alternative examples could be developed in which the true average treatment effect equals alternative values, and in which the range of estimates varies just as widely. Although the calculations behind Table 11.2 are simple, the point of the table is to show that one can obtain any estimate of an average treatment effect by making different assumptions about the appropriate adjustment factor.[16]

In view of this problem, what should a researcher do? If there are strong theoretical reasons for arguing that a particular adjustment factor is correct, and others agree, then analysis is straightforward. If not, which we suspect is generally the case, then it may be possible to argue for a range of adjustment factors. In this case, a researcher may be able to bound the causal effect to a region on which all researchers can agree.

In general, the assumption that a particular adjustment factor is correct must be based on a set of assumptions about how $E[Y_{it}^0]$ for those in the treatment and control groups evolves over time. This leads naturally to a more explicit model-based approach, which we present in the next section. As we will also see, with data from more than one pretreatment time period, it may be possible to test the adequacy of the assumptions and thus the appropriateness of a particular adjustment factor.

---

[16]Recall Panel Data Demonstration 2. There, the divergent-trajectory scenario with no self-selection yielded a change score estimate that was biased upward by 1. Adopting the logic of this explanation, the correct adjustment factor was 5/3, so that $(110.5 − 98) − (5/3)(99.5 − 98) = 10$. However, the change score estimator assumed that the correct adjustment factor was 1, resulting in $(110.5 − 98) − (1)(99.5 − 98) = 11$.

11.3.2   Model-Based Approaches

In discussing panel data models we have until now considered only traditional methods of estimating a causal effect. There is, however, much merit to considering explicit models of the evolution of Y_{it}^1 and Y_{it}^0 and asking, "Under what model assumptions do different methods give consistent estimates?" In this section, we take this approach and address four questions:

1. What is the dynamic structure of the outcome? In particular, how are future values of the outcome related to previous values of the outcome? Answering this question is critical if our goal is to estimate counterfactual values. In the potential outcome framework for the sort of examples we will consider, we are interested primarily in the dynamic structure of Y_{it}^0, which is the potential outcome variable under the control state.

2. How is assignment to the treatment determined? As in cross-sectional attempts to estimate causal effects, modeling treatment assignment/selection is crucial if a researcher hopes to generate consistent estimates of a particular causal effect.

3. What are the alternative methods of estimation that can be used to consistently estimate average effects, given a valid set of assumptions?

4. How can the estimated model be tested against the data?

We will consider these four questions in this order.

Dynamic Structure

As shown in any advanced time series textbook, the dynamic structure of the outcome can be infinitely complex. In the context of panel data models, researchers have typically considered fairly simple structures, often because of the limited number of waves of data that are available. Rather than trying to provide an exhaustive account – which would take a book in and of itself – we primarily focus on conceptual issues.

The broad statistics and econometric literature on panel data models is quite distinct from the estimation of treatment effects from a counterfactual perspective. Implicit in much of this literature is the assumption that causal effects are constant across individuals, such that causes/treatments simply shift the outcome by fixed amounts. From a counterfactual perspective, such assumptions are overly rigid. A necessary component of estimating a treatment effect is the consideration of the hypothetical evolution of Y_{it}^0 for the treatment group after the treatment occurs. If treatment effects are heterogeneous and selection is on the treatment effect itself, then the ATT is usually the parameter of interest, as it is often the only one that can be identified by any model (and, fortunately, it is also often of inherent substantive interest).

Consider the following possible two equations for the generation of Y_{it}^0:

$$Y_{it}^0 = \lambda_i + e_{it}, \tag{11.18}$$

$$e_{it} = \rho e_{it-1} + v_{it-1}, \tag{11.19}$$

Table 11.3 Alternative Trajectories of the Outcome Under the Control State for Different Assumptions About Its Dynamic Structure

Model	Assumed Constraints		Evolution of Y_{it}^0
A	$\rho = 0$	$\mathrm{Var}(\lambda_i) \neq 0$	Immediate regression of individual values to separate group expectations
B	$\rho \neq 0$	$\mathrm{Var}(\lambda_i) = 0$	Regression over time of individual values to a common expectation
C	$\rho \neq 0$	$\mathrm{Var}(\lambda_i) \neq 0$	Regression over time of individual values to separate group expectations

where λ_i is a time-constant, individual-varying fixed effect, v_{it-1} is pure random noise (that is, uncorrelated with everything), and ρ is the correlation between e_{it} over time (not the correlation between Y_{it}^0 over time, which we labeled as r earlier). Equation (11.19) specifies an autoregressive process of order (1). It is order (1) because the current e_{it} is dependent on only the last e_{it-1}, not e_{it-2} or any errors from prior time periods. There are many possible ways that the current error could be dependent on past errors. These define what are known as the class of autoregressive moving average (ARMA) models.

Within the current context (i.e., assuming that we know that Equations (11.18) and (11.19) are capable of representing the full dynamic structure of Y_{it}^0), determining the dynamic portion of the model for Y_{it}^0 amounts to asking whether $\mathrm{Var}(\lambda_i) = 0$, $\rho = 0$, or both. Multiple tests are available to evaluate these restrictions (see, again, texts such as Hamilton 1994 and Hendry 1995 for comprehensive details). Most standard data analysis programs allow a researcher to estimate a full model on the pretreatment values of Y_{it}, assuming that neither $\mathrm{Var}(\lambda_i) = 0$ nor $\rho = 0$, and then to reestimate various constrained versions. Thereafter, a researcher can then use standard likelihood ratio tests of these model constraints. Such tests on the pretreatment data are not full tests of how Y_{it}^0 evolves for the treatment group in the absence of treatment (here again, we are back to the issue for the ITS model in Figure 11.1). Thus, a researcher most likely will need to make some untestable assumptions.

Consider the following scenarios. If both $\mathrm{Var}(\lambda_i)$ and ρ are nonzero (and, furthermore, selection into the treatment is on λ_i only), how then do the values of Y_{it}^0 in the treatment and control group evolve? When asking this question, we are implicitly considering how $E[Y_{it}^0|D_i^* = 1]$ and $E[Y_{it}^0|D_i^* = 0]$ evolve over time toward one or more values, even though the evolution of these two conditional expectations represent average trajectories of individual-specific patterns of evolution in trajectories of Y_{it}^0. Consider a summary of these different situations in Table 11.3, which are then depicted as pairs of lines in Figure 11.7.

Note that Model A is consistent with the assumptions of the change score model. Model B is consistent with the assumptions of the analysis of covariance model. Model C is consistent with neither, but we suspect that it is the most common scenario in empirical research.

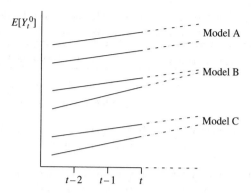

Figure 11.7 Depictions of possible trajectories, as specified by the models in Table 11.3, for $E[Y_{it}^0|D_i^* = 1]$ (the upper line of each pair, corresponding to the treatment group) and $E[Y_{it}^0|D_i^* = 0]$ (the lower line of each pair, corresponding to the control group).

Many of the most common models in panel data analysis assume that there is virtually no dynamic structure to the process that generates the outcome. In fact, most versions of the model that generally goes by the name of a "fixed effects" model are based on the following implicit model for the generation of Y^0 and Y^1:

$$Y_{it}^0 = \lambda_i + T\tau + v_{it}, \tag{11.20}$$
$$Y_{it}^1 = Y_{it}^0 + \delta_i, \tag{11.21}$$

where λ_i is a fixed time constant, individual-level determinant of the outcome in the absence of treatment, $T\tau$ is a time trend common to all (because T is a variable measuring time, τ is a constant coefficient that does not vary over individuals or time), v_{it} is random noise, and δ_i is an individual-specific additive causal effect that is assumed to be independent of λ_i and v_{it}. The assumed data generation process that motivates the most standard form of a fixed effect model is equivalent to the assumption that each individual has his or her own intercept, but there is neither serial correlation in v_{it} nor individual-specific trajectories in time.

The motivation for the standard fixed effects model can be generalized by allowing each individual to have his or her own slope with respect to time, which is indicated by subscripting τ by i in the assumed data generation model in Equations (11.20) and (11.21). This more general model is then

$$Y_{it}^0 = \lambda_i + T\tau_i + v_{it}, \tag{11.22}$$
$$Y_{it}^1 = Y_{it}^0 + \delta_i, \tag{11.23}$$

which, apart from the stochastic term v_{it}, was considered already in the previous section; see Equations (11.14) and (11.15). There, we showed that allowing for differences between $E[\tau_i|D_i^* = 1]$ and $E[\tau_i|D_i^* = 0]$ could lead to the necessity of adjustment factors ranging from negative to positive infinity. The attractiveness of this model, of course, is that it allows $E[Y_{it}^0]$ for the treatment and control groups to evolve in parallel, diverge, or converge. This will depend, respectively, on whether the difference in

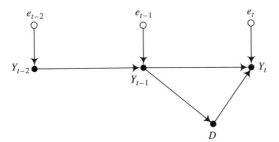

Figure 11.8 A model of endogenous treatment assignment in which selection is on the pretreatment outcome, Y_{t-1}.

the expected slopes for the treatment and control groups is zero, positive, or negative. But substantial amounts of data are needed to estimate it, and certainly from more than just one pretreatment time period and one posttreatment time period.

Determining the Assignment Process

As we have argued throughout this book, the key to estimating a treatment effect is understanding the process of treatment assignment/selection. One of the advantages of conceptualizing and then analyzing the dynamic process of the outcome is that it may provide evidence about the factors that structure the assignment process. Two general cases are of particular interest and lead to quite different estimation strategies. The issue, however, is potentially tricky in that we may want to condition on one or more endogenous variables. As we have discussed at many points throughout this book, conditioning on an endogenous variable that is a collider along a back-door path will unblock an already blocked back-door path, thus creating a new source of confounding.

Consider first the case in which assignment is directly a function of previous values of Y as in Figure 11.8. In this graph, the association between Y_t and D does not identify the causal effect of D on Y_t because they are connected by a back-door path through Y_{t-1}: $D \leftarrow Y_{t-1} \rightarrow Y_t$. However, this back-door path can be blocked by conditioning on Y_{t-1}.

Note that Y_{t-1} is a collider on the path $e_{t-2} \rightarrow Y_{t-2} \rightarrow Y_{t-1} \leftarrow e_{t-1}$. Thus, conditioning on Y_{t-1} will induce associations between Y_{t-2} and e_{t-1} as well as between e_{t-2} and e_{t-1}. These new associations are unproblematic, however, because they do not create any new as-if back-door paths between D and Y_t. Note also that if we thought that treatment assignment D was determined by earlier values of Y, we could condition on these Y's without creating as-if back-door paths that confound the effect of interest.

Consider an alternative and much more complex model, presented in Figure 11.9, where treatment assignment D is determined by λ instead of Y_{t-1}. For this model, there is an unblocked back-door path connecting D to Y_t: $D \leftarrow \lambda \rightarrow Y_t$. What happens if we condition on Y_{t-1}? Obviously, the unblocked back-door path $D \leftarrow \lambda \rightarrow Y_t$ remains unblocked because Y_{t-1} does not lie along it. In addition, conditioning on Y_{t-1} unblocks a set of already blocked back-door paths because Y_{t-1} is a collider on the previously

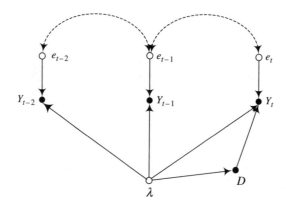

Figure 11.9 A model of endogenous treatment assignment in which selection is on a fixed effect that also determines the outcome.

blocked back-door paths represented collectively by $D \leftarrow \lambda \rightarrow Y_{t-1} \leftarrow e_{t-1} \dashleftarrow\dashrightarrow e_t \rightarrow Y_t$. In combination, conditioning on Y_{t-1} has only made things worse. We failed to block the original back-door path that was of concern, and we unblocked already blocked back-door paths.

Consider this problem from another perspective. Suppose that the standard motivation of a fixed effect model is in place, such that one has reason to believe that Y_{it}^0 and Y_{it}^1 are generated in the simple way specified for Equations (11.20) and (11.21). Suppose, therefore, that we wish to estimate the following regression equation:

$$Y_{it} = l_i + D_{it}c + e_{it}, \tag{11.24}$$

where the l_i are individual-specific intercepts, and where we then hope that the estimated coefficient c on the treatment exposure variable in time period t will then be equal to the ATE (or the ATT if self-selection is present).

If we had a measure of λ_i, then estimating the model in Equation (11.24) would be straightforward. We could put in a dummy variable specification to parameterize the l_i intercepts. For the graph in Figure 11.9, and from a standard structural equation perspective, Y_{it-1} can be thought of as a measure of λ_i. This then suggests that the following regression equation can be estimated instead of Equation (11.24):

$$Y_{it} = a + Y_{it-1}b + D_{it}c + e_{it}. \tag{11.25}$$

If we think of Y_{it-1} as a measure of λ_i, then it contains measurement error because it is also a function of e_{it-1}. As a result, the coefficient on Y_{it-1} will be downwardly biased. In fact, the estimate of b will be a consistent estimate of r, which is the reliability of Y. This is not surprising because Equation (11.25) is the analysis of covariance model. Thus, the analysis of covariance model can be interpreted as a fixed effect model in which we have used a noisy measure of λ_i as an adjustment variable.

In the last section, we saw that the choice of either the analysis of covariance model or the change score model can be consequential. Accordingly, it is critical to determine whether assignment to D is function of Y_{it-1} or λ_i. If we have two or more pretreatment observations, this is easy to do. The first step is to determine whether $b = c$ or $c = 0$ in the following model:

$$\text{Logit}(D_i) = a + Y_{it-1}b + Y_{it-2}c. \tag{11.26}$$

In Figure 11.8, D is dependent only on Y_{t-1}. Thus, c should equal 0. In Figure 11.9, D is associated with only Y_{t-1} and Y_{t-2} through their joint dependence on λ. As a result, $b = c$.[17] Obviously, with this test, we may also include additional observed variables that we believe also determine D. And the test generalizes in the obvious way when we observe Y at more than two pretreatment time periods.

Effect Estimation

Extensive discussions exist in the econometrics literature on the estimation of the fixed effects model and its generalizations. Basically, there are two different estimation strategies. For the differencing approach, the specification is

$$\begin{aligned}
\text{differencing: } Y_{it} - Y_{it-1} &= (\lambda_i - \lambda_i) + (D_{it} - D_{it-1})d \\
&\quad + (X_{it} - X_{it-1})b + (e_{it} - e_{it-1}) \\
&= (D_{it} - D_{it-1})d + (X_{it} - X_{it-1})b + (e_{it} - e_{it-1}),
\end{aligned} \tag{11.27}$$

where treatment exposure occurs between time period $t - 1$ and t. In contrast, the dummy variable approach is

$$\text{individual dummies: } Y_{it} = P_i l_i + D_{it}d + X_{it}b + e_{it}, \tag{11.28}$$

where P_i is a dummy variable for person i and l_i is its associated coefficient. This second method amounts to estimating a separate intercept for each individual. It is also equivalent to differencing Y and X from their respective individual-level means. If one wants to estimate the generalized fixed effect model in which there is an interaction between an individual effect and time, one can do this by either differencing additional times or by interacting the individual dummies, P_i, with time and estimating a separate interaction term for each dummy.

To understand the differences between these two approaches, evolving conventions in data analysis must be noted. For the representation of the change score model and the analysis of covariance model in Equations (11.4) and (11.5), we conformed to the

---

[17]These assumptions can more easily be tested by estimating the model

$$\text{Logit}(D_i) = a + Y_{it-1}b + (Y_{it-1} + Y_{it-2})c$$

and testing whether $b = 0$ or $c = 0$. Here, $(Y_{it-1} + Y_{it-2})$ is essentially acting as a measure of λ_i. This strategy is based on the following trick often used to test for the equality of coefficients for two variables X and Z. Let the coefficient on X be m and on Z be $m + n$. Run the following regression equation:

$$\begin{aligned}
Y &= Xm + Z(m + p) + u \\
&= (X + Z)m + Zp + u
\end{aligned}$$

and use a standard statistical test to evaluate the null hypothesis that $p = 0$.

convention in the literature wherein the models are written out so that they can be estimated easily with a cross-sectional dataset. In other words, time is encoded in the variables, so that time-varying outcome variables, Y_{it} and Y_{it-1}, are regressed on a cross-sectional treatment group dummy variable D_i^* in an implicit dataset with one record for each of N individuals. This data setup is also the implicit background for the differencing specification of the fixed effect estimator in Equation (11.27).

As can be seen in virtually any panel data textbook (e.g., Baltagi 2005), the convention is to now structure one's dataset in person–time records, which results in $N \times T$ records rather than N records. Forcing time to be constant within each data record allows for variables such as D_{it} and D_i^* to be cleanly parameterized in subsequent analysis. This data setup is the implicit background for the individual dummy specification of the fixed effect estimator in Equation (11.28), and that is why there is no reference to time $t-1$ versus time t in Equation (11.28). Throughout the remainder of this section, we will write with such an $N \times T$ dataset in mind. The ideas, however, do not depend on such an implicit structuring, as one can switch back and forth between both setups based on analogs to Equations (11.27) and (11.28).

As for the alternative fixed effect specifications in Equations (11.27) and (11.28), the two methods give identical answers when there are only two points in time. When the errors are correlated across time and there are more than two time periods, the two methods will give somewhat different estimates, although both estimators are consistent. Which estimator is preferable depends on the nature of the correlation structure. We need not be concerned about this issue here; see Baltagi (2005), Hsiao (2003), and Wooldridge (2010) for discussion.

Traditional fixed effect and differencing methods are generally inefficient, however, if the goal is only to estimate the effect of a treatment. These methods simply eliminate unobserved individual effects from the data. Doing so powerfully eliminates all associations between the treatment variable D_{it} and unobserved time-constant, individual-level variables. If the coefficients of all observed variables are of interest, then this is appropriate.

In the present case, however, our focus is simply on estimating the effect of treatment exposure, D_{it}. As pointed out repeatedly in previous chapters, one approach to consistently estimate the effect of D_{it} is to balance the data with respect to all systematic determinants of D_{it}. As discussed in the last section, our interest in the case of linear models (the situation with nonlinear models being more complicated) is in differences in the expected trajectories of Y_{it}^0 for those in the treatment group and those in the control group. If we want to use the average observed values of Y_{it} in the control group in the posttreatment time periods in order to predict the average counterfactual values of Y_{it}^0 in the treatment group in the posttreatment time periods, then it is essential that differences in the average trajectories of these two groups be modeled effectively.

To be more concrete, suppose that we have three time points. If the only difference in the expected trajectories of Y_{it}^0 for the two groups is in their average levels, then all we need to do is allow for differences in group-level intercepts by estimating

$$Y_{it} = a + D_i^* b + D_{it} d + e_{it}. \tag{11.29}$$

Here, the coefficient b captures differences in the expected trajectories for the treatment and control groups, such that the intercept for the control group is a and the intercept for the treatment group is $a + b$.[18]

If the expected trajectories also differ in their slopes, then we need to include a term for time and an interaction term between group membership and time, as in

$$Y_{it} = a + D_i^* b + Tc + (D_i^* \times T)c' + D_{it}d + e_{it}. \tag{11.30}$$

The coefficient b again captures differences in the expected trajectories, and c' now captures differences in the slopes of the expected trajectories. Interactions between D^* and higher-order polynomials of T (or any other function of time) can also be introduced, assuming sufficient pretreatment data are available.

Estimating the model in Equation (11.30) is equivalent to differencing out the treatment/control group expectations of λ_i and τ_i in the following assumed data generation model for Y^0 and Y^1, based on an augmentation of Equations (11.22) and (11.23). Expanding a standard fixed effect model separately for the treatment and control groups using the time-constant indicator of the treatment group D^* yields

for $D_i^* = 0$: $\hspace{5cm}$ (11.31)
$$Y_{it,D^*=0}^0 = (\mu_{\lambda,D^*=0} + v_{i,D^*=0}) + T(\mu_{\tau,D^*=0} + \tau_{i,D^*=0}'),$$
$$Y_{it}^1 = Y_{it}^0 + \delta_i,$$

and

for $D_i^* = 1$: $\hspace{5cm}$ (11.32)
$$Y_{it,D^*=1}^0 = (\mu_{\lambda,D^*=1} + v_{i,D^*=1}) + T(\mu_{\tau,D^*=1} + \tau_{i,D^*=1}'),$$
$$Y_{it}^1 = Y_{it}^0 + \delta_i.$$

Here, $\mu_{\lambda,D^*=0}$ is the expectation of λ_i in the control group, and $\lambda_i = \mu_{\lambda,D^*=0} + v_{i,D^*=0}$ for those in the control group. Likewise, $\mu_{\tau,D^*=0}$ is the expectation of τ_i in the control group, and $\tau_i = \mu_{\tau,D^*=0} + \tau_{i\lambda,D^*=0}'$ for those in the control group. The terms for the treatment group are defined analogously.

In this setup, the terms $v_{i,D^*=0}$, $v_{i,D^*=1}$, $T\tau_{i,D^*=0}'$, and $T\tau_{i,D^*=1}'$ become components of the error term e_{it} of Equation (11.30) as constant individual-level differences v_i and time-varying individual differences $T\tau_i'$. Because $v_{i,D^*=0}$, $v_{i,D^*=1}$, $T\tau_{i,D^*=0}'$, and $T\tau_{i,D^*=1}'$ are all by construction uncorrelated with D_i^*, e_{it} is uncorrelated with D_i^*, assuming any extra individual or time-varying noise embedded within e_{it} is completely random. Furthermore, the coefficient a in Equation (11.30) is equal to $\mu_{\lambda,D^*=0}$, and the coefficient b is equal to $\mu_{\lambda,D^*=1} - \mu_{\lambda,D^*=0}$. Thus, b captures the difference in the expected intercept for individuals in the treatment and control groups. Likewise, the coefficient c in Equation (11.30) is equal to $\mu_{\tau,D^*=0}$, and the coefficient c' is then equal to $\mu_{\tau,D^*=1} - \mu_{\tau,D^*=0}$. And thus c' captures the difference in expected slope of the time trends for individuals in the treatment and control groups.

---

[18]Notice that we can include both D_{it} and D_i^* in the same regression equation because we have multiple records for each individual over time in our dataset. In posttreatment time periods, $D_i^* = D_{it}$ for all individuals (assuming that no one leaves the treatment state before the end of the study), but in pretreatment time periods, $D_i^* = D_{it}$ only for individuals in the control group.

The coefficient d on D_{it} is a consistent estimate of the ATE because the expectations of λ_i and τ_i are balanced across the treatment and control groups. All of their systematic components related to treatment group membership are parameterized completely by a, b, c, and c'. This leaves all remaining components of the distributions of λ_i and τ_i safely relegated to the error term.[19]

There are two advantages of estimating Equation (11.30), as opposed to using traditional methods for estimating fixed effect models and their generalizations. First, conceptually, the model makes clear that if the goal is consistent estimation, then the expected trajectories of Y_{it}^0 for the treatment and control groups must be correctly modeled, not necessarily all individual-specific trajectories. Later, we show how this principle leads to a general specification test. Second, there are potential efficiency gains. For example, in a standard fixed effect model, half of the overall degrees of freedom are lost by specifying individual-specific fixed effects (when there are only two time periods of data). In estimating Equation (11.30), only one degree of freedom is lost to an estimate of the difference in the intercept for the mean of Y_{it}^0. We should note, however, that this minimization in the loss of degrees of freedom is moderated by the fact that the errors in Equation (11.30) are likely to be highly correlated within individuals (because the errors within individuals include a common fixed effect). An estimation procedure should be used for the standard errors that accounts for this correlation.

In the situation in which Y_{it} is determined only by Y_{it-1} and D_{it}, estimation is simpler. As already discussed with reference to Figure 11.8, conditioning on Y_{it-1} is sufficient to block all back-door paths connecting D to Y_{it}. How the necessary conditioning should be performed will depend on the details of the application. Conditioning could be done by matching, as in the analysis of Dehejia and Wahba (1999) for the National Supported Work data (although there is no evidence that they attempt to test the suitability of this specification as opposed to a fixed effect specification).[20] Alternatively, Y_{it-1} could be conditioned on by a regression model, as in an analysis of covariance model. These models will not yield the same results, and it may be difficult to choose between them. More complicated specifications in which Y_{it} is a function of both Y_{it-1} and individual-level variables are also possible. Halaby (2004) provides a clear introduction to these methods.

Model Testing

By now, we hope to have convinced the reader that maintained modeling assumptions can have large consequences. Given this dependence, it is critical that researchers be explicit about the assumptions that they have made and be able to defend those assumptions. Assumptions can be defended either theoretically or on empirical grounds. Often neither is done. In fact, they are made often without any explicit recognition. Fortunately, if pretreatment observations are available for multiple time periods, it is

---

[19]Moreover, because this model is set up as a linear specification of the treatment effect, a lack of balance in higher-order (centered) moments of λ_i and τ_i does not affect the estimation of d.

[20]For detailed discussions of the appropriate model specification for these data, see Smith and Todd (2005) and associated comment and reply.

possible in many circumstances to test the assumptions against the data. Here, we describe two conceptually different, but mathematically closely related, approaches.

In discussing strategies to increase confidence in a causal effect estimate from an ITS model, we suggested that a researcher could either use a dependent variable for which no effect should occur or estimate the effect for a group for which no treatment effect should occur. Evidence of a treatment effect in either case is evidence that the model is misspecified.

Heckman and Hotz (1989) suggest applying this same principle to panel data when two or more pretreatment waves of data are available. Specifically, they suggest taking one of the pretreatment outcomes and analyzing it as if it occurred posttreatment. A researcher then simply applies the estimation method to the new pseudo-posttreatment data and tests for whether there is evidence of a "treatment effect." Because neither the treatment nor the control group has experienced a treatment in the data under consideration, evidence of a treatment effect is evidence that the model is misspecified (i.e., that the model has failed to fully adjust for differences between the treatment and control groups).

In the analysis of covariance model, care must be taken. Here, it is implicitly assumed that selection is on Y_{it-1}. For example, for a logit specification of the probability of treatment selection, it is implicitly assumed that

$$\text{Logit}(D_i^*) = a + Y_{it-1}b. \tag{11.33}$$

In this model, D_i^* is a function of Y_{it-1}. This is mathematically equivalent to maintaining that D_i^* is a function of $Y_{it-2} + (Y_{it-1} - Y_{it-2})$. Generally, Y_{it-1} will be correlated with $(Y_{it-1} - Y_{it-2})$. Consider the following model:

$$Y_{it-1} = a + Y_{it-2}r + D_i^*c + u_i. \tag{11.34}$$

Because D_i^* is a function of both Y_{it-1} and $(Y_{it-1} - Y_{it-2})$, and Y_{it-1} is correlated with the latter term, in general c will not equal 0. The basic point is that D_i^* is partially a function of a component of Y_{it-1} that is not contained in Y_{it-2}. In general, the coefficient c on D_i^* is a function of this dependence.

We can, however, run time backwards. Accordingly, we can estimate

$$Y_{it-2} = a + Y_{it-1}r + v_i. \tag{11.35}$$

If we are going to use the testing strategy of Heckman and Hotz (1989) to evaluate an analysis of covariance model, we should then test whether $c = 0$ in the following related model:

$$Y_{it-2} = a + Y_{it-1}r + D_i^*c + e_i. \tag{11.36}$$

Because there is no component of D_i^* that depends on Y_{it-2} conditional on Y_{it-1}, c should equal 0 in this model if the analysis of covariance model has correctly adjusted for treatment group differences in Y_{it-2}.

Heckman and Hotz's test also indicates how two-period, pretreatment-posttreatment data can be used. What we should do is fit a cross-sectional model. We should then treat the pretreatment outcome as if it were a posttreatment outcome and then test for a treatment effect. Evidence of a treatment effect is evidence that our cross-sectional model has failed to fully adjust for pretreatment differences between the treatment and control groups.

To better understand these procedures, consider a more general specification of this type of test. Recall that, net of our adjustments for various functions of time and other variables, we seek to evaluate for these tests whether the trajectories of Y_{it}^0 are equivalent in the pretreatment data for the treatment and control groups. A variety of different models can be assessed. A fixed effect model allows for differences in the intercepts for two groups. A model with individual- or group-specific time coefficients allows for differences in slopes. If we have enough pretreatment data, we can add additional functions of time to our model and thereby allow for even more heterogeneity for the trajectories of Y_{it}^0.

The most general specification would be to choose one time period as the base, create a dummy variable for all other time periods, and allow these dummy variables to fully interact with our treatment group indicator D^*. This is what is known as the saturated model in this tradition. It is a completely flexible functional form and allows for completely separate time trajectories for Y_{it}^0 for the treatment and control groups. It is of little use in estimating the true causal effect.

Using just the pretreatment data, we can, however, compare the saturated model with any more restrictive model – such as a fixed effects model – using an F-test or likelihood ratio test. Evidence that the more restrictive model does not fit is evidence that the more restrictive model fails to fully model the differences between treatment and control groups in the trajectories of Y_{it}^0.

Consider the results of Heckman and Hotz (1989), a portion of which is presented in Table 11.4.[21] For their analysis, Heckman and Hotz estimated a wide range of alternative models of the effect of the National Supported Work program on the 1978 earnings of participants who were high school dropouts. The first column reports selected estimated effects from their study. The experimental estimate suggests that there is no evidence that the program has an effect (given a point estimate of $-\$48$ with a standard error of \$144). The regression and fixed effect models show large negative effects, which are statistically significant by conventional standards. The random-growth models, which allow for individual slope coefficients for the trajectories of earnings, suggest a modest but still nonsignificant negative effect.

The second column of Table 11.4 reports the p values for tests of no treatment effect, in which the preprogram 1975 earnings are used as if they were in fact posttreatment earnings. If the models that are tested adequately adjust for underlying differences between the treatment and control groups in the trajectories of earnings, one would expect the treatment effect estimate to be nonsignificant (i.e., have a high p value). In the case of the regression and fixed effects models, the faux-treatment-effect estimate is highly significant, indicating a lack of model fit. In the case of the random-growth model, however, it appears to fit the data.

---

[21]Although Heckman and Hotz (1989) is an exemplary early example of this sort of analysis, the basic specification test approach is used in one form or another in other work as well (e.g., Petersen, Penner, and Høgsnes 2011).

Table 11.4 Specification Tests from the Analysis of Heckman and Hotz (1989) of the Effect of the National Supported Work Program on the Earnings of High School Dropouts

	Estimated effect, in dollers	p values for specification tests	
		Preprogram 1975 earnings	Postprogram 1978 earnings
Experiment	−48 (144)		
Regression	−1884 (247)	.000	.000
Fixed effect model (pre-1972)	−1886 (242)	.000	.000
Random-growth model (pre-1972 and 1973)	−231 (414)	.375	.329

Note: Results are from tables 3 and 5 of Heckman and Hotz (1989).

The third column reports results from a similar test, where Heckman and Hotz analyze the valid 1978 posttreatment data as if one time period was in fact pretreatment data. Again, they test for whether there is evidence of a treatment effect. As with the tests reported in the second column, if the model is properly specified, then there should be no evidence of a treatment effect. But here also the p values for the regression and fixed effects models suggest a significant treatment effect, indicating that these models do not fit the data. And, as before, the p value for the random-growth model indicates that it fits the data.[22]

The National Supported Work data have been analyzed by many different social scientists, and they are perhaps the most widely used data to assess the relative explanatory power of alternative types of panel data estimators. There has been considerable debate about whether or not researchers need methods that take account of unobservables or whether adjusting only for observables is sufficient. Smith and Todd (2005; see also the associated comment and reply) show clearly how sensitive estimates can be to the sample that is chosen. Their results support Heckman's position that there are important situations for which treatment selection is likely to be a function of unobserved variables.

---

[22]A note of caution is warranted here. As in all situations, the power of these tests is a function of sample size. In this case, there are only 566 treatment cases. From the data, it is impossible to tell whether the random-growth model fits the data because it is a sufficiently correct model of the underlying individual-specific trajectories or rather because the sample is small.

11.4 Conclusions

Longitudinal data may be helpful for estimating causal effects, but longitudinal data do not constitute a magic bullet. Defendable assumptions about the treatment assignment process must be specified. And, to use longitudinal data to its maximum potential, researchers must carefully consider the dynamic process that generates the outcome, clearly define the causal effect of interest, and then use constrained models only when there is reason to believe that they fit the underlying data.

With this chapter, we have completed our consideration of identification and analysis strategies that can be relied upon to estimate the ATE, ATT, and ATC. We make no claim to have considered anything but a subset of all strategies that are available, but we have attempted to cover the material that receives the most attention in the literature. In the next chapter, we consider how analysis can proceed when the prospects are low for point identification of a causal effect of interest. Informative empirical analysis is still possible, even though strong causal conclusions cannot be developed.

11.5 Appendix to Chapter 11: Time-Varying Treatment Regimes

The longitudinal data models and analysis strategies that we have considered up until this point make the strong assumptions that the treatment is administered only once and that the timing of the treatment is fixed. Models that relax these assumptions are needed in order to consider research questions where they cannot be sustained, or where maintaining them alters the structure of the questions that are of genuine interest. Social scientists regularly encounter systems of causal effects where the timing of the treatment varies across individuals and where the treatment is repeated across multiple time periods in dynamic fashion.

Dealing with data where the timing of the treatment varies and where there are multiple treatments at different times complicates analysis considerably. First, individuals may differ not only in whether they receive the treatment or not, but they also may differ on when the treatment is received. These additional sources of differences between individuals lead to new identification challenges. Second, new questions about treatment effects may be of interest: Does it matter when the treatment occurs? If there are repeated treatments, what are the effects of different combinations of treatments? Third, new estimation requirements arise. Even in the seemingly ideal case where ignorability holds (i.e., there are no unobserved confounders of the treatments and the outcomes, so that all causal effects are identified), standard conditioning methods such as matching and regression do not always provide consistent estimates. As a result, new specialized methods must be utilized.

Robins and his collaborators have developed several different approaches to modeling data where treatments vary in these ways (see Robins 1997, 1998, 1999, 2000; Robins and Hernán 2009). Unfortunately, these methods do not solve the problem of selection on the unobservables that the main body of this chapter addresses. In fact, as we will discuss below, serious estimation challenges exist even in the absence of selection on the unobservables, with the main challenge being the dependency between treatment states and intermediate outcomes even when unconfoundedness holds. Even so, the methods we present in this appendix are an exciting frontier of methodological scholarship that social scientists need to learn. And directed graphs elucidate the crucial issues.

In this appendix, we will follow the presentation of Robins and Hernán (2009). We will examine methods that are appropriate when the data have been generated by what is known as a "dynamic treatment regime" – a treatment exposure at one point in time is potentially a function of past treatment history and/or current and/or past time-varying covariates. However, we will only discuss the estimation of the effects of what are called "static regimes": the difference in an outcome between individuals when all individuals follow one regime versus another regime. Thus, although we will assume that the data have been generated by a dynamic regime, we will focus on the effects of static regimes.

Types of Regimes

Fixed Treatment Regimes. Situations where the timing of a single treatment or multiple treatments is determined at the beginning of the study are considered to have fixed treatment times. The case where treatment occurs at the same time for all individuals provides the simplest case. The Operation Ceasefire study discussed in Section 11.1 is an example. A college or university where promotion to associate professor always occurs in the sixth year after initial appointment is another. More complicated examples are also possible. A randomized experiment for the evaluation of a worker training program (LaLonde 1986) would be considered to have fixed treatment times, even though participants in the experiment have work histories of different lengths, because for each individual the timing of treatment assignment is known at baseline and thus fixed. A medical treatment regime where different doses or types of drugs are taken for fixed periods of time would be considered to have fixed treatment times as long as which drugs are taken and how long they are taken are determined at baseline.

Nondynamic Regimes. When treatment assignment is endogenous in the restricted sense that treatment status at one point in time is a (nondeterministic) function of treatment status at prior points in time, the treatment regime is labeled "nondynamic" by Robins and Hernán (2009). We do not find this label particularly helpful. The classic example is a sequentially randomized experiment where individuals are randomized at different stages to different treatment possibilities, such that the probability of receiving a treatment at a later stage is solely a function of the outcome of a previous randomization. A simple example would be a worker training experiment where (1) individuals are first randomized to either training or no training and then (2) individuals who are assigned to training are then randomized into specific training programs for either computer skills or construction skills. The key feature of a nondynamic regime is that all randomization probabilities are fixed at baseline before randomization is enacted, so that the randomization probabilities are not functions of either time-varying covariates or actual prior treatment statuses. When such additional dependence does exist, the regime is considered to be dynamic.

Dynamic Regimes. Fixed and nondynamic treatment regimes represent constrained forms of time-varying treatment structures, and they are the ones that present no new analysis challenges beyond those explained in prior chapters of this book. Dynamic treatment regimes are a more general class of models where treatment status at time t, D_t, is determined by a covariate or set of time-varying covariates that may be functions of earlier treatments. The classic example comes from medicine, where treatment at time t is determined as a function of the patient's observed symptoms at time t, which are, in turn, a function of whether or not the patient received a treatment in a prior time period. There are many social science examples as well, and we next consider an example that follows from others we have considered in this book.

The Catholic School Example as a Dynamic Treatment Regime

Consider the effect of Catholic schooling on twelfth grade test scores. For a dynamic treatment regime version of this effect, we allow students to do what some of them are actually observed to do: change the type of school – Catholic or public – that they are enrolled in between the tenth and twelfth grades.[23] In order to keep the example from becoming too complex, we assume that students can only change type of school at the end of tenth grade (i.e., students are assumed to be in the same type of school in both the eleventh and the twelfth grades).

The indicator variables for enrollment type in the tenth grade and in the eleventh and twelfth grades are D_{10} and D_{12}, respectively. Students may follow any of the following four regimes:

1. $D_{10} = D_{12} = 0$ (public school throughout),

2. $D_{10} = D_{12} = 1$ (Catholic school throughout),

3. $D_{10} = 1$, $D_{12} = 0$ (Catholic school, then public school),

4. $D_{10} = 0$, $D_{12} = 1$ (public school, then Catholic school).

The questions for analysis are how twelfth grade tests scores are affected by these alternative treatment regimes, not simply what the effect of twelfth grade enrollment status is on test scores in the twelfth grade. The effect of Catholic schooling is likely to differ across students who have been enrolled in Catholic schools continuously from the tenth grade through the twelfth grade in comparison to students who switch from public schools after the tenth grade and are then enrolled in Catholic schools in the eleventh and twelfth grades.

Because of the complexity of treatment regime patterns, we therefore need to understand the effects that generate tenth grade test scores, even if our primary goal is to estimate effects that can be measured at the end of high school in the twelfth grade. Accordingly, we must model test scores in both the tenth and twelfth grades, Y_{10} and Y_{12}, respectively. For this appendix, we will reduce the complexity of our discussion and the worked example we offer below by analyzing Y_{10} and Y_{12} as two dummy variables that indicate whether students receive high rather than low test scores.

---

[23]For Panel Data Demonstration 1 (see page 273), we considered the effect of Catholic schooling on tenth and twelfth grade test scores, but we restricted attention, as in the existing literature, to students who remained in the same types of schools in both grades. For Panel Data Demonstration 2 (see page 365), we considered only the effects of Catholic schooling on tenth grade test scores, which we modeled using pretreatment outcome measures from the eighth grade.

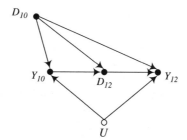

Figure 11.10 The Catholic school effect in the tenth and twelfth grades as a dynamic treatment regime.

Consider the directed graph in Figure 11.10. In order to keep the discussion as simple as possible, Figure 11.10 does not include any observed variables other than the two treatment indicator variables, D_{10} and D_{12}, and the two outcome indicator variables, Y_{10} and Y_{12}. Additional observed variables would affect any or all of these four variables, as in the other demonstrations in this book that analyze the Catholic school effect. It is appropriate to think of Figure 11.10 as the graph that applies within one stratum defined by some combination of other variables. This simplification implies that, unlike in the other demonstrations, we ignore the determinants of D_{10} (see, for example, Equations (5.24) and (5.25) in Matching Demonstration 4). This means that we are holding aside all of the other complications shown in prior demonstrations and now considering an additional set of complications that would still remain if we could solve all of the other complications explained in prior demonstrations. A more realistic model would include variables that would allow us to demonstrate all of the complications at once, but our presentation would then lose focus on the issues we wish to explain in this appendix.

For Figure 11.10, we also assume that whether a student is enrolled in a Catholic school by the twelfth grade, D_{12}, is solely determined by whether the student was enrolled in a Catholic school in tenth grade, D_{10}, and by how successful the student was on the test at the end of tenth grade, Y_{10}. Thus, the directed graph allows for persistence effects (i.e., students may be more likely to be in the same type of school in twelfth grade as in the tenth grade) and also the possibility that some students change school type based on test performance at the end of the tenth grade. For example, the directed graph is consistent with scenarios where students in public schools who perform poorly at the end tenth grade switch to Catholic schools thereafter.

Finally, note the restricted way in which the unobserved variable U structures Y_{10} and Y_{12} and nothing else in the graph. Because we have assumed that we are analyzing Figure 11.10 within a stratum defined by variables that may also determine treatment assignment, the unobserved variable U in Figure 11.10 is not one of the variables that define these strata. U might reflect the portions of innate intelligence, motivation, or other variables that determine the test score outcomes but that have no role in structuring treatment selection decisions. The critical assumption in the model is that selection into D_{12} is solely a function of observed variables, in this case D_{10} and Y_{10}. Unobserved variables that determine D_{12} and either Y_{10} or Y_{12} are assumed not to exist. As we discuss below, this assumption is the key to identification.

To clarify the language of treatment regimes, note that Figure 11.10 represents a dynamic treatment regime because the second treatment, D_{12}, is determined by an intermediate outcome, Y_{10}, that is itself determined by the initial treatment, D_{10}. If Y_{10} were omitted from Figure 11.10, we would then have a nondynamic treatment regime because the second treatment, D_{12}, would be determined only by the first treatment, D_{10}, in addition to idiosyncratic determinants unrelated to all else. And, if the dependence of D_{12} on D_{10} did not exist, we would have a fixed treatment regime consisting of the two treatments, D_{10} and D_{12}, that could be analyzed separately without concern for the implications of one analysis on the other.

Independent of whether Figure 11.10 contains Y_{10} or not, or whether D_{10} is a cause of D_{12} or not, the directed graph represents a structure of effects that is equivalent to a sequentially randomized experiment. Because D_{10} is a function of no other variables, it is akin to a randomized treatment (again, within strata of observed and unobserved variables that structure treatment assignment in the worlds considered for prior demonstrations). Analogously, D_{12} can be thought of as having been determined by a randomization scheme where the probability of receiving the treatment in the twelfth grade is determined by two specific observed variables, D_{10} and Y_{10}. Since the directed graph in Figure 11.10 represents a sequentially randomized experiment, ignorability holds with the result that the total causal effects of D_{10} on Y_{10}, D_{10} on Y_{12}, and D_{12} on Y_{12} are all identified. We explain this result, and related results for direct causal effects, in considerable detail below.

General Identification Conditions

As noted in the introduction to this appendix, our concern now is whether it is possible to identify the causal effects of a fixed regime (i.e., the effects of combinations across time of Catholic and public school enrollment) from data generated as part of a dynamic treatment regime. Identification of the causal effect of a fixed regime of treatments holds under three conditions, two of which are demanding and worth explaining.[24] The most important condition is that sequential ignorability holds (i.e., conditional on observables, treatment assignment in each time period is independent of the potential outcomes). This assumption is equivalent to maintaining that no unobserved confounders exist that determine both the treatments and outcomes. In general, this condition is untestable and must be defended by appeals to substantive knowledge. We have discussed this issue extensively in previous chapters.

For this appendix, we are assuming away this problem by focusing on an analysis within a stratum where we can assume that sequential ignorability holds. Return to the directed graph in Figure 11.10. The fact that it has the same structure as the directed graph that would be an appropriate representation of a sequentially randomized experiment means that ignorability holds for both D_{10} and D_{12} separately and together. This result can be confirmed by noting that neither D_{10} nor D_{12} is affected by any unobserved confounders that also affect either Y_{10} or Y_{12}.

The second condition is known as positivity. For this condition, it must be the case that at least some individuals have followed each logically possible treatment

---

[24]The third condition is what is known as consistency, which states that the value of the realized Y under a specific treatment condition is equal to the potential outcome under that treatment. Hernán (2005) discusses situations where consistency may not hold.

regime within each stratum. When there are a large number of possible treatment combinations and/or strata, nonpositivity can be a serious problem because there may well be too few individuals (indeed, perhaps none) in particular combinations of treatments and the strata defined by observed confounders. As with any stratification procedure, one may face the "curse of dimensionality." As we discuss below, the method of G-computation, which is a stratification-type procedure, is often impractical because of the curse of dimensionality (even though G-computation is a very useful way to think through identification results).

A Specific Setup of the Dynamic Treatment Regime Variant of the Catholic School Example

In order to demonstrate how various methods do or do not work, we now introduce a hypothetical empirical dataset that is consistent with the directed graph in Figure 11.10. We do so in two steps. First, we posit a set of equations consistent with Figure 11.10. To keep matters simple, we use linear equations, although we will allow for interactions. Second, we generate a table of the expected values for the endogenous variables in the directed graph in Figure 11.10. We use expected values as opposed to a fully simulated dataset because this setup makes it easier to demonstrate that a particular estimation strategy will generate consistent estimates of effects of interest.

First, we assume that $E[D_{10}] = .20$ because 20 percent of the students are enrolled in Catholic schools in the tenth grade. U is a dummy indicator variable that measures whether a student is high, as opposed to low, on unobserved characteristics, such as motivation, effort, and mental ability. We assume that $E[U] = .60$ because 60 percent of students are in the high category for U. With these distributions for D_{10} and U, we set the expected value of test scores at the end of tenth grade as

$$E[Y_{10}|D_{10}, U] = .1 + .2D_{10} + .4U. \tag{11.37}$$

Enrolling in a Catholic school increases the probability that a student will be in the high-test-score group by .2. In addition, being in the high category of the unobserved variable ($U = 1$) increases the probability of being in the high-test-score category by .4.[25]

---

[25]Note that we have not defined $E[Y_{10}|D_{10}, U]$ directly in values of underlying potential outcomes. We could do so by working toward a conditional expectation function analogous to Equation (11.37) by starting with three equations:

$$Y_{12}^0 = .1 + .4U + v^0,$$

$$Y_{10}^1 = Y_{10}^0 + .2 + v^1,$$

$$Y_{10} = D_{10}Y_{10}^1 + (1 - D_{10})Y_{10}^0.$$

We will not use this type of setup for this example because we will not focus on individual-level variability of causal effects and because the Markov structure of Figure 11.10, along with binary variables, allows us to fit saturated models to recover all conditional expectations that match those we could define explicitly with potential outcomes. This is consistent with the dynamic treatment regime literature, where Markov assumptions for (sometimes implicit) causal graphs are often used to allow for a consideration of causal effects that are structured by how observed treatment variables have effects on observed outcome variables. These observed outcome variables could be defined in terms of underlying potential outcomes.

Having set the distributions for D_{10} and Y_{10}, we then set enrollment in Catholic schools in the twelfth grade as

$$E[D_{12}|D_{10}, Y_{10}] = .5 + .4D_{10} - .4Y_{10} + .4(D_{10} \times Y_{10}), \qquad (11.38)$$

which specifies a cross-product interaction between D_{10} and Y_{10}. Accordingly, Equation (11.38) indicates that, for students enrolled in public schools in the tenth grade ($D_{10} = 0$) and in the low-test-score group ($Y_{10} = 0$), their probability of being in a Catholic school in the twelfth grade is .5. However, for students enrolled in a public school in tenth grade ($D_{10} = 0$) and in the high-test-score group ($Y_{10} = 1$), their probability of being in Catholic school in the twelfth grade is only .1. We specify this pattern to give the treatment regime a dynamic structure where students doing well in public schools at the end of the tenth grade are much less likely to decide to switch enrollment status to enter into a Catholic school in the twelfth grade. Finally, for students enrolled in Catholic schools in the tenth grade ($D_{10} = 1$), the probability of being in a Catholic school in the twelfth grade is .9, regardless of whether or not these students are in the high-test-score or the low-test-score groups at the end of the tenth grade.

To complete the specification of the example, we set the expected value of the test scores at the end of the twelfth grade as

$$E[Y_{12}|D_{10}, D_{12}, U] = .2 + .2D_{12} + .1(D_{10} \times D_{12}) + .4U. \qquad (11.39)$$

The unobserved variable U has the largest effect on the expected value of Y_{12} (but, at .4, has the same effect on the expected value of Y_{12} as for Y_{10}). Being in a Catholic school in the twelfth grade increases the probability of being in the high-test-score group by .2. However, being in a Catholic school in the tenth grade only increases the chances of being in the high-test-score group in the twelfth grade if one was also enrolled in a Catholic school in the twelfth grade. This repeated treatment effect generates a boost of .1 for students enrolled in Catholic schools in both the tenth and twelfth grades.[26]

Table 11.5 presents expected values for the endogenous variables, Y_{10}, D_{12}, and Y_{12} based on this setup. Note that there are four variables, D_{10}, Y_{10}, D_{12}, Y_{12}, and that each takes on the value 0 and 1. The table reports the conditional expected values for the three endogenous variables: $E[Y_{10}|D_{10}]$, $E[D_{12}|D_{10}, Y_{10}]$, and $E[Y_{12}|D_{10}, Y_{10}, D_{12}]$. The final column of Table 11.5 presents the proportion of the sample in the 16 strata defined across the 4 dichotomous variables. These values can be used to calculate additional conditional expectations defined by combinations of values for the variables in the first four columns.

Identification of Total and Direct Effects for the Example   For total causal effects, which have been the focus of this book, the key identification results in the dynamic treatment regime literature are given by the back-door criterion. In cases where the total causal effect is not equivalent to the direct causal effect (e.g., the effect of D_{10} on Y_{12} in Figure 11.10, which has both a direct effect, $D_{10} \rightarrow Y_{12}$, and two

---

[26]In other words, there is an implicit term in Equation (11.39) of $0 \times D_{10}$ because enrollment in a Catholic school in the tenth grade has no effects on twelfth grade test scores, which implies that students do not carry with them a lagged effect of tenth grade Catholic schooling when they switch from Catholic schools to public schools for the eleventh and twelfth grades.

Table 11.5 Expected Values for the Endogenous Variables in the Directed Graph in Figure 11.10

D_{10}	Y_{10}	D_{12}	Y_{12}	$E[Y_{10}|D_{10}]$	$E[D_{12}|D_{10},Y_{10}]$	$E[Y_{12}|D_{10},D_{12},Y_{10}]$	Proportion
1	1	1	1	.460	.900	.811	.079
1	1	1	0	.460	.900	.811	.018
1	1	0	1	.460	.900	.511	.006
1	1	0	0	.460	.900	.511	.005
1	0	1	1	.460	.900	.657	.054
1	0	1	0	.460	.900	.657	.028
1	0	0	1	.460	.900	.357	.003
1	0	0	0	.460	.900	.357	.006
0	1	1	1	.260	.100	.753	.020
0	1	1	0	.260	.100	.753	.007
0	1	0	1	.260	.100	.553	.135
0	1	0	0	.260	.100	.553	.109
0	0	1	1	.260	.500	.582	.154
0	0	1	0	.260	.500	.582	.110
0	0	0	1	.260	.500	.382	.101
0	0	0	0	.260	.500	.382	.163

indirect effects, $D_{10} \to Y_{10} \to D_{12} \to Y_{12}$ and $D_{10} \to D_{12} \to Y_{12}$), the key identification results are given by a related literature on causal mediation; see Pearl (2009), Wang and Sobel (2013), and VanderWeele (in press). Accordingly, the first step in an identification analysis, which we offer below for the Catholic school example, is to consider the total causal effects in Figure 11.10 by consulting the back-door criterion. The results that we offer below will be consistent with many others offered in Chapters 4 through 7. We then give sustained attention to the identification of direct effects, concentrating on what have been labeled the "controlled direct effects" in the literature on causal mediation. These effects have not been considered in this book up until now in any explicit way, although the careful reader will have seen references to them in the details of Chapter 10. Our identification analysis will make use of the conditional expectations reported in Table 11.5, with reference to Equations (11.37)–(11.39) that generate them. In the section that follows, we will then discuss how to estimate these identified effects from a single sample of data.

Total Effects. Table 11.6 indicates which of the eight possible total causal effects in Figure 11.10 are identified, as well as the method that would need to be used in order to consistently estimate those that are identified.[27] Most of the identification results in Table 11.6 are not surprising, although the last three differ from others that

---

[27]The fact that all of the total effects of observed variables in Figure 11.10 are identified means that we can test whether they are equal to 0 or not and thus at least partially consider the appropriateness of the directed graph as a whole. Elwert (2013, table 13.1) gives an example of how such testing can be organized.

Table 11.6 Identification Status of the Total Causal Effects in Figure 11.10

Total Effect	Identified?	Method for Estimation
1. $U \rightarrow Y_{10}$	No	
2. $U \rightarrow Y_{12}$	No	
3. $D_{10} \rightarrow Y_{10}$	Yes	Unconditional association
4. $Y_{10} \rightarrow D_{12}$	Yes	Condition on D_{10}
5. $D_{12} \rightarrow Y_{12}$	Yes	Condition on D_{10} and Y_{10}
6. $(Y_{10} \rightarrow D_{12} \rightarrow Y_{12})$	Yes	Front-door combination of already identified effects $Y_{10} \rightarrow D_{12}$ and $D_{12} \rightarrow Y_{12}$
7. $(D_{10} \rightarrow D_{12}) + (D_{10} \rightarrow Y_{10} \rightarrow D_{12})$	Yes	Unconditional association
8. $(D_{10} \rightarrow Y_{12}) + (D_{10} \rightarrow D_{12} \rightarrow Y_{12})$ $\quad + (D_{10} \rightarrow Y_{10} \rightarrow D_{12} \rightarrow Y_{12})$	Yes	Unconditional association

we have considered explicitly in this book. We will examine these eight effects in the order in which they are listed Table 11.6, and they fall into five distinct identification patterns according to the order in which they are listed.

First, because is U is unobserved, the total effects of U on both Y_{10} and Y_{12} are not identified by the observed data. Second, because D_{10} and Y_{10} are not connected by any confounders that generate a back-door path, it follows that their unconditional association identifies $D_{10} \rightarrow Y_{10}$. For our hypothetical data, the effect is equal to the difference in the expected values of Y_{10} for the two values of D_{10},

$$E[Y_{10}|D_{10}=1] - E[Y_{10}|D_{10}=0] = .460 - .260$$
$$= .200,$$

given in Table 11.5. Nonetheless, recall that this result follows from the construction of the example, where we have assumed that we are analyzing the Catholic school effect within a stratum where Figure 11.10 can be accepted as reasonable.

Third, the next two total effects are identified by conditioning that is warranted by the back-door criterion. For the effect $Y_{10} \rightarrow D_{12}$, three back-door paths are present:

1. $Y_{10} \leftarrow D_{10} \rightarrow D_{12}$,

2. $Y_{10} \leftarrow D_{10} \rightarrow Y_{12} \leftarrow D_{12}$,

3. $Y_{10} \leftarrow U \rightarrow Y_{12} \leftarrow D_{12}$.

Without any conditioning, paths 2 and 3 are blocked by the collider Y_{12}. However, path 1 remains unblocked and generates a noncausal association between Y_{10} and D_{12}. When D_{10} is conditioned on, all three back-door paths are blocked and a consistent estimate of the effect $Y_{10} \rightarrow D_{12}$ can be obtained. Similarly, for the effect $D_{12} \rightarrow Y_{12}$, four back-door paths are present:

1. $D_{12} \leftarrow D_{10} \rightarrow Y_{12}$,

2. $D_{12} \leftarrow Y_{10} \leftarrow D_{10} \rightarrow Y_{12}$,

3. $D_{12} \leftarrow Y_{10} \leftarrow U \rightarrow Y_{12}$,

4. $D_{12} \leftarrow D_{10} \rightarrow Y_{10} \leftarrow U \rightarrow Y_{12}$.

In the absence of conditioning, path 4 is blocked by the collider Y_{10}, while paths 1, 2, and 3 remain unblocked. If only D_{10} is conditioned on, paths 1, 2, and 4 are blocked, but path 3 remains unblocked. If only Y_{10} is conditioned on, paths 2, 3, and 4 are blocked, but path 1 remains unblocked. If both D_{10} and Y_{10} are conditioned on, then all four back-door paths are blocked and a consistent estimate of the effect $D_{12} \rightarrow Y_{12}$ can be obtained.

Fourth, the total causal effect of Y_{10} on Y_{12}, which is a two-edge directed path $Y_{10} \rightarrow D_{12} \rightarrow Y_{12}$, is also identified. This result follows from a double consideration of the back-door criterion, which is the front-door identification strategy presented in Chapter 10. The edge-by-edge identification results that constitute the front-door identification strategy are given in the corresponding rows of the table immediately above the row for the total effect of Y_{10} on Y_{12}, and as discussed already.

Fifth, the final two total effects are identified but are different because they are composed of both a direct effect and one or more indirect effects through chains of mediation. The total causal effect of D_{10} on D_{12} (which is composed of the two directed paths $D_{10} \rightarrow Y_{10} \rightarrow D_{12}$ and $D_{10} \rightarrow D_{12}$) is identified by the unconditional association between D_{10} and D_{12} because no back-door paths between D_{10} and Y_{12} are present that generate confounding. Likewise, the total causal effect of D_{10} on Y_{12}, which is composed of the three directed paths ($D_{10} \rightarrow Y_{10} \rightarrow D_{12} \rightarrow Y_{12}$, $D_{10} \rightarrow D_{12} \rightarrow Y_{12}$, and $D_{10} \rightarrow Y_{12}$) is identified by the unconditional association between D_{10} and Y_{12}, again because no back-door paths are present that generate confounding. Note, however, that we have established these last two identification results by construction of the example, just as for our prior consideration of the total effect of D_{10} on Y_{10}.

Overall, the directed graph in Figure 11.10 has a simple structure, allowing for the estimation of all total causal effects of the observed variables, including the total effects of D_{10} on both Y_{10} and Y_{12} as well as the total causal effect of D_{12} on Y_{12}. The simple structure yields straightforward identification results because Figure 11.10 is equivalent to a sequential randomized experiment.

Direct Effects. Consider now the identification of direct effects, ignoring those that are equal to their total effects (i.e., $D_{10} \rightarrow Y_{10}$). Two such direct effects are present in Figure 11.10, one of which is straightforward and of secondary interest ($D_{10} \rightarrow D_{12}$) and one of which is not straightforward but of primary interest ($D_{10} \rightarrow Y_{12}$). These effects are identified, but they are not identified using the back-door criterion to warrant conditioning variables (because no back-door paths confound these two direct effects).

To begin to appreciate the complications, it is useful to consider how the conditional expectations in Equations (11.38) and (11.39) structure these direct effects. Consider first the direct effect $D_{10} \rightarrow D_{12}$. Here, the direct effect varies with the two values of Y_{10} because of the nature of the dynamic treatment regime. We can use Equation (11.38) to see how these effects were set by construction. The first step is to generate the conditional expectations that will define the direct effects:

$$E[D_{12}|D_{10} = 1, Y_{10} = 1] = .5 + .4(1) - .4(1) + .4(1 \times 1) \tag{11.40}$$
$$= .9,$$

$$E[D_{12}|D_{10} = 0, Y_{10} = 1] = .5 + .4(0) - .4(1) + .4(0 \times 1) \tag{11.41}$$
$$= .1,$$

$$E[D_{12}|D_{10} = 1, Y_{10} = 0] = .5 + .4(1) - .4(0) + .4(1 \times 0) \tag{11.42}$$
$$= .9,$$

$$E[D_{12}|D_{10} = 0, Y_{10} = 0] = .5 + .4(0) - .4(0) + .4(0 \times 0) \tag{11.43}$$
$$= .5.$$

These values are also given in the sixth column of Table 11.5.

The second step is take the values yielded by Equations (11.40)–(11.43) and then calculate what have become known as "controlled direct effects" in the literature on mediation, direct, and indirect effects (see Pearl 2001, 2009, 2012a, 2012b; VanderWeele 2009a, 2010; Wang and Sobel 2013):

$$\text{CDE}_{D_{10} \to D_{12}}(Y_{10} = 1)$$
$$= E[D_{12}|D_{10} = 1, Y_{10} = 1] - E[D_{12}|D_{10} = 0, Y_{10} = 1] \tag{11.44}$$
$$= .9 - .1$$
$$= .8,$$

$$\text{CDE}_{D_{10} \to D_{12}}(Y_{10} = 0)$$
$$= E[D_{12}|D_{10} = 1, Y_{10} = 0] - E[D_{12}|D_{10} = 0, Y_{10} = 0] \tag{11.45}$$
$$= .9 - .5$$
$$= .4.$$

These two controlled direct effects for $D_{10} \to D_{12}$ are two distinct components of the direct effect, each of which can be calculated when Y_{10} is set to one of its two values of 0 or 1. The label "controlled" refers to the action of setting the value for the variable Y_{10} before calculating the effect of the primary causal variable on the outcome variable of interest.

When controlled direct effects are identified, then the direct effect can be considered identified. In other words, the direct effect is identified because all of its component effects are identified.[28] To understand this result, notice first that all of the calculations

---

[28]Nonetheless, one may prefer to have a single value for a direct effect, rather than multiple values for each controlled direct effect. The most common single-value direct effect is what has been labeled both

carried out for Equations (11.40)–(11.46) use values for conditional expectations and probabilities that are functions in observed variables only. As such, if a sample of infinite size were available, we could exactly calculate these effects for such a sample because the sample values would be exactly equal to these population values. This explanation, however, is too shallow. The deepest explanation can be found in the primary literature on results that have been established for graphs that have a Markov structure; see Pearl (2009), after reading our appendix to Chapter 3. The core idea is that the simple structure of the graph in Figure 11.10, where no unblocked back-door paths are present between the two treatment variables, ensures that we can define the conditional expectations using Equations (11.37)–(11.39) and then assert that differences in the estimated values of the conditional expectations on the left-hand sides of Equations (11.40)–(11.43) are causal contrasts that identify controlled direct effects. Consider a counterexample for insight. If the graph in Figure 11.10 cannot be defended for the substantive example because an unobserved confounder of D_{10} and D_{12} exists, such that a back-door path $D_{10} \leftarrow C \rightarrow D_{12}$ exists where C is unobserved, then the conditional expectations that define the controlled direct effects in Equations (11.44) and (11.45) would not identify causal effects. The difference between the sample analogs to these conditional expectations, such as

$$E_N[d_{i12} = 1 | d_{i10} = 1, y_{i10} = 1] - E[d_{i12} = 1 | d_{i10} = 0, y_{i10} = 1],$$

would not converge to the relevant controlled direct effect, $\text{CDE}_{D_{10} \rightarrow D_{12}}(Y_{10} = 1)$ because the confounder C generates additional noncausal dependence between D_{10} and D_{12} within strata defined by Y_{10}.[29]

We find another explanation of this identification result helpful as well because of the way it connects to the sort of thinking that we have used extensively when

---

the "pure direct effect" and the "natural direct effect" in the causal mediation literature. Whatever label chosen, it is a weighted average of the controlled direct effects that correspond to the distribution of the mediator that exists in the control group. In this case, the natural direct effect is

$$\begin{aligned} \text{NDE}_{D_{10} \rightarrow D_{12}} &= \text{CDE}_{D_{10} \rightarrow D_{12}}(Y_{10} = 1) \times \Pr[Y_{10} = 1 | D_{10} = 0] \\ &\quad + \text{CDE}_{D_{10} \rightarrow D_{12}}(Y_{10} = 0) \times \Pr[Y_{10} = 0 | D_{10} = 0] \\ &= (.8 \times .34) + (.4 \times .66) \\ &= .536, \end{aligned} \quad (11.46)$$

where the values of .34 and .66 for $\Pr[Y_{10} = 1 | D_{10} = 0]$ and $\Pr[Y_{10} = 0 | D_{10} = 0]$ are calculated from the final column of Table 11.5. In this case, the natural direct effect is a counterfactual quantity: the Catholic school persistence effect, net of the dynamic nature of the treatment regime that we have assumed by construction exists. Thus, apart from how test score performance is determined in the tenth grade, the estimate suggests that the probability of entering a Catholic school in the twelfth grade is higher by .536 if a student was enrolled in a Catholic school in the tenth grade. We do not find this counterfactual single-value direct effect to provide any additional information beyond the controlled direct effects. In the broader literature on causal mediation, natural direct effects yield effects that are more informative, typically when the mediator represents the primary substantive mechanism that generates the total effect.

[29]In the causal graph literature, this identification result follows from the Markov structure of the graph, which allows all differences in the conditional expectations of observed variables in the pre-intervention graph to be equal to their under-intervention differences. These equalities could be made more explicit by using either potential outcome notation or Pearl's $do(.)$ operator. When the graph has a Markov structure, such representations are redundant, as we explained in the appendix to Chapter 3.

invoking the back-door criterion in this book. Consider the following graphical explanation for the role of conditioning in calculating controlled direct effects. In order to estimate the controlled direct effects, we need to condition on Y_{10}. The basic idea here is that we need to set the indirect effect of D_{10} on Y_{12} to 0 by blocking the directed path $D_{10} \rightarrow Y_{10} \rightarrow D_{12}$ in order to then calculate the separable controlled direct effects. As a result, conditioning is essential to the calculation of these effects. Yet, note also that Y_{10} is a collider on a path that begins at D_{10} and ends at D_{12}, which is $D_{10} \rightarrow Y_{10} \leftarrow U \rightarrow Y_{12} \leftarrow D_{12}$. Although this path is not a back-door path between these two variables (because it does not begin with $D_{10} \leftarrow$), conditioning on Y_{10} does nonetheless induce an association between D_{10} and U. Fortunately, this induced association is blocked by a second collider on the same path, Y_{12}. Accordingly, we can see that conditioning on Y_{10} effectively sets the indirect path to 0 without inducing any unwanted noncausal associations between D_{10} and D_{12}. This result suggests that a straightforward approach for the estimation of the direct effect $D_{10} \rightarrow D_{12}$ is to calculate the conditional associations between D_{10} and D_{12} for each value of Y_{10}.

To fully appreciate the depth of the issues involved in the dynamic treatment regime literature, we need to consider the direct effect of D_{10} on Y_{12}, which is $D_{10} \rightarrow Y_{12}$ in Figure 11.10. This direct effect is identified according to some of the same reasoning just laid out for the direct effect of D_{10} on D_{12}, with some very important differences we will fully explain below. First note that the direct effect $D_{10} \rightarrow Y_{12}$ is a much more important substantive effect to estimate because it is a net average treatment effect for the treatment introduced in the first time period on the final outcome observed. It is also a more complicated direct effect to consider because the indirect effect of D_{10} on Y_{12} traverses two separate directed paths, $D_{10} \rightarrow D_{12} \rightarrow Y_{12}$ and $D_{10} \rightarrow Y_{10} \rightarrow D_{12} \rightarrow Y_{12}$. Fortunately, because the directed path through Y_{10} lies on top of the directed path through D_{12}, the indirect effect through Y_{10} is fully absorbed into the distribution of D_{12}. Accordingly, and following the reasoning above, we can set both indirect effects to 0 by conditioning in D_{12}.

As was the case for the direct effect $D_{10} \rightarrow D_{12}$, the first step to developing an explanation for the identification results is to generate the conditional expectations that define the controlled direct effects, plugging all combinations of values for D_{10} and D_{12} into Equation (11.39) while allowing U to vary:

$$E[Y_{12}|D_{10}=1, D_{12}=1, U] = .2 + .2(1) + .1(1 \times 1) + .4U$$
$$= .5 + .4U, \tag{11.47}$$

$$E[Y_{12}|D_{10}=0, D_{12}=1, U] = .2 + .2(1) + .1(0 \times 1) + .4U$$
$$= .4 + .4U, \tag{11.48}$$

$$E[Y_{12}|D_{10}=1, D_{12}=0, U] = .2 + .2(0) + .1(1 \times 0) + .4U$$
$$= .2 + .4U, \tag{11.49}$$

$$E[Y_{12}|D_{10}=0, D_{12}=0, U] = .2 + .2(0) + .1(0 \times 0) + .4U$$
$$= .2 + .4U. \tag{11.50}$$

Note that, even though Y_{10} lies on a path that is part of the indirect effect of D_{10} on Y_{12}, Y_{10} is absent from Equation (11.39). According to the graph, this effect is fully absorbed into the effect of D_{12} on Y_{12}. The controlled direct effects are then

$$
\begin{aligned}
\text{CDE}_{D_{10} \to Y_{12}} & (D_{12} = 1) \\
& = E[Y_{12}|D_{10} = 1, D_{12} = 1, U] - E[Y_{12}|D_{10} = 0, D_{12} = 1, U] \\
& = (.5 + .4U) - (.4 + .4U) \\
& = .1,
\end{aligned} \tag{11.51}
$$

$$
\begin{aligned}
\text{CDE}_{D_{10} \to Y_{12}} & (D_{12} = 0) \\
& = E[Y_{12}|D_{10} = 1, D_{12} = 0, U] - E[Y_{12}|D_{10} = 0, D_{12} = 0, U] \\
& = (.2 + .4U) - (.2 + .4U) \\
& = 0,
\end{aligned} \tag{11.52}
$$

where the effects of U cancel.[30] These controlled direct effects match the values of Equation (11.39) by construction, and they indicate that being in a Catholic school in the tenth grade has no direct effect on test scores in the twelfth grade unless one is in a Catholic school in the twelfth grade.[31]

We again must ask: How do we know that the controlled direct effects are all identified? As for the direct effect $D_{10} \to D_{12}$, the core of the answer is the same for the direct effect $D_{10} \to Y_{12}$. It is still the case that the controlled directed effects are identified by the observed data based of the results established for graphs that have a Markov structure (again, see Pearl 2009 after reading our appendix to Chapter 3).

---

[30] Although we did not specify a value for U in Equations (11.47)–(11.50), and simply allowed U to cancel in the calculation of the controlled direct effects, we could have developed eight separate conditional expectations because we know the values of U and the probability distribution of U. We do not do so because this information is not typically available to the analyst. It would also require us to then average over these eight conditional expectations to get to the four values for the conditional expectations that can be used to directly calculate the true controlled direct effects. Instead, we show how to solve for these values using other methods in the next section.

[31] For completeness, the natural direct effect is then

$$
\begin{aligned}
\text{NDE}_{D_{10} \to Y_{12}} & = \text{CDE}_{D_{10} \to Y_{12}}(D_{12} = 1) \times \Pr[D_{12} = 1|D_{10} = 0] \\
& \quad + \text{CDE}_{D_{10} \to Y_{12}}(D_{12} = 0) \times \Pr[D_{12} = 0|D_{10} = 0] \\
& = (.1 \times .36) + (0 \times .64) \\
& = .036,
\end{aligned} \tag{11.53}
$$

where the values of .36 and .64 for $\Pr[D_{12} = 1|D_{10} = 0]$ and $\Pr[D_{12} = 0|D_{10} = 0]$ are calculated from the final column of Table 11.5. The natural direct effect is small and is again a counterfactual quantity: the effect of Catholic schooling in the tenth grade on test scores in the twelfth grade, net of the dynamic nature of the treatment regime that we have assumed by construction exists. Thus, apart from how Catholic school attendance is determined in the twelfth grade, the estimate suggests that the probability of being in the high-test-score group increases by .036 if a student was enrolled in a Catholic school in the tenth grade. This counterfactual effect is an example of a natural direct effect that probably does not deserve attention. The weighting in Equation (11.53) is a direct function of the number of students who enter Catholic schools in the twelfth grade, having been in public schools in the tenth grade, and this is precisely the group for whom the controlled direct effect is equal to 0. In this case, the controlled direct effects are sensible and have clear interpretations.

The key feature that establishes identification is again the absence of confounders that would generate noncausal associations through unblocked back-door paths between the treatment and outcome variables (in this case between D_{10} and Y_{12} and between D_{12} and Y_{12}).

Although identification is again positive, all of the shallower explanations we offered for the direct effect $D_{10} \rightarrow D_{12}$ no longer easily apply. Instead, the same explanatory strategies reveal complexities that suggest why a clever set of techniques has been developed to estimate direct effects of these types. When we discussed the identification of the direct effect $D_{10} \rightarrow D_{12}$, we were able to point out that sample analogs to the conditional expectations on the left-hand sides of Equations (11.40)–(11.43) would converge to the true values for those conditional expectations as the sample size approaches infinity. When considering the direct effect $D_{10} \rightarrow Y_{12}$, an equivalent claim is true for the conditional expectations on the left-hand sides of Equations (11.47)–(11.50), but with one debilitating caveat: We cannot form the sample analogs to the true conditional expectations because U is an unobserved variable. And, if we try to estimate controlled direct effects by recklessly substituting in sample analogs to $E[Y_{12}|D_{12},D_{10}]$ for what would be the proper sample analogs to $E[Y_{12}|D_{12},D_{10},U]$ that are impossible to generate from the observed data, we will obtain inconsistent and biased estimates of the controlled direct effects. The usual culprit produces this bias: a collider. This complication can be seen in the graph, as we now explain.

Recall our prior graphical explanation for the crucial role that conditioning on the mediator played in the identification of the controlled direct effects for $D_{10} \rightarrow D_{12}$. We explained that conditioning on Y_{10} effectively sets the indirect effect path $D_{10} \rightarrow Y_{10} \rightarrow D_{12}$ to 0, enabling identification of the controlled direct effects within strata defined by the mediator, Y_{10}. For the direct effect $D_{10} \rightarrow Y_{12}$, the situation is more complicated. In order to estimate the controlled direct effects in this case, we again need to condition in a way that sets the indirect effect to 0. But, now we need to condition in a way that blocks two paths, $D_{10} \rightarrow D_{12} \rightarrow Y_{12}$ and $D_{10} \rightarrow Y_{10} \rightarrow D_{12} \rightarrow Y_{12}$. The obvious candidate conditioning variable is D_{12} because it lies on both of these paths. At the same time, it should also be obvious that Y_{10} is not a good candidate for conditioning. Not only would conditioning on Y_{10} fail to block the path $D_{10} \rightarrow D_{12} \rightarrow Y_{12}$, Y_{10} is a collider on the path $D_{10} \rightarrow Y_{10} \leftarrow U \rightarrow Y_{12}$ that begins at D_{10} and ends at Y_{12}. Conditioning on Y_{10} would induce an association between D_{10} and U, which would then generate a noncausal association between D_{10} and Y_{12} because U is a direct cause of Y_{12}.

Upon closer inspection, however, we can see that D_{12} is a descendant of Y_{10}. As a result, conditioning on D_{12} will also induce an association between D_{10} and U, which then generates a noncausal association between D_{10} and Y_{12} within strata defined by the mediator D_{12}. Given this predicament, it is not obvious how we can condition on D_{12} to set the indirect effect to 0 without triggering a new source of confounding. Yet, without conditioning in a way that will set the indirect effect to 0, we cannot estimate the controlled direct effects that according to the causal graph literature are, in fact, identified because of the absence of unobserved confounders. This predicament is often referred to colloquially in this literature as "damned if you do and damned if you don't."

Before too much despair accumulates, note that if we could find a way to condition on D_{12} in order to set the indirect effects to 0 and then also adjust away the induced bias that travels by way of U, we would be able to estimate the controlled direct effects. The signal contribution of the literature on dynamic treatment regimes is a solution to this predicament, which we will explain below when presenting estimation methods in the next section. The key innovation is to use the relationships that constitute the indirect effects (i.e., the joint probability distribution of D_{10}, Y_{10}, and D_{12}) to adjust away the induced confounding that travels by way of the unobserved variable U when we condition on D_{12}, Y_{10}, or both.

Estimation Strategies

We will not discuss how to estimate total effects in detail because the strategies should be obvious. As shown in Table 11.6, many of these effects can be estimated using the naive estimator because unconditional associations identify the effects. Even when this is not the case, standard back-door conditioning estimators can be used to estimate the others. Instead, we will focus in this section on the two direct effects, $D_{10} \rightarrow D_{12}$ and $D_{10} \rightarrow Y_{12}$. Estimation of the first of these effects is straightforward, once the identification result is known. Estimation of the second of these effects is not straightforward and is the focus of the dynamic treatment regime literature.

Estimating the Direct Effect of D_{10} on D_{12}. How would one estimate the controlled direct effects of D_{10} on D_{12} in a finite sample? Assuming that conditions such as positivity are met for the available data, our identification explanation in the section above suggests one straightforward method. The researcher estimates sample analogs to the conditional expectations on the left-hand sides of Equations (11.40)–(11.43), which would be

$$E_N[d_{i12} = 1 | d_{i10} = 1, y_{i10} = 1] \quad \text{for} \quad E[D_{12} | D_{10} = 1, Y_{10} = 1],$$
$$E_N[d_{i12} = 1 | d_{i10} = 0, y_{i10} = 1] \quad \text{for} \quad E[D_{12} | D_{10} = 0, Y_{10} = 1],$$
$$E_N[d_{i12} = 1 | d_{i10} = 1, y_{i10} = 0] \quad \text{for} \quad E[D_{12} | D_{10} = 1, Y_{10} = 0],$$
$$E_N[d_{i12} = 1 | d_{i10} = 0, y_{i10} = 0] \quad \text{for} \quad E[D_{12} | D_{10} = 0, Y_{10} = 0].$$

Differences between these four estimated conditional expectations can then be taken to estimate the controlled direct effects in Equations (11.44) and (11.45).[32]

In the literature on dynamic treatment regimes, this straightforward estimation strategy is labeled G-computation (Robins 1986; Robins and Hernán 2009), where the "G" is an abbreviation of "General." G-computation is one of three related approaches to the estimation of treatment effects for dynamic treatment regimes. For examples such as this one, where we have three binary variables, positivity, and no emergent conditioning bias from colliders, G-computation takes a particularly simple form and is nearly certain to be feasible. As such, the other two estimation methods, which we detail in the next section, would not need to be used.

Estimating the Direct Effect of D_{10} on Y_{12}. To estimate the direct effect of D_{10} on Y_{12}, the dynamic treatment effect literature offers three methods. The first, G-Computation (Robins 1986), was just presented, and it conveys the core identification

---

[32]In addition, these estimated controlled direct effects can then be combined into a weighted average, using the sample analog to $\Pr[Y_{10} = 1 | D_{10} = 0]$, which would then yield a consistent estimate of the natural direct effect.

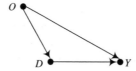

Figure 11.11 An illustrative directed graph for G-computation.

challenge and how it is resolved. As we will show in this section, it is more complicated for the direct effect of D_{10} on Y_{12}. The second approach, known as the estimation of marginal structural models (MSMs), is feasible for research scenarios in which
G-Computation is infeasible because of the curse of dimensionality (Robins 1998, 1999,
2000). We will also briefly discuss a third approach, G-Estimation, and the intuition
behind it.

G-Computation. Consider the simple example in Figure 11.11 of a directed graph
where D is a single fixed treatment and Y is an outcome of interest. As in many other
examples in this book, the association between D and Y does not identify the causal
effect of D on Y because of the back-door path $D \leftarrow O \rightarrow Y$. However, because O is
observed, we can condition on it and generate a consistent estimate of $D \rightarrow Y$ for the
reasons discussed extensively in Chapters 4 through 7.

If D and O are discrete, then the overall causal effect of D on Y is easily estimated
via stratification (see Matching Demonstration 1 on page 145). Two different strategies,
however, are possible. The standard approach would be to first estimate the causal
effect of Y within strata of O and then calculate the overall causal effect of D on
Y as the weighted average of the causal effects across strata where the weights are
proportional to stratum size; see Equations (5.5) and (5.6). A second approach would
be to estimate the expectation of each of the potential outcomes, Y^1 and Y^0, within
each stratum of O as

$$E_N[y_i|d_i = 1, o_i = o] \xrightarrow{p} E[Y^1|O = o], \tag{11.54}$$

$$E_N[y_i|d_i = 0, o_i = o] \xrightarrow{p} E[Y^0|O = o].$$

Weighted sums of these stratified estimates can then be taken, which will be consistent
estimates of $E[Y^1]$ and $E[Y^0]$ because

$$E[Y^1] = \sum_O E[Y^1|O = o]\Pr(O = o), \tag{11.55}$$

$$E[Y^0] = \sum_O E[Y^0|O = o]\Pr(O = o). \tag{11.56}$$

To then estimate the ATE, the researcher takes the difference between the sample
analogs to the expectations in Equations (11.55) and (11.56).

This general approach to estimation is G-computation, and Equations (11.55) and
(11.56) are an example of what are labeled G-formulas. The "G" for "General" is
meant to indicate that this procedure represents a general approach to the estimation
of a causal effect. If a causal effect is identified, and ignorability holds, then in principle

a causal effect can be estimated using G-computation. A simple but key requirement is that we are considering average causal effects defined using expectations. In this case, causal effects calculated as weighted averages of differences within strata are equal to causal effects calculated as differences between weighted outcomes across strata.

Consider what G-computation does. G-computation stratifies the sample in order to estimate the expected potential outcome for each possible treatment regime under the assumption that all individuals have the same treatment status. With consistent estimates of these expected potential outcomes, estimating the effect of any treatment regime relative to any other treatment regime then only requires that one calculate the differences between these expectations. An important point in this literature is that controlled direct effects can be thought of as differences between compound effects for different combinations of treatments, as we will demonstrate below.

Now consider again the directed graph in Figure 11.10. From a G-computation perspective, what we want to estimate are the outcomes of Y_{12} for the four different regimes (i.e., the four possible combinations of values of D_{10} and D_{12}). Because suspense offers no explanatory value for this appendix, we will reveal these values before we show how to estimate them:

$$E[Y_{12}|D_{10} = 1, D_{12} = 1] = .74, \tag{11.57}$$

$$E[Y_{12}|D_{10} = 0, D_{12} = 1] = .64, \tag{11.58}$$

$$E[Y_{12}|D_{10} = 1, D_{12} = 0] = .44, \tag{11.59}$$

$$E[Y_{12}|D_{10} = 0, D_{12} = 0] = .44. \tag{11.60}$$

From the values in Equations (11.57)–(11.60), it is trivial to calculate the causal effect of one regime relative to another.[33] Notice this shift in language: from total and direct effects to alternative treatment regimes. In particular, we can calculate the expected effect of enrollment in Catholic school in the tenth and twelfth grade as

$$E[Y_{12}|D_{10} = 1, D_{12} = 1] - E[Y_{12}|D_{10} = 0, D_{12} = 0] = .74 - .44 = .3.$$

Most important for our consideration of direct effects, the values in Equations (11.57)–(11.60) can be used to calculate the difference in Y_{12} produced by D_{10}, separately by the value of D_{12}. These are, in fact, what we labeled the controlled direct effects above:

$$
\begin{aligned}
\mathrm{CDE}_{D_{10} \to Y_{12}} &(D_{12} = 1) \\
&= E[Y_{12}|D_{10} = 1, D_{12} = 1] - E[Y_{12}|D_{10} = 0, D_{12} = 1] \\
&= .74 - .64 \\
&= .1,
\end{aligned}
\tag{11.61}
$$

---

[33]In addition, two equivalences are worth noting at this point. Given the Markov structure of the graph, the values in Equations (11.57)–(11.60) are equivalent to expected potential outcomes. They are also equivalent to the conditional expectations in Equations (11.47)–(11.50), averaged over the distribution of U.

$$\text{CDE}_{D_{10} \to Y_{12}}(D_{12} = 0)$$
$$= E[Y_{12}|D_{10} = 1, D_{12} = 0] - E[Y_{12}|D_{10} = 0, D_{12} = 0] \quad (11.62)$$
$$= .44 - .44$$
$$= 0,$$

but now U is no longer present for reasons we will explain below; see Equations (11.51) and (11.52) for comparison. In the literature on dynamic treatment regimes, these differences are simply the effect of Catholic schooling in the tenth grade on test scores in the twelfth grade calculated first for the regime in which students also attend Catholic schools in the twelfth grade and then second for the regime in which students do not attend Catholic schools in the twelfth grade.[34] In short, with the values in Equations (11.57)–(11.60), we can calculate all of the treatment effects we want for all contrasts across the permissible treatment regimes.

How do we estimate these values with G-computation? We use the appropriate G-formula:

$$E[Y_{12}|D_{10}, D_{12}] = E[Y_{12}|D_{10}, D_{12}, Y_{10} = 1] \times \Pr[Y_{10} = 1|D_{10}] \quad (11.63)$$
$$+ E[Y_{12}|D_{10}, D_{12}, Y_{10} = 0] \times \Pr[Y_{10} = 0|D_{10}].$$

Although the structure of this G-formula is given by the graph and the identifying assumptions that it represents, the computed expectation can be interpreted as the expected outcome for Y_{12} for the particular combination of values set by interventions on D_{10} and D_{12}. The right-hand side is a sum of two products, where each product includes the appropriate conditional expectation weighted by whether Y_{10} equals 1 or 0. This weight is conditional on D_{10}, but not D_{12} (because Y_{10} is determined by D_{10} but not by D_{12}).[35]

Operationally, we take eight strata defined across all two-way combinations of D_{10}, D_{12}, and Y_{10} and then calculate the mean values for Y_{12}. We then average over strata defined by Y_{10}, conditional on patterns of D_{12} and D_{10}, to generate the four values in Equations (11.57)–(11.60).

For this example, the expectations for the relevant eight strata are presented in the seventh column of Table 11.5. Sample analogs to the conditional expectations will converge to the true values for these conditional expectations:

$$E_N[y_{12}|d_{10} = 1, d_{12} = 1, y_{10} = 1] \xrightarrow{p} .811,$$
$$E_N[y_{12}|d_{10} = 1, d_{12} = 1, y_{10} = 0] \xrightarrow{p} .657,$$
$$E_N[y_{12}|d_{10} = 0, d_{12} = 1, y_{10} = 1] \xrightarrow{p} .753,$$
$$E_N[y_{12}|d_{10} = 0, d_{12} = 1, y_{10} = 0] \xrightarrow{p} .582, \quad (11.64)$$

---

[34] In this literature, controlled direct effects are considered differences in compound effects for alternative combinations of treatments.

[35] For completeness, we should note that the G-formula for the direct effect $D_{10} \to D_{12}$ is much simpler. It is $E[D_{12}|D_{10}, Y_{10}]$ and does not require averaging over any underlying strata.

$$E_N[y_{12}|d_{10} = 1, d_{12} = 0, y_{10} = 1] \xrightarrow{p} .511,$$

$$E_N[y_{12}|d_{10} = 1, d_{12} = 0, y_{10} = 0] \xrightarrow{p} .357,$$

$$E_N[y_{12}|d_{10} = 0, d_{12} = 0, y_{10} = 1] \xrightarrow{p} .553,$$

$$E_N[y_{12}|d_{10} = 0, d_{12} = 0, y_{10} = 0] \xrightarrow{p} .382.$$

Inserting these conditional expectations into the G-formula in Equation (11.63), and while assuming an infinite sample, we can calculate the desired four values in Equations (11.57)–(11.60) as

$$
\begin{aligned}
E_N[y_{12}|d_{10} &= 1, d_{12} = 1] \\
&= .811 \times \Pr_N[y_{10} = 1|d_{10} = 1] + .657 \times \Pr_N[y_{10} = 0|d_{10} = 1] \\
&= (.811 \times .55) + (.657 \times .45) \\
&= .74,
\end{aligned}
\tag{11.65}
$$

$$
\begin{aligned}
E_N[y_{12}|d_{10} &= 0, d_{12} = 1] \\
&= .753 \times \Pr_N[y_{10} = 1|d_{10} = 0] + .582 \times \Pr_N[y_{10} = 0|d_{10} = 0] \\
&= (.753 \times .34) + (.582 \times .66) \\
&= .64,
\end{aligned}
\tag{11.66}
$$

$$
\begin{aligned}
E_N[y_{12}|d_{10} &= 1, d_{12} = 0] \\
&= .511 \times \Pr_N[y_{10} = 1|d_{10} = 1] + .357 \times \Pr_N[y_{10} = 0|d_{10} = 1] \\
&= (.511 \times .55) + (.357 \times .45) \\
&= .44,
\end{aligned}
\tag{11.67}
$$

$$
\begin{aligned}
E_N[y_{12}|d_{10} &= 0, d_{12} = 0] \\
&= .553 \times \Pr[y_{10} = 1|d_{10} = 0] + .382 \times \Pr_N[y_{10} = 0|d_{10} = 0] \\
&= (.553 \times .34) + (.382 \times .66) \\
&= .44,
\end{aligned}
\tag{11.68}
$$

where the conditional probabilities $\Pr_N[y_{10} = 1|d_{10} = 1]$, $\Pr_N[y_{10} = 0|d_{10} = 1]$, $\Pr_N[y_{10} = 1|d_{10} = 0]$, and $\Pr_N[y_{10} = 0|d_{10} = 0]$ are calculated from the final column of Table 11.5.

With the four values produced by Equations (11.65)–(11.68), we can estimate the causal effect for any two contrasting treatment regimes, as shown above with reference to the values in Equations (11.57)–(11.60). In effect, the structure of the graph allows us to collapse the strata defined by Y_{10} as long as the strata are weighted in accordance with the conditional probability distributions encoded by the graph. As we will explain in the next section, there are alternative methods available to achieve this type of weighted collapsing.

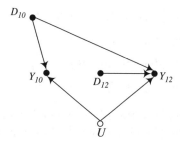

Figure 11.12 A directed graph for a pseudo-population produced using inverse probability of treatment weighting.

Alternatives to G-Computation. If there is sufficient data and the appropriate ignorability assumptions hold, estimating a saturated model via G-computation will always provide consistent estimates of the causal effect of any one regime relative to another. From these treatment regime differences, one can then calculate the relevant controlled direct effects.

However, because of the curse of dimensionality, G-computation-based estimates can be very imprecise because some conditional expectations will be estimated for strata with very small samples. In this final section, we consider two alternative approaches to G-computation that are meant to deal with the curse of dimensionality: marginal structural models (MSMs) estimated via inverse probability of treatment weighting (IPTW), and structural nested mean models (SNMMs) estimated via a G-estimation. Our discussion of the later approach will be brief.

Marginal Structural Models (MSMs). These models are attracting interest in sociology and have been used in several published papers (Wimer, Sampson, and Laub 2008; Sharkey and Elwert 2011; Wodtke, Harding, and Elwert 2011). They are labeled structural because they estimate a causal effect and marginal because the effect that is estimated is the marginal effect over a set of collapsed strata. MSMs are attracting interest because they are feasible when G-computation is not, because they are comparatively easy to understand relative to G-estimation (see below), and because they can be estimated using available software.

Consider the directed graph in Figure 11.12. Notice that D_{12} is not determined by either D_{10} or Y_{10}. As a result, D_{10} no longer has an indirect effect on Y_{12} through Y_{10} and D_{12}. As a result, there is no reason to condition on either Y_{10} or D_{12} when estimating the causal effect of D_{10} on Y_{12} for this graph. Notice also that the path $D_{10} \rightarrow Y_{10} \leftarrow U \rightarrow Y_{12}$ that connects D_{10} to Y_{12} is blocked by the collider Y_{10}. As long as we do not condition on Y_{10}, this path will remain blocked. If the directed graph in Figure 11.12 described our data, the unconditional association between D_{10} and Y_{12} would identify the (direct) effect of D_{10} on Y_{12}. It would also be the case that the unconditional association between D_{12} and Y_{12} would identify the effect of D_{12} on Y_{12}.

Of course, the causal dependence of D_{12} on D_{10} and Y_{10} cannot just be assumed away. Nonetheless, Robins (1999) showed how one can create a pseudo-population in which the directed graph in Figure 11.12 can be substituted for the directed graph in

Figure 11.10. The pseudo-population construction is achieved using the inverse probability weighting methods we presented in Chapter 7. In particular, after examining the directed graph in Figure 11.10, one estimates a propensity score model predicting assignment to D_{12} as a function of D_{10} and Y_{10}:

$$\Pr(D_{12} = 1 | D_{10}, Y_{10}) = F(D_{10}, Y_{10}). \tag{11.69}$$

For each individual, the probability of enrolling in a Catholic school in the twelfth grade can either be estimated nonparametrically from the data if there are a sufficient number of cases or using a general linear model, such as a logit or probit. We can then define the weights in two different ways:

$$\text{For } d_{12i} = 1: \quad w_{i,\,\text{MSM}} = \frac{1}{\hat{p}_i},$$

$$\text{For } d_{12i} = 0: \quad w_{i,\,\text{MSM}} = \frac{1}{1 - \hat{p}_i},$$

or

$$\text{For } d_{12i} = 1: \quad sw_{i,\,\text{MSM}} = \frac{E_N[d_{12} = 1]}{\hat{p}_i},$$

$$\text{For } d_{12i} = 0: \quad sw_{i,\,\text{MSM}} = \frac{1 - E_N[d_{12} = 1]}{1 - \hat{p}_i},$$

where $\hat{p}$ is the estimated probability for each individual of entering Catholic schooling in the twelfth grade based on Equation (11.69). Robins and Hernán (2009) refer to the w_i weights as "unstabilized weights" and the sw_i weights as "stabilized weights." Unstabilized weights generally have greater variance and typically lead to wider confidence intervals. As such, stabilized weights are usually recommended, although there are special circumstances where stabilized weights will produce inconsistent estimates of causal effects (Robins and Hernán 2009:576).

Table 11.7 presents the pseudo-population proportions that would result from nonparametric estimation of the weights for the dynamic treatment regime version of the Catholic school effect we have been analyzing in this appendix. The final column of this table can be directly compared to the observed population proportions in Table 11.5 that we set by construction for the hypothetical example. If we use these estimated proportions to calculate weighted means of Y_{12} conditional on values for all four combinations of D_{10} and D_{12}, we can then calculate differences to recover the total effect of D_{10} on Y_{12} as well as the two controlled direct effects.

The advantage of MSMs, relative to G-computation, is that we do not have to calculate the means of Y_{12} within each stratum (i.e., across Y_{10} as in our example). Instead, we estimate weights and then calculate contrasts for Y_{12} with the weighted data. This is the same advantage that the propensity-score-based weighting estimators presented in Chapter 7 have relative to the full stratification estimators presented in Chapter 5. MSMs still require positivity with respect to all possible treatment regimes, but they permit sparseness in observed variables that determine treatment assignment, just as for propensity-score estimators for a fixed-time treatment regime.

Table 11.7 Pseudo-Population Proportions for
the Directed Graph in Figure 11.12

D_{10}	Y_{10}	D_{12}	Y_{12}	Pseudo-Population Proportion
1	1	1	1	.041
1	1	1	0	.010
1	1	0	1	.093
1	1	0	0	.037
1	0	1	1	.031
1	0	1	0	.036
1	0	0	1	.021
1	0	0	0	.025
0	1	1	1	.082
0	1	1	0	.020
0	1	0	1	.011
0	1	0	0	.064
0	0	1	1	.145
0	0	1	0	.104
0	0	0	1	.107
0	0	0	0	.173

Accordingly, MSMs can be a very useful approach to dealing with "damned if you do, damned if you don't" situations. MSMs, however, are not a panacea and have several weaknesses. First, and most importantly, even stabilized weights can produce unusually large weights resulting in imprecise estimates. In Chapter 7, we have discussed various methods, such as trimming, for dealing with this situation, noting that these methods will lead to biased estimates. Second, MSMs cannot be used when an instrumental variable is available. Third, the types of sensitivity analysis (see Chapter 12) that can be done with MSMs are limited (Robins and Hernán 2009:592–93). We next consider a method, G-estimation (not to be confused with G-computation) that does not share these problems. Unfortunately, G-estimation is both more difficult to understand and more difficult to implement.

G-estimation. Our discussion of G-estimation will be brief because our goal is to give the reader an intuitive understanding of how G-estimation works, with the hope that when readers encounter these methods, they will have a basic understanding of the procedure. G-estimation is closely related to the method of generalized estimating equations (GEE) and other methods-of-moments estimators. G-estimation seeks a set of estimates that satisfies an orthogonality condition. As we have discussed above, identification occurs when ignorability holds. Ignorability holds if the potential outcomes are independent of an individual's past treatment, conditional on their past and present covariate history. The intuition behind G-estimation is that one wants to create a set of predicted counterfactual potential outcomes by modeling the observed outcomes. This goal is pursued by searching for a set of parameters that results in orthogonality of both observed and the predicted potential outcomes with respect to treatment assignment, conditional on treatment history and current and past covariate values. Unfortunately, in most cases it is necessary to use a grid search procedure to find the desired set of parameters, which can require substantial computing power and time to generate results. Also, as far as we are aware, no software routines have been shared for use with either commercial or freely available data analysis programs. For more details on G-estimation, consult (Robins and Hernán 2009:577–92).

Conclusions

We have chosen a simple example where the data are generated from a hypothetical dynamic treatment regime, and we have then shown how it is possible to identify static treatment effects from such data. This example is rich enough to convey both the basic identification results and the clever ways in which estimation is rendered feasible with the methods developed by Robins and his colleagues. If one's interest is in identifying dynamic treatment effects, then identification and estimation are considerably more challenging (see Robins and Hernán 2009). If sequential ignorability does not hold because confounders such as U in our example determine either or both of the treatment exposure variables, then the models developed in this literature are not identified, just as in the case for fixed treatment regimes. In addition, not all estimation issues have been resolved. G-computation is beyond reproach, but it is often infeasible because of insufficient data. Marginal structural models can be fragile for the same reasons as the weighted regression estimators discussed in Chapter 7. The specification of the model that generates the weights must be correct, and the concern with extreme weights will be present for many applications, especially if some patterns of treatment exposure are comparatively rare. Finally, G-estimation also requires that the model predicting potential outcomes be correct, something which may be difficult to test. In addition to Robins and Hernán (2009), we recommend that readers seek additional guidance in Chakraborty and Moodie (2013).

Part V: Estimation When Causal Effects Are Not Point-Identified by Observables

Chapter 12

Distributional Assumptions, Set Identification, and Sensitivity Analysis

In this chapter, we consider how analysts can proceed when no observed variables are available to point-identify and then estimate causal effects using the procedures explained in prior chapters. We discuss three complementary approaches. First, we will review the early literature on selection-bias adjustment, which shows clearly how distributional assumptions about unobservables, when harnessed from within a structural model, can identify causal parameters of interest. Point identification and estimation utilizing this strategy was frequent before it became clear to researchers in the 1990s how rarely the required distributional assumptions were warranted for their applications. The more recent selection-bias adjustment literature offers less restrictive semiparametric methods, but it also reemphasizes the relative value of instrumental variables in contrast to distributional assumptions for unobservables.

We will then consider the exact opposite strategy. Rather than place strong and usually untestable assumptions on the distributional characteristics of unobservables, the set-identification approach asks what can be learned about particular causal parameters by asserting only weak but defendable assumptions about unobserved variables. Instead of attempting to point-identify average causal effects, the set-identification approach suggests that it is more credible to try to limit the interval within which an average treatment effect must fall.

Finally, we will consider the related approach known as sensitivity analysis. Here, the analyst offers an estimate based on the provisional maintenance of an identifying assumption – most often, ignorability or selection on the observables – and then assesses the extent to which the estimate would vary as violations of the identifying assumption increase in severity. These final two modeling strategies are complementary, and each of them may be an effective path forward when an analyst has low confidence that any of the methods presented in prior chapters will suffice.

12.1 Distributional Assumptions and Latent Variable Selection-Bias Models

Heckman's early work in the 1970s on selection bias, particularly his lambda method, has received a great deal of attention in the social sciences. His closely related work on dummy endogenous variables, pursued at the same time, has received far less attention (see Heckman 1978, 1979). Although completed before most of Rubin and Rosenbaum's work on propensity scores, Heckman's work on dummy endogenous variables can be understood in relation to the propensity score approach.

Suppose that we are again interested in estimating the average treatment effect (ATE) for the effect of a binary cause D on an interval-scaled outcome Y. Recall the treatment selection model from the econometric tradition and Equation (4.7) in particular:

$$\tilde{D} = Z\phi + U. \tag{12.1}$$

For this equation, $\tilde{D}$ is a latent continuous variable, Z represents observed variables that determine treatment selection (either instruments for D or determinants of treatment selection that are also determinants of Y), and ϕ is a coefficient (or a vector of coefficients if Z represents more than one variable). Most importantly for these methods, U represents both systematic unobserved determinants of treatment selection as well as completely random idiosyncratic determinants of treatment selection. It is the unobserved systematic determinants in U that generate confounding that cannot be removed by conditioning on Z, and therefore this model is used to represent selection on the unobservables. Finally, as in Section 4.3.2, the latent continuous variable $\tilde{D}$ in Equation (4.7) is related to the treatment selection dummy variable, D, by

$$D = 1 \quad \text{if} \quad \tilde{D} \geq 0,$$
$$D = 0 \quad \text{if} \quad \tilde{D} < 0,$$

where the threshold 0 is arbitrary because U has no inherent metric.

Rather than focus on the conditional probability of treatment selection (i.e., directly on the propensity score $\Pr[D = 1|Z]$), Heckman instead focused on the conditional expectation function of the latent variable $\tilde{D}$. Using the linearity of Equation (12.1), he wrote the conditional expectation function as

$$E[\tilde{D}|Z\phi, D] = Z\phi + E[U|Z\phi, D]. \tag{12.2}$$

Accordingly, the expectation of $\tilde{D}$ is a function in $Z\phi$ and D, which therefore varies across individuals according to characteristics measured by z_i, and whether or not individuals select into the treatment, d_i. Heckman's insight was the recognition that, although one cannot observe u_i directly for each individual (unlike z_i and d_i), one can formulate an expression for a conditional expectation function of U if one is willing to assert distributional assumptions that imply values for the expectation of U conditional on $Z\phi$ and D. The standard assumption for selection-bias adjustment is that U is normally distributed with mean 0 and variance 1.[1] If $f(.)$ is the normal density function

---

[1] In the treatment effects context, it is also common to assert trivariate normality for the distributions of u_i, v_i^0, and v_i^1, where the latter two are the individual-level variation in the potential

and $F(.)$ is the corresponding cumulative distribution function, then standard results for truncated normal distributions yield the following expressions for the conditional expectation functions for U:

$$E[U|Z\phi, D=1] = \frac{f(Z\phi)}{F(Z\phi)}, \tag{12.3}$$

$$E[U|Z\phi, D=0] = \frac{-f(Z\phi)}{[1-F(Z\phi)]}. \tag{12.4}$$

Equation (12.3) simply gives the formula for Heckman's λ in a standard sample selection problem. In the treatment effects context, this expression is the λ for the treatment group $(D=1)$. The full dummy endogenous variable model requires a second λ for those in the control group $(D=0)$, which is calculated from Equation (12.4). These two λ's are then used to form two new regressors,

$$d_i \times \left(\frac{f(z_i\hat{\phi})}{F(z_i\hat{\phi})}\right) \quad \text{and} \quad (1-d_i) \times \left(\frac{-f(z_i\hat{\phi})}{[1-F(z_i\hat{\phi})]}\right), \tag{12.5}$$

that are specified as adjustment variables alongside other conditioning variables in the regression model that estimates the effect of D on Y. Thus, the procedure here is identical to Heckman's general lambda method for correcting for selection bias, except that two distinct λ's are utilized, each of which applies either to the treatment group or to the control group. Because these supplementary adjustment terms are functions in $z_i\hat{\phi}$ and d_i, the model is often referred to as a control function estimator.

As many researchers have come to appreciate, estimates from these models can be very sensitive to assumptions about the distribution of U. This recognition prompted the development of a second generation of selection-bias estimators that introduced semiparametric estimation of the treatment selection equation (see Honor and Powell 1994; Pagan and Ullah 1999). And it led to a greater appreciation of a point that Heckman made clearly in the original development of the approach: sample selection-bias models are most effectively estimated when the variables in Z include instrumental variables, so that the added regressors in Equation (12.5) have an exogenous source of variation unrelated to the distributional assumption placed on U.

More recently, Heckman and Vytlacil (2005, 2007) have offered a latent index model for heterogeneous treatment effects that can represent latent variable selection-bias estimators, their semiparametric extensions, and instrumental variable (IV) estimators that satisfy monotonicity assumptions. Although the latent index model accommodates the original control function strategy and is consistent with the basic structure of the potential outcome model, it emphasizes the core identifying power of local instrumental variables (LIVs) rather than the distributional assumptions about unobservables that were heavily relied upon in the applied literature; see our discussion of local IVs and MTEs in Section 9.3.4. As such, one can read the latent index model as an appeal for piecewise IV identification of heterogeneous treatment effects, not for structural identification of average treatment effects by assumptions about the

---

outcomes; see Section 4.3.2. In addition to the two-stage models explained in this section, these models can also be estimated by maximum likelihood and nonlinear least squares. See Winship and Mare (1984, 1992) and Fu, Winship, and Mare (2004) for further discussion.

distributions of unobservables. In response to this most recent literature, the control function approach to selection-bias adjustment appears likely to continue to decline in usage in the applied literature, except in cases where instrumental variables for treatment selection equations are available.

12.2 Set Identification with Minimal Assumptions

If one cannot (or does not) impose a strong assumption to point-identify and then estimate an average treatment effect, it is natural to wonder about an opposing strategy for analysis: What can be learned about average causal effects while imposing weaker but defendable assumptions? In a series of articles and books that have built on less formalized work of the past, Manski (1994, 1995, 1997, 2003, 2013b) has investigated plausible values for treatment effect parameters, such as the ATE, that are consistent with the data when weak assumptions alone are maintained. In this section, we introduce Manski's work to show that, in most research situations, the observed data themselves provide some information on the size of average treatment effects without any assumptions. We then introduce some of the additional assumptions considered by Manski to show how they bound average causal effects to permissible intervals. We use the label of "set identification" in contrast to point identification of single values. Manski often uses the alternative phrase "partial identification," and many writers refer to this strategy as an "analysis of bounds."

12.2.1 No-Assumptions Bounds

To see why the collection of data can bound the range of permissible values for the ATE, consider a hypothetical example in which we know that both Y^1 and Y^0 are bounded by 0 and 1, and thus that Y is also bounded by 0 and 1. Examples could be dichotomous potential outcomes for graduating high school or not in response to two causal states, such as Catholic schooling or public schooling, discussed extensively already. Or the potential outcomes could be whether one voted for Al Gore, depending on whether or not one received a butterfly ballot. This latter example (see Section 1.3.2) would be especially appropriate to analyze from a bounds and set-identification perspective because the goal of such a causal analysis would be to determine whether the causal effect is plausibly large enough to have flipped the election results. With that goal clearly in focus, concerns over putative statistical significance are of secondary concern.

In this case, we know from the definitions of Y^1 and Y^0, even without collecting data, that the ATE, $E[\delta]$, cannot be greater than 1 or less than -1. This is a straightforward implication of the obvious point that no individual treatment effect can be greater than 1 or less than -1. Accordingly, the maximum ATE of 1 would occur if $E[Y^1|D=1] = E[Y^1|D=0] = 1$ and $E[Y^0|D=1] = E[Y^0|D=0] = 0$, whereas the minimum ATE of -1 would occur instead if $E[Y^1|D=1] = E[Y^1|D=0] = 0$ and $E[Y^0|D=1] = E[Y^0|D=0] = 1$. Thus, we know that $E[\delta]$ must be contained in an interval of length 2, which can be stated as a known bound on $E[\delta]$, using the notation

$$-1 \leq E[\delta] \leq 1 \tag{12.6}$$

Table 12.1 A Hypothetical Example of the Calculation of Bounds
for the ATE

	$E[Y^1\mid .]$	$E[Y^0\mid .]$
Naive estimator suggests $E[\delta] = .4$		
Treatment group	$E[Y^1\mid D=1]=.7$	$E[Y^0\mid D=1]=\;?$
Control group	$E[Y^1\mid D=0]=\;?$	$E[Y^0\mid D=0]=.3$
Largest possible $E[\delta] = .7$		
Treatment group	$E[Y^1\mid D=1]=.7$	$E[Y^0\mid D=1]=0$
Control group	$E[Y^1\mid D=0]=1$	$E[Y^0\mid D=0]=.3$
Smallest possible $E[\delta] = -.3$		
Treatment group	$E[Y^1\mid D=1]=.7$	$E[Y^0\mid D=1]=1$
Control group	$E[Y^1\mid D=0]=0$	$E[Y^0\mid D=0]=.3$

or that $E[\delta]$ lies in an interval with closed bounds, as in

$$E[\delta] \in [-1,1]. \tag{12.7}$$

Manski has shown that we can improve on these bounded intervals considerably
without making additional assumptions but by collecting data on Y and D. By following a systematic sampling strategy from a well-defined population, we can assert
the specific convergence results stated earlier in Equations (2.9)–(2.11). These would
ensure that, for an infinite sample, $E_N[d_i] = E[D]$, $E_N[y_i|d_i = 1] = E[Y^1|D = 1]$, and
$E_N[y_i|d_i = 0] = E[Y^0|D=0]$. Knowledge of these three quantities allows one to narrow
the bound of width 2 in Equations (12.6) and (12.7) to one with a width of only 1.

Consider the hypothetical example depicted in Table 12.1, where, as shown in
the first panel, we stipulate that $E[Y^1|D = 1] = .7$ and $E[Y^0|D = 0] = .3$. The naive
estimator, $E_N[y_i|d_i = 1] - E_N[y_i|d_i = 0]$, would yield values that converge to .4 as the
sample size increases. Note that the naive estimator does not use the information
about the distribution of the sample across the observed values of D. And, we leave
question marks to stand in for the unknown values of the counterfactual conditional
expectations $E[Y^1|D=0]$ and $E[Y^0|D=1]$.

Suppose that $E[D] = .5$ and that our sample is infinite such that sampling error
is 0. And, for simplicity of notation and consistency with the decomposition in Chapter 2, allow π to again stand in for $E[D]$. For the second and third panels of Table
12.1, the minimum and maximum values of 0 and 1 for the counterfactual conditional
expectations $E[Y^1|D=0]$ are $E[Y^0|D=1]$ and substituted for the question marks in
the first panel.

If half of the sample is in the treatment group, and if $E[Y^1|D = 1] = .7$ and
$E[Y^0|D = 0] = .3$, then the largest possible treatment effect is .7, whereas the smallest possible treatment effect is $-.3$. This result is a straightforward calculation using
what-if values for the decomposition of the ATE presented in Equation (2.10), which

was used earlier in Chapter 2 and elsewhere to discuss the bias of the naive estimator:

$$E[\delta] = \{\pi E[Y^1|D=1] + (1-\pi)E[Y^1|D=0]\} \tag{12.8}$$
$$- \{\pi E[Y^0|D=1] + (1-\pi)E[Y^0|D=0]\}.$$

Plugging in the values from the second panel of Table 12.1 into the decomposition in Equation (12.8) yields the largest possible treatment effect as

$$E[\delta] = \{(.5)(.7) + (1-.5)(1)\} - \{(.5)(0) + (1-.5)(.3)\}$$
$$= .85 - .15$$
$$= .7,$$

whereas plugging in the values from the third panel yields the smallest possible treatment effect as

$$E[\delta] = \{(.5)(.7) + (1-.5)(0)\} - \{(.5)(1) + (1-.5)(.3)\}$$
$$= .35 - .65$$
$$= -.3.$$

Thus, the constraints implied by the observed data alone guarantee, for this example, that $E[\delta] \in [-.3, .7]$, which is an interval of length 1 and is half the length of the maximum interval calculated before estimates of π, $E[Y^1|D=1]$, and $E[Y^0|D=0]$ were obtained from the data.

Manski labels this interval the "no-assumptions bound" because it requires knowledge only of the bounds on Y^1 and Y^0, as well as collection of data on Y and D from a systematic random sample of a well-defined population. Consider now a more general development of these same ideas.

The treatment effect can be bounded by finite values only when the potential outcomes Y^1 and Y^0 are bounded by finite values. In other words, because $E[Y^1|D=0]$ and $E[Y^0|D=1]$ are both unobserved, they can take on any values from $-\infty$ to ∞ in the absence of known restrictions on the ranges of Y^1 and Y^0. Thus, in the absence of any known restrictions on $E[Y^1|D=0]$ and $E[Y^0|D=1]$, $E[\delta]$ is contained in the completely uninformative interval between $-\infty$ and ∞.

Manski (1994, 1995, 2003) shows that with potential outcome variables bounded by 1 and 0, the no-assumptions bound will always be of length 1. This result is obtained by a more general manipulation of Equation (12.8), for which the lower bound is derived as

$$\{\pi E[Y^1|D=1] + (1-\pi)(0)\} - \{\pi(1) + (1-\pi)E[Y^0|D=0]\}, \tag{12.9}$$

and the upper bound is derived as

$$\{\pi E[Y^1|D=1] + (1-\pi)(1)\} - \{\pi(0) + (1-\pi)E[Y^0|D=0]\}. \tag{12.10}$$

Simplifying and then combining Equations (12.9) and (12.10) yields the no-assumptions bound for potential outcomes bounded by 1 and 0:

$$\pi E[Y^1|D=1] - (1-\pi)E[Y^0|D=0] - \pi \tag{12.11}$$
$$\leq E[\delta]$$
$$\leq \pi E[Y^1|D=1] - (1-\pi)E[Y^0|D=0] + (1-\pi).$$

The length of the bound is 1 because the upper and lower bounds differ only by two complementary probabilities π and $1-\pi$ that sum to 1. The location of the bound in the $[-1,1]$ interval is set by the common term $\pi E[Y^1|D=1] - (1-\pi)E[Y^0|D=0]$, to which $-\pi$ and $1-\pi$ are then added to form the bound.

More generally, if Y^1 and Y^0 are bounded by any finite values a and b, where $b > a$, such as theoretical minimum and maximum scores on a standardized test, then Equation (12.6) can be written more generally as

$$-b+a \leq E[\delta] \leq b-a, \tag{12.12}$$

which then generates the more general no-assumptions bound for the observed data:

$$\pi E[Y^1|D=1] - (1-\pi)E[Y^0|D=0] + a(1-\pi) - b\pi \tag{12.13}$$
$$\leq E[\delta]$$
$$\leq \pi E[Y^1|D=1] - (1-\pi)E[Y^0|D=0] + b(1-\pi) - a\pi.$$

The same basic results hold as before. The no-assumptions bound always includes 0, and it is only half as wide as the bound in Equation (12.12) implied only by the bounds on the potential outcomes. In this case, the no-assumptions bound is of length $b-a$ rather than of length $2(b-a)$.[2] As with the case for potential outcomes bounded by 1 and 0, the particular location of the interval of length $b-a$ in $[-b+a, b-a]$ is again determined by the same term, $\pi E[Y^1|D=1] - (1-\pi)E[Y^0|D=0]$.

12.2.2 Bounds Under Additional Weak Assumptions

For Manski, calculation of the no-assumptions bound is only the starting point of a set-identification analysis. The primary goal is to analyze how additional assumptions can narrow the no-assumptions bound. Manski's basic perspective is summarized nicely in the following passage:

> Empirical researchers should be concerned with both the logic and the credibility of their inferences. Credibility is a subjective matter, yet I take there to be wide agreement on a principle I shall call:
> *The Law of Decreasing Credibility*: The credibility of inference decreases with the strength of the assumptions maintained.
> This principle implies that empirical researchers face a dilemma as they decide what assumptions to maintain: Stronger assumptions yield inferences that may be more powerful but less credible. (Manski 2003:1)

Manski has shown that weak and often plausible assumptions can substantially narrow the no-assumptions bound. Consider the following simple assumptions about the direction of individual-level treatment effects and the direction of self selection.[3]

---

[2]To see this result, note that $[b-a] + [-1(-b+a)]$ simplifies to $2(b-a)$, and $[b(1-\pi) - a\pi] + [-1(a(1-\pi) - b\pi)]$ simplifies to $b-a$.

[3]For empirical examples with the same structure, see first the applications offered at the time Manski was developing the set-identification perspective (e.g., Manski and Nagin 1998; Manski, Sandefur, McLanahan, and Powers 1992). More recently, Manski has used the same basic approach to critique the "dueling certitudes" and other noncredible claims that pervade public policy pronouncements in academic research and beyond (Manski 2013b). See also Rein and Winship (1999).

Monotone Treatment Response

In many situations, it may be reasonable to assume that the individual-level treatment effect cannot be negative, such that $\delta \geq 0$ for every individual i. Manski labels this assumption the monotone treatment response (MTR) assumption.

Under MTR (in the direction where $\delta \geq 0$), the lower bound for the ATE must be 0. MTR implies that members of the control group have counterfactual values of y_i^1 that are at least as high as their observed values of y_i. The reasoning here is simple: If $y_i = y_i^0$ for members of the control group, then the MTR assumption that $y_i^1 \geq y_i^0$ for all individuals i implies that $y_i^1 \geq y_i$ for members of the control group. The opposite is likewise true for members of the treatment group; their counterfactual values for y_i^0 are no higher than their observed values of y_i. Under MTR for the hypothetical example presented in Table 12.1, one can therefore replace the extreme values of 1 and 0 in the no-assumptions lower bound:

$$\{(.5)(.7) + (1-.5)(0)\} - \{(.5)1 + (1-.5)(.3)\} = -.3$$

with less extreme values of .7 and .3. As a result, one obtains a new lower bound:

$$\{(.5)(.7) + (1-.5)(.3)\} - \{(.5)(.7) + (1-.5)(.3)\} = 0,$$

implying that the bound for the ATE, assuming MTR in this direction, is $[0, .7]$ rather than $[-.3, .7]$.

Monotone Treatment Selection

It may also be reasonable to assume that those who receive the treatment have higher average outcomes under potential exposure to both the treatment and control, which Manski labels the monotone treatment selection (MTS) assumption. In most cases, one would assume jointly that $E[Y^1|D=1] \geq E[Y^1|D=0]$ and $E[Y^0|D=1] \geq E[Y^0|D=0]$, which is traditionally thought of as positive self-selection. (But one could flip the direction of the assumption, just as we also noted for MTR.) Manski and Pepper (2000) present this type of MTS assumption for the example of the effect of education on earnings, for which it is equivalent to assuming that individuals with higher education would on average receive higher wages than individuals with lower education, under counterfactual conditions in which they had the same levels of education.

For the hypothetical example presented in Table 12.1, MTS implies that the naive estimator would be an upper bound for the ATE. In short, MTS in this direction stipulates that members of the treatment group could not have done any worse in the control state than those observed in the control group (and, vice versa, that members of the control group could not have done any better in the treatment state than those observed in the treatment group). Maintaining MTS allows one to replace the extreme values of 1 and 0 in the no-assumptions upper bound,

$$\{(.5)(.7) + (1-.5)(1)\} - \{(.5)0 + (1-.5)(.3)\} = .7,$$

with the less extreme values of .7 and .3, yielding for this example,

$$\{(.5)(.7) + (1-.5)(.7)\} - \{(.5)(.3) + (1-.5)(.3)\} = .4.$$

MTS thereby implies that the bound for the ATE is $[-.3, .4]$ rather than $[-.3, .7]$.

Monotone Treatment Response and Treatment Selection Together

The power of Manski's approach comes from the ability to apply multiple assumptions at the same time in order to narrow the no-assumptions bound to a bound that is considerably more informative. For the hypothetical example presented in Table 12.1, one can narrow the no-assumptions bound from $[-.3, .7]$ to $[0, .4]$ by invoking the MTR and MTS assumptions together. The resulting bound still includes 0, but in some applications such narrowing may well be helpful in ruling out some unreasonable and extreme causal claims.

Monotone Instrumental Variables

Whereas the recent work of Heckman and his colleagues has demonstrated how powerful the conclusions of a study can be if a perfect and finely articulated IV is available (see Section 9.3.4), Manski's work on IVs probes the opposite territory, following on the consideration in his early work of the capacity of traditional IVs to narrow the no-assumptions bound (see Manski 1994, 1995). Manski and Pepper (2000) investigate what can be learned about the ATE when both standard and weaker IV assumptions are maintained. Their results are then presented as a component of the general set-identification methodology laid out in Manski (2003; see especially chapter 9).

Manski and Pepper (2000) first define the traditional IV assumption in terms of mean independence, after conditioning on a covariate X for generality. In our notation, the standard IV assumption of no direct effect of Z on Y (by either Y^1 or Y^0) is written out as

$$E[Y^1|X, Z = z'] = E[Y^1|X, Z = z''], \tag{12.14}$$

$$E[Y^0|X, Z = z'] = E[Y^0|X, Z = z''], \tag{12.15}$$

for any two values z' and z'' of Z (and separately for strata defined by X). In other words, Equations (12.14) and (12.15) require that the expectations of the potential outcomes be equal within strata defined by Z (conditional on X). Accordingly, the bounds analysis presented above then applies within each stratum of Z, and the bound on the ATE can then be defined as the intersection of the bounds across the strata defined by Z. This result implies that the no-assumptions bound can only be narrowed by an IV if the no-assumptions bounds that could be calculated within each stratum z of Z also vary across the strata defined by Z.

Manski and Pepper (2000) then consider a weaker assumption for an IV analysis, known as a monotone IV (MIV) assumption. It states that for all values of the instrument Z, in which $z'' \geq z'$, the variable Z is an MIV if

$$E[Y^1|X, Z = z''] \geq E[Y^1|X, Z = z'], \tag{12.16}$$

$$E[Y^0|X, Z = z''] \geq E[Y^0|X, Z = z']. \tag{12.17}$$

In Equations (12.16) and (12.17), the expected values of both potential outcomes are weakly increasing in Z. Note that this usage of the concept of monotonicity is quite different than for a local average treatment effect (LATE) analysis or an LIV analysis (see Section 9.3, beginning on page 305). For an MIV, monotonicity refers to the relationship between the instrument and the potential outcomes, not the relationship between the treatment and the instrument. The latter type of assumption is referred to as monotone treatment selection (MTS) by Manski and his colleagues (see our discussion above).

To what extent does an MIV narrow the bound for the ATE (or, in the words of Manski 2003, shrink the identification region for it)? The short answer: No more than a traditional IV (and usually considerable less), but still enough to have some identifying power.

It is easier to explain how an MIV bounds the expectation of each potential outcome than it is to demonstrate directly how an MIV bounds the ATE that is a function of these expectations. Consider just the determination of the upper bound for $E[Y^1]$.

Under the standard IV assumption of mean independence, the upper bound for $E[Y^1]$ is equal to the smallest upper bound across the different subpopulations defined by the instrument Z. More precisely, the upper bound is the smallest value that $E[Y^1|Z=z]$ takes on across all values z of Z.

In contrast, under the weaker MIV assumption, the upper bound for $E[Y^1]$ is a weighted average of subpopulation upper bounds, for which each subpopulation upper bound is defined across the values of Z. The implicit calculation of this upper bound on $E[Y^1]$ can be described in a series of as-if algorithmic steps. First, a value of Z is selected as z'. The upper bound for $E[Y^1]$, with respect to z', is set as the smallest value that $E[Y^1|Z=z]$ takes on across all values z'' of Z where $z'' \geq z'$.[4] After this smallest upper bound is found with z' fixed, the next value of z' is selected, and a smallest upper bound is found with respect to this next z'. After all values of Z have been selected as z', the upper bound for $E[Y^1]$ is set as the weighted average of the subpopulation upper bounds with respect to each value of Z, where the weights are the marginal distribution of Z attached pointwise to the smallest upper bounds set with respect to all z' of Z. The determination of the lower bound on $E[Y^1]$ under MIV is the opposite of this procedure, in the sense that the greatest lower bound is first sought across subpopulations of Z, restricting the range of Z over which one searches in each step to be larger than the selected anchoring point z' of Z.

To find the bounds implied by an MIV for the ATE, the bounds on $E[Y^0]$ must be calculated with the same basic procedure. These bounds can then be substituted into the same basic framework introduced in the last section, with the marginal distribution of D used to characterize the known distribution across treatment states.

Of course, as we noted above, the goal of a set-identification analysis is to invoke combinations of weak assumptions.[5] As with the more general set-identification

---

[4]Because this smallest upper bound relative to z' is selected from a smaller subset of possible upper bounds (only those greater than z' rather than all values of Z), the resulting weighted upper bound across z' is by definition no smaller (and usually larger) than would be the upper bound suggested by a more stringent mean-independence IV assumption. As a result, MIVs never shrink the identification region more than traditional IVs (and usually considerable less).

[5]In a prior version of their manuscript, Manski and Pepper used the MTR, MTS, and MIV assumptions together to determine the bounds on the effect of education on the logged wages of respondents for the National Longitudinal Survey of Youth. When they invoked MTR and MTS assumptions, they found that the bound for the effect of a twelfth year of schooling was [0, .199], that the bound for the effect of a fifteenth year of schooling was [0, .255], and that the bound for the effect of a sixteenth year of schooling was [0, .256]. When they then used the Armed Forces Qualification Test as an MIV while still maintaining the MTR and MTS assumptions, they obtained narrower bounds, respectively, of [0, .126], [0, .162], and [0, .167].

approach, it is generally difficult to eliminate values of 0 for average treatment effects in this tradition, but the methodology can be used, as Manski shows, to convincingly reject extreme causal effect assertions and build regions of credible inference to move the literature forward in any given area. Moreover, it is presumably much easier to find MIVs in any substantive area than IVs that allow for the identification of LATEs or full schedules of MTEs.

In sum, the set-identification approach is a principled and strict framework for developing warranted causal claims. In part because it is so strict, but more fundamentally because scholars and reviewers value point estimates so highly, the usage rate of set-identification strategies in empirical research remains quite low. For exceptions, it is well worth consulting the applications offered at the time Manski was developing his set-identification perspective (e.g., Manski and Nagin 1998; Manski et al. 1992). More recently, Blundell, Gosling, Ichimura, and Meghir (2007) use a bounds approach to analyze wage trajectories for men and women. Gunderson, Kreider, and Pepper (2012) offer an MIV application to generate support for the claim that reduced-price lunches in public schools have positive effects on health outcomes of children. Manski and Pepper (2013) offer a full analysis of the deterrence effect of the death penalty. And, following on the general orientation reflected in this last piece, Manski (2013b) offers an extended critique of the "dueling certitudes" and other noncredible claims that pervade public policy pronouncements in academic research and beyond.

12.3 Sensitivity Analysis for Provisional Causal Effect Estimates

Consider a situation where a researcher suspects, based on substantive knowledge from other studies or a particular theoretical perspective, that treatment selection is non-ignorable because selection is, in part, on unobserved variables. Yet, the researcher is unwilling to adopt the set-identification perspective because she feels that the violations of the point-identifying assumptions are very small. In this case, she may want to offer a causal effect estimate under the assumption that treatment selection is ignorable, but she should then judge how wrong the results may be and offer interpretations that are appropriately cautious.

Although Blalock is often accused of inspiring naive causal analysis (and we have engaged, as well, in our own share of Blalock criticism in Chapter 1), he did warn against overconfident causal claims based on conditioning with regression methods long before most of the methods we have presented in this book were developed. At the end of his influential 1961 book *Causal Inferences in Nonexperimental Research*, he wrote:

> It seems safe to conclude that the problem of making causal inferences on the basis of nonexperimental data is by no stretch of the imagination a simple one. A number of simplifying assumptions must be made, perhaps so many that the goal would seem almost impossible to achieve. The temptation may very well be to give up the entire venture. But, as we have

suggested at a number of points, there may not be any satisfactory alternatives. Most likely, social scientists will continue to make causal inferences, either with or without an explicit awareness of these problems. (Blalock 1964[1961]:184)

His recommendation was to report a range of plausible results, and he lamented the pressures to settle in on only one favored set of results:

> it would be extremely helpful if social scientists would develop the habit of contrasting the results of several *different* procedures. ... If conclusions or results differed, one might gain valuable insights as to why specific differences have occurred. ... Present practices and publication policies are undoubtedly unfavorable to such a strategy. ... there are undoubtedly pressures on the individual investigator to report only his "best" results in instances where different techniques have not given the same results. (Blalock 1964[1961]:184–85)

In the decades since scholars such as Blalock appealed for the multimodel reporting of results, a number of scholars have attempted to systematize this approach.

In the counterfactual tradition, the approach known as sensitivity analysis has been the most influential.[6] Based largely on the work of Rosenbaum and his colleagues (see Rosenbaum 1991, 1992, 2002, 2010), the guiding principle of the approach is simple: When reporting and interpreting an estimate of a causal effect, researchers should analyze and report how sensitive the estimates and interpretations are to the maintained assumptions of the analysis.

For Rosenbaum, sensitivity analysis of this form is quite distinct from other types of robustness checks on one's results because of its concern with estimates of causal effects. Rosenbaum (1999:275) writes, "Assumptions are of three kinds: (i) the scientific model or hypothesis, which is the focus of scientific interest, (ii) incidental assumptions, needed for statistical inference but of little or no scientific interest and (iii) pivotal assumptions which rival the scientific hypothesis and are the focus of scientific controversy." Sensitivity analysis is usually focused narrowly on the specific and pivotal assumption of ignorability of treatment assignment. Here, the question of interest is usually, How sensitive are estimates of an average causal effect to the potential effects of unobservable treatment selection patterns?

Rosenbaum (2002; see also Rosenbaum 2010) devotes a large portion of his excellent book on observational data analysis to strategies for performing sensitivity analysis, especially for matching designs.[7] Consider for reference the simple directed graph in

---

[6]There is a related literature on sensitivity analysis that is tied to simulation methods and deterministic modeling more generally. In this tradition, for example, the parameters of a deterministic simulation of an outcome are varied systematically. If variation in nuisance parameters of no inherent interest does not change the basic results of the model, then the model is robust. And, as described in Saltelli, Tarantola, Campolongo, and Ratto (2004), such methods can be used to assess the uncertainty of model predictions based on the uncertainty of model inputs. Such results can direct future effort optimally, so as to reduce the uncertainty of those inputs that generate the most uncertainty of prediction.

[7]For expositions of the basic perspective written for social scientists, see DiPrete and Gangl (2004), Frank (2000), and Gangl (2013).

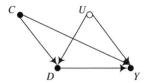

Figure 12.1 A graph in which the causal effect of D on Y is confounded by an observed variable C and an unobserved variable U.

Figure 12.1, where an analyst seeks to estimate the effect of D on Y. In this case, the researcher observes only one of two known confounders, C but not U. For Rosenbaum, adjusting for C would remove "overt bias" but would leave the "hidden bias" produced by $D \leftarrow U \rightarrow Y$ untouched. For a sensitivity analysis, the typical steps would be

1. Offer a provisional point estimate of the ATE for the effect of D on Y by conditioning on C and assuming that ignorability holds.

2. Choose multiple pairs of values for the effects of U on D and of U on Y.

3. Using either analytic methods or simulation methods, present results that show how much the provisional point estimate of the ATE would be expected to change assuming each pair of values.

4. Identify and report the pairs of values that would push the provisional point estimate of the ATE below the critical level that would prevent one from rejecting the null hypothesis of no effect.

5. Use external information on what is known about U and its relationships to D and Y to assess whether or not the pairs of values that would prevent one from rejecting the null hypothesis of no effect are reasonably likely.[8]

If, after after executing these steps, the researcher can argue that all plausible violations of ignorability (i.e., the pairs of values for the effects of U on D and of U on Y) would not change the qualitative conclusions, then one can report the levels of sensitivity from step 3 and confidently stick to the main substantive conclusion that the ATE is sufficiently unlikely to be equal to 0. If, however, plausible violations of ignorability would change the main substantive conclusion, because 0 cannot be regarded as implausible, then one should step back from strong conclusions and consider other methods of analysis.

Examples in the literature demonstrate the utility of the sensitivity analysis approach, although for obvious reasons it is hard to find published examples where causal assertions have been abandoned in response to sensitivity analyses. Instead, sensitivity analysis is usually used to defend conclusions against the reasonable objections of either

---

[8]Note how this strategy differs from the set-identification approach presented above. From this alternative orientation, the researcher would first calculate no-assumptions bounds for the conditional average effect of D on Y within strata of C. Then, the researcher would introduce additional credible assumptions about the individual-level and/or average effects of U on D and U on Y in an attempt to narrow the no-assumptions bounds to intervals that all researchers can accept.

fictitious fair critics or the specific assertions of real critics. Consider an example of each type.

In response to fictitious critics, Harding (2003) assesses the strength of the relationship that a composite of unobserved variables would have to have with a treatment variable and an outcome variable in order to challenge the causal interpretations of his estimates of the effect of neighborhood context on the odds of completing high school. He uses Rosenbaum's specific pair-matching thresholds for violations of ignorability to demonstrate that convincing support for his conclusions is available. He is able to conclude, "Thus, to drive the [estimated] neighborhood effect to nonsignificance, parental involvement would have to be more powerful than family income or having poorly educated parents, net of these variables" (Harding 2003:712).

For a response to real critics, VanderWeele (2011b) defends the claim of Christakis and Fowler (2007) that obesity is contagious and spreads through social networks. The critics claimed that the results of Christakis and Fowler did not provide support for the causal effects of interest and instead reflected both homophily bias and unmeasured effects of shared environments (see, e.g., Cohen-Cole and Fletcher 2008; Shalizi and Thomas 2011). VanderWeele (2011b) performed a sensitivity analysis, also based on Rosenbaum's general procedures, and concluded that his sensitivity analysis suggested that the contagion effects on obesity "were reasonably robust to latent homophily" (VanderWeele 2011b:252). The results, however, were not definitive, even if they were encouraging to the original authors (see Christakis and Fowler 2013). Shalizi (2012) raised objections to the realism of the sensitivity analysis, while also making the more fundamental point that the original models were themselves less suited to the estimation task than Christakis and Fowler recognized when first offering them.

Given the close connections between set identification and sensitivity analysis, are there reasons to favor one or the other approach when both are equally feasible? To consider this question, it is useful to first review the writing of Manski and Rosenbaum. Manski's identification focus is clear, as explained in the quotations that we used in Section 3.1 (see page 78). In contrast, Rosenbaum sees a targeted focus on identification to be considerably less helpful:

> The idea of identification draws a bright red line between two types of problems. Is this red line useful? ... In principle, in a problem that is formally not identified, there may be quite a bit of information about β [a causal parameter of interest], perhaps enough for some particular practical decision .... Arguably, a bright red line relating assumptions to asymptotics is less interesting than an exact confidence interval describing what has been learned from the evidence actually at hand. (Rosenbaum 2002: 185–86)

Rosenbaum's objection to the bright red line of identification is issued in a discussion of an IV estimator, which we explained in Chapter 9 can offer an estimate of a formally identified parameter that may be so noisy in a dataset of finite size that one cannot possibly learn anything from the estimate. However, an alternative estimator in the same context – usually a least squares regression estimator – that does not formally identify a parameter because it remains asymptotically biased even in an infinite sample may nonetheless provide sufficiently systematic information so as to remain useful, especially if one has a sense from other aspects of the analysis of the likely direction and size of any remaining bias.

We accept Rosenbaum's perspective to some extent. It is undeniable that an empirical researcher who forsakes all concern with statistical inference could be led astray by considering only estimates that are formally identified. But, for this book, our perspective has been much closer to that of Manski. We have focused on identification problems almost exclusively because our primary goal has been to help researchers to determine what assumptions must be maintained in order to identify causal effects and then estimate them. Accordingly, we do not take a particular position on whether set identification or sensitivity analysis is a better strategy when point estimation of a causal effect of interest is infeasible. We give more space with a worked example for an analysis of bounds only because it fits with the identification focus of this book.

Our position notwithstanding, Manski does have a principled reason for favoring set identification, even though he sees it as "mathematically complementary" to sensitivity analysis (Manski 2003:5). He regards Rosenbaum's sensitivity analysis approach as too narrow, writing:

> Where Rosenbaum and I differ is that I do not view the assumption of ignorable treatment selection to have a special status in observational studies of treatment effects. As an economist, I usually am inclined to think that treatments are purposefully selected and that comparison of outcomes plays an important role in the selection process. Perhaps the departures from ignorable treatment selection that Rosenbaum entertains in his sensitivity analysis can be interpreted behaviorally in terms of some model of purposeful treatment selection, but for now I do not see how. (Manski 1999:281)

Other social scientists may agree with this position, based on whether they also feel that it is too difficult to discuss departures from ignorability of treatment assignment in a coherent way without reference to the behavior of individuals. To some degree, we found ourselves in the same situation when discussing the charter school example in Section 8.3, where we found it necessary to elaborate the back-door path that Rosenbaum would regard as a source of hidden bias. This orientation likely also underlies the critique that Shalizi (2012) offers of VanderWeele's sensitivity analysis in defense of Christakis and Fowler:

> [VanderWeele 2011b] is a truly ingenious paper, which advanced the field ... However, it did so under very strong parametric and substantive assumptions, such as, e.g., all latent homophily being due to a single binary variable, which interacts with observables in very specific and limiting ways. Proving results under these restrictions is more than anyone else has done, but before one appeals to the results in empirical problems, one needs to either have some scientific reason to think the restrictions hold, or a mathematical reason to think that the conclusions are robust to substantial departures from those assumptions. Since those mathematical reasons are, at least for now, unavailable, we are forced to rely on scientific knowledge. Is anyone prepared to argue that we ought, on biological or sociological

grounds, to think that everything relevant to friendship formation and obesity (in suburban Massachusetts) boils down to one binary variable? (Shalizi 2012:2)

It remains to be determined, therefore, whether sensitivity analysis can serve as an effective defense against real critics.

Nonetheless, sensitivity analysis can be and is being generalized, to the extent that the boundary between sensitivity analysis and set identification seems likely to gradually disappear. Consider, for example, savvy applied work such as Berk and de Leeuw (1999), where simulation is used to insert uncertainty into conclusions that may result from violations of assumptions. The prospects for such work are bright, especially when used to extend interpretations of findings by exploring alternative structures for outcomes, rather than simply trying to defend against the claims of critics who challenge identifying assumptions.

12.4   Conclusions

In this chapter, we have considered three separate strategies for how analysts can proceed when they are unable to use observed data to point-identify average causal effects of interest. After using the classic selection-bias literature as an example of how distributional assumptions for unobservables can identify some estimates of interest, we then took the position of the more recent literature that distributional assumptions may be particularly hard to justify. We then considered the set-identification literature, which is motivated by a similar type of skepticism. Rather than issue daring assumptions that may be hard to justify, the goal is to offer easily defendable assumptions that are not open to doubt. This literature has shown that through combinations of assumptions, some causal claims can be eliminated. Yet, these methods have not been widely used because they typically cannot eliminate 0 as a permissible value for estimated causal effects and, even more so, because social scientists continue to very much prefer single-value estimates. Finally, we considered an approach that is complementary to set identification, where analysts offer provisional point estimates of average treatment effects, with full recognition that they are likely inconsistent and biased. Analysts then examine how wrong these estimates may be by assessing their sensitivity to violations of the maintained assumptions.

Part VI: Conclusions

Chapter 13

Counterfactuals and the Future of Empirical Research in Observational Social Science

What role should the counterfactual approach to observational data analysis play in causal analysis in the social sciences? Some scholars see its elaboration as a justification for experimental methodology as an alternative to observational data analysis. We agree that by laying bare the challenges that confront causal analysis with observational data, the counterfactual approach does indirectly support experimentation as an alternative to observation. But, because experiments are often (perhaps usually) infeasible for most of the causal questions that practicing social scientists appear to want to answer, this implication, when considered apart from others, is understandably distressing.

We see the observational data analysis methods associated with the potential outcome model, motivated using directed graphs, as useful tools that can help to improve the investigation of causal relationships within the social sciences, *especially* when experiments are infeasible. Accordingly, we believe that the methods associated with the counterfactual approach complement and extend older approaches to causal analysis with observational data by shaping the goals of an analysis, requiring explicit consideration of individual-level heterogeneity of causal effects, encouraging a wider consideration of available identification strategies, and clarifying standards for credible interpretations.

In this chapter, we first shore up our presentation of the counterfactual approach by considering several critical perspectives on its utility. We weigh in with the arguments that we find most compelling, and it will not be surprising to the reader that we find these objections less serious than do those who have formulated them. However, we use quotations from the published literature in order to demonstrate the richness of these debates and to encourage a full reading of the material that we are able to summarize only briefly here.

We conclude the chapter with a discussion of causal modeling practices where we delineate the multiple modes of causal inquiry that exist in observational social science. We argue that the counterfactual approach is useful because it facilitates movement between modes of causal inquiry, guiding analysis to the deepest level that theory and data can enable at any given point in time.

13.1   Objections to Adoption of the Counterfactual Approach

The counterfactual approach is not without serious objections from thoughtful critics. In this section, we will present three objections from the published literature. If these objections are accepted, then a justification for eschewing much of what we have presented in prior chapters is available. Such a decision, however, would presumably require the adoption of the counterposition(s) of those who raise the following objections. We will not lay out these alternative approaches to causal analysis here, but their main features are implicit in the objections.

We will not further discuss two objections that we have already addressed in detail in prior chapters. First, for a response to the supposed limitations for practice implied by the typical adoption of stable unit treatment value assumption (SUTVA), see Section 2.5. As argued there, the potential outcome model cannot be considered problematic or inappropriate for the social sciences because it more clearly demonstrates a challenge to the definition and estimation of causal effects than other less complete frameworks for causal inference. In addition, counterfactual analysis can proceed when SUTVA does not hold, although the potential outcome model then shows how difficult it can be to identify average causal effects of various forms.

Second, for a response to the claim that the counterfactual approach does not give sufficient attention to generative mechanisms, see Chapter 10. While the counterfactual approach takes no general position on the necessary depth of a causal claim, we argued there that (1) the counterfactual approach is suitable for use when evaluating levels of support for conjectured alternative mechanisms that generate observed outcomes and (2) the counterfactual approach is entirely consistent with the theory construction agenda of the generative mechanisms movement, as long as such theory construction is not itself passed off as causal explanation.[1]

---

[1] We will also not discuss the claim that, by adopting a notion of causality that grants currency to counterfactuals, the analyst is forced to include in any explanation an abundance of ancillary causal claims, typically those grounded in causality by omission ("because the moon did not crash into the earth") and those grounded in causality by prehistory ("because of the birth of the universe"). Proponents of this red-herring objection then argue that, because such ancillary claims have little or no explanatory value, but the counterfactual approach requires that they be admitted into the explanation, it must be the case that the counterfactual approach to causation is indefensible because it is incomplete. See Martin (2011) for the style of argument, where he writes, for example, "[W]e cannot ask the question, What causes B's death? and bring in anything less than an infinite number of causes, with little way of telling them apart. Almost everything that led our world to be our world (and not some other) was a cause of B's death.... Committed counterfacualists have learned to live

Objection 1: Nonmanipulable Causes Cannot Be Analyzed When Using the Potential Outcome Model

The counterfactual approach is most natural when one can conceive of a specific procedure for manipulating the cause of interest, or in the words of Berk (2004:91), "in the sense that conscious human action can change things." If the effects of a cause that cannot be manipulated by "human action" are of interest, then the potential outcome model loses some of its transparency because the counterfactuals become harder to conceptualize. The statistician Paul Holland is usually cited as offering the most eloquent statement of this position, using the motto "no causation without manipulation" that he developed in a prior discussion with Rubin (see Holland 1986:959).[2] The objection we consider in this section maintains that, because one must agree with the 1986 Holland-Rubin position in order to utilize the potential outcome model, scholars who are interested in the effects of (typically) immutable characteristics – such as gender and race – must instead find some other approach to causal analysis to use in their research. And, given that so many social scientists are interested in these sorts of effects, the potential outcome model must therefore be regarded as one that has quite limited utility for observational social science (see, e.g., Goldthorpe 2001).

We find this objection unpersuasive for the following reasons. Even if one adopts the counterfactual approach while also accepting the rigid position of "no causation without manipulation," the entailed restrictions on permissible analysis are not as limiting as is often claimed. Woodward (2003) has argued this position more completely than anyone else (see, in particular, his section 3.4, "In What Sense Must Interventions Be Possible?"). In developing his conception of an intervention, Woodward concludes:

> The sorts of counterfactuals that cannot be legitimately used to elucidate the meaning of causal claims will be those for which we cannot coherently describe what it would be like for the relevant intervention to occur at all or for which there is no conceivable basis for assessing claims about what would happen under such interventions because we have no basis for disentangling, even conceptually, the effect of changing the cause variable alone from the effects of other sorts of changes that accompany changes in the cause variable. (Woodward 2003:132)

In this regard, what matters is not the ability for humans to manipulate the cause through some form of actual physical intervention but rather that we be able, as observational analysts, to conceive of the conditions that would follow from a hypothetical (but perhaps physically impossible) intervention.[3]

---

with this strange conclusion" (38). The position that we have maintained in this book is that the simple and direct explanatory goals at hand are valid guides for the delineation of causal states and, furthermore, that claims about the sizes of estimated causal effects will be sufficiently informative for the purposes at hand if the causal states that define them are local and reasonable. Simply put, claims for the relevance of causality by omission, and especially causality by prehistory, can be held aside by scientific judgment alone because they are sufficiently remote relative to the goals of the analysis.

[2] Even so, Holland's position is frequently misinterpreted. He did not argue that nonmanipulable causes cannot be analyzed using the potential outcome model but rather that the effects of nonmanipulable causes are ill-defined and therefore cannot be estimated effectively with any methods. For an update to his argument, see Holland (2008).

[3] Woodward (2003:122) also discusses how the focus on conceivable interventions allows one to avoid having to contemplate preposterously unreasonable counterfactuals. Holland (2008) still supports the

In addition, the manipulability criterion is not as relevant as is sometimes supposed. For example, if discrimination is the topic of study, the attributes of individuals do not need to be manipulated, only the perception of them by potential discriminators. Deception then can be used to gain leverage on the causal effects of interest, while the attributes of the subjects of potential discrimination remain fixed. Greiner and Rubin (2011) offer a discussion of appropriate research designs, consistent with the potential outcome model, where the causal effects of immutable characteristics can be studied by manipulations of perceptions.

Even if Woodward's position is rejected, and thus if a researcher does not feel comfortable using counterfactuals to define the causal effects for hard-to-conceive-of-actually-manipulating attributes, the counterfactual approach can still be used in an as-if mode in order to sharpen the goals and contribution of an analysis. This is the position of Glymour (1986), who responds to the original Holland-Rubin position by arguing that counterfactuals can be used to elucidate effects that are presumed to have been caused by a set of potentially manipulable factors but are merely referred to collectively by nominal labels. The goal of research is to push the analysis further down from the nominal labels to the separable manipulable factors. In pursuit of this goal, it can be efficient to first define the counterfactuals associated with nonmanipulable attributes as if they could be easily and naturally manipulated by human action.[4]

Consider studying the effect of race on socioeconomic outcomes, one of the most common topics of study in the literature on social inequality. Although race cannot be manipulated easily, the framework can still be used to motivate an attempt to estimate average gains that employed black adults working full-time, full-year would expect to capture if all prospective employers believed them to be white. This effect would differ from an attempt to estimate what the black–white gap in earnings would be if black adults had grown up in families with the same average level of income as whites, and hence would have had the available resources with which to purchase higher-quality secondary and postsecondary education. By helping to frame such fine distinctions between alternative causal states, the potential outcome model helps to sharpen research questions and then shape reasonable interpretations of whatever results can be distilled from the data that are at hand. Such a basic framing of differentials is a necessary first step, and only thereafter can one begin to systematically investigate the particular mechanisms that generate the causal relationships and that can explain the sources of the differences that motivate the inquiry in the first place.

In this regard, stating merely that an attribute D causes an outcome Y is simply the crudest form of a mechanism sketch (as we discussed in Chapter 10). To further investigate this mechanism sketch, counterfactuals must then be defined for whatever conjectured process is thought to generate the causal effect of D on Y. Reskin (2003), for example, provides a clear exposition of how the investigation of such effects

---

original 1986 motto, but his position is more nuanced than much of the secondary literature implies and appears to have softened to some small degree.

[4]We should also note, consistent with our prior discussion in Section 2.1, that some of the causes that we can easily conceive of manipulating are also nominal, in the sense that the causal capacities inherent in them may be attributable to only some of their constituent features. From this perspective, perhaps most of the causes that we can conceive of manipulating – obtaining another year of schooling, enrolling in a worker training program, moving to an alternative neighborhood – are nominal starting points as well.

of attributes should proceed. She argues that researchers should lay out alternative mechanisms that generate ascriptive inequality – primarily those based on the discriminatory motives of gatekeepers versus those based on unequal access in structures of opportunity – and then evaluate the relative of importance of each set of mechanisms in targeted empirical analysis.

Objection 2: The Counterfactual Approach Is Appropriate Only for Modeling the Effects of Causes, Not the Causes of Effects

If an investigator is interested in estimating the effect of a particular cause, D, on an outcome of interest, Y, then the potential outcome model offers a strong foundation for empirical inquiry. Causal analysis of this type is often labeled an *effects-of-causes* analysis, and we have explained the counterfactual approach to such analysis at great length in this book.

If, instead, an investigator is interested in answering the all-encompassing question "What causes Y?" then the investigator is interested in conducting what is often labeled a *causes-of-effects* analysis. No progress in such an investigation is possible without consulting extant theory or abducting relevant hypotheses that suggest that the effects of one or more candidate causes – A, B, C, D, and so on – need to be modeled to determine whether evidence exists that one or more of these causes contributes to the full account that explains the pattern of all observations of Y. For this type of investigation, the potential outcome model is less natural. In fact, when potential outcomes are defined for each distinct combination of states across all candidate causes, the resulting notation quickly becomes an impediment to practice when more than two or three candidate causal variables must be considered. Fortunately, the causal graphs that we have also utilized in this book offer corresponding definitions for causal effects and, as a result, permit efficient representations of full causal systems. These systems, when fully written out, can effectively guide model specifications that can, in theory, estimate the effects of an unlimited number of causes of an outcome of interest. For this reason, it is false to claim that the counterfactual approach can only be used to motivate studies that estimate the effects of causes. This claim can only be considered reasonable if one restricts the counterfactual approach to the potential outcome model and if one is concerned unduly with the practicality of using potential outcomes to define effects for complex patterns of causal configurations.

The issues worth debating, as we see them, are whether social scientists pose too many causes-of-effects questions relative to effects-of-causes questions and whether the counterfactual approach differs from other approaches in its capacity to enable the analysis of causes-of-effects questions. Michael Sobel has provided perhaps the most vigorous indictment of the common practice of attempting to estimate simultaneously large numbers of causes of an effect of interest with observational data. To cite one example of his perspective, he writes in the overview essay "Causal Inference in the Social Sciences" for a millennium issue of the *Journal of the American Statistical Association*:

> much of quantitative political science and sociology may be characterized as a highly stylized search for new causes of effects. Researchers typically begin with one or more outcomes and a list of causes identified by previous

workers. Potentially new causes are then listed; if these account ("explain") for additional variability of the response, then the new causes are held to affect the outcome, and the significant coefficients in the new model are endowed with a causal interpretation. The process is repeated by subsequent workers, resulting in further "progress." When researchers realize that they are merely adding more and more variables to a predictive conditioning set, one wonders what will take the place of the thousands of purported (causal) effects that currently fill the journals. (Sobel 2000:650)

If Sobel is correct (and we are inclined to agree with him), then the social sciences in the past few decades have given too much attention to the estimation of the causes of effects and too little attention to the simpler and more tractable goal of estimating the effects of particular causes. But even if one disagrees with this position, maintaining instead that causes-of-effects questions should always be in the foreground, it is hard to deny that the quest for a full account of the causes of any outcome is aided (and perhaps, at times, best advanced) by the pursuit of well-defined questions that focus narrowly on the effects of particular causes. As knowledge of these effects accumulates, we can then attempt to build full causal accounts of all of the effects on the outcome of interest. If the counterfactual approach is then justified only as an instrumental path toward the ultimate goal of sustaining full causal accounts, little of its appeal is diminished in our view.[5]

The philosopher Paul Humphreys offers an elegant appeal to this pragmatic realist–empiricist position in the concluding paragraph of his 1989 book, *The Chances of Explanation: Causal Explanation in the Social, Medical, and Physical Sciences*. After reflecting on the demise of the covering law model of explanation and some of its failed alternatives, Humphreys writes:

> What, then, do we have to replace the traditional account of explanatory knowledge? It is not propositional knowledge.... Nor is it *de dicto* knowledge.... It is *de re* knowledge of the causes that contributed to the effect, gained in a cumulative manner, coupled with the discovery of previously unknown structures that pushes the moving boundary of the observable even further from ordinary experience and allows us to become acquainted with finer and finer details of how the world works. When we know of all of the causes, then we shall know all there is to know of the kind with which we are concerned, for all that we will be left is pure chance, and chance is, as I have said, literally nothing. (Humphreys 1989:140–41)

---

[5]One frontier of current methodological scholarship, consistent with the counterfactual approach, is the elaboration of sufficient component cause models that bridge classic epidemiological models with causal graph methodology (see VanderWeele and Robins 2007b, 2008, 2009; VanderWeele, Vansteelandt, and Robins 2010; Vansteelandt, VanderWeele, and Robins 2012). These causes-of-effects models adopt a threshold-crossing framework where combinations of risk factors are modeled as jointly sufficient to generate the outcome of interest (typically, onset of a health condition of some form). Pearl (2009, chapters 9 and 10) has long recognized the value in using his structural equation approach to develop "actual cause" models. We expect that in the coming years recognition that these sorts of models exist will decrease the frequency with which others raise the causes-of-effects objection discussed in this section.

The idea here is that we should strive to explain phenomena by estimating the effects of putative causes on particular outcomes. And, in practice for observational data analysis, most such attempts require a consideration of at least two types of other causes of the outcome: (1) those that may lie along back-door paths in a causal graph and (2) those that constitute the mechanisms that link the putative cause to the outcome. Over time, we collect more data and estimate the effects of the same causes on the same outcomes in a variety of conditions, which deepens our understanding of the additional causes that were considered unobservables initially.

The counterfactual approach we have presented in this book is well suited to this pragmatic account of social science research, where progress results from credible advances rather than grand claims. Even so, the counterfactual approach is not inconsistent with the development and analysis of full causal systems that deliver answers for causes-of-effects questions, even though it encourages skepticism of results that are warranted only by assumptions of dubious validity.

Objection 3: Causal Inference Should Not Depend on Metaphysical Quantities Such as Potential Outcomes

In a wide-ranging article titled "Causal Inference Without Counterfactuals," the statistician A. Philip Dawid argues that potential outcomes are inherently metaphysical, and thus the counterfactual model for causal inference is "generally unhelpful and frequently misleading" (Dawid 2000:409). He argues that causal claims must rest only on inherently testable ideas, and he presents an alternative that he argues succeeds in realizing the goal of estimating the effects of causes.[6]

Although some aspects of Dawid's critique are rather involved, some of its central features are quite simple. He first argues that the potential outcome model embraces "fatalism" because it implicitly assumes that the potential outcomes of individuals (i.e., y_i^1 and y_i^0 for a binary cause) are fixed values that are regarded as "predetermined attributes" of individuals "waiting only to be uncovered by suitable experimentation" (Dawid 2000:412). For his alternative, Dawid considers a basic Bayesian decision model that, without reference to potential outcomes, allows a statistician to (1) design a randomized experiment to compare the effects of a treatment on an outcome (in comparison with a base state of either no treatment or an alternative treatment) and (2) convey the results of this experiment to inform a relevant decision maker of the expected effect of applying the treatment to an additional subject similar to those on whom the experiment was conducted.[7] After contrasting his model with the counterfactual model, he concludes:

> I have argued that the counterfactual approach to causal inference is essentially metaphysical, and full of temptation to make "inferences" that cannot be justified on the basis of empirical data and are thus unscientific.

---

[6]For brevity, we do not cover Dawid's position on modeling the causes of effects here. We just addressed the causes-of-effects versus effects-of-causes position in the last section.

[7]Following in the Bayesian tradition, the decision maker (who could also be the statistician) can then assess the consequences of applying the treatment to the subject, with reference to cost considerations and other determinants of her loss function. These additional points are not stressed by Dawid, in part because they are not incompatible with the counterfactual model.

> An alternative approach based on decision analysis, naturally appealing and fully scientific, has been presented. This approach is completely satisfactory for addressing the problem of inference about the effects of causes, and the familiar "black box" approach of experimental statistics is perfectly adequate for this purpose. (Dawid 2000:423)

The response to Dawid's critique of the potential outcome model has been largely negative, as revealed in the comments published alongside it. The crux of the counter-argument is the following. If perfect experiments for every causal question of interest could be designed and then implemented, then potential outcomes are unnecessary (even though they might still be useful for some purposes, such as to think through the results that might have emerged from alternative experimental protocols). Furthermore, no one seems to disagree with the claim that we should use study designs that can rescue us from having to apply assumptions to what-if quantities. In this regard, it would of course be preferable to be able to use a crossover design in all situations, which Rothman et al. (2008) describe as follows:

> The classic *crossover* study is a type of experiment in which two (or more) treatments are compared, as in any experimental study. In a crossover study, however, each subject receives both treatments, with one following the other. Preferably, the order in which the two treatments are applied is randomly chosen for each subject. Enough time should be allocated between the two administrations so that the effect of each treatment can be measured and can subside before the other treatment is given. A persistent effect of the first intervention is called a *carryover effect*. A crossover study is only valid to study treatments for which effects occur within a short induction period and do not persist, i.e., carryover effects must be absent, so that the effect of the second intervention is not intermingled with the effect of the first. (Rothman et al. 2008:125)

The appeal of this sort of a study design, in view of Dawid's critique, is that each ostensibly metaphysical potential outcome would become an observed outcome. Unfortunately, as Rothman et al. (2008) note, the crossover design is only feasible when one has control over the allocation of the treatment(s), is only effective when the treatment effects of interest do not leave carryover effects behind, and is only possible if the effects have a short enough duration that one can allocate and assess each of them during the available observation window. These conditions exist very rarely for the causal questions that concern social scientists, and as a result crossover studies are most common in clinical settings where the goal is to examine "the efficacy of therapies intended to reduce the frequency or severity of chronic, recurrent problems, such as seizures" (Rothman et al. 2008:649).

For observational data analysis, which Dawid barely mentions in his article, we see no way to escape having to assert what-if assumptions about potential outcomes in order to move forward. Robins and Greenland (2000), when commenting on the Dawid critique, consider many of the applied examples for which counterfactual models have been used. They conclude:

> By narrowly concentrating on randomized experiments with complete com-
> pliance, Dawid, in our opinion, incorrectly concludes that an approach to
> causal inference based on "decision analysis" and free of counterfactuals is
> completely satisfactory for addressing the problem of inference about the
> effects of causes. We argue that when attempting to estimate the effects of
> causes in observational studies or in randomized experiments with noncom-
> pliance ... [a] reliance on counterfactuals or their logical equivalent cannot
> be avoided. (Robins and Greenland 2000:431)

This basic point is echoed by most of the other commentators on the article. Cox
(2000:424) asks this question: "And yet: has the philosophical coherence [of Dawid's
position], if not thrown the baby out with the bathwater, at least left the baby seriously
bruised in some vital organs?" Casella and Schwartz (2000:425–26) conclude, "Dawid
insists that such choices [about how to conduct a causal analysis] ... must be based
on strict principles that can be verified empirically. We believe that such a program is
so overly rigid that, in the end, science is not served."

Even so, Dawid's objection to the metaphysical nature of potential outcomes does
bring up a deeper question: What is the epistemological status of the potential outcome model?[8] It is certainly not positivist, as it breaks radically from the positivist–
empiricist prescription that analysis must be based only on observed quantities. The
reliance on what-if potential outcomes and the consideration of unobservables – in the
treatment assignment process and for the presumed generative mechanism – consigns
the counterfactual approach to the postpositivist model of science generally labeled
realism (see Psillos 1999; Putnam 1975; see also Godfrey-Smith 2003 for an overview).
But, because each unobserved potential outcome could have become an observed outcome if the unobservables in the treatment assignment process had been configured
in an alternative way, and the variables that constitute the generative mechanism
could be observed as well, the entire framework aspires to help us become reductive
empiricists.

In this regard, we suspect that most social scientists who adopt the counterfactual
approach would find the enigmatic but enabling position of Donald Campbell similar
to their own, as he laid it out in 1977:

> I am a fallibilist and antifoundationalist. No part of our system of knowl-
> edge is immune to correction. There are no firm building blocks, either
> as indubitable and therefore valid axioms, or in any set of posits that are
> unequivocal once made. Nor are there even any unequivocal experiences or
> explicit operations on which to found certainty of communication in lieu
> of a certainty of knowledge.

---

[8]When wading into the philosophy literature, one point leaps out in regard to Dawid's charge. The
literature on causality is considered by most philosophers to be properly situated in the subfield of
metaphysics. Thus, Dawid's charge that the potential outcome framework is metaphysical is, from the
perspective of philosophers, both accurate and untroubling. See also the position of Pearl (2009:33–34),
who argues that, even if counterfactuals are metaphysical in some sense because they are structures
in the mind, individuals are all too happy to use them in their daily lives. For this reason alone, it
seems natural to build notions of causality around them.

> I am some kind of a realist, some kind of a critical, hypothetical, corri-
> gible, scientific realist. But I am against direct realism, naive realism, and
> epistemological complacency. (Campbell 1988[1977]:444–45)

We read this passage as indicating that Campbell recognized, grudgingly to be sure, that the pursuit of valid knowledge necessitates the invocation of unobservable quantities, such as potential outcomes. This necessity tips one into the realm of realism, wherein such unobservables are given provisional truth status for pragmatic purposes. But, for any corrigible realist like Campbell, the ultimate aspiration is to bring conjectured unobservables out in the open, drawing analysis down to its most generic form as an as-if positivist endeavor.

For scholars who find the counterfactual approach to observational data analysis unappealing, perhaps because of the persuasiveness of some of the objections presented in this section, a variety of responses exists. Some scholars argue, simply, that we should devote much more attention to descriptive modeling because causal analysis can be so intractable. We begin the next section with this position, after which we then move on to consider alternative modes of causal inquiry within which the counterfactual approach can be utilized.

13.2   Modes of Causal Inquiry in the Social Sciences

Some scholars argue that many long-standing questions in the social sciences can and should be pursued without any reference to causality. When the sole interest is a parsimonious account of "what the data show" or "what the historical record reveals," then all strategies for causal analysis are largely irrelevant, including the counterfactual approach. But, there are also good reasons to consider descriptive analysis more generally. Berk (2004:218) notes that "good description is the bread and butter of good science and good policy research." Sobel (1996:376) declares that many questions "neither require nor benefit from the introduction of causal considerations, and the tendency to treat such questions as if they are causal only leads to confusion."

We agree with this position to a great extent. But the appeal for descriptive research cannot be taken too far. Epistemological questions still arise in descriptive research, as not all descriptive accounts can be considered equally worthy of our attention. Perhaps more importantly, social scientists may not want to backpedal from causal inquiry, lest journalists, partisan think-tank scholars, and politicians take it over entirely. We would most prefer to see both judicious descriptive analysis and careful causal analysis jointly crowd out poor causal analysis in the social sciences and beyond.

As we have noted at several points, when considering the causal controversies that pervade the literature, many social scientists have echoed the point summarized most clearly for sociology by John Goldthorpe, who writes, "sociologists have to find their own ways of thinking about causation, proper to the kinds of research that they can realistically carry out and the problems that they can realistically address" (Goldthorpe 2001:8). With this appeal in mind, we conclude this book with a discussion of the primary modes of causal inquiry that exist in observational social science.

We have two goals for this concluding presentation of complementary modes of causal inquiry: (1) to affirm that all of these modes of inquiry are valuable and (2) to argue that adopting a counterfactual approach can help us to move productively between them. Even so, we would not want to argue for the hegemony of counterfactual thinking. The specter of widespread mechanical adoption of the potential outcome model is truly frightening to us: No one wishes to have to review journal submissions in which scholars define the average treatment effect for the treated (ATT), offer up a fairly arbitrary set of matching estimates, and then a disingenuous sensitivity analysis that purportedly bolsters the results. This prospect alone is sufficient for us to recommend considerable caution. But, perhaps more important, we cannot envision the pursuit of empirical analysis in our own research areas without engaging in analysis that includes all of the modes below. In other words, we can see many future profitable lines of inquiry in our own research in which counterfactual thinking will be used only as a background tool to encourage clear thinking.

With that backing, we offer the following (necessarily oversimplified) representation of complementary modes of causal inquiry in observational social science:

Mode 1: Associational Analysis. In practice, most causal inquiry begins with an assessment, sometimes unreported, of whether putative observed causes are associated with an outcome of interest in some way. It is often stated that establishing such associations are a precondition for subsequent causal analysis, as reflected in the aphorism "no causation without association."[9]

Mode 2: Conditional Associational Analysis. After associations have been established, it is customary to then reestimate the associations after conditioning on values of other observed variables. Although quite general, the typical approach usually follows one of two implementations: (1) conditioning on other variables that are thought to determine the outcome and that may also be related to the cause(s) and (2) conditioning on variables that determine the cause(s) and that may also be related to the outcome. Although conceptually quite different (as we noted in our discussion in Section 4.4), the goal of such conditioning is to eliminate obvious sources of confounding in order to build basic support for causal relevance.

Mode 3a: Targeted Analysis of the Effects of One or More Focal Causes. After an assessment of basic causal relevance for a set of putative causes, the effects of one or more particular causes are then estimated in targeted analysis. This mode of causal inquiry can be (or, we would argue, should be) undertaken after first elaborating a directed graph entailed by extant theory. In any case, careful attention to model specification is needed, adopting assumptions, for example, of ignorability if identification is by back-door conditioning or exclusion restrictions if identification is by instrumental variables.

---

[9]Formally, of course, there are cases for which this may not be true, such as when individual-varying causal effects perfectly cancel out each other or when suppression effects exist. We ignore exceptions such as these here, as they are rare. The sorts of physical equilibria that rigidly generate balancing of responses to causal shocks have no clear analogs in the social sciences that would generate similar perfect cancellation of unit-specific causal effects. And, although suppression effects exist in the social sciences, we cannot think of examples for which they have been shown to completely erase bivariate associations that are at least partly causal.

Mode 3b: Mechanism-Based Analysis. After (or while) targeted estimation of the effects of one or more causes is undertaken, a common practice in causal inquiry is to then introduce intervening variables between the focal causal variable(s) subjected to targeted analysis and the outcome variable in an effort to develop a mechanistic explanation of the process that generates the causal effect. Such a form of causal inquiry can proceed, as is often the case, even though it may be unclear whether or not all forms of confounding have been eliminated.

Mode 4: All-Cause Structural Analysis. At its most ambitious level, causal inquiry is pursued as an attempt to identify all causes in chains of causality from the causal variables to the outcome variable, eliminating or neutralizing all confounding in order to identify all parameters of full causal systems. This approach is best represented in the forms of structural equation modeling that prevail in economics, wherein all specifications are justified with appeals to microeconomic theory and purport to explain the entire "who, when, where, and how" of all of the causes of the outcome of interest.[10]

Consider now how the counterfactual approach can be related to these complementary but increasingly ambitious modes of causal inquiry and how adoption of the approach facilitates movement between them as advances in theory construction and data collection are achieved. For Mode 1 (associational analysis), the counterfactual approach encourages the consideration of causes for which one can conceive of the reasonable and theoretically possible conditions that would result from an intervention that changes the causal variable from one value to another. When causal inquiry is stuck by convention on the consideration of the effects of attributes for which the conditions of as-if interventions are unclear, the approach encourages additional effort to break the associational analysis into pieces that are more amenable to analysis by consideration of specific counterfactual dependencies. The hope of such a refined and expansive strategy is that the analysis can be brought down to a level where manipulations are more easily conceivable, which may then allow for a redirection of the extant associational inquiry. Such redirection may be necessary in order to open up the possibility of advancement beyond Modes 1 and 2.

For the transition from Mode 1 to Mode 2 (that is, from associational to conditional associational analysis), the counterfactual model encourages separate but comparative consideration of the determinants of the causes and the determinants of the outcome. Determinants of causal variables lost some deserved attention as regression methods became dominant in observational social science in the 1980s and 1990s, but the matching literature associated with the counterfactual model has restored some of this attention by focusing analysis on the goal of balancing the data with respect to the determinants of the cause. For the subsequent transition to Mode 3a (targeted analysis of one or more focal causes), the joint consideration of both types of variables then allows for a determination, by carefully defined assumptions grounded in theory, of which sufficient subset of such determinants may be used in a conditioning

---

[10]In addition, all-cause structural analysis may drill down to model directly the separable causal effects of the constituent features that compose the causal states. In other words, the causal variables analyzed for Modes 1, 2, and 3a may come to be regarded as too broad to motivate sensible structural models. This decision may emerge following analysis within Mode 3b when the analyst comes to recognize that the relevant generative mechanisms apply to only a subset of the causal state's constitutive features.

strategy to achieve identification of the causal effect by eliminating all confounding. If the assumptions cannot be sustained for any set of observed variables, then point identification of the focal causal effects by back-door conditioning is impossible.

If no other research design is possible, such as those that exemplify Modes 3b and 4, then the counterfactual model suggests clearly why analysis should then expand beyond efforts to point-identify causal effects to alternative forms of set identification and sensitivity analysis. In addition, the counterfactual approach also encourages especially careful examination of all back-door paths from the causal variable to the outcome variable (rather than simply a one-by-one consideration of all variables that lie along any back-door paths). This form of systematic consideration of model identification can prevent researchers from mistakenly generating new confounding between the causal variables and the outcome variable, as occurs when the analyst conditions on a collider variable that lies along an already blocked back-door path.

For the transition to Mode 3b (mechanism-based analysis), the counterfactual approach shows that such a transition is substantive, not methodological. No new estimation techniques beyond those used for Modes 2 and 3a are required. All that is required to pursue Mode 3b is a theory that suggests which variables (and in which configuration) represent the causal mechanism that brings about the causal effect. Moreover, the counterfactual approach suggests that typical mechanism-based analyses must be clearly separated into two very different varieties: those that merely point to possible causal pathways and those that fully identify the causal effects of interest. For analysis of the former type, some explanation can be achieved. But, to realize the more ambitious goals of the latter, the mechanistic variables that are specified must be exhaustive and isolated (or made so by suitable conditioning). Mechanistic variables that are not independent of unblocked back-door paths after conditioning on observed variables cannot be used to identify a causal effect.

For the transition to Mode 4 (all-cause structural analysis), the counterfactual approach sensitizes researchers to the stringency of the required assumptions. Pearl (2009) is surely correct that our ambition should always be to get as close as possible to all-cause structural models where accounts can be offered for all of the causes of an outcome. Adopting the counterfactual approach helps to clarify the dubious nature of the maintained assumptions that pervade the published literature where all-cause models are offered. For example, most such models are based on combinations of two identification strategies – basic conditioning and instrumental variable techniques – under maintained assumptions that causal effects are homogeneous (or at least conditionally random). As the matching literature associated with the counterfactual approach has shown, such homogeneity is rarely justified empirically even for singular causes, and most parametric forms of conditioning average systematic causal effect heterogeneity in arcane ways. Moreover, the instrumented variable (IV) literature associated with the counterfactual approach has shown that, in the presence of such heterogeneity, IVs estimate marginal causal effects that cannot then be extrapolated to other segments of the population of interest without introducing unsupportable assumptions of homogeneity of effects.

By making these issues clear, the counterfactual approach shows how high the demands on theory and on data must be in order to sustain the all-cause mode of causal inquiry: Answers to all of the "who, when, where, and how" questions for the causal relationships of interest must be provided. Dodging these questions by introducing homogeneity assumptions that have only a dubious grounding in theory (or a solid grounding in dubious theory) can undermine causal claims, at least in the eyes of a fair critic. In this sense, the counterfactual approach encourages modesty of causal inquiry.

Finally, such modesty can be pursued outside of these modes of inquiry, after which analysis can then shift into whichever of these modes seems most appropriate. As shown perhaps most clearly when adopting Manski's set-identification perspective (see Chapter 12), the range that a causal effect may take on can be defined and then examined with the potential outcome model before any data are considered. Thereafter, an assessment can be undertaken of the data that are available to estimate average causal effects of various forms. For these steps, assumptions about why a causal relationship may exist need not be introduced, nor in fact assumptions that there is any systematic relationship between the processes that generate causal relationships at the individual level. Such flexibility leaves analysis wide open at the outset, focusing attention on clear definitions of effects of interest, independent of data availability issues.

But, once an analysis has begun, a theoretical position must be adopted in order to provide answers to at least some of the "who, when, where, and how" questions for the causal relationship of interest. Given provisional answers to these questions, the counterfactual approach helps one to consider which modes of causal inquiry are feasible. Analysis may have to stop at the associational level, yielding nothing other than analogs to the naive estimator. If some of the determinants of the cause and/or outcome are systematic and observed, then analysis can move down to conditional variants of associational analysis, followed by an analysis of bounds (and perhaps a sensitivity analysis). If a directed graph can be drawn that suggests that ignorability assumptions or exclusion restrictions are available for focal causal effects, then it may be feasible to generate consistent estimates of the effects of these specific causes. If theory and data are available to examine what brings about the effect of the cause, then analysis can move down toward a mechanism-based mode of causal inquiry. Finally, if theory is finely articulated and supported by past substantive scholarship, and if all data requirements are met, then full structural modeling can be attempted. Such all-cause models represent a standard that is rarely achieved but properly valued, for they provide complete explanations not only of the causes of an outcome but of every linkage in the causal chains between them. Counterfactual modeling guides us as close to this standard as is appropriate, given the constraints of theory and of data that prevail at any given time.

References

Abadie, Alberto and Guido W. Imbens. 2006. "Large Sample Properties of Matching Estimators for Average Treatment Effects." *Econometrica* 74:235–67.

—. 2008. "On the Failure of the Bootstrap for Matching Estimators." *Econometrica* 76:1537–57.

—. 2009. "Matching on the Estimated Propensity Score." National Bureau of Economic Research, Cambridge, Massachusetts.

—. 2011. "Bias-Corrected Matching Estimators for Average Treatment Effects." *Journal of Business and Economic Statistics* 29:1–11.

—. 2012. "A Martingale Representation for Matching Estimators." *Journal of the American Statistical Association* 107:833–43.

Abbott, Andrew D. 1988. "Transcending General Linear Reality." *Sociological Theory* 6:169–86.

—. 2001[1998]. *Time Matters: On Theory and Method.* Chicago: University of Chicago Press.

Abdulkadiroglu, Atila, Joshua D. Angrist, Susan M. Dynarski, Thomas J. Kane, and Parag Pathak. 2011. "Accountability and Flexibility in Public Schools: Evidence from Boston's Charters and Pilots." *Quarterly Journal of Economics* 126:699–748.

Adler, Nancy E. and Katherine Newman. 2002. "Socioeconomic Disparities in Health: Pathways and Policies." *Health Affairs* 21:60–76.

Agresti, Alan. 2002. *Categorical Data Analysis.* New York: Wiley.

Alexander, Jeffrey C. 2003. *The Meanings of Social Life: A Cultural Sociology.* New York: Oxford University Press.

Alexander, Karl L. and Aaron M. Pallas. 1983. "Private Schools and Public Policy: New Evidence on Cognitive Achievement in Public and Private Schools." *Sociology of Education* 56:170–82.

—. 1985. "School Sector and Cognitive Performance: When Is a Little a Little?" *Sociology of Education* 58:115–28.

Allison, Paul D. 1990. "Change Scores as Dependent Variables in Regression Analysis." *Sociological Methodology* 20:93–114.

—. 2009. *Fixed Effects Regression Models*. Los Angeles: Sage.

Althauser, Robert P. and Donald B. Rubin. 1970. "The Computerized Construction of a Matched Sample." *American Journal of Sociology* 76:325–46.

—. 1971. "Measurement Error and Regression to the Mean in Matched Samples." *Social Forces* 50:206–14.

Altonji, Joseph G., Todd E. Elder, and Christopher Taber. 2005a. "An Evaluation of Instrumental Variable Strategies for Estimating the Effects of Catholic Schooling." *Journal of Human Resources* 40:791–821.

—. 2005b. "Selection on Observed and Unobserved Variables: Assessing the Effectiveness of Catholic Schools." *Journal of Political Economy* 113:151–83.

Alwin, Duane F. and Robert M. Hauser. 1975. "The Decomposition of Effects in Path Analysis." *American Sociological Review* 40:37–47.

An, Weihua. 2010. "Bayesian Propensity Score Estimators: Incorporating Uncertainties in Propensity Scores into Causal Inference." *Sociological Methodology* 40: 151–89.

—. 2013. "Models and Methods to Identify Peer Effects." Pp. 514–32 in *The Sage Handbook of Social Network Analysis*, edited by J. Scott and P. J. Carrington. Thousand Oaks: Sage.

Angrist, Joshua D. 1990. "Lifetime Earnings and the Vietnam Era Draft Lottery: Evidence from Social Security Administrative Records." *American Economic Review* 80:313–36.

—. 1998. "Estimating the Labor Market Impact of Voluntary Military Service Using Social Security Data on Military Applicants." *Econometrica* 66:249–88.

Angrist, Joshua D. and Stacey H. Chen. 2011. "Schooling and the Vietnam-Era GI Bill: Evidence from the Draft Lottery." *American Economic Journal: Applied Economics* 3:96–119.

Angrist, Joshua D., Susan M. Dynarski, Thomas J. Kane, Parag A. Pathak, and Christopher R. Walters. 2010. "Inputs and Impacts in Charter Schools: KIPP Lynn." *American Economic Review: Papers & Proceedings* 100:1–5.

Angrist, Joshua D. and Ivan Fernandez-Val. 2013. "ExtrapoLATE-ing: External Validity and Overidentification in the LATE Framework." Pp. 401–36 in *Advances in Economics and Econometrics*, vol. 3, edited by D. Acemoglu, M. Arellano, and D. Eddie. Cambridge: Cambridge University Press.

Angrist, Joshua D. and Guido W. Imbens. 1995. "Two-Stage Least Squares Estimation of Average Causal Effects in Models with Variable Treatment Intensity." *Journal of the American Statistical Association* 90:431–42.

Angrist, Joshua D., Guido W. Imbens, and Donald B. Rubin. 1996. "Identification of Causal Effects Using Instrumental Variables." *Journal of the American Statistical Association* 87:328–36.

Angrist, Joshua D. and Alan B. Krueger. 1991. "Does Compulsory School Attendance Affect Schooling and Earnings?" *Quarterly Journal of Economics* 106:979–1014.

—. 1992. "The Effect of Age at School Entry on Educational Attainment: An Application of Instrumental Variables with Moments from Two Samples." *Journal of the American Statistical Association* 87:328–36.

—. 1994. "Why Do World War II Veterans Earn More Than Nonveterans?" *Journal of Labor Economics* 12:74–97.

—. 1999. "Empirical Strategies in Labor Economics." Pp. 1277–366 in *Handbook of Labor Economics*, vol. 3, edited by O. C. Ashenfelter and D. Card. Amsterdam: Elsevier.

—. 2001. "Instrumental Variables and the Search for Identification: From Supply and Demand to Natural Experiments." *Journal of Economic Perspectives* 15:65–83.

Angrist, Joshua D. and Victor Lavy. 1999. "Using Maimonides' Rule to Estimate the Effect of Class Size on Scholastic Achievement." *Quarterly Journal of Economics* 114:533–75.

Angrist, Joshua D. and Jörn-Steffen Pischke. 2009. *Mostly Harmless Econometrics: An Empiricist's Companion.* Princeton: Princeton University Press.

—. 2010. "The Credibility Revolution in Empirical Economics: How Better Research Design Is Taking the Con Out of Econometrics." *The Journal of Economic Perspectives* 24:3–30.

Arminger, Gerhard, Clifford C. Clogg, and Michael E. Sobel, Eds. 1995. *Handbook of Statistical Modeling for the Social and Behavioral Sciences.* New York: Plenum.

Armor, David J. 1995. *Forced Justice: School Desegregation and the Law.* New York: Oxford University Press.

Aronow, Peter M. and Allison Carnegie. 2013. "Beyond LATE: Estimation of the Average Treatment Effect with an Instrumental Variable." *Political Analysis* 21:492–506.

Ashenfelter, Orley C. 1978. "Estimating the Effect of Training Programs on Earnings." *Review of Economics and Statistics* 60:47–57.

Ashenfelter, Orley C. and David Card. 1985. "Using the Longitudinal Structure of Earnings to Estimate the Effect of Training Programs." *Review of Economics and Statistics* 67:648–60.

Austin, Peter C. 2009a. "Balance Diagnostics for Comparing the Distribution of Baseline Covariates between Treatment Groups in Propensity-Score Matched Samples." *Statistics in Medicine* 28:3083–107.

—. 2009b. "Some Methods of Propensity-Score Matching Had Superior Performance to Others: Results of an Empirical Investigation and Monte Carlo Simulations." *Biometrical Journal* 51:171–84.

Baltagi, Badi H. 2005. *Econometric Analysis of Panel Data.* Chichester: J. Wiley & Sons.

Bang, Heejung and James M. Robins. 2005. "Doubly Robust Estimation in Missing Data and Causal Inference Models." *Biometrics* 61:962–72.

Barnow, Burt S., Glen G. Cain, and Arthur S. Goldberger. 1980. "Issues in the Analysis of Selectivity Bias." *Evaluation Studies Review Annual* 5:43–59.

Barringer, Sondra N., Erin Leahey, and Scott R. Eliason. 2013. "A History of Causal Analysis in the Social Sciences." Pp. 9–26 in *Handbook of Causal Analysis for Social Research*, edited by S. L. Morgan. Dordrecht: Springer.

Becker, Gary S. 1993[1964]. *Human Capital: A Theoretical and Empirical Analysis with Special Reference to Education.* Chicago: University of Chicago Press.

Becker, Sascha O. and Andrea Ichino. 2005. "Pscore: Stata Programs for ATT Estimation Based on Propensity Score Matching." Software Routine for Stata, Department of Economics, University of Warwick, Coventry, United Kingdom.

Behrens, Angela, Christopher Uggen, and Jeff Manza. 2003. "Ballot Manipulation and the 'Menace of Negro Domination': Racial Threat and Felon Disenfranchisement in the United States, 1850–2002." *American Journal of Sociology* 109:559–605.

Berger, Mark C. and Barry T. Hirsch. 1983. "The Civilian Earnings Experience of Vietnam-Era Veterans." *Journal of Human Resources* 18:455–79.

Berinsky, Adam J. and Gabriel S. Lenz. 2011. "Education and Political Participation: Exploring the Causal Link." *Political Behavior* 33:357–73.

Berk, Richard, Geoffrey Barnes, Lindsay Ahlman, and Ellen Kurtz. 2010. "When Second Best Is Good Enough: A Comparison between a True Experiment and a Regression Discontinuity Quasi-Experiment." *Journal of Experimental Criminology* 6:191–208.

Berk, Richard A. 1988. "Causal Inference for Sociological Data." Pp. 155–72 in *Handbook of Sociology*, edited by N. J. Smelser. Newbury Park: Sage.

—. 2004. *Regression Analysis: A Constructive Critique*. Thousand Oaks: Sage.

—. 2006. "An Introduction to Ensemble Methods for Data Analysis." *Sociological Methods and Research* 34:263–95.

—. 2008. *Statistical Learning from a Regression Perspective*. New York: Springer.

Berk, Richard A. and Jan de Leeuw. 1999. "An Evaluation of California's Inmate Classification System Using a Generalized Discontinuity Design." *Journal of the American Statistical Association* 94:1045–52.

Berk, Richard A. and Phyllis J. Newton. 1985. "Does Arrest Really Deter Wife Battery? An Effort to Replicate the Findings of the Minneapolis Spouse Abuse Experiment." *American Sociological Review* 50:253–62.

Berk, Richard A., Phyllis J. Newton, and Sarah Fenstermaker Berk. 1986. "What a Difference a Day Makes: An Empirical Study of the Impact of Shelters for Battered Women." *Journal of Marriage and the Family* 48: 481–90.

Berk, Richard A. and David Rauma. 1983. "Capitalizing on Nonrandom Assignment to Treatments: A Regression-Discontinuity Evaluation of a Crime-Control Program." *Journal of the American Statistical Association* 78:21–27.

Berzuini, Carlo, Philip Dawid, and Luisa Bernardinelli, Eds. 2012. *Causality: Statistical Perspectives and Applications*. Hoboken: Wiley.

Bhaskar, Roy. 1998[1997]. "Philosophy and Scientific Realism." Pp. 16–47 in *Critical Realism: Essential Readings*, edited by M. S. Archer, R. Bhaskar, A. Collier, T. Lawson, and A. Norrie. London: Routledge.

Black, Sandra E. 1999. "Do Better Schools Matter? Parental Valuation of Elementary Education." *Quarterly Journal of Economics* 114:577–99.

Blalock, Hubert M. 1964[1961]. *Causal Inferences in Nonexperimental Research*. Chapel Hill: University of North Carolina Press.

Blau, Peter M. and Otis Dudley Duncan. 1967. *The American Occupational Structure*. New York: Wiley.

Bloom, Howard S. 2012. "Modern Regression Discontinuity Analysis." *Journal of Research on Educational Effectiveness* 5:43–82.

Blossfeld, Hans-Peter. 2009. "Causation as a Generative Process: The Elaboration of an Idea for the Social Sciences and an Application to an Analysis of an Interdependent Social System." Pp. 83–109 in *Causal Analysis in Population Studies: Concepts, Methods, Applications*, edited by H. Engelhardt, H.-P. Kohler, and A. Prskawetz. Dordrecht: Springer.

Blundell, Richard, Amanda Gosling, Hidehiko Ichimura, and Costas Meghir. 2007. "Changes in the Distribution of Male and Female Wages Accounting for Employment Composition Using Bounds." *Econometrica* 75:323–63.

Bollen, Kenneth A. 1989. *Structural Equations with Latent Variables*. New York: Wiley.

—. 1995. "Structural Equation Models That Are Nonlinear in Latent Variables: A Least-Squares Estimator." *Sociological Methodology* 25:223–51.

—. 1996a. "An Alternative Two-Stage Least Squares (2SLS) Estimator for Latent Variable Equations." *Psychometrika* 61:109–21.

—. 1996b. "A Limited-Information Estimator for LISREL Models with or Without Heteroscedastic Errors." Pp. 227–41 in *Advanced Structural Equation Modeling: Issues and Techniques*, edited by G. A. Marcoulides and R. E. Schumacker. Mahwah: Lawrence Erlbaum.

—. 2001. "Two-Stage Least Squares and Latent Variable Models: Simultaneous Estimation and Robustness to Misspecifications." Pp. 119–38 in *Structural Equation Modeling: Present and Future – a Festschrift in Honor of Karl Jöreskog*, edited by R. Cudeck, S. du Toit, and D. Sörbom. Lincolnwood: Scientific Software International.

—. 2012. "Instrumental Variables in Sociology and the Social Sciences." *Annual Review of Sociology* 38:37–72.

Bollen, Kenneth A. and Judea Pearl. 2013. "Eight Myths About Structural Equation Models." Pp. 301–28 in *Handbook of Causal Analysis for Social Research*, edited by S. L. Morgan. Dordrecht: Springer.

Boudon, Raymond. 1998. "Social Mechanisms Without Black Boxes." Pp. 172–203 in *Social Mechanisms: An Analytical Approach to Social Theory*, edited by P. Hedström and R. Swedberg. Cambridge: Cambridge University Press.

Bound, John, David A. Jaeger, and Regina M. Baker. 1995. "Problems with Instrumental Variables Estimation When the Correlation between the Instruments and the Endogenous Explanatory Variable Is Weak." *Journal of the American Statistical Association* 90:443–50.

Bourdieu, Pierre. 1973. "Cultural Reproduction and Social Reproduction." Pp. 71–112 in *Knowledge, Education, and Cultural Change: Papers in the Sociology of Education*, edited by R. K. Brown. London: Tavistock.

Bowden, Roger J. and Darrell A. Turkington. 1984. *Instrumental Variables*. Cambridge: Cambridge University Press.

Braakmann, Nils. 2011. "The Causal Relationship between Education, Health and Health Related Behaviour: Evidence from a Natural Experiment in England." *Journal of Health Economics* 30:753–63.

Brady, Henry E. and David Collier, Eds. 2010. *Rethinking Social Inquiry: Diverse Tools, Shared Standards*. Lanham: Rowman and Littlefield.

Braga, Anthony A., David Hureau, and Christopher Winship. 2008. "Losing Faith? Police, Black Churches, and the Resurgence of Youth Violence in Boston." *Ohio State Journal of Criminal Law* 6:141–72.

Braga, Anthony A., David M. Kennedy, Elin J. Waring, and Anne Morrison Piehl. 2001. "Problem-Oriented Policing, Deterrence, and Youth Violence: An Evaluation of Boston's Operation Ceasefire." *Journal of Research in Crime and Delinquency* 38:195–225.

Brand, Jennie E. and Dwight Davis. 2011. "The Impact of College Education on Fertility: Evidence for Heterogeneous Effects." *Demography* 48:863–87.

Brand, Jennie E. and Juli Simon Thomas. 2013. "Causal Effect Heterogeneity." Pp. 189–213 in *Handbook of Causal Analysis for Social Research*, edited by S. L. Morgan. Dordrecht: Springer.

Brand, Jennie E. and Yu Xie. 2010. "Who Benefits Most from College? Evidence for Negative Selection in Heterogeneous Economic Returns to Higher Education." *American Sociological Review* 75:273–302.

Breen, Richard and Jan O. Jonsson. 2005. "Inequality of Opportunity in Comparative Perspective: Recent Research on Educational Attainment and Social Mobility." *Annual Review of Sociology* 31:223–43.

Brewster, Karin L. and Ronald R. Rindfuss. 2000. "Fertility and Women's Employment in Industrialized Nations." *Annual Review of Sociology* 26:271–96.

Browning, Harley L., Sally C. Lopreato, and Dudley L. Poston Jr. 1973. "Income and Veteran Status: Variations among Mexican Americans, Blacks and Anglos." *American Sociological Review* 38:74–85.

Bryk, Anthony S., Valerie E. Lee, and Peter B. Holland. 1993. *Catholic Schools and the Common Good*. Cambridge: Harvard University Press.

Bumpass, Larry L., Ronald R. Rindfuss, and Richard B. Janosik. 1978. "Age and Marital Status at First Birth and the Pace of Subsequent Fertility." *Demography* 15:75–86.

Bunge, Mario. 2004. "How Does It Work? The Search for Explanatory Mechanisms." *Philosophy of Social Science* 34:182–210.

Bush, Robert R. and Frederick Mosteller. 1955. *Stochastic Models for Learning*. New York: Wiley.

—. 1959. "A Comparison of Eight Models." Pp. 335–49 in *Studies in Mathematical Learning Theory*, edited by R. R. Bush and W. K. Estes. Stanford: Stanford University Press.

Campbell, Donald T. 1957. "Factors Relevant to the Validity of Experiments in Social Settings." *Psychological Bulletin* 54:297–312.

—. 1988[1977]. *Methodology and Epistemology for Social Science: Selected Papers.* Chicago: University of Chicago Press.

Campbell, Donald T. and Julian C. Stanley. 1966[1963]. *Experimental and Quasi-Experimental Designs for Research.* Chicago: Rand McNally.

Carbonaro, William and Elizabeth Covay. 2010. "School Sector and Student Achievement in the Era of Standards Based Reforms." *Sociology of Education* 83: 160–82.

Card, David. 1999. "The Causal Effect of Education on Earnings." Pp. 1801–63 in *Handbook of Labor Economics*, vol. 3A, edited by O. C. Ashenfelter and D. Card. Amsterdam: Elsevier.

Card, David and Enrico Moretti. 2007. "Does Voting Technology Affect Election Outcomes? Touch-Screen Voting and the 2004 Presidential Election." *Review of Economics and Statistics* 89:660–73.

Carneiro, Pedro, James J. Heckman, and Edward J. Vytlacil. 2010. "Evaluating Marginal Policy Changes and the Average Effect of Treatment for Individuals at the Margin." *Econometrica* 78:377–94.

—. 2011. "Estimating Marginal Returns to Education." *American Economic Review* 101:2754–81.

Cartwright, Nancy S. 2007a. "Counterfactuals in Economics: A Commentary." Pp. 191–216 in *Causation and Explanation*, edited by J. K. Campbell, M. O'Rourke, and H. Silverstein. Cambridge: MIT Press.

—. 2007b. *Hunting Causes and Using Them: Approaches in Philosophy and Economics.* Cambridge: Cambridge University Press.

Casella, George and Stephen P. Schwartz. 2000. "Comment on 'Causal Inference without Counterfactuals' by A. P. Dawid." *Journal of the American Statistical Association* 95:425–27.

Caughey, Devin and Jasjeet S. Sekhon. 2011. "Elections and the Regression Discontinuity Design: Lessons from Close U.S. House Races, 1942–2008." *Political Analysis* 19:385–408.

Cawley, John and Christopher J. Ruhm. 2012. "The Economics of Risky Health Behaviors." Pp. 95–199 in *Handbook of Health Economics*, vol. 2, edited by M. V. Pauly, T. G. Mcguire, and P. P. Barros. Amsterdam: Elsevier.

Cawley, John H., Ed. 2011. *The Oxford Handbook of the Social Science of Obesity.* New York: Oxford University Press.

Cawley, John, John Moran, and Kosali Simon. 2010. "The Impact of Income on the Weight of Elderly Americans." *Health Economics* 19:979–93.

Ceci, Stephen J. 1991. "How Much Does Schooling Influence General Intelligence and Its Cognitive Components? A Reassessment of the Evidence." *Developmental Psychology* 27:703–22.

Center for Research on Education Outcomes. 2009. "Multiple Choice: Charter School Performance in 16 States." Stanford University, Stanford, California.

Chakraborty, Bibhas and Erica E. M. Moodie. 2013. *Statistical Methods for Dynamic Treatment Regimes: Reinforcement Learning, Causal Inference, and Personalized Medicine.* New York: Springer.

Chapin, F. Stuart. 1932. "The Advantages of Experimental Sociology in the Study of Family Group Patterns." *Social Forces* 11:200–207.

—. 1947. *Experimental Designs in Sociological Research.* New York: Harper.

Christakis, Nicholas A. and James H. Fowler. 2007. "The Spread of Obesity in a Large Social Network over 32 Years." *New England Journal of Medicine* 357:370–79.

—. 2013. "Social Contagion Theory: Examining Dynamic Social Networks and Human Behavior." *Statistics in Medicine* 32:556–77.

Chubb, John E. and Terry M. Moe. 1990. *Politics, Markets, and America's Schools.* Washington, DC: Brookings Institution.

Clotfelter, Charles T. 2004. *After Brown: The Rise and Retreat of School Desegregation.* Princeton: Princeton University Press.

Cochran, William G. 1968. "The Effectiveness of Adjustment by Subclassification in Removing Bias in Observational Studies." *Biometrics* 24:295–313.

Cochran, William G. and Gertrude M. Cox. 1950. *Experimental Designs.* New York: Wiley.

Cochran, William G. and Donald B. Rubin. 1973. "Controlling Bias in Observational Studies: A Review." *Sankhya* 35:417–46.

Cohen-Cole, Ethan and Jason M. Fletcher. 2008. "Detecting Implausible Social Network Effects in Acne, Height, and Headaches: Longitudinal Analysis." *British Medical Journal* 337:a2533.

Cohen-Zada, Danny and Todd Elder. 2009. "Historical Religious Concentrations and the Effects of Catholic Schooling." *Journal of Urban Economics* 66: 65–74.

Coleman, James S. 1961. *The Adolescent Society: The Social Life of the Teenager and Its Impact on Education.* New York: Free Press.

—. 1964. *Introduction to Mathematical Sociology.* New York: Free Press.

—. 1981. *Longitudinal Data Analysis*. New York: Basic Books.

—. 1990. *Foundations of Social Theory*. Cambridge: Harvard University Press.

Coleman, James S. and Thomas Hoffer. 1987. *Public and Private Schools: The Impact of Communities*. New York: Basic Books.

Coleman, James S., Thomas Hoffer, and Sally Kilgore. 1982. *High School Achievement: Public, Catholic, and Private Schools Compared*. New York: Basic Books.

Collier, Andrew. 2005. "Philosophy and Critical Realism." Pp. 327–45 in *The Politics of Method in the Human Sciences: Positivism and Its Epistemological Others*, edited by G. Steinmetz. Durham: Duke University Press.

Collins, John, Ned Hall, and L. A. Paul, Eds. 2004. *Causation and Counterfactuals*. Cambridge: MIT Press.

Collins, Randall. 2004. *Interaction Ritual Chains*. Princeton: Princeton University Press.

Cook, Michael D. and William N. Evans. 2000. "Families or Schools? Explaining the Convergence in White and Black Academic Performance." *Journal of Labor Economics* 18:729–54.

Cook, Thomas D. and Donald T. Campbell. 1979. *Quasi-Experimentation: Design & Analysis Issues for Field Settings*. Chicago: Rand McNally.

Cook, Thomas D., Peter M. Steiner, and Steffi Pohl. 2009. "How Bias Reduction Is Affected by Covariate Choice, Unreliability, and Mode of Data Analysis: Results from Two Types of Within-Study Comparisons." *Multivariate Behavioral Research* 44:828–47.

Cox, David R. 1958. *Planning of Experiments*. New York: Wiley.

—. 1992. *Planning of Experiments*. New York: Wiley.

Cox, D. R. 2000. "Comment on 'Causal Inference Without Counterfactuals' by A. P. Dawid." *Journal of the American Statistical Association* 95:424–25.

Cox, David R. and Nancy Reid. 2000. *The Theory of the Design of Experiments*. Boca Raton: Chapman & Hall/CRC.

Cox, David R. and Nanny Wermuth. 1996. *Multivariate Dependencies: Models, Analysis and Interpretation*. New York: Chapman Hall.

—. 2001. "Some Statistical Aspects of Causality." *European Sociological Review* 17: 65–74.

Crain, Robert L. and Rita E. Mahard. 1983. "The Effect of Research Methodology on Desegregation-Achievement Studies: A Meta-Analysis." *American Journal of Sociology* 88:839–54.

Craver, Carl F. 2007. *Explaining the Brain: Mechanisms and the Mosaic Unity of Neuroscience.* Oxford: Oxford University Press.

Crump, Richard K., V. Joseph Hotz, Guido W. Imbens, and Oscar A. Mitnik. 2009. "Dealing with Limited Overlap in Estimation of Average Treatment Effects." *Biometrika* 96:187–99.

Cunha, Flavio and James J. Heckman. 2007. "Identifying and Estimating the Distributions of Ex Post and Ex Ante Returns to Schooling." *Labour Economics* 14: 870–93.

Cutright, Phillips. 1974. "The Civilian Earnings of White and Black Draftees and Nonveterans." *American Sociological Review* 39:317–27.

Davenport, Tiffany C., Alan S. Gerber, Donald P. Green, Christopher W. Larimer, Christopher B. Mann, and Costas Panagopoulos. 2010. "The Enduring Effects of Social Pressure: Tracking Campaign Experiments over a Series of Elections." *Political Behavior* 32:423–30.

Davis, Kingsley. 1945. "The World Demographic Transition." *Annals of the American Academy of Political and Social Science* 237:1–11.

Dawid, A. Philip. 2000. "Causal Inference Without Counterfactuals." *Journal of the American Statistical Association* 95:407–24.

—. 2002. "Influence Diagrams for Causal Modelling and Inference." *International Statistical Review* 70:161–89.

—. 2010. "Beware of the DAG!" *Journal of Machine Learning Research* 6:59–86.

—. 2012. "The Decision-Theoretic Approach to Causal Inference." Pp. 25–42 in *Causality: Statistical Perspectives and Applications*, edited by C. Berzuini, P. Dawid, and L. Bernardinelli. Hoboken: Wiley.

de Luna, Xavier, Ingeborg Waernbaum, and Thomas S. Richardson. 2011. "Covariate Selection for the Nonparametric Estimation of an Average Treatment Effect." *Biometrika* 98:861–75.

De Tray, Dennis. 1982. "Veteran Status as a Screening Device." *American Economic Review* 72:133–42.

Deaton, Angus. 2010. "Instruments, Randomization, and Learning about Development." *Journal of Economic Literature* 48:424–55.

Dechter, Rina, Hector Geffner, and Joseph Y. Halpern, Eds. 2010. *Heuristics, Probability, and Causality: A Tribute to Judea Pearl.* London: College Publications.

Dehejia, Rajeev H. and Sadek Wahba. 1999. "Causal Effects in Nonexperimental Studies: Reevaluating the Evaluation of Training Programs." *Journal of the American Statistical Association* 94:1053–62.

Devlin, Bernie, Stephen E. Fienberg, Daniel P. Resnick, and Kathryn Roeder, Eds. 1997. *Intelligence, Genes, and Success: Scientists Respond to the Bell Curve.* New York: Springer.

Diamond, Alexis and Jasjeet S. Sekhon. 2013. "Genetic Matching for Estimating Causal Effects: A General Multivariate Matching Method for Achieving Balance in Observational Studies." *Review of Economics and Statistics* 95:932–45.

Dinkel, Robert M. 1952. "Occupation and Fertility in the United States." *American Sociological Review* 17:178–83.

DiPrete, Thomas A. and Markus Gangl. 2004. "Assessing Bias in the Estimation of Causal Effects: Rosenbaum Bounds on Matching Estimators and Instrumental Variables Estimation with Imperfect Instruments." *Sociological Methodology* 34:271–310.

Draper, Norman R. and Harry Smith. 1998. *Applied Regression Analysis.* New York: Wiley.

Druckman, James N., James H. Kuklinski, and Arthur Lupia, Eds. 2011. *Cambridge Handbook of Experimental Political Science.* Cambridge: Cambridge University Press.

DuMouchel, William H. and Greg J. Duncan. 1983. "Using Sample Survey Weights in Multiple Regression Analyses of Stratified Samples." *Journal of the American Statistical Association* 78:535–43.

Duncan, Greg J. 2008. "When to Promote, and When to Avoid, a Population Perspective." *Demography* 45:763–84.

Duncan, Otis Dudley. 1966. "Path Analysis: Sociological Examples." *American Journal of Sociology* 72:1–16.

—. 1975. *Introduction to Structural Equation Models.* New York: Academic Press.

—. 1984. *Notes on Social Measurement: Historical and Critical.* New York: Russell Sage Foundation.

Duncan, Otis Dudley, David L. Featherman, and Beverly Duncan. 1972. *Socioeconomic Background and Achievement.* New York: Seminar Press.

Duncan, Otis Dudley, Archibald O. Haller, and Alejandro Portes. 1968. "Peer Influences on Aspirations: A Reinterpretation." *American Journal of Sociology* 74:119–37.

Dunning, Thad. 2012. *Natural Experiments in the Social Sciences: A Design-Based Approach.* Cambridge: Cambridge University Press.

Ehrenberg, Ronald G. 2004. "Econometric Studies of Higher Education." *Journal of Econometrics* 121:19–37

Elster, Jon. 1989. *Nuts and Bolts for the Social Sciences.* Cambridge: Cambridge University Press.

—. 1998. "A Plea for Mechanisms." Pp. 45–73 in *Social Mechanisms: An Analytical Approach to Social Theory*, edited by P. Hedström and R. Swedberg. Cambridge: Cambridge University Press.

Elwert, Felix. 2013. "Graphical Causal Models." Pp. 245–73 in *Handbook of Causal Analysis for Social Research*, edited by S. L. Morgan. Dordrecht: Springer.

Elwert, Felix and Christopher Winship. 2010. "Effect Heterogeneity and Bias in Main-Effects-Only Regression Models." Pp. 327–36 in *Heuristics, Probability and Causality: A Tribute to Judea Pearl*, edited by R. Dechter, H. Geffner, and J. Y. Halpern. London: College Publications.

—. 2014. "Endogenous Selection Bias: The Problem of Conditioning on a Collider Variable." *Annual Review of Sociology* 40:31–53.

Engelhardt, Henriette, Hans-Peter Kohler, and Alexia Prskawetz, Eds. 2009. *Causal Analysis in Population Studies: Concepts, Methods, Applications*. Dordrecht: Springer.

Evans, William and Robert M. Schwab. 1995. "Finishing High School and Starting College: Do Catholic Schools Make a Difference?" *Quarterly Journal of Economics* 110:41–74.

Firebaugh, Glenn. 2008. *Seven Rules for Social Research*. Princeton: Princeton University Press.

Fiscella, Kevin, Peter Franks, Marthe R. Gold, and Carolyn M. Clancy. 2000. "Inequality in Quality: Addressing Socioeconomic, Racial, and Ethnic Disparities in Health Care." *Journal of the American Medical Association* 283:2579–84.

Fisher, Ronald A. 1935. *The Design of Experiments*. Edinburgh: Oliver and Boyde.

Fligstein, Neil and Doug McAdam. 2012. *A Theory of Fields*. New York: Oxford University Press.

Fox, John. 1984. *Linear Statistical Models and Related Methods: With Applications to Social Research*. New York: Wiley.

—. 2008. *Applied Regression Analysis and Generalized Linear Models*. Los Angeles: Sage.

Frangakis, Constantine E. and Donald B. Rubin. 2002. "Principal Stratification in Causal Inference." *Biometrics* 58:21–29.

Frank, Kenneth A. 2000. "Impact of a Confounding Variable on the Inference of a Regression Coefficient." *Sociological Methods and Research* 29:147–94.

Freedman, David A. 1983. "A Note on Screening Regression Equations." *American Statistician* 37:152–55.

—. 2005. *Statistical Models: Theory and Practice*. Cambridge: Cambridge University Press.

—. 2010. *Statistical Models and Causal Inference: A Dialogue with the Social Sciences.* New York: Cambridge University Press.

Freedman, David A. and Richard A. Berk. 2008. "Weighting Regressions by Propensity Scores." *Evaluation Review* 32:392–409.

Freedman, Ronald and Amos H. Hawley. 1949. "Unemployment and Migration in the Depression." *Journal of the American Statistical Association* 44:260–72.

Freese, Jeremy and J. Alex Kevern. 2013. "Types of Causes." Pp. 27–41 in *Handbook of Causal Analysis for Social Research*, edited by S. L. Morgan. Dordrecht: Springer.

Frumento, Paolo, Fabrizia Mealli, Barbara Pacini, and Donald B. Rubin. 2012. "Evaluating the Effect of Training on Wages in the Presence of Noncompliance, Nonemployment, and Missing Outcome Data." *Journal of the American Statistical Association* 107:450–66.

Fu, Vincent Kang, Christopher Winship, and Robert D. Mare. 2004. "Sample Selection Bias Models." Pp. 409–30 in *Handbook of Data Analysis*, edited by M. A. Hardy and A. Bryman. Thousand Oaks: Sage.

Fuller, Bruce and Richard F. Elmore. 1996. *Who Chooses? Who Loses? Culture, Institutions, and the Unequal Effects of School Choice.* New York: Teachers College Press.

Gangl, Markus. 2010. "Causal Inference in Sociological Research." *Annual Review of Sociology* 36:21–47.

—. 2013. "Partial Identification and Sensitivity Analysis." Pp. 377–402 in *Handbook of Causal Analysis for Social Research*, edited by S. L. Morgan. Dordrecht: Springer.

Garen, John. 1984. "The Returns to Schooling: A Selectivity Bias Approach with a Continuous Choice Variable." *Econometrica* 52:1199–218.

Garfinkel, Irwin, Charles F. Manski, and C. Michalopoulos. 1992. "Micro Experiments and Macro Effects." Pp. 253–73 in *Evaluating Welfare and Training Programs*, edited by C. F. Manski and I. Garfinkel. Cambridge: Harvard University Press.

Gelman, Andrew and Jennifer Hill. 2007. *Applied Regression and Multilevel/ Hierarchical Models.* Cambridge: Cambridge University Press.

Gelman, Andrew and Gary King. 1990. "Estimating Incumbency Advantage Without Bias." *American Journal of Political Science* 34:1142–64.

Gelman, Andrew and Xiao-Li Meng, Eds. 2004. *Applied Bayesian Modeling and Causal Inference from Incomplete-Data Perspectives: An Essential Journey with Donald Rubin's Statistical Family.* New York: Wiley.

Gennetian, Lisa A., Lisa Sanbonmatsu, Lawrence F. Katz, Jeffrey R. Kling, Matthew Sciandra, Jens Ludwig et al. 2012. "The Long-Term Effects of Moving to Opportunity on Youth Outcomes." *Cityscape* 14:137–67.

Gerber, Alan S. and Donald P. Green. 2012. *Field Experiments: Design, Analysis, and Interpretation.* New York: W. W. Norton.

Gerber, Alan S., Donald P. Green, and Christopher W. Larimer. 2008. "Social Pressure and Voter Turnout: Evidence from a Large-Scale Field Experiment." *American Political Science Review* 102:33–48.

Gerring, John. 2007. *Case Study Research: Principles and Practices.* New York: Cambridge University Press.

Glymour, Clark. 1986. "Statistics and Metaphysics: Comment on Holland's 'Statistics and Causal Inference.'" *Journal of the American Statistical Association* 81:964–66.

Glymour, Clark, Richard Scheines, and Peter Spirtes. 2001. *Causation, Prediction, and Search,* 2nd ed. Cambridge: MIT Press.

Godfrey-Smith, Peter. 2003. *Theory and Reality: An Introduction to the Philosophy of Science.* Chicago: University of Chicago Press.

Goertz, Gary. 2006. *Social Science Concepts: A User's Guide.* Princeton: Princeton University Press.

Goertz, Gary and James Mahoney. 2012. *A Tale of Two Cultures: Qualitative and Quantitative Research in the Social Sciences.* Princeton: Princeton University Press.

Goldberger, Arthur S. 1972. "Structural Equation Methods in the Social Sciences." *Econometrica* 40:979–1001.

—. 1991. *A Course in Econometrics.* Cambridge: Harvard University Press.

Goldberger, Arthur S. and Glen G. Cain. 1982. "The Causal Analysis of Cognitive Outcomes in the Coleman, Hoffer, and Kilgore Report." *Sociology of Education* 55:103–22.

Goldin, Claudia and Lawrence F. Katz. 2002. "The Power of the Pill: Oral Contraceptives and Women's Career and Marriage Decisions." *Journal of Political Economy* 110:730–70.

Goldthorpe, John H. 2001. "Causation, Statistics, and Sociology." *European Sociological Review* 17:1–20.

—. 2007. *On Sociology.* 2 vols. Stanford: Stanford University Press.

Gorski, Philip S. 2004. "The Poverty of Deductivism: A Constructive Realist Model of Sociological Explanation." *Sociological Methodology* 34:1–33.

Greene, William H. 2000. *Econometric Analysis.* Upper Saddle River: Prentice-Hall.

Greenwood, Ernest. 1945. *Experimental Sociology: A Study in Method*. New York: King's Crown Press.

Greiner, D. James and Donald B. Rubin. 2011. "Causal Effects of Perceived Immutable Characteristics." *Review of Economics and Statistics* 93:775–85.

Gundersen, Craig, Brent Kreider, and John Pepper. 2012. "The Impact of the National School Lunch Program on Child Health: A Nonparametric Bounds Analysis." *Journal of Econometrics* 166:79–91.

Guo, Shenyang and Mark W. Fraser. 2010. *Propensity Score Analysis: Statistical Methods and Applications*. Thousand Oaks: Sage.

Haavelmo, Trygve. 1943. "The Statistical Implications of a System of Simultaneous Equations." *Econometrica* 11:1–12.

Hahn, Jinyong. 1998. "On the Role of the Propensity Score in Efficient Semiparametric Estimation of Average Treatment Effects." *Econometrica* 66:315–31.

Hahn, Jinyong and Jerry Hausman. 2003. "Weak Instruments: Diagnosis and Cures in Empirical Economics." *American Economic Review* 93:118–25.

Hahn, Jinyong, Petra Todd, and Wilbert Van der Klaauw. 2001. "Identification and Estimation of Treatment Effects with a Regression-Discontinuity Design." *Econometrica* 69:201–9.

Hainmueller, Jens. 2012. "Entropy Balancing for Causal Effects: A Multivariate Reweighting Method to Produce Balanced Samples in Observational Studies." *Political Analysis* 20:25–46.

Halaby, Charles N. 2004. "Panel Models in Sociological Research: Theory into Practice." *Annual Review of Sociology* 30:507–44.

Hallinan, Maureen T. and Warren N. Kubitschek. 2012. "A Comparison of Academic Achievement and Adherence to the Common School Ideal in Public and Catholic Schools." *Sociology of Education* 85:1–22.

Halloran, M. Elizabeth, Ira M. Longini, and Claudio J. Struchiner. 2010. *Design and Analysis of Vaccine Studies*. New York: Springer.

Ham, John C., Xianghong Li, and Patricia B. Reagan. 2011. "Matching and Semi-Parametric IV Estimation, a Distance-Based Measure of Migration, and the Wages of Young Men." *Journal of Econometrics* 161:208–27.

Hamilton, James D. 1994. *Time Series Analysis*. Princeton: Princeton University Press.

Hanmer, Michael J., Won-Ho Park, Michael W. Traugott, Richard G. Niemi, Paul S. Herrnson, Benjamin B. Bederson et al. 2010. "Losing Fewer Votes: The Impact of

Changing Voting Systems on Residual Votes." *Political Research Quarterly* 63:129–42.

Hansen, Ben B. 2004. "Full Matching in an Observational Study of Coaching for the SAT." *Journal of the American Statistical Association* 99:609–18.

—. 2008. "The Prognostic Analogue of the Propensity Score." *Biometrika* 95:481–88.

Hansen, Ben B. and Jake Bowers. 2008. "Covariate Balance in Simple, Stratified and Clustered Comparative Studies." *Statistical Science* 23:219–36.

Hansen, Ben B., Mark Fredrickson, and Josh Buckner. 2013. "Optmatch: Functions for Optimal Matching." Software Routine for R, Department of Statistics, University of Michigan, Ann Arbor, Michigan.

Harder, Valerie S., Elizabeth A. Stuart, and James C. Anthony. 2010. "Propensity Score Techniques and the Assessment of Measured Covariate Balance to Test Causal Associations in Psychological Research." *Psychological Methods* 15:234–49.

Harding, David J. 2003. "Counterfactual Models of Neighborhood Effects: The Effect of Neighborhood Poverty on Dropping Out and Teenage Pregnancy." *American Journal of Sociology* 109:676–719.

Harding, David J., Lisa Gennetian, Christopher Winship, Lisa Sanbonmatsu, and Jeffrey Kling. 2011. "Unpacking Neighborhood Influences on Education Outcomes: Setting the Stage for Future Research." Pp. 277–96 in *Social Inequality and Educational Disadvantage*, edited by G. Duncan and R. Murnane. New York: Russell Sage.

Harding, David J. and Kristin Seefeldt. 2013. "Mixed Methods and Causal Analysis." Pp. 91–110 in *Handbook of Causal Analysis for Social Research*, edited by S. L. Morgan. Dordrecht: Springer.

Hastie, Trevor, Robert Tibshirani, and Jerome Friedman. 2001. *The Elements of Statistical Learning: Data Mining, Inference, and Prediction.* New York: Springer.

Hauser, Philip M. 1962. "On Stouffer's Social Research to Test Ideas." *Public Opinion Quarterly* 26:329–34.

Hauser, Robert M. and Arthur S. Goldberger. 1971. "The Treatment of Unobservable Variables in Path Analysis." *Sociological Methodology* 3:81–117.

Hauser, Robert M., John Robert Warren, Min-Hsiung Huang, and Wendy Y. Carter. 2000. "Occupational Status, Education, and Social Mobility in the Meritocracy." Pp. 179–229 in *Meritocracy and Economic Inequality*, edited by K. J. Arrow, S. Bowles, and S. N. Durlauf. Princeton: Princeton University Press.

Hayashi, Fumio. 2000. *Econometrics.* Princeton: Princeton University Press.

Hayford, Sarah R. and S. Philip Morgan. 2008. "Religiosity and Fertility in the United States: The Role of Fertility Intentions." *Social Forces* 86:1163–88.

Heckman, James J. 1974. "Shadow Prices, Market Wages, and Labor Supply." *Econometrica* 42:679–94.

—. 1978. "Dummy Endogenous Variables in a Simultaneous Equation System." *Econometrica* 46:931–59.

—. 1979. "Selection Bias as a Specification Error." *Econometrica* 47:153–61.

—. 1989. "Causal Inference and Nonrandom Samples." *Journal of Educational Statistics* 14:159–68.

—. 1992. "Randomization and Social Policy Evaluation." Pp. 201–30 in *Evaluating Welfare and Training Programs*, edited by C. F. Manski and I. Garfinkel. Cambridge: Harvard University Press.

—. 1996. "Randomization as an Instrumental Variable." *Review of Economics and Statistics* 77:336–41.

—. 1997. "Instrumental Variables: A Study of Implicit Behavioral Assumptions Used in Making Program Evaluations." *Journal of Human Resources* 32:441–62.

—. 2000. "Causal Parameters and Policy Analysis in Economics: A Twentieth Century Retrospective." *Quarterly Journal of Economics* 115:45–97.

—. 2005. "The Scientific Model of Causality." *Sociological Methodology* 35:1–97.

—. 2008a. "Econometric Causality." *International Statistical Review* 76:1–27.

—. 2008b. "Schools, Skills, and Synapses." *Economic Inquiry* 46:289–324.

—. 2011. "Building Bridges Between Structural and Program Evaluation Approaches to Evaluating Policy." *Journal of Economic Literature* 48:356–98.

Heckman, James J. and V. Joseph Hotz. 1989. "Choosing among Alternative Non-experimental Methods for Estimating the Impact of Social Programs: The Case of Manpower Training." *Journal of the American Statistical Association* 84:862–74.

Heckman, James J., Hidehiko Ichimura, Jeffrey A. Smith, and Petra Todd. 1998. "Characterizing Selection Bias Using Experimental Data." *Econometrica* 66:1017–98.

Heckman, James J., Hidehiko Ichimura, and Petra Todd. 1997. "Matching as an Econometric Evaluation Estimator: Evidence from Evaluating a Job Training Programme." *Review of Economic Studies* 64:605–54.

—. 1998. "Matching as an Econometric Evaluation Estimator." *Review of Economic Studies* 65:261–94.

Heckman, James J., Robert J. LaLonde, and Jeffrey A. Smith. 1999. "The Economics and Econometrics of Active Labor Market Programs." Pp. 1865–2097 in *Handbook of Labor Economics*, vol. 3, edited by O. C. Ashenfelter and D. Card. Amsterdam: Elsevier.

Heckman, James J. and Richard Robb. 1985. "Alternative Methods for Evaluating the Impact of Interventions." Pp. 156–245 in *Longitudinal Analysis of Labor Market Data*, edited by J. J. Heckman and B. Singer. Cambridge: Cambridge University Press.

——. 1986. "Alternative Methods for Solving the Problem of Selection Bias in Evaluating the Impact of Treatments on Outcomes." Pp. 63–113 in *Drawing Inferences from Self-Selected Samples*, edited by H. Wainer. New York: Springer.

——. 1989. "The Value of Longitudinal Data for Solving the Problem of Selection Bias in Evaluating the Impact of Treatment on Outcomes." Pp. 512–38 in *Panel Surveys*, edited by D. Kasprzyk, G. Duncan, G. Kalton, and M. P. Singh. New York: Wiley.

Heckman, James J., Jeffrey Smith, and Nancy Clements. 1997. "Making the Most out of Programme Evaluations and Social Experiments: Accounting for Heterogeneity in Programme Impacts." *Review of Economic Studies* 64:487–535.

Heckman, James J., Justin L. Tobias, and Edward J. Vytlacil. 2003. "Simple Estimators for Treatment Parameters in a Latent-Variable Framework." *Review of Economics and Statistics* 85:748–55.

Heckman, James J. and Sergio Urzua. 2010. "Comparing IV with Structural Models: What Simple IV Can and Cannot Identify." *Journal of Econometrics* 156:27–37.

Heckman, James J., Sergio Urzua, and Edward J. Vytlacil. 2006. "Understanding Instrumental Variables in Models with Essential Heterogeneity." *Review of Economics and Statistics* 88:389–432.

Heckman, James J. and Edward J. Vytlacil. 1999. "Local Instrumental Variables and Latent Variable Models for Identifying and Bounding Treatment Effects." *Proceedings of the National Academy of Sciences of the United States of America* 96:4730–34.

——. 2000. "The Relationship between Treatment Parameters within a Latent Variable Framework." *Economics Letters* 66:33–39.

——. 2005. "Structural Equations, Treatment Effects, and Econometric Policy Evaluation." *Econometrica* 73:669–738.

——. 2007. "Econometric Evaluation of Social Programs, Part I: Causal Models, Structural Models and Econometric Policy Evaluation." Pp. 4779–874 in *Handbook of Econometrics*, vol. 6, part b, edited by J. J. Heckman and E. E. Leamer. Amsterdam: Elsevier.

Hedström, Peter. 2005. *Dissecting the Social: On the Principles of Analytical Sociology.* Cambridge: Cambridge University Press.

Hedström, Peter and Peter Bearman, Eds. 2009. *The Oxford Handbook of Analytical Sociology.* Oxford: Oxford University Press.

Hedström, Peter and Richard Swedberg, Eds. 1998. *Social Mechanisms: An Analytical Approach to Social Theory.* Cambridge: Cambridge University Press.

Hedström, Peter and Lars Udehn. 2009. "Analytical Sociology and Theories of the Middle Range." Pp. 25–47 in *The Oxford Handbook of Analytical Sociology,* edited by P. Hedström and P. Bearman. Oxford: Oxford University Press.

Hedström, Peter and Petri Ylikoski. 2010. "Causal Mechanisms in the Social Sciences." *Annual Review of Sociology* 36:49–67.

Heller, Ruth, Paul R. Rosenbaum, and Dylan S. Small. 2010. "Using the Cross-Match Test to Appraise Covariate Balance in Matched Pairs." *American Statistician* 64:299–309.

Henderson, John and Sara Chatfield. 2011. "Who Matches? Propensity Scores and Bias in the Causal Effects of Education on Participation." *Journal of Politics* 73:646–58.

Hendry, David F. 1995. *Dynamic Econometrics.* Oxford: Oxford University Press.

Henig, Jeffrey R. 2008. *Spin Cycle: How Research Is Used in Policy Debates: The Case of Charter Schools.* New York: Russell Sage.

Hernán, Miguel A. 2005. "Invited Commentary: Hypothetical Interventions to Define Causal Effects – Afterthought or Prerequisite?" *American Journal of Epidemiology* 162:618–20.

Hernán, Miguel A., Sonia Hernandez-Diaz, and James M. Robins. 2004. "A Structural Approach to Selection Bias." *Epidemiology* 15:615–25.

Hernán, Miguel A. and James M. Robins. 2006a. "Estimating Causal Effects from Epidemiological Data." *Journal of Epidemiology and Community Health* 60:578–86.

—. 2006b. "Instruments for Causal Inference: An Epidemiologist's Dream?" *Epidemiology* 17:360–72.

Herrnstein, Richard J. and Charles A. Murray. 1994. *The Bell Curve: Intelligence and Class Structure in American Life.* New York: Free Press.

Herron, Michael C. and Jasjeet S. Sekhon. 2003. "Overvoting and Representation: An Examination of Overvoted Presidential Ballots in Broward and Miami-Dade Counties." *Electoral Studies* 22:21–47.

Highton, Benjamin. 2009. "Revisiting the Relationship between Educational Attainment and Political Sophistication." *Journal of Politics* 71:1564–76.

Hill, Jennifer and Jerome P. Reiter. 2006. "Interval Estimation for Treatment Effects Using Propensity Score Matching." *Statistics in Medicine* 25:2230–56.

Hill, Jennifer, Christopher Weiss, and Fuhua Zhai. 2011. "Challenges with Propensity Score Strategies in a High-Dimensional Setting and a Potential Alternative." *Multivariate Behavioral Research* 46:477–513.

Hill, Jennifer L. 2011. "Bayesian Nonparametric Modeling for Causal Inference." *Journal of Computational and Graphical Statistics* 20:217–40.

Hirano, Keisuke and Guido W. Imbens. 2001. "Estimation of Causal Effects Using Propensity Score Weighting: An Application to Data on Right Heart Catheterization." *Health Services & Outcomes Research Methodology* 2:259–78.

—. 2004. "The Propensity Score with Continuous Treatments." Pp. 73–84 in *Applied Bayesian Modeling and Causal Inference from Incomplete-Data Perspectives: An Essential Journey with Donald Rubin's Statistical Family*, edited by A. Gelman and X.-L. Meng. New York: Wiley.

Hirano, Keisuke, Guido W. Imbens, and Geert Ridder. 2003. "Efficient Estimation of Average Treatment Effects Using the Estimated Propensity Score." *Econometrica* 71:1161–89.

Hitchcock, Christopher. 2007. "What's Wrong with Neuron Diagrams?" Pp. 69–92 in *Causation and Explanation*, edited by J. K. Campbell, M. O'Rourke, and H. Silverstein. Cambridge: MIT Press.

Ho, Daniel, Kosuke Imai, Gary King, and Elizabeth Stuart. 2011. "MatchIt: Nonparametric Preprocessing for Parametric Causal Inference." Software Routine for R, Department of Government, Harvard University, Cambridge, Massachusetts.

Ho, Daniel E., Kosuke Imai, Gary King, and Elizabeth A. Stuart. 2007. "Matching as Nonparametric Preprocessing for Reducing Model Dependence in Parametric Causal Inference." *Political Analysis* 15:199–236.

Hodge, Robert W. and Naohiro Ogawa. 1991. *Fertility Change in Contemporary Japan.* Chicago: University of Chicago Press.

Hoffer, Thomas, Andrew M. Greeley, and James S. Coleman. 1985. "Achievement Growth in Public and Catholic Schools." *Sociology of Education* 58:74–97.

Holland, Paul W. 1986. "Statistics and Causal Inference." *Journal of the American Statistical Association* 81:945–70.

—. 2008. "Causation and Race." Pp. 93–109 in *White Logic, White Methods: Racism and Methodology*, edited by T. Zuberi and E. Bonilla-Silva. Lanham: Rowman and Littlefield.

Holmlund, Helena, Mikael Lindahl, and Erik Plug. 2011. "The Causal Effect of Parents' Schooling on Children's Schooling: A Comparison of Estimation Methods." *Journal of Economic Literature* 49:615–51.

Hong, Guanglei. 2010. "Marginal Mean Weighting Through Stratification: Adjustment for Selection Bias in Multilevel Data." *Journal of Educational and Behavioral Statistics* 35:499–531.

—. 2012. "Marginal Mean Weighting Through Stratification: A Generalized Method for Evaluating Multivalued and Multiple Treatments with Nonexperimental Data." *Psychological Methods* 17:44–60.

Hong, Guanglei and Stephen W. Raudenbush. 2006. "Evaluating Kindergarten Retention Policy: A Case Study of Causal Inference for Multilevel Observational Data." *Journal of the American Statistical Association* 101:901–10.

—. 2013. "Heterogeneous Agents, Social Interactions, and Causal Inference." Pp. 331–52 in *Handbook of Causal Analysis for Social Research*, edited by S. L. Morgan. Dordrecht: Springer.

Honoré, Bo E. and James L. Powell. 1994. "Pairwise Difference Estimators of Censored and Truncated Regression Models." *Journal of Econometrics* 64:241–78.

Hood, William C. and Tjalling C. Koopmans, Eds. 1953. *Studies in Econometric Method.* New York: Wiley.

Hout, Michael. 2012. "Social and Economic Returns to College Education in the United States." *Annual Review of Sociology* 38:379–400.

Howell, William G. and Paul E. Peterson. 2002. *The Education Gap: Vouchers and Urban Schools.* Washington, DC: Brookings Institution.

Hoxby, Caroline M. 1996. "The Effects of Private School Vouchers on Schools and Students." Pp. 177–208 in *Holding Schools Accountable: Performance-Based Reform in Education*, edited by H. F. Ladd. Washington, DC: Brookings Institution.

Hoxby, Caroline M, Ed. 2003. *The Economics of School Choice.* Chicago: University of Chicago Press.

—, Ed. 2004. *College Choices: The Economics of Where to Go, When to Go, and How to Pay for It.* Chicago: University of Chicago Press.

Hoxby, Caroline M., Sonali Murarka, and Jenny Kang. 2009. "How New York City's Charter Schools Affect Achievement." National Bureau of Economic Research, Cambridge, Massachusetts.

Hsiao, Cheng. 2003. *Analysis of Panel Data.* Cambridge: Cambridge University Press.

Hudgens, Michael G. and M. Elizabeth Halloran. 2008. "Toward Causal Inference with Interference." *Journal of the American Statistical Association* 103:832–42.

Hume, David. 1977[1772]. *An Enquiry Concerning Human Understanding.* Indianapolis: Hackett.

Humphreys, Paul. 1989. *The Chances of Explanation: Causal Explanation in the Social, Medical, and Physical Sciences.* Princeton: Princeton University Press.

—. 2004. *Extending Ourselves: Computational Science, Empiricism, and Scientific Method.* New York: Oxford University Press.

Hyman, Herbert. 1955. *Survey Design and Analysis: Principles, Cases and Procedures.* Glencoe: Free Press.

Hyman, Herbert H. 1962. "Samuel A. Stouffer and Social Research." *Public Opinion Quarterly* 26:323–28.

Iacus, Stefano M. and Gary King. 2012. "How Coarsening Simplifies Matching-Based Causal Inference Theory." Department of Economics, Business and Statistics, University of Milan, Milan, Italy.

Iacus, Stefano M., Gary King, and Giuseppe Porro. 2011. "Multivariate Matching Methods That Are Monotonic Imbalance Bounding." *Journal of the American Statistical Association* 106:345–61.

—. 2012a. "Causal Inference Without Balance Checking: Coarsened Exact Matching." *Political Analysis* 20:1–24.

—. 2012b. "CEM: Coarsened Exact Matching Software." Software Routine for Stata, Department of Government, Harvard University, Cambridge, Massachusetts.

Imai, Kosuke, Gary King, and Elizabeth A. Stuart. 2008. "Misunderstandings between Experimentalists and Observationalists about Causal Inference." *Journal of the Royal Statistical Society* 171:481–502.

Imai, Kosuke and Marc Ratkovic. 2014. "Covariate Balancing Propensity Score." *Journal of the Royal Statistical Society*, Series B 76:243–63.

Imai, Kosuke and David A. van Dyk. 2004. "Causal Inference with General Treatment Regimes: Generalizing the Propensity Score." *Journal of the American Statistical Association* 99:854–66.

Imbens, Guido W. 2000. "The Role of the Propensity Score in Estimating Dose-Response Functions." *Biometrika* 87:706–10.

—. 2004. "Nonparametric Estimation of Average Treatment Effects under Exogeneity: A Review." *Review of Economics and Statistics* 86:4–29.

—. 2010. "Better LATE Than Nothing: Some Comments on Deaton (2009) and Heckman and Urzua (2009)." *Journal of Economic Literature* 48:399–423.

Imbens, Guido W. and Joshua D. Angrist. 1994. "Identification and Estimation of Local Average Treatment Effects." *Econometrica* 62:467–75.

Imbens, Guido W. and Thomas Lemieux. 2008. "Regression Discontinuity Designs: A Guide to Practice." *Journal of Econometrics* 142:1–21.

Imbens, Guido W. and Donald B. Rubin. 1997. "Estimating Outcome Distributions for Compliers in Instrumental Variables Models." *Review of Economic Studies* 64: 555–74.

Imbens, Guido W. and Jeffrey M. Wooldridge. 2009. "Recent Developments in the Econometrics of Program Evaluation." *Journal of Economic Literature* 47:5–86.

Jackson, Michelle V. 2013. *Determined to Succeed? Performance versus Choice in Educational Attainment.* Stanford: Stanford University Press.

Jencks, Christopher S. and Susan E. Mayer. 1990. "The Social Consequences of Growing Up in a Poor Neighborhood." Pp. 111–86 in *Inner-City Poverty in the United States*, edited by L. E. Lynn and M. G. H. McGeary. Washington, D.C.: National Academy Press.

Jepsen, Christopher. 2003. "The Effectiveness of Catholic Primary Schooling." *Journal of Human Resources* 38:928–41.

Jin, Hui, John Barnard, and Donald B. Rubin. 2010. "A Modified General Location Model for Noncompliance with Missing Data: Revisiting the New York City School Choice Scholarship Program Using Principal Stratification." *Journal of Educational and Behavioral Statistics* 35:154–73.

Jin, Hui and Donald B. Rubin. 2009. "Public Schools versus Private Schools: Causal Inference with Partial Compliance." *Journal of Educational and Behavioral Statistics* 34:24–45.

Joffe, Marshall M. 2011. "Principal Stratification and Attribution Prohibition: Good Ideas Taken Too Far." *International Journal of Biostatistics* 7:1–22.

Joffe, Marshall M. and Paul R. Rosenbaum. 1999. "Propensity Scores." *American Journal of Epidemiology* 150:327–31.

Judd, Charles M. and David A. Kenny. 1981. *Estimating the Effects of Social Interventions.* New York: Cambridge University Press.

Kam, Cindy D. and Carl L. Palmer. 2008. "Reconsidering the Effects of Education on Political Participation." *Journal of Politics* 70:612–31.

—. 2011. "Rejoinder: Reinvestigating the Causal Relationship between Higher Education and Political Participation." *Journal of Politics* 73:659–63.

Karpinos, Bernard D. 1938. "The Differential True Rates of Growth of the White Population in the United States and Their Probable Effects on the General Growth of the Population." *American Journal of Sociology* 44:251–73.

Keane, Michael P. 2010. "A Structural Perspective on the Experimentalist School." *Journal of Economic Perspectives* 24:47–58.

Kempthorne, Oscar. 1948. "Review of Experimental Designs in Sociological Research by F. Stuart Chapin." *Journal of the American Statistical Association* 43:489–92.

—. 1952. *The Design and Analysis of Experiments*. New York: Wiley.

Kemptner, Daniel, Hendrik Jürges, and Steffen Reinhold. 2011. "Changes in Compulsory Schooling and the Causal Effect of Education on Health: Evidence from Germany." *Journal of Health Economics* 30:340–54.

Kendall, Patricia L. and Paul F. Lazarsfeld. 1950. "Problems of Survey Analysis." Pp. 133–96 in *Continuities in Social Research: Studies in the Scope and Method of "The American Soldier,"* edited by R. K. Merton and P. F. Lazarsfeld. Glencoe: Free Press.

Kennedy, David M. 1997. "Pulling Levers: Chronic Offenders, High-Crime Settings, and a Theory of Prevention." *Valparaiso University Law Review* 31:449–84.

Keyfitz, Nathan. 1948. "Review of Experimental Designs in Sociological Research by F. Stuart Chapin." *American Journal of Sociology* 54:259–60.

King, Gary, Robert O. Keohane, and Sidney Verba. 1994. *Designing Social Inquiry: Scientific Inference in Qualitative Research*. Princeton: Princeton University Press.

Kish, Leslie. 1965. *Survey Sampling*. New York: Wiley.

—. 1987. *Statistical Design for Research*. New York: Wiley.

Kling, Jeffrey R., Jeffrey B. Liebman, and Lawrence F. Katz. 2007. "Experimental Analysis of Neighborhood Effects." *Econometrica* 75:83–119.

Kling, Jeffrey R., Jens Ludwig, and Lawrence F. Katz. 2005. "Neighborhood Effects on Crime for Female and Male Youth: Evidence from a Randomized Housing Voucher Experiment." *Quarterly Journal of Economics* 120:87–130.

Knight, Carly R. and Christopher Winship. 2013. "The Causal Implications of Mechanistic Thinking: Identification Using Directed Acyclic Graphs (DAGs)." Pp. 275–99 in *Handbook of Causal Analysis for Social Research*, edited by S. L. Morgan. Dordrecht: Springer.

Koller, Daphne and Nir Friedman. 2009. *Probabilistic Graphical Models: Principles and Techniques*. Cambridge: MIT Press.

Koopmans, Tjalling C. and Olav Reiersøl. 1950. "The Identification of Structural Characteristics." *Annals of Mathematical Statistics* 21:165–81.

Krasno, Jonathan S. and Donald P. Green. 2008. "Do Televised Presidential Ads Increase Voter Turnout? Evidence from a Natural Experiment." *Journal of Politics* 70:245–61.

Krueger, Alan B. and Pei Zhu. 2004. "Another Look at the New York City School Voucher Experiment." *American Behavioral Scientist* 47:658–98.

Krueger, Thomas M. and William F. Kennedy. 1990. "An Examination of the Super Bowl Stock Market Predictor." *Journal of Finance* 45:691–97.

Ladd, Helen F. 2002. "School Vouchers: A Critical View." *Journal of Economic Perspectives* 16:3–24.

LaLonde, Robert J. 1986. "Evaluating the Econometric Evaluations of Training Programs with Experimental Data." *American Economic Review* 76:604–20.

—. 1995. "The Promise of Public Sector-Sponsored Training Programs." *Journal of Economic Perspectives* 9:149–68.

Lauritzen, Steffen L. 1996. *Graphical Models.* Oxford: Clarendon.

Lazarsfeld, Paul F., Bernard Berelson, and Hazel Gaudet. 1955[1948]. "Political Interest and Voting Behavior." Pp. 155–8 in *The Language of Social Research: A Reader in the Methodology of Social Research*, edited by P. F. Lazarsfeld and M. Rosenberg. Glencoe: Free Press.

Leamer, Edward E. 1978. *Specification Searches: Ad Hoc Inference with Nonexperimental Data.* New York: Wiley.

—. 1983. "Let's Take the Con Out of Econometrics." *American Economic Review* 73:31–43.

—. 2010. "Tantalus on the Road to Asymptopia." *Journal of Economic Perspectives* 24:31–46.

Lechner, Michael. 2002a. "Program Heterogeneity and Propensity Score Matching: An Application to the Evaluation of Active Labor Market Policies." *Review of Economics and Statistics* 84:205–20.

—. 2002b. "Some Practical Issues in the Evaluation of Heterogeneous Labour Market Programmes by Matching Methods." *Journal of Royal Statistical Society* 165:59–82.

Lee, Brian K., Justin Lessler, and Elizabeth A. Stuart. 2009. "Improving Propensity Score Weighting Using Machine Learning." *Statistics in Medicine* 29:337–46.

—. 2011. "Weight Trimming and Propensity Score Weighting." *PLoS ONE* 6:e18174.

Lee, Myoung-jae. 2005. *Micro-Econometrics for Policy, Program, and Treatment Effects.* Oxford: Oxford University Press.

Lee, Ronald. 2003. "The Demographic Transition: Three Centuries of Fundamental Change." *Journal of Economic Perspectives* 17:167–90.

Leuven, Edwin and Barbara Sianesi. 2012. "PSMATCH2: Stata Module to Perform Full Mahalanobis and Propensity Score Matching, Common Support Graphing, and Covariate Imbalance Testing." Software Routine for Stata, Department of Economics, Boston College, Boston, Massachusetts.

Levy, Paul S., Stanley Lemeshow, Paul P. Biemer, and Sharon L. Christ. 2008. "Constructing Survey Weights." Pp. 489–516 in *Sampling of Populations: Methods and Applications*. Hoboken: Wiley.

Lewis, David. 1973. "Causation." *Journal of Philosophy* 70:556–67.

Lieberson, Stanley. 1985. *Making It Count: The Improvement of Social Research and Theory*. Berkeley: University of California Press.

Lieberson, Stanley and Freda B. Lynn. 2002. "Barking Up the Wrong Branch: Scientific Alternatives to the Current Model of Sociological Science." *Annual Review of Sociology* 28:1–19.

Lindahl, Mikael. 2005. "Estimating the Effect of Income on Health and Mortality Using Lottery Prizes as an Exogenous Source of Variation in Income." *Journal of Human Resources* 40:144–68.

Link, Bruce G. and Jo C. Phelan. 1995. "Social Conditions as Fundamental Causes of Disease." *Journal of Health and Social Behavior* 35:80–94.

Little, Roderick J. A. 1982. "Models for Nonresponse in Sample Surveys." *Journal of the American Statistical Association* 77:237–50.

Little, Roderick J. A. and Donald B. Rubin. 2002. *Statistical Analysis with Missing Data*. Hoboken: Wiley.

Long, J. Scott. 1997. *Regression Models for Categorical and Limited Dependent Variables*. Thousand Oaks: Sage Publications.

Long, J. Scott and Laurie H. Ervin. 2000. "Using Heteroscedasticity Consistent Standard Errors in the Linear Regression Model." *American Statistician* 54:217–24.

Lu, Bo, Elaine Zanutto, Robert Hornik, and Paul R. Rosenbaum. 2001. "Matching with Doses in an Observational Study of a Media Campaign Against Drug Abuse." *Journal of the American Statistical Association* 96:1245–53.

Lubienski, Sarah T. and Christopher Lubienski. 2003. "School Sector and Academic Achievement: A Multilevel Analysis of NAEP Mathematics Data." *American Educational Research Journal* 43:651–98.

Lumley, Thomas and Achim Zeileis. 2013. "Sandwich: Robust Covariance Matrix Estimators." Software Routine for R, Department of Statistics, University of Washington, Seattle, Washington.

Lutfey, Karen and Jeremy Freese. 2005. "Toward Some Fundamentals of Fundamental Causality: Socioeconomic Status and Health in the Routine Clinic Visit for Diabetes." *American Journal of Sociology* 110:1326–72.

Machamer, Peter, Lindley Darden, and Carl F. Craver. 2000. "Thinking About Mechanisms." *Philosophy of Science* 67:1–25.

Machin, Stephen, Olivier Marie, and Sunčica Vujić. 2011. "The Crime Reducing Effect of Education." *Economic Journal* 121:463–84.

MacKinnon, David P. 2008. *Introduction to Statistical Mediation Analysis.* New York: Lawrence Erlbaum.

Manski, Charles F. 1994. "The Selection Problem." Pp. 143–70 in *Advances in Econometrics: Sixth World Congress*, vol. 1, edited by C. A. Sims. Cambridge: Cambridge University Press.

—. 1995. *Identification Problems in the Social Sciences.* Cambridge: Harvard University Press.

—. 1997. "Monotone Treatment Response." *Econometrica* 65:1311–34.

—. 1999. "Comment on 'Choice as an Alternative to Control in Observational Studies' by Rosenbaum." *Statistical Science* 14:279–81.

—. 2003. *Partial Identification of Probability Distributions.* New York: Springer.

—. 2013a. "Identification of Treatment Response with Social Interactions." *Econometrics Journal* 16:S1–S23.

—. 2013b. *Public Policy in an Uncertain World: Analysis and Decisions.* Cambridge: Harvard University Press.

Manski, Charles F. and Irwin Garfinkel, Eds. 1992. *Evaluating Welfare and Training Programs.* Cambridge: Harvard University Press.

Manski, Charles F. and Daniel S. Nagin. 1998. "Bounding Disagreements about Treatment Effects: A Case Study of Sentencing and Recidivism." *Sociological Methodology* 28:99–137.

Manski, Charles F. and John V. Pepper. 2000. "Monotone Instrumental Variables: With an Application to the Returns to Schooling." *Econometrica* 68:997–1010.

—. 2013. "Deterrence and the Death Penalty: Partial Identification Analysis Using Repeated Cross Sections." *Journal of Quantitative Criminology* 29:123–41.

Manski, Charles F., Gary D. Sandefur, Sara McLanahan, and Daniel Powers. 1992. "Alternative Estimates of the Effect of Family Structure During Adolescence on High School Graduation." *Journal of the American Statistical Association* 87:25–37.

Manza, Jeff and Christopher Uggen. 2004. "Punishment and Democracy: Disenfranchisement of Nonincarcerated Felons in the United States." *Perspectives on Politics* 2:491–505.

Manzo, Gianluca. 2011. "Relative Deprivation in Silico: Agent-Based Models and Causality in Analytical Sociology." Pp. 266–308 in *Analytical Sociology and Social Mechanisms*, edited by P. Demeulenaere. Cambridge: Cambridge University Press.

Marcantonio, Richard J. and Thomas D. Cook. 1994. "Convincing Quasi-Experiments: The Interrupted Time Series and Regression-Discontinuity Designs." Pp. 133–54 in *Handbook of Practical Program Evaluation*, edited by J. S. Wholey, H. P. Hatry, and K. E. Newcomer. San Francisco: Jossey-Bass.

Marini, Margaret M. and Burton Singer. 1988. "Causality in the Social Sciences." *Sociological Methodology* 18:347–409.

Mark, Melvin M. and Steven Mellor. 1991. "Effect of Self-Relevance of an Event on Hindsight Bias: The Foreseeability of a Layoff." *Journal of Applied Psychology* 76:569–77.

Martin, John Levi. 2011. *The Explanation of Social Action*. New York: Oxford University Press.

Mayer, Albert J. and Sue Marx. 1957. "Social Change, Religion, and Birth Rates." *American Journal of Sociology* 62:383–90.

Mayer, Alexander K. 2011. "Does Education Increase Political Participation?" *Journal of Politics* 73:633–45.

McCaffrey, Daniel F., Greg Ridgeway, and Andrew R. Morral. 2004. "Propensity Score Estimation with Boosted Regression for Evaluating Causal Effects in Observational Studies." *Psychological Methods* 9:403–25.

McDowall, David, Richard McCleary, Errol E. Meidinger, and Richard A Hay Jr. 1980. *Interrupted Time Series Analysis*. Beverly Hills: Sage.

McLanahan, Sara. 2004. "Diverging Destinies: How Children Are Faring under the Second Demographic Transition." *Demography* 41:607–27.

—. 2009. "Fragile Families and the Reproduction of Poverty." *Annals of the American Academy of Political and Social Science* 621:111–31.

McLanahan, Sara and Christine Percheski. 2008. "Family Structure and the Reproduction of Inequalities." *Annual Review of Sociology* 34:257–76.

McLanahan, Sara and Gary D. Sandefur. 1994. *Growing Up with a Single Parent: What Hurts, What Helps*. Cambridge: Harvard University Press.

McLanahan, Sara, Laura Tach, and Daniel Schneider. 2013. "The Causal Effects of Father Absence." *Annual Review of Sociology* 39:399–427.

Mebane, Walter R. 2004. "The Wrong Man Is President! Overvotes in the 2000 Presidential Election in Florida." *Perspectives on Politics* 2:525–35.

Merton, Robert K. 1968. "The Matthew Effect in Science." *Science* 159:56–63.

Messner, Steven F., Eric P. Baumer, and Richard Rosenfeld. 2004. "Dimensions of Social Capital and Rates of Criminal Homicide." *American Sociological Review* 69:882–903.

Moffitt, Robert. 2005. "Remarks on the Analysis of Causal Relationships in Population Research." *Demography* 42:91–108.

Moffitt, Robert A. 1996. "Comment on 'Identification of Causal Effects Using Instrumental Variables' by Angrist, Imbens, and Rubin." *Journal of the American Statistical Association* 91:462–65.

—. 2003. "Causal Analysis in Population Research: An Economist's Perspective." *Population and Development Review* 29:448–58.

Morgan, Stephen L. 2001. "Counterfactuals, Causal Effect Heterogeneity, and the Catholic School Effect on Learning." *Sociology of Education* 74:341–74.

—. 2005. *On the Edge of Commitment: Educational Attainment and Race in the United States*. Stanford: Stanford University Press.

—, Ed. 2013. *Handbook of Causal Analysis for Social Research*. Dordrecht: Springer.

Morgan, Stephen L. and David J. Harding. 2006. "Matching Estimators of Causal Effects: Prospects and Pitfalls in Theory and Practice." *Sociological Methods and Research* 35:3–60.

Morgan, Stephen L., Theodore S. Leenman, Jennifer J. Todd, and Kim A. Weeden. 2013. "Occupational Plans, Beliefs about Educational Requirements, and Patterns of College Entry." *Sociology of Education* 86:197–217.

Morgan, Stephen L. and Jennifer J. Todd. 2008. "A Diagnostic Routine for the Detection of Consequential Heterogeneity of Causal Effects." *Sociological Methodology* 38:231–81.

Morgan, Stephen L. and Christopher Winship. 2012. "Bringing Context and Variability Back into Causal Analysis." Pp. 319–54 in *The Oxford Handbook of Philosophy of Social Science*, edited by H. Kincaid. Oxford: Oxford University Press.

Morgan, S. Philip and Ronald R. Rindfuss. 1999. "Reexamining the Link of Early Childbearing to Marriage and to Subsequent Fertility." *Demography* 36:59–75.

Morton, Rebecca B. and Kenneth C. Williams. 2010. *Experimental Political Science and the Study of Causality: From Nature to the Lab*. Cambridge: Cambridge University Press.

Mumford, Stephen and Rani L. Anjum. 2011. *Getting Causes from Powers*. Oxford: Oxford University Press.

Murnane, Richard J., Stephen E. Newstead, and Randall J. Olsen. 1985. "Comparing Public and Private Schools: The Puzzling Role of Selectivity Bias." *Journal of Business and Economic Statistics* 3:23–35.

Murnane, Richard J. and John B. Willett. 2011. *Methods Matter: Improving Causal Inference in Educational and Social Science Research*. Oxford: Oxford University Press.

Musick, Kelly. 2002. "Planned and Unplanned Childbearing among Unmarried Women." *Journal of Marriage and Family* 64:915–29.

Musick, Kelly, Paula England, Sarah Edgington, and Nicole Kangas. 2009. "Education Differences in Intended and Unintended Fertility." *Social Forces* 88:543–72.

Neal, Derek. 1997. "The Effects of Catholic Secondary Schooling on Educational Achievement." *Journal of Labor Economics* 14:98–123.

—. 2002. "How Vouchers Could Change the Market for Education." *Journal of Economic Perspectives* 16:25–44.

Neyman, Jerzy Splawa. 1935. "Statistical Problems in Agricultural Experimentation (with Discussion)." *Journal of the Royal Statistical Society*, Series B 2:107–80.

—. 1990[1923]. "On the Application of Probability Theory to Agricultural Experiments. Essay on Principles. Section 9." *Statistical Science* 5:465–80.

Ní Bhrolcháin, Máire and Tim Dyson. 2007. "On Causation in Demography: Issues and Illustrations." *Population and Development Review* 33:1–36.

Nie, Norman H., Jane Junn, and Kenneth Stehlik-Barry. 1996. *Education and Democratic Citizenship in America*. Chicago: University of Chicago Press.

Noell, Jay. 1982. "Public and Catholic Schools: A Reanalysis of 'Public and Private Schools.'" *Sociology of Education* 55:123–32.

Notestein, Frank W. 1933. "The Differential Rate of Increase among the Social Classes of the American Population." *Social Forces* 12:17–33.

—. 1950. "The Population of the World in the Year 2000." *Journal of the American Statistical Association* 45:335–45.

Oreopoulos, Philip. 2006. "Estimating Average and Local Average Treatment Effects of Education When Compulsory Schooling Laws Really Matter." *American Economic Review* 96:152–75.

Pagan, Adrian and Aman Ullah. 1999. *Nonparametric Econometrics*. New York: Cambridge University Press.

Pampel, Fred C., Patrick M. Krueger, and Justin T. Denney. 2010. "Socioeconomic Disparities in Health Behaviors." *Annual Review of Sociology* 36:349–70.

Parsons, Talcott. 1937. *The Structure of Social Action*. New York: McGraw-Hill.

—. 1951. *The Social System*. Glencoe: Free Press.

Paul, L. A. and Ned Hall. 2013. *Causation: A User's Guide.* Oxford: Oxford University Press.

Pearl, Judea. 2000. *Causality: Models, Reasoning, and Inference.* Cambridge: Cambridge University Press.

—. 2001. "Direct and Indirect Effects." Pp. 411–20 in *Proceedings of the Seventeenth Conference on Uncertainty in Artificial Intelligence.* San Francisco: Morgan Kaufmann.

—. 2009. *Causality: Models, Reasoning, and Inference,* 2nd ed. Cambridge: Cambridge University Press.

—. 2011. "Principal Stratification – a Goal or a Tool?" *International Journal of Biostatistics* 7:1–13.

—. 2012a. "The Causal Mediation Formula: A Guide to the Assessment Pathways and Mechanisms." *Prevention Science* 13:426–36.

—. 2012b. "The Mediation Formula: A Guide to the Assessment of Causal Pathways in Nonlinear Models." Pp. 151–79 in *Causality: Statistical Perspectives and Applications,* edited by C. Berzuini, P. Dawid, and L. Bernardinelli. Hoboken: Wiley.

Petersen, Trond, Andrew M. Penner, and Geir Høgsnes. 2011. "The Male Marital Wage Premium: Sorting vs. Differential Pay." *Industrial and Labor Relations Review* 64:283–304.

Peterson, Paul E. and William G. Howell. 2004. "Efficiency, Bias, and Classification Schemes: A Response to Alan B. Krueger and Pei Zhu." *American Behavioral Scientist* 47:699–717.

Phelan, Jo C., Bruce G. Link, Ana Diez-Roux, Ichiro Kawachi, and Bruce Levin. 2004. "'Fundamental Causes' of Social Inequalities in Mortality: A Test of the Theory." *Journal of Health and Social Behavior* 45:265–85.

Phelan, Jo C., Bruce G. Link, and Parisa Tehranifar. 2010. "Social Conditions as Fundamental Causes of Health Inequalities: Theory, Evidence, and Policy Implications." *Journal of Health and Social Behavior* 51:S28–S40.

Powers, Daniel A. and Yu Xie. 2000. *Statistical Methods for Categorical Data Analysis.* San Diego: Academic Press.

Pratt, John W. and Robert Schlaifer. 1984. "On the Nature and Discovery of Structure." *Journal of the American Statistical Association* 79:9–33.

—. 1988. "On the Interpretation and Observation of Laws." *Journal of Econometrics* 39:23–52.

Psillos, Stathis. 1999. *Scientific Realism: How Science Tracks Truth.* London: Routledge.

Putnam, Hilary. 1975. *Mind, Language, and Reality.* New York: Cambridge University Press.

Quandt, Richard E. 1972. "A New Approach to Estimating Switching Regression." *Journal of the American Statistical Association* 67:306–10.

Raftery, Adrian E. 1995. "Bayesian Model Selection in Social Research." *Sociological Methodology* 25:111–63.

Ragin, Charles C. 1987. *The Comparative Method: Moving Beyond Qualitative and Quantitative Strategies.* Berkeley: University of California Press.

—. 2008. *Redesigning Social Inquiry: Fuzzy Sets and Beyond.* Chicago: University of Chicago Press.

Raudenbush, Stephen W. and Anthony S. Bryk. 2002. *Hierarchical Linear Models: Applications and Data Analysis Methods.* Thousand Oaks: Sage.

Reardon, Sean F., Jacob E. Cheadle, and Joseph P. Robinson. 2009. "The Effect of Catholic Schooling on Math and Reading Development in Kindergarten Through Fifth Grade." *Journal of Research on Educational Effectiveness* 2:45–87.

Reed, Isaac. 2011. *Interpretation and Social Knowledge: On the Use of Theory in the Human Sciences.* Chicago: University of Chicago Press.

Reiersøl, Olav. 1941. "Confluence Analysis by Means of Lag Moments and Other Methods of Confluence Analysis." *Econometrica* 9:1–24.

Rein, Martin and Christopher Winship. 1999. "The Dangers of 'Strong' Causal Reasoning in Social Policy." *Society* 36:38–46.

Reskin, Barbara F. 2003. "Including Mechanisms in Our Models of Ascriptive Inequality." *American Sociological Review* 68:1–21.

Retherford, Robert D. and Naohiro Ogawa. 2006. "Japan's Baby Bust: Causes, Implications, and Policy Responses." Pp. 5–47 in *The Baby Bust: Who Will Do the Work? Who Will Pay the Taxes?*, edited by F. R. Harris. Lanham: Rowman and Littlefield.

Robins, James M. 1986. "A New Approach to Causal Inference in Mortality Studies with a Sustained Exposure Period: Application to Control of the Healthy Worker Survivor Effect." *Mathematical Modelling* 7:1393–512.

—. 1997. "Causal Inference from Complex Longitudinal Data." Pp. 69–117 in *Latent Variable Modeling and Applications to Causality: Lecture Notes in Statistics*, edited by M. Berkane. New York: Springer.

—. 1998. "Marginal Structural Models." Pp. 1–10 in *1997 Proceedings of the American Statistical Association, Section on Bayesian Statistical Science.* Alexandria, Virginia.

—. 1999. "Association, Causation, and Marginal Structural Models." *Synthese* 121: 151–79.

—. 2000. "Marginal Structural Models versus Structural Nested Modes as Tools for Causal Inference." Pp. 95–134 in *Statistical Models in Epidemiology: The Environment and Clinical Trials*, edited by M. E. Halloran and D. A. Berry. New York: Springer.

Robins, James M. and Sander Greenland. 2000. "Comment on 'Causal Inference Without Counterfactuals' by A. P. Dawid." *Journal of the American Statistical Association* 95:431–35.

Robins, James M. and Miguel A. Hernán. 2009. "Estimation of the Causal Effects of Time-Varying Exposures." Pp. 553–99 in *Longitudinal Data Analysis*, edited by G. Fitzmaurice, M. Davidian, G. Verbeke, and G. Molenberghs. Boca Raton: Chapman & Hall/CRC.

Robins, James M. and Thomas Richardson. 2010. "Alternative Graphical Causal Models and the Identification of Direct Effects." Center for Statistics and the Social Sciences, Working Paper 100, University of Washington, Seattle, Washington.

Robins, James M. and Ya'acov Ritov. 1997. "Toward a Curse of Dimensionality Appropriate (CODA) Asymptotic Theory for Semi-Parametric Models." *Statistics in Medicine* 16:285–319.

Robins, James M. and Andrea Rotnitzky. 2001. "Comment on 'Inference for Semiparametric Models: Some Questions and an Answer' by Bickel and Kwon." *Statistica Sinica* 11:920–36.

Robins, James M., Andrea Rotnitzky, and Lue Ping Zhao. 1994. "Estimation of Regression Coefficients When Some Regressors Are Not Always Observed." *Journal of the American Statistical Association* 89:846–66.

Rohwer, Götz. 2010. *Models in Statistical Social Research*. London: Routledge.

Rose, Arnold. 1962. "*Review of Social Research to Test Ideas: Selected Writings by Samuel A. Stouffer*." *American Sociological Review* 27:720–21.

Rose, Arnold M. 1942. "A Research Note on the Influence of Immigration on the Birth Rate." *American Journal of Sociology* 47:614–21.

Rosen, Sherwin and Paul Taubman. 1982. "Changes in Life-Cycle Earnings: What Do Social Security Data Show?" *Journal of Human Resources* 17:321–38.

Rosenbaum, Paul R. 1987. "Model-Based Direct Adjustment." *Journal of the American Statistical Association* 82:387–94.

—. 1989. "Optimal Matching for Observational Studies." *Journal of the American Statistical Association* 84:1024–32.

—. 1991. "Sensitivity Analysis for Matched Case Control Studies." *Biometrics* 47: 87–100.

—. 1992. "Detecting Bias with Confidence in Observational Studies." *Biometrika* 79:367–74.

—. 1999. "Choice as an Alternative to Control in Observational Studies." *Statistical Science* 14:259–304.

—. 2002. *Observational Studies.* New York: Springer.

—. 2010. *Design of Observational Studies.* New York: Springer.

—. 2012. "Optimal Matching of an Optimally Chosen Subset in Observational Studies." *Journal of Computational and Graphical Statistics* 21:57–71.

Rosenbaum, Paul R. and Donald B. Rubin. 1983a. "Assessing Sensitivity to an Unobserved Covariate in an Observational Study with Binary Outcome." *Journal of the Royal Statistical Society* 45:212–18.

—. 1983b. "The Central Role of the Propensity Score in Observational Studies for Causal Effects." *Biometrika* 70:41–55.

—. 1984. "Reducing Bias in Observational Studies Using Subclassification on the Propensity Score." *Journal of the American Statistical Association* 79:516–24.

—. 1985a. "The Bias Due to Incomplete Matching." *Biometrics* 41:103–16.

—. 1985b. "Constructing a Control Group Using Multivariate Matched Sampling Methods." *American Statistician* 39:33–38.

Rosenzweig, Mark R. and Kenneth I. Wolpin. 2000. "Natural 'Natural Experiments' in Economics." *Journal of Economic Literature* 38:827–74.

Rossell, Christine H., David J. Armor, and Herbert J. Walberg, Eds. 2002. *School Desegregation in the 21st Century.* Westport: Praeger.

Rothman, Kenneth J., Sander Greenland, and Timothy L. Lash. 2008. *Modern Epidemiology.* Philadelphia: Wolters Kluwer Health/Lippincott Williams & Wilkins.

Roy, A. D. 1951. "Some Thoughts on the Distribution of Earnings." *Oxford Economic Papers* 3:135–46.

Rubin, Donald B. 1973a. "Matching to Remove Bias in Observational Studies." *Biometrics* 29:159–83.

—. 1973b. "The Use of Matched Sampling and Regression Adjustment to Remove Bias in Observational Studies." *Biometrics* 29:185–203.

—. 1974. "Estimating Causal Effects of Treatments in Randomized and Nonrandomized Studies." *Journal of Educational Psychology* 66: 688–701.

—. 1976. "Multivariate Matching Methods That Are Equal Percent Bias Reducing, I: Some Examples." *Biometrics* 32:109–20.

—. 1977. "Assignment to Treatment Group on the Basis of a Covariate." *Journal of Educational Statistics* 2:1–26.

—. 1978. "Bayesian Inference for Causal Effects: The Role of Randomization." *Annals of Statistics* 6:34–58.

—. 1979. "Using Multivariate Matched Sampling and Regression Adjustment to Control Bias in Observational Studies." *Journal of the American Statistical Association* 74:318–28.

—. 1980a. "Bias Reduction Using Mahalanobis-Metric Matching." *Biometrics* 36: 293–98.

—. 1980b. "Comment on 'Randomization Analysis of Experimental Data in the Fisher Randomization Test' by Basu." *Journal of the American Statistical Association* 75:591–93.

—. 1981. "Estimation in Parallel Randomized Experiments." *Journal of Educational Statistics* 6:377–400.

—. 1986. "Which Ifs Have Causal Answers (Comment on 'Statistics and Causal Inference' by Paul W. Holland)." *Journal of the American Statistical Association* 81: 961–62.

—. 1990. "Formal Modes of Statistical Inference for Causal Effects." *Journal of Statistical Planning and Inference* 25:279–92.

—. 1991. "Practical Implications of Modes of Statistical Inference for Causal Effects and the Critical Role of the Assignment Mechanism." *Biometrics* 47:1213–34.

—. 2005. "Causal Inference Using Potential Outcomes: Design, Modeling, Decisions." *Journal of the American Statistical Association* 100:322–31.

—. 2006. *Matched Sampling for Causal Effects.* New York: Cambridge University Press.

Rubin, Donald B. and Neal Thomas. 1992. "Characterizing the Effect of Matching Using Linear Propensity Score Methods with Normal Distributions." *Biometrika* 79:787–809.

—. 1996. "Matching Using Estimated Propensity Scores: Relating Theory to Practice." *Biometrics* 52:249–64.

—. 2000. "Combining Propensity Score Matching with Additional Adjustments for Prognostic Covariates." *Journal of the American Statistical Association* 95:573–85.

Ruppert, David, M. P. Wand, and Raymond J. Carroll. 2003. *Semiparametric Regression.* Cambridge: Cambridge University Press.

Ruud, Paul A. 2000. *An Introduction to Classical Econometric Theory.* New York: Oxford University Press.

Salmon, Wesley C. 1984. *Scientific Explanation and the Causal Structure of the World.* Princeton: Princeton University Press.

—. 1989. *Four Decades of Scientific Explanation.* Minneapolis: University of Minnesota Press.

Saltelli, Andrea, Stefano Tarantola, Francesca Campolongo, and Marco Ratto. 2004. *Sensitivity Analysis in Practice: A Guide to Assessing Scientific Models.* Hoboken: Wiley.

Sampson, Robert J. 2008. "Moving to Inequality: Neighborhood Effects and Experiments Meet Social Structure." *American Journal of Sociology* 114:189–231.

—. 2012. *Great American City: Chicago and the Enduring Neighborhood Effect.* Chicago: University of Chicago Press.

Sampson, Robert J., Patrick Sharkey, and Stephen W. Raudenbush. 2008. "Durable Effects of Concentrated Disadvantage on Verbal Ability among African-American Children." *Proceedings of the National Academy of Sciences* 105: 845–52.

Scharfstein, Daniel O., Andrea Rotnitzky, and James M. Robins. 1999. "Adjusting for Nonignorable Drop-out Using Semiparametric Nonresponse Models." *Journal of the American Statistical Association* 94:1096–120.

Schofield, Janet Ward. 1995. "Review of Research on School Desegregation's Impact on Elementary and Secondary School Students." Pp. 597–617 in *Handbook of Research on Multicultural Education,* edited by J. A. Banks and C. A. M. Banks. New York: Macmillan.

Schultz, Henry. 1938. *The Theory and Measurement of Demand.* Chicago: University of Chicago Press.

Schwartz, Saul. 1986. "The Relative Earnings of Vietnam and Korean-Era Veterans." *Industrial and Labor Relations Review* 39:564–72.

Sekhon, Jasjeet S. 2004. "The 2004 Florida Optical Voting Machine Controversy: A Causal Analysis Using Matching." Working Paper, Department of Government, Harvard University, Cambridge, Massachusetts.

—. 2007. "Alternative Balance Metrics for Bias Reduction in Matching Methods for Causal Inference." Working Paper, Survey Research Center, University of California, Berkeley, California.

—. 2009. "Opiates for the Matches: Matching Methods for Causal Inference." *Annual Review of Political Science* 12:487–508.

—. 2013. "Matching." Software Routine for R, Travers Department of Political Science, University of California, Berkeley, California.

Sekhon, Jasjeet S. and Rocio Titiunik. 2012. "When Natural Experiments Are Neither Natural Nor Experiments." *American Political Science Review* 106:35–57.

Seltzer, Judith A., Christine A. Bachrach, Suzanne M. Bianchi, Caroline H. Bledsoe, Lynne M. Casper, P. Lindsay Chase-Lansdale et al. 2005. "Explaining Family Change and Variation: Challenges for Family Demographers." *Journal of Marriage and Family* 67:908–25.

Sewell, William H. 1964. "Community of Residence and College Plans." *American Sociological Review* 29:24–38.

Sewell, William H., Archibald O. Haller, and George W. Ohlendorf. 1970. "The Educational and Early Occupational Status Attainment Process: Replication and Revision." *American Sociological Review* 35:1014–24.

Sewell, William H., Archibald O. Haller, and Alejandro Portes. 1969. "The Educational and Early Occupational Attainment Process." *American Sociological Review* 34: 82–92.

Sewell, William H., Robert M. Hauser, Kristen W. Springer, and Taissa S. Hauser. 2004. "As We Age: A Review of the Wisconsin Longitudinal Study, 1957–2001." *Research in Social Stratification and Mobility* 20:3–111.

Shadish, William R., Thomas D. Cook, and Donald Thomas Campbell. 2001. *Experimental and Quasi-Experimental Designs for Generalized Causal Inference*. Boston: Houghton Mifflin.

Shadish, William R., Rodolfo Galindo, Vivian C. Wong, Peter M. Steiner, and Thomas D. Cook. 2011. "A Randomized Experiment Comparing Random and Cutoff-Based Assignment." *Psychological Methods* 16:179–91.

Shalizi, Cosma Rohilla. 2012. "Comment on 'Why and When "Flawed" Social Network Analyses Still Yield Valid Tests of No Contagion."' *Statistics, Politics, and Policy* 3:1–3.

Shalizi, Cosma Rohilla and Andrew C. Thomas. 2011. "Homophily and Contagion Are Generically Confounded in Observational Social Network Studies." *Sociological Methods and Research* 40:211–39.

Sharkey, Patrick and Felix Elwert. 2011. "The Legacy of Disadvantage: Multigenerational Neighborhood Effects on Cognitive Ability." *American Journal of Sociology* 116:1934–81.

Shpitser, Ilya. 2012a. "Graph-Based Criteria of Identifiability of Causal Questions." Pp. 59–70 in *Causality: Statistical Perspectives and Applications*, edited by C. Berzuini, P. Dawid, and L. Bernardinelli. Hoboken: Wiley.

—. 2012b. "Structural Equations, Graphs and Interventions." Pp. 15–24 in *Causality: Statistical Perspectives and Applications*, edited by C. Berzuini, P. Dawid, and L. Bernardinelli. Hoboken: Wiley.

Shpitser, Ilya and Judea Pearl. 2007. "What Counterfactuals Can Be Tested." Pp. 352–59 in *Proceedings of the Twenty-Third Conference on Uncertainty in Artificial Intelligence*, edited by R. Parr and L. van der Gaag. Corvallis: Association for Uncertainty in Artificial Intelligence Press.

Shpitser, Ilya, Tyler J. VanderWeele, and James M. Robins. 2010. "On the Validity of Covariate Adjustment for Estimating Causal Effects." Pp. 527–36 in *Proceedings of the Twenty-Sixth Conference on Uncertainty in Artificial Intelligence*, edited by P. Grünwald and P. Spirtes. Catalina Island: Association for Uncertainty in Artificial Intelligence Press.

Singer, Burton and Margaret M. Marini. 1987. "Advancing Social Research: An Essay Based on Stanley Lieberson's Making It Count." *Sociological Methodology* 17: 373–91.

Small, Dylan S. and Paul R. Rosenbaum. 2008. "War and Wages: The Strength of Instrumental Variables and Their Sensitivity to Unobserved Biases." *Journal of the American Statistical Association* 103:924–33.

Smith, Herbert L. 1989. "Integrating Theory and Research on the Institutional Determinants of Fertility." *Demography* 26:171–84.

—. 1990. "Specification Problems in Experimental and Nonexperimental Social Research." *Sociological Methodology* 20:59–91.

—. 1997. "Matching with Multiple Controls to Estimate Treatment Effects in Observational Studies." *Sociological Methodology* 27:325–53.

—. 2003. "Some Thoughts on Causation as It Relates to Demography and Population Studies." *Population and Development Review* 29:459–69.

—. 2009. "Causation and Its Discontents." Pp. 233–42 in *Causal Analysis in Population Studies: Concepts, Methods, Applications*, edited by H. Engelhardt, H.-P. Kohler, and A. Prskawetz. Dordrecht: Springer.

—. 2013. "Research Design: Toward a Realistic Role for Causal Analysis." Pp. 45–73 in *Handbook of Causal Analysis for Social Research*, edited by S. L. Morgan. Dordrecht: Springer.

Smith, Jeffery A. and Petra Todd. 2005. "Does Matching Overcome Lalonde's Critique of Nonexperimental Estimators?" *Journal of Econometrics* 125:305–53.

Sobel, Michael E. 1995. "Causal Inference in the Social and Behavioral Sciences." Pp. 1–38 in *Handbook of Statistical Modeling for the Social and Behavioral Sciences*, edited by G. Arminger, C. C. Clogg, and M. E. Sobel. New York: Plenum.

—. 1996. "An Introduction to Causal Inference." *Sociological Methods and Research* 24:353–79.

—. 2000. "Causal Inference in the Social Sciences." *Journal of the American Statistical Association* 95:647–51.

—. 2006. "What Do Randomized Studies of Housing Mobility Demonstrate? Causal Inference in the Face of Interference." *Journal of the American Statistical Association* 101:1398–407.

Sondheimer, Rachel M. and Donald P. Green. 2010. "Using Experiments to Estimate the Effects of Education on Voter Turnout." *American Journal of Political Science* 54:174–89.

Sørensen, Aage B. 1998. "Theoretical Mechanisms and the Empirical Study of Social Processes." Pp. 238–66 in *Social Mechanisms: An Analytical Approach to Social Theory, Studies in Rationality and Social Change*, edited by P. Hedström and R. Swedberg. Cambridge: Cambridge University Press.

Sørensen, Aage B. and Stephen L. Morgan. 2000. "School Effects: Theoretical and Methodological Issues." Pp. 137–60 in *Handbook of the Sociology of Education*, edited by M. T. Hallinan. New York: Kluwer/Plenum.

Sovey, Allison J. and Donald P. Green. 2011. "Instrumental Variables Estimation in Political Science: A Readers' Guide." *American Journal of Political Science* 55: 188–200.

Staiger, Douglas and James H. Stock. 1997. "IV Regression with Weak Instruments." *Econometrica* 65:557–86.

Steiner, Peter M., Thomas D. Cook, William R. Shadish, and M. H. Clark. 2010. "The Importance of Covariate Selection in Controlling for Selection Bias in Observational Studies." *Psychological Methods* 15:250–67.

Stevens, Mitchell L., Elizabeth A. Armstrong, and Richard Arum. 2008. "Sieve, Incubator, Temple, Hub: Empirical and Theoretical Advances in the Sociology of Higher Education." *Annual Review of Sociology* 34:127–51.

Stewart, Susan T., David M. Cutler, and Allison B. Rosen. 2009. "Forecasting the Effects of Obesity and Smoking on U.S. Life Expectancy." *New England Journal of Medicine* 361:2252–60.

Stock, James H. and Mark W. Watson. 2007. *Introduction to Econometrics*. Boston: Pearson/Addison Wesley.

Stolzenberg, Ross M. 2004. "Multiple Regression Analysis." Pp. 165–207 in *Handbook of Data Analysis*, edited by M. A. Hardy and A. Bryman. Thousand Oaks: Sage.

Stouffer, Samuel A. 1949. *The American Soldier*. Princeton: Princeton University Press.

—. 1950. "Some Observations on Study Design." *American Journal of Sociology* 55:355–61.

—. 1955. *Communism, Conformity, and Civil Liberties: A Cross-Section of the Nation Speaks Its Mind.* Garden City: Doubleday.

—. 1962[1948]. *Social Research to Test Ideas.* Glencoe: Free Press.

Strevens, Michael. 2008. *Depth: An Account of Scientific Explanation.* Cambridge: Harvard University Press.

Stuart, Elizabeth A. 2010. "Matching Methods for Causal Inference: A Review and a Look Forward." *Statistical Science* 25:1–21.

Stuart, Elizabeth A. and Nicholas S. Ialongo. 2010. "Matching Methods for Selection of Participants for Follow-Up." *Multivariate Behavioral Research* 45:746–65.

Swinburn, Boyd A., Gary Sacks, Kevin D. Hall, Klim McPherson, Diane T. Finegood, Marjory L. Moodie et al. 2011. "The Global Obesity Pandemic: Shaped by Global Drivers and Local Environments." *Lancet* 378:804–14.

Tchetgen Tchetgen, Eric J. and Tyler J. VanderWeele. 2010. "On Causal Inference in the Presence of Interference." *Statistical Methods in Medical Research* 21:55–75.

Tenn, Steven. 2005. "An Alternative Measure of Relative Education to Explain Voter Turnout." *Journal of Politics* 67:271–82.

Thistlewaite, D. L. and Donald T. Campbell. 1960. "Regression-Discontinuity Analysis: An Alternative to the Ex Post Facto Experiment." *Journal of Educational Psychology* 51:309–17.

Thompson, Steven K. 2002. *Sampling.* New York: Wiley.

Thompson, Warren S. 1948. "Differentials in Fertility and Levels of Living in the Rural Population of the United States." *American Sociological Review* 13:516–34.

—. 1949. "The Demographic Revolution in the United States." *Annals of the American Academy of Political and Social Science* 262:62–69.

Thornton, Arland, Georgina Binstock, Kathryn M. Yount, Mohammad Jalal Abbasi-Shavazi, Ghimire Dirgha, and Yu Xie. 2012. "International Fertility Change: New Data and Insights from the Developmental Idealism Framework." *Demography* 49:677–98.

Treiman, Donald J. 2009. *Quantitative Data Analysis: Doing Social Research to Test Ideas.* San Francisco: Jossey-Bass.

Trochim, William M. K. 1984. *Research Design for Program Evaluation: The Regression-Discontinuity Approach.* Beverly Hills: Sage Publications.

Tu, Wanzhu and Xiao-Hua Zhou. 2002. "A Bootstrap Confidence Interval Procedure for the Treatment Effect Using Propensity Score Subclassification." *Health Services & Outcomes Research Methodology* 3:135–47.

Tuttle, Christina C., Brian Gill, Philip Gleason, Virginia Knechtel, Ira Nichols-Barrer, and Alexandra Resch. 2013. "KIPP Middle Schools: Impacts on Achievement and Other Outcomes." Mathematica Policy Research, Washington, DC.

Tyack, David B. 1974. *The One Best System: A History of American Urban Education.* Cambridge: Harvard University Press.

Uggen, Christopher, Angela Behrens, and Jeff Manza. 2005. "Criminal Disenfranchisement." *Annual Review of Law and Social Science* 1:307–22.

Uggen, Christopher and Jeff Manza. 2002. "Democratic Contraction? Political Consequences of Felon Disenfranchisement in the United States." *American Sociological Review* 67:777–803.

Valliant, Richard, Jill A. Dever, and Frauke Kreuter. 2013. *Practical Tools for Designing and Weighting Survey Samples.* New York: Springer.

Van der Klaauw, Wilbert. 2002. "Estimating the Effect of Financial Aid Offers on College Enrollment: A Regression-Discontinuity Approach." *International Economic Review* 43:1249–87.

van der Laan, M. J. and James M. Robins. 2003. *Unified Methods for Censored Longitudinal Data and Causality.* New York: Springer.

VanderWeele, Tyler J. 2008. "Simple Relations Between Principal Stratification and Direct and Indirect Effects." *Statistics and Probability Letters* 78:2957–62.

——. 2009a. "Marginal Structural Models for the Estimation of Direct and Indirect Effects." *Epidemiology* 20:18–26.

——. 2009b. "On the Distinction Between Interaction and Effect Modification." *Epidemiology* 20:863–71.

——. 2010. "Bias Formulas for Sensitivity Analysis for Direct and Indirect Effects." *Epidemiology* 21:540–51.

——. 2011a. "Principal Stratification – Uses and Limitations." *International Journal of Biostatistics* 7:1–14.

——. 2011b. "Sensitivity Analysis for Contagion Effects in Social Networks." *Sociological Methods and Research* 40:240–55.

——. 2012. "Mediation Analysis with Multiple Versions of the Mediator." *Epidemiology* 23:454–63.

——. In press. *Explanation in Causal Inference: Methods for Mediation and Interaction.* Oxford: Oxford University Press.

VanderWeele, Tyler J. and Weihua An. 2013. "Social Networks and Causal Analysis." Pp. 353–74 in *Handbook of Causal Analysis for Social Research*, edited by S. L. Morgan. Dordrecht: Springer.

VanderWeele, Tyler J. and James M. Robins. 2007a. "Four Types of Effect Modification: A Classification Based on Directed Acyclic Graphs." *Epidemiology* 18:561–68.

—. 2007b. "The Identification of Synergism in the Sufficient-Component-Cause Framework." *Epidemiology* 18:329–39.

—. 2008. "Empirical and Counterfactual Conditions for Sufficient Cause Interactions." *Biometrika* 95:49–61.

—. 2009. "Minimal Sufficient Causation and Directed Acyclic Graphs." *Annals of Statistics* 37:1437–65.

VanderWeele, Tyler J. and Ilya Shpitser. 2011. "A New Criterion for Confounder Selection." *Biometrics* 67:1406–13.

VanderWeele, Tyler J., Stijn Vansteelandt, and James M. Robins. 2010. "Marginal Structural Models for Sufficient Cause Interactions." *American Journal of Epidemiology* 171:506–14.

Vansteelandt, Stijn, Tyler J. VanderWeele, and James M. Robins. 2012. "Semiparametric Tests for Sufficient Cause Interaction." *Journal of the Royal Statistical Society* 74:223–44.

Verba, Sidney and Norman H. Nie. 1972. *Participation in America: Political Democracy and Social Equality.* New York: Harper.

Verba, Sidney, Kay Lehman Schlozman, and Henry E. Brady. 1995. *Voice and Equality: Civic Voluntarism in American Politics.* Cambridge: Harvard University Press.

Vytlacil, Edward J. 2002. "Independence, Monotonicity, and Latent Index Models: An Equivalence Result." *Econometrica* 70:331–41.

Wald, Abraham. 1940. "The Fitting of Straight Lines If Both Variables Are Subject to Error." *Annals of Mathematical Statistics* 11:284–300.

Wand, Jonathan, Kenneth W. Shotts, Jasjeet S. Sekhon, Walter R. Mebane Jr., Michael C. Herron, and Henry E. Brady. 2001. "The Butterfly Did It: The Aberrant Vote for Buchanan in Palm Beach County, Florida." *American Political Science Review* 95:793–810.

Wang, Xiaolu and Michael E. Sobel. 2013. "New Perspectives on Causal Mediation Analysis." Pp. 215–42 in *Handbook of Causal Analysis for Social Research*, edited by S. L. Morgan. Dordrecht: Springer.

Wells, Amy Stuart and Robert L. Crain. 1994. "Perpetuation Theory and the Long-Term Effects of School Desegregation." *Review of Educational Research* 64:531–55.

West, Martin R. and Ludger Woessmann. 2010. "'Every Catholic Child in a Catholic School': Historical Resistance to State Schooling, Contemporary Private Competition and Student Achievement across Countries." *Economic Journal* 120:F229–55.

Westoff, Charles F. and Larry Bumpass. 1973. "The Revolution in Birth Control of U.S. Roman Catholics." *Science* 179:41–44.

Westoff, Charles F. and Elise F. Jones. 1979. "The End of 'Catholic' Fertility." *Demography* 16:209–17.

Westoff, Charles F. and Norman B. Ryder. 1977. *The Contraceptive Revolution.* Princeton: Princeton University Press.

Whelpton, Pascal K. 1932. "Trends in Age Composition and in Specific Birth-Rates, 1920–30." *American Journal of Sociology* 37:855–61.

White, Halbert and Karim Chalak. 2009. "Settable Systems: An Extension of Pearl's Causal Model with Optimization, Equilibrium, and Learning." *Journal of Machine Learning Research* 10:1759–99.

Willis, Robert and Sherwin Rosen. 1979. "Education and Self-Selection." *Journal of Political Economy* 87:S7–S35.

Willms, J. Douglas. 1985. "Catholic-School Effects on Academic Achievement: New Evidence from the High School and Beyond Follow-up Study." *Sociology of Education* 58:98–114.

Wilson, William Julius. 2011. "Reflections on a Sociological Career That Integrates Social Science with Social Policy." *Annual Review of Sociology* 37:1–18.

Wimer, Christopher, Robert J. Sampson, and John Laub. 2008. "Estimating Time-Varying Causes and Outcomes, with Application to Incarceration and Crime." Pp. 37–59 in *Applied Data Analytic Techniques for Turning Points Research*, edited by P. Cohen. New York: Routledge.

Winship, Christopher and David J. Harding. 2008. "A General Strategy for the Identification of Age, Period, Cohort Models: A Mechanism-Based Approach." *Sociological Methods and Research* 36:362–401.

Winship, Christopher and Sanders Korenman. 1997. "Does Staying in School Make You Smarter? The Effect of Education on IQ in the Bell Curve." Pp. 215–34 in *Intelligence, Genes, and Success: Scientists Respond to the Bell Curve*, edited by B. Devlin, S. E. Fienberg, D. P. Resnick, and K. Roeder. New York: Springer.

Winship, Christopher and Robert D. Mare. 1984. "Regression Models with Ordinal Variables." *American Sociological Review* 49:512–25.

——. 1992. "Models for Sample Selection Bias." *Annual Review of Sociology* 18: 327–50.

Winship, Christopher and Stephen L. Morgan. 1999. "The Estimation of Causal Effects from Observational Data." *Annual Review of Sociology* 25:659–706.

Winship, Christopher and Michael E. Sobel. 2004. "Causal Analysis in Sociological Studies." Pp. 481–503 in *Handbook of Data Analysis*, edited by M. A. Hardy and A. Bryman. Thousand Oaks: Sage.

Winship, Scott and Christopher Winship. 2013. "The Permanent and Transitory Effect of Schooling on Mental Ability." Working Paper, Department of Sociology, Harvard University, Cambridge, Massachusetts.

Wodtke, Geoffrey T., David J. Harding, and Felix Elwert. 2011. "Neighborhood Effects in Temporal Perspective: The Impact of Long-Term Exposure to Concentrated Disadvantage on High School Graduation." *American Sociological Review* 76:713–36.

Wong, Vivian C., Peter M. Steiner, and Thomas D. Cook. 2013. "Analyzing Regression-Discontinuity Designs with Multiple Assignment Variables: A Comparative Study of Four Estimation Methods." *Journal of Educational and Behavioral Statistics* 38:107–41.

Woodward, James. 2003. *Making Things Happen: A Theory of Causal Explanation*. New York: Oxford University Press.

Wooldridge, Jeffrey M. 2010. *Econometric Analysis of Cross Section and Panel Data*. Cambridge: MIT Press.

Working, E. J. 1927. "What Do Statistical 'Demand Curves' Show?" *Quarterly Journal of Economics* 41:212–35.

Working, Holbrook. 1925. "The Statistical Determination of Demand Curves." *Quarterly Journal of Economics* 39:503–45.

Wright, Erik Olin. 1997. *Class Counts: Comparative Studies in Class Analysis*. Cambridge: Cambridge University Press.

Wright, Sewall. 1921. "Correlation and Causation." *Journal of Agricultural Research* 20:557–85.

—. 1925. "Corn and Hog Correlations." U.S. Department of Agriculture, Washington, DC.

—. 1934. "The Method of Path Coefficients." *Annals of Mathematical Statistics* 5:161–215.

Wu, Lawrence and Barbara L. Wolfe, Eds. 2001. *Out of Wedlock: Causes and Consequences of Nonmarital Fertility*. New York: Russell Sage Foundation.

Wu, Lawrence L. 1996. "Effects of Family Instability, Income, and Income Instability on the Risk of a Premarital Birth." *American Sociological Review* 61:386–406.

—. 2008. "Cohort Estimates of Nonmarital Fertility for U.S. Women." *Demography* 45:193–207.

Xie, Yu. 2007. "Otis Dudley Duncan's Legacy: The Demographic Approach to Quantitative Reasoning in Social Science." *Research in Social Stratification and Mobility* 25:141–56.

—. 2011. "Population Heterogeneity and Causal Inference." Population Studies Center, Institute for Social Research, University of Michigan, Ann Arbor, Michigan.

Xie, Yu, Jennie E. Brand, and Ben Jann. 2012. "Estimating Heterogeneous Treatment Effects with Observational Data." *Sociological Methodology* 42:314–47.

Yamamoto, Teppei. 2012. "Understanding the Past: Statistical Analysis of Causal Attribution." *American Journal of Political Science* 56:237–56.

Yang, Dan, Dylan S. Small, Jeffrey H. Silber, and Paul R. Rosenbaum. 2012. "Optimal Matching with Minimal Deviation from Fine Balance in a Study of Obesity and Surgical Outcomes." *Biometrics* 68:628–36.

Yinger, Milton J., Kiyoshi Ikeda, and Frank Laycock. 1967. "Treating Matching as a Variable in a Sociological Experiment." *American Sociological Review* 32:801–12.

Zhang, Junni L., Donald B. Rubin, and Fabrizia Mealli. 2008. "Evaluating the Effects of Job Training Programs on Wages Through Principal Stratification." *Advances in Econometrics* 21:117–45.

—. 2009. "Likelihood-Based Analysis of Causal Effects of Job-Training Programs Using Principal Stratification." *Journal of the American Statistical Association* 104:166–76.

Zhao, Shandong, David A. van Dyk, and Kosuke Imai. 2012. "Causal Inference in Observational Studies with Non-Binary Treatments." Department of Statistics, University of California, Irvine, California.

Zubizarreta, José R. 2012. "Using Mixed Integer Programming for Matching in an Observational Study of Kidney Failure After Surgery." *Journal of the American Statistical Association* 107:1360–71.

Index